12th International Spring Seminar on Nuclear Physics 2017

Current Problems and Prospects for Nuclear Structure

Journal of Physics: Conference Series Volume 966

Ischia, Italy
15 - 19 May 2017

Editors:

Angela Gargano
Luigi Coraggio
Nunzio Itaco

ISBN: 978-1-5108-5896-1
ISSN: 1742-6588

Printed from e-media with permission by:

Curran Associates, Inc.
57 Morehouse Lane
Red Hook, NY 12571

Some format issues inherent in the e-media version may also appear in this print version.

Printed by Curran Associates, Inc. (2018)

For permission requests, please contact the Institute of Physics at the address below.

Institute of Physics
Dirac House, Temple Back
Bristol BS1 6BE UK

Phone: 44 1 17 929 7481
Fax: 44 1 17 920 0979

techtracking@iop.org

Additional copies of this publication are available from:

Curran Associates, Inc.
57 Morehouse Lane
Red Hook, NY 12571 USA
Phone: 845-758-0400
Fax: 845-758-2633
Email: curran@proceedings.com
Web: www.proceedings.com

Table of contents

Volume 966

12th International Spring Seminar on Nuclear Physics: Current Problems and Prospects for Nuclear Structure

15–19 May 2017, Sant'Angelo d'Ischia, Italy

Accepted papers received: 31 January 2018
Published online: 22 February 2018

Preface

Papers

IOP Conf. Series: Journal of Physics: Conf. Series **966** (2018) 011001 doi:10.1088/1742-6596/966/1/011001

Foreword

The 12th International Seminar on Nuclear Physics was held in Ischia from May 15 to May 19, 2017. This Seminar is part of a series of topical meetings started in Sorrento in 1986 on the initiative of Aldo Covello and continued with meetings held every two or three years in the Naples area, with the aim to create a lively and informal environment for a fruitful debate, between theoretical and experimental researchers, on current achievements and future developments of nuclear structure.
The present Seminar, devoted to *Current Problems and Prospects for Nuclear Structure,* has maintained the tradition and the original spirit of the Spring Seminar series. The topics covered during the week represent the main lines of investigations in the field of nuclear structure including its role in the context of fundamental physics:

- Advanced methods for nuclear structure studies
- Nuclear structure studies through RIB experiments
- Nuclear structure & fundamental physics

A comprehensive overview of these three subjects was given through general talks, more specialized seminars, and posters, as well as through lively debates we had due the very active participation of all the attendants. A specific discussion in concluding the Seminar was dedicated to *The role of nuclear structure towards the new frontier in nuclear physics.*

The Seminar had about 100 participants from some 20 countries [please see the list of participants]. This is in line with the tradition of these meetings, as is the fact that more than 50% of the present participants attended one or more of the previous Seminars.

The present volume contains 63 manuscripts out of the 81 invited talks and posters presented at the Seminar. All of these manuscripts have been peer reviewed. We would like to thank all the colleagues who have acted as referees to assess the suitability of the various articles for publication in *Journal of Physics: Conference Series.* We are confident that the high quality of all papers contained in these Proceedings will be appreciated by the nuclear physics community.

We are deeply grateful to the members of the Advisory Committee for their valuable cooperation in preparing the scientific program.

We also warmly acknowledge the financial support of the Istituto Nazionale di Fisica Nucleare, the Dipartimento di Fisica and the Università di Napoli "Federico II", the Università della Campania "Luigi Vanvitelli" who helped make the Seminar possible.

Angela Gargano
Luigi Coraggio
Nunzio Itaco
Editors

LOCAL ORGANIZING COMMITTEE

L. Coraggio
L. De Angelis
G. De Gregorio
T. Fukui
A. Gargano
N. Itaco

INTERNATIONAL ADVISORY COMMITTEE

T. Aumann *(GSI)*
C. Barbieri *(Surrey)*
M.J.G. Borge *(Madrid&CERN)*
B.A. Brown *(MSU)*
G. Colo' *(Milano)*
J. Dobaczewski *(Warsaw)*
G. de France *(GANIL)*
M. Hjorth-Jensen *(MSU&Oslo)*
F. Iachello *(Yale)*
T.T.S. Kuo *(Stony Brook)*
T. Otsuka *(Tokyo)*
R. Raabe *(KU Leuven)*
I. Talmi *(Rehovot)*
J. Vary *(Iowa)*
S. Bacca *(TRIUMF)*
A. Bonaccorso *(INFN Pisa)*
A. Bracco *(Milano)*
F. Cappuzzello *(Catania)*
G. de Angelis *(INFN LNL)*
A. Faessler *(Tübingen)*
A. Gade *(MSU)*
M. Hjorth-Jensen *(MSU&Oslo)*
A. Jungclaus *(CSIC Madrid)*
R. Machleidt *(Idaho)*
A. Poves *(Madrid)*
A. Schwenk *(TU Darmstadt)*
U. van Kolck *(IPN Orsay)*
F.R. Xu *(PKU Beijing)*

SPONSORS OF THE SEMINAR

Dipartimento di Fisica "Ettore Pancini", Università di Napoli "Federico II"
Istituto Nazionale di Fisica Nucleare
Università di Napoli "Federico II"
Università della Campania "Luigi Vanvitelli"

IOP Conf. Series: Journal of Physics: Conf. Series **966** (2018) 011002 doi:10.1088/1742-6596/966/1/011002

12th INTERNATIONAL SPRING SEMINAR ON NUCLEAR PHYSICS

CURRENT PROBLEMS AND PROSPECTS FOR NUCLEAR STRUCTURE

SANT'ANGELO D'ISCHIA, MAY 15-19, 2017

PROGRAM

Monday, May 15

9:00 Opening Address

Chairperson: **C. H. Dasso** *(Sevilla)*

9:20	**A. Faessler** *(Tübingen)*:	Can one determine the neutrino mass by electron capture?
9:45	**M. J. Borge** *(Madrid):*	Recent results from ISOLDE and HIE-ISOLDE
10:10	**H. Sagawa** *(Wako):*	Three-body model for nuclei near and beyond neutron drip line
10.35	**J. Kvasil** *(Praha):*	Elimination of the spurious admixtures in the QRPA E0, E1, E2 and M1 excitations of deformed nuclei

10:55 Coffee Break

Chairperson: **S. Lunardi** *(Padova)*

11:20	**F. Sammarruca** *(Moscow)*:	Chiral effective field theory and nucleonic matter
11:45	**P. Reiter** *(Köln):*	Nuclear-structure physics with MINIBALL at HIE-ISOLDE
12:10	**A. Gal** *(Jerusalem):*	Charge symmetry breaking in Λ hypernuclei
12:35	**C. Michelagnoli** *(Grenoble):*	FIPPS - A new instrument for the spectroscopy of n-rich nuclei at ILL
12:55	**G. De Gregorio** *(Napoli):*	Odd nuclei spectroscopy within a self-consistent multiphonon approach

13:15 Session close

Chairperson: **G. de Angelis** *(Legnaro)*

15:00	**J.D. Holt** *(Vancouver)*:	Extending ab initio nuclear structure to all light- and medium-mass nuclei
15:25	**J.J. Valiente Dobón** *(Legnaro)*:	Nuclear structure of moderately exotic nuclei populated via multinucleon transfer reactions
15:50	**F. Cappuzzello** *(Catania)*:	Recent results on nuclear structure at the INFN-LNS laboratory
16:10	**H. Heylen** *(Heidelberg):*	The structure of exotic nuclei studied using collinear laser spectroscopy: Recent results from COLLAPS

16:30 Coffee Break

Chairperson: **N. Itaco** *(Napoli)*

16:55	**B.R. Barrett** *(Tucson):*	The no core shell model in an effective field theory approach
17:20	**M. Horoi** *(Mount Pleasant)*:	Nuclear structure for double beta decay
17:45	**M. La Commara** *(Napoli):*	$^{8}B+^{208}Pb$ elastic scattering at Coulomb barrier energies

18:05 Session close

IOP Conf. Series: Journal of Physics: Conf. Series **966** (2018) 011002 doi:10.1088/1742-6596/966/1/011002

Tuesday, May 16

Chairperson: **P. Reiter** *(Köln):*

9:00	**R. Machleidt** (*Moscow):*	Consistent high-quality two-nucleon potentials up to fifth order of the chiral expansion
9:25	**T. Aumann** *(Darmstadt):*	Quasifree (p,2p) knockout reactions with radioactive beams in inverse and complete kinematics at R3B
9:50	**G. de France** *(Caen)*:	The EXILL campaign
10:15	**S. Shlomo** *(College Station)*:	Short-range correlations effect on the single particle proton $3s_{1/2}$ wave function in ^{206}Pb

10:40 Coffee break

Chairperson: **M. J. Borge** *(Madrid)*

11:05	**U. van Kolck** *(Orsay)*:	Nuclear physics from an expansion around the unitarity limit
11:30	**C. Barbieri** *(Guildford)*:	Self-consistent Green's function computations of medium mass isotopes
11:55	**D. Carbone** *(Catania):*	Nuclear structure studies performed using the (^{18}O,^{16}O) two-neutron transfer reactions
12:15	**P. Pragati** *(Roorkee)*:	Competition between spherical and octupole deformation in ^{216}Fr

12:35 Session close

IOP Conf. Series: Journal of Physics: Conf. Series **966** (2018) 011002 doi:10.1088/1742-6596/966/1/011002

Wednesday, May 17

Chairperson: **R. Machleidt** *(Moscow)*

8:30	**B.A. Brown** *(East Lansing)*:	Nuclear matrix elements for double-beta decay of ^{76}Ge
8:55	**D. Balabanski** *(Bucharest)*:	Experiments with brilliant gamma beams at ELI-NP
9:20	**S. Bacca** *(Vancouver)*:	Electromagnetic reactions from coupled-cluster theory
9:45	**B. Fernández-Domínguez** *(Santiago de Compostela)*:	How sharp is the transition into the N=20 island of inversion for Mg isotopes?
10:05	**M. Cavallaro** *(Catania)*:	Nuclear reactions for neutrinoless double beta decay
10:25	**Yu. M. Tchuvil'sky** *(Moskva)*:	Ab initio calculations of one-nucleon halo states

10:45 Coffee Break

Chairperson: **L. Coraggio** *(Napoli)*

11:05	**F. Nowacki** *(Strasbourg)*:	Shell model far from stability: Islands of inversion mergers
11:30	**G. Benzoni** *(Milano)*:	Evolution of deformation in Ba isotopic chain
12:55	**L.S. Ferreira** *(Lisboa)*:	New proton drip-line nuclei relevant to nuclear astrophysics
12:15	**W.A. Richter** *(Somerset West)*:	Shell-model studies of the ^{34g,m}Cl(p,γ) ^{35}Ar rp reaction rates
12:35	**P. Gulshani** *(Mississauga)*:	A microscopic crancking model for uni-axial rotation and its connection to other collective models

12:55 Session close

Chairperson: **H. Witala** *(Kraków)*

14:40	**T. Otsuka** (*Tokyo):*	Quantum self-organization and nuclear collectivities
15:05	**A. Andrighetto** *(Legnaro)*:	SPES: An intense source of neutron-rich radioactive beams at LEGNARO
15:30	**L. M. Fraile** *(Madrid):*	Spying on intruders in the ^{68}Ni region
15:50	**K. Arnswald** *(Köln)*:	Enhanced collectivity along the N=Z line: Lifetime measurements in ^{44}Ti, ^{48}Cr, and ^{52}Fe
16:10	**T. Suzuki** *(Tokyo)*:	Structure of two-neutron halo in light drip-line nuclei
16:30	**S. Karampagia** *(East Lansing)*:	Low-energy magnetic radiation enhancement within the nuclear shell model

16:50 Coffee Break

Chairperson: **U. van Kolck** *(Orsay)*

17:15	**G. Hagen** *(Oak Ridge)*:	Coupled-cluster computations of atomic nuclei
17:40	**V. Zelevinsky** *(East Lansing)*:	Nuclear level density, thermalization, chaos and collectivity
18:05	**S.M. Lenzi** *(Padova)*:	The AGATA campaign in GANIL: Results and perspectives
18:30	**I. N. Gratchev** *(Grenoble)*:	The identification of excited states and collectivity of ^{88}Se
18:50	**J. Wilhelmy** *(Köln):*	Investigation of the γ-decay behavior of ^{48,50}Ti and ^{52}Cr with the $γ^3$ setup at HIγS

19:10 Session close

IOP Conf. Series: Journal of Physics: Conf. Series **966** (2018) 011002 doi:10.1088/1742-6596/966/1/011002

Thursday, May 18

Chairperson: **J. N. Ginocchio** *(Los Alamos)*

9:00	**F. Iachello** *(New Haven)*:	Cluster structure of light nuclei
9:25	**M. Gai** *(Groton):*	Experimental tests of the algebraic cluster model
9:50	**A. Heusler** *(Heidelberg)*:	Rotating and vibrating α-clusters in heavy nuclei
10:10	**V. Della Rocca** *(L'Aquila):*	Cluster-shell model: Structure of ^{9}Be, ^{9}B
10:30	**A. Di Pietro** *(Catania)*	Experimental investigation of exotic clustering in ^{13}B and ^{14}C using resonance scattering method

10:50 Coffee Break

Chairperson: **B.A. Brown** *(East Lansing)*

11:15	**P. Ring** *(München)*:	On the way to a microscopic derivation of covariant density functionals in nuclei
11:40	**F.R. Xu** *(Beijing)*:	Resonances of many-body nuclear systems
12:05	**P.E. Georgoudis** *(Jerusalem)*:	Aspects of shape coexistence in the geometric collective model
12:25	**N. Sandulescu** *(Bucharest)*:	Proton-neutron pairing and alpha-like quartet correlations in nuclei
12:45	**M. Sambataro** *(Catania):*	Quartetting in even-even and odd-odd N=Z nuclei

13:05 Session close

Chairperson: **F. Sammarruca** *(Moscow)*:

14:40	**A. Vitturi** *(Padova):*	Interplay of break-up and transfer processes in reactions involving weakly-bound systems: A simple one-dimensional model
15:05	**A. Jungclaus** *(Madrid)*:	Selected results from in-beam and decay γ-ray spectroscopy in the ^{132}Sn region performed at RIKEN
15:30	**V.O. Nesterenko** *(Dubna)*:	Dipole toroidal resonance: Vortical properties, possible origins of pygmy mode, anomalous deformation features
15:50	**Y. F. Niu** *(Bucharest)*:	Beyond mean-field description of Gamow-Teller resonance and β-decay in superfluid nuclei
16:10	**P. Vesely** *(Praha)*:	Recent applications of the equation of motion phonon method to even-even nuclei and exotic nuclear systems

16:30 Coffee Break

Chairperson: **G. Benzoni** *(Milano)*

16:55	**F. Recchia** *(Padova)*:	Neutron knockout from ^{68,70}Ni ground and isomeric states
17:15	**C. Parascandolo** *(Napoli)*:	Dynamical dipole mode in heavy-ion fusion reactions
17:35	**D. Gambacurta** *(Bucharest)*:	Subtraction method in the SRPA approximation with Skyrme and Gogny interactions
17:55	**J. Pei** *(Beijing)*:	Probing surface flows in deformed soft dipole modes

18:15 **Poster session**

Friday, May 19

Chairperson: **F. R. Xu** (*Beijing*)

09:00	**K. Langanke** *(Darmstadt)*:	The FAIR chance for nuclear astrophysics
09:25	**D. Hasell** *(Cambridge)*:	Hard two-photon contribution to elastic lepton-proton scatteringdetermined by the OLYMPUS experiment
09:50	**N. Dinh Dang** *(Wako)*:	Role of exact pairing in the description of nuclear level density and radiative strength function
10:15	**S. Pastore** *(Los Alamos)*:	Fundamental physics with electroweak probes of nuclei
10:35	**R. Crespo** *(Lisboa)*:	Merging first principle studies and few-body reaction formalism for interpretation of RIB experiments

10:55 Coffee Break

Chairperson: **M. Gai** (*Groton*)

11:20	**A.A. Voinov** *(Dubna)*:	Study of the $^{249-251}Cf+^{48}Ca$ reactions: Recent results and outlook
11:40	**N. Pietralla** *(Darmstadt)*:	Shell evolution and E2 collectivity: New spectroscopic information
12:05	**Y. Alhassid** *(New Haven)*:	The shell model Monte Carlo method: Recent developments and applications
12:30	**N. van Giai** *(Orsay)*:	Relativistic Hartree-Fock-Bogoliubov theory: Applications to spherical nuclei
12:55	**A. Zilges** *(Köln)*:	The low lying electric multipole response of nuclei: New experimental approaches
13:15	**J.M.G. Gómez** *(Madrid)*:	Experimental evidence of chaos in the bound states of ^{208}Pb

13:35 Session close

Chairperson: **A. Di Pietro** (*Catania*)

15:00	**C.M. Petrache** *(Orsay)*:	Exotic rotations in lanthanides
15:25	**H. Naidja** *(Strasbourg)*:	Spectroscopic properties and collectivity in the nuclei above ^{132}Sn core

15:45 Coffee Break

16:10 **Discussion:** The role of nuclear structure towards the new frontier in nuclear physics
S. Bacca *(Vancouver)*, **G. de Angelis** *(Legnaro)*, **F. Iachello** *(New Haven)*

18:10 Session close

IOP Conf. Series: Journal of Physics: Conf. Series **966** (2018) 011003 doi:10.1088/1742-6596/966/1/011003

12th INTERNATIONAL SPRING SEMINAR ON NUCLEAR PHYSICS

CURRENT PROBLEMS AND PROSPECTS FOR NUCLEAR STRUCTURE

SANT'ANGELO D'ISCHIA, MAY 15-19, 2017

List of participants

Y. Alhassid	Center for Theoretical Physics, Sloane Physics Laboratory, Yale University New Haven, Connecticut 06520, USA *yoram.alhassid@yale.edu*
A. Andrighetto	Istituto Nazionale di Fisica Nucleare, Laboratori di Legnaro Viale dell'Università 2, 35020 Legnaro (PD), Italy *alberto.andrighetto@lnl.infn.it*
K. Arnswald	Institut für Kernphysik, Universität zu Köln, D-50937 Köln, Germany *karnswald@ikp.uni-koeln.de*
T. Aumannn	Technische Universität Darmstadt, and GSI Helmholtzzentrum für Schwerionenforschung, D-64291 Darmstadt, Germany *t.aumann@gsi.de*
S. Bacca	Theory Department, Science Division, TRIUMF, 4004 Wesbrook Mall, Vancouver, BC, Canada V6T 2A3 *bacca@triumf.ca*
D. L. Balabanski	ELI-NP, IFIN-HH 30 Reactorului Str., RO-077125 Magurele-Bucharest, Romania *dimiter.balabanski@eli-np.ro*
C. Barbieri	Department of Physics, University of Surrey, Guildford GU2 7XH, UK *C.Barbieri@surrey.ac.uk*
B. R. Barrett	Department of Physics, PO Box 21008, University of Arizona, Tucson, Arizona 85721-0081, USA *bbarrett@Physics.arizona.edu*
G. Benzoni	Istituto Nazionale di Fisica Nucleare, Sezione di Milano Via Celoria 16, I-20133 Milano, Italy *giovanna.benzoni@mi.infn.it*
M. J. Garcia Borge	ISOLDE-EP, CERN, 1121 Genève 23, Switzerland and Instituto de Estructura de la Materia, CSIC, Serrano 113bis, E-28006-Madrid, Spain *mgb@cern.ch*
B. A. Brown	Department of Physics and Astronomy, and National Superconducting Cyclotron Laboratory, Michigan State University East Lansing, Michigan 48824-1321, USA *brown@nscl.msu.edu*
F. Cappuzzello	Dipartimento di Fisica ed Astronomia, Università di Catania, and INFN-LNS Via S. Sofia 64, 95123 Catania, Italy *cappuzzello@lns.infn.it*
D. Carbone	Istituto Nazionale di Fisica Nucleare, Laboratori Nazionali del Sud Via S. Sofia 64, 95123 Catania, Italy *carboned@lns.infn.it*
	I

M. Cavallaro	Istituto Nazionale di Fisica Nucleare, Laboratori Nazionali del Sud Via S. Sofia 64, 95123 Catania, Italy *manuela.cavallaro@lns.infn.it*
M. Cicerchia	Istituto Nazionale di Fisica Nucleare, Laboratori di Legnaro Viale dell'Università 2, 35020 Legnaro (PD), Italy *cicerchia@lnl.infn.it*
L. Coraggio	Istituto Nazionale di Fisica Nucleare, Sezione di Napoli Complesso Universitario di Monte S. Angelo, Via Cintia, 80126 Napoli, Italy *coraggio@na.infn.it*
R. Crespo	Departamento de Fisica, Instituto Superior Técnico, Universidade de Lisboa Av. Rovisco Pais 1, 1049-001, Lisboa, Portugal *respo@tecnico.ulisboa.pt*
C. H. Dasso	Departamento de Física Atómica, Molecular y Nuclear, Universidad de Sevilla Apartado 1065, E-41080 Sevilla, Spain *dasso@us.es*
G. de Angelis	Istituto Nazionale di Fisica Nucleare, Laboratori Nazionali di Legnaro Viale dell'Università 2, 35020 Legnaro (PD), Italy *deangelis@lnl.infn.it*
L. De Angelis	Istituto Nazionale di Fisica Nucleare, Sezione di Napoli Complesso Universitario di Monte S. Angelo, Via Cintia, 80126 Napoli, Italy *deangelis@na.infn.it*
G. de France	GANIL, CEA/DRF-CNRS/IN2P3, BP 55027, 14076 Caen Cedex 5, France gilles.defrance@ganil.fr
G. De Gregorio	Dipartimento di Fisica, Università di Napoli Federico II Complesso Universitario di Monte S. Angelo, Via Cintia, 80126 Napoli, Italy *degregorio@na.infn.it*
V. Della Rocca	Gran Sasso Science Institute Viale Francesco Crispi 7, 67100 L'Aquila, Italy *valeria.dellarocca@gssi.it*
N. Dinh Dang	Quantum Hadron Physics Laboratory, RIKEN Nishina Center for Accelerator-Based Science 2-1 Hirosawa, Wako city, 351-0198 Saitama, Japan *dang@riken.jp*
A. Di Pietro	Istituto Nazionale di Fisica Nucleare, Laboratori Nazionali del Sud Via S. Sofia 64, 95123 Catania, Italy dipietro@lns.infn.it
A. Faessler	Institut für Theoretische Physik, Universität Tübingen Auf der Morgenstelle 14, D-72076 Tübingen, Germany *faessler@uni-tuebingen.de*
B. Fernández Domínguez	Universidade de Santiago de Compostela 15754 Santiago de Compostela, Spain *beatriz.fernandez.dominguez@usc.es*
L. S. Ferreira	Center of Physics and Engineering of Advanced Materials, CeFEMA, Instituto Superior Técnico, Universidade de Lisboa Av. Rovisco Pais, P1049 001, Lisboa, Portugal *flidia@ist.utl.pt*
L. M. Fraile	Universidad Complutense de Madrid Av. Séneca, 2, 28040 Madrid, Spain *mfraile@ucm.es*

T. Fukui	Istituto Nazionale di Fisica Nucleare, Sezione di Napoli Complesso Universitario di Monte S. Angelo, Via Cintia, 80126 Napoli, Italy *fukui@na.infn.it*
M. Gai	Laboratory for Nuclear Science at Avery Point, University of Connecticut 1084 Shennecossett Rd, Groton, CT 06340-6097, USA *moshe.gai@uconn.edu*
A. Gal	Racah Institute of Physics, The Hebrew University Jerusalem 91904, Israel *avragal@savion.huji.ac.il*
D. Gambacurta	Extreme Light Infrastructure - Nuclear Physics (ELI-NP) Măgurele, Romania *danilo.gambacurta@eli-np.ro*
P. E. Georgoudis	Racah Institute of Physics, The Hebrew University Jerusalem 91904, Israel *panos@phys.huji.ac.il*
A. Gargano	Istituto Nazionale di Fisica Nucleare, Sezione di Napoli Complesso Universitario di Monte S. Angelo, Via Cintia, 80126 Napoli, Italy *gargano@na.infn.it*
J. N. Ginocchio	Theoretical Division, Los Alamos National Laboratory MS B283 Los Alamos, NM 87545, USA *gino@lanl.gov*
J. M. G. Gómez	Grupo de Física Nuclear, Facultad de Ciencias Físicas, Universidad Complutense de Madrid E-28040 Madrid, Spain *gomezk@nuc1.fis.ucm.es*
I. N. Gratchev	LPSC, Université Grenoble-Alpes, CNRS/IN2P3, Institut National Polytechnique de Grenoble, F-38026 Grenoble Cedex, France *grachev@lpsc.in2p3.fr*
P. Gulshani	NUTECH Services 3313 Fenwick Crescent, Mississauga, Ontario, Canada L5L 5N1 *matlap@bell.net*
G. Hagen	Physics Division, Oak Ridge National Laboratory, Oak Ridge, TN 37831, USA and Department of Physics and Astronomy, University of Tennessee, Knoxville, TN 37996, USA *hageng@ornl.gov*
D. Hasell	Massachusetts Institute of Technoology Cambridge, MA, 02139, USA *hasell@mit.edu*
A. Heusler	Gustav-Kirchhoff-Strasse 7/1 D-69120 Heidelberg, Germany *A.Heusler@mpi-hd.mpg.de*
H. Heylen	Max-Planck-Institut für Kernphysik D-69117 Heidelberg, Germany *hanne.heylen@cern.ch*
J. D. Holt	TRIUMF 4004 Wesbrook Mall, Vancouver BC V6T 2A3 Canada *jholt@triumf.ca*
M. Horoi	Department of Physics, Central Michigan University Mount Pleasant, Michigan 48859, USA *mihai.horoi@cmich.edu*

F. Iachello	Center for Theoretical Physics, Sloane Laboratory, Yale University, New Haven, Connecticut 06520-8120, USA *Francesco.Iachello@yale.edu*
N. Itaco	Università della Campania and INFN- Sezione di Napoli Viale Lincoln 5, 81100 Caserta, Italy *itaco@na.infn.it*
A. Jungclaus	Consejo Superior de Investigaciones Cientificas C/ Serrano 113bis, E-28006 Madrid, Spain *andrea.jungclaus@csic.es*
S. Karampagia	National Superconducting Cyclotron Laboratory, Michigan State University East Lansing, MI 48824-1321, USA *karampag@nscl.msu.edu*
H. Kaur	Department of Physics, Guru Nanak Dev University Amritsar-143 005, India *harjeet.phy@gndu.ac.in*
J. Kvasil	Institute of Particle and Nuclear Physics MFF UK, Charles University V Holesovickach 2, 180 00, Prague, Czech Republic *kvasil@ipnp.troja.mff.cuni.cz*
M. Laassiri	LPNR Laboratory, Faculty of Sciences, Mohammed V University Rabat, Morocco *mounia.laassiri@gmail.com*
M. La Commara	Università di Napoli Federico II and INFN-Sezione di Napoli, Complesso Universitario di Monte S. Angelo, Via Cintia, 80126 Napoli, Italy *lacommara@na.infn.it*
K. Langanke	GSI Helmholtzzentrum für Schwerionenforschung, D-64291 Darmstadt, Germany *k.langanke@gsi.de*
S. M. Lenzi	Dipartimento di Fisica e Astronomia dell'Università e INFN-Sezione di Padova Via Marzolo 8, 35131 Padova, Italy *lenzi@pd.infn.it*
J. Li	Peking University, Beijing 100871, China *jianguo_li@pku.edu.cn*
S. Lunardi	Dipartimento di Fisica e Astronomia dell'Università e INFN-Sezione di Padova Via Marzolo 8, 35131 Padova, Italy *lunardi@pd.infn.it*
Y. Ma	Peking University, Beijing 100871, China *yuanzhuoma@pku.edu.cn*
R. Machleidt	Department of Physics, University of Idaho Moscow, ID 83844-0903, USA *machleid@uidaho.edu*
C. Michelagnoli	Institut Laue-Langevin 71 Avenue des Martyrs, 38000 Grenoble, France *michelagnolic@ill.fr*
A. Muoio	Istituto Nazionale di Fisica Nucleare, Laboratori Nazionali del Sud Via S. Sofia 64, 95123 Catania, Italy *muoio@lns.infn.it*
H. Naidja	Université de Strasbourg, Strasbourg, France Constantine 1 University, 25000 Constantine, Algeria *houda.naidja@iphc.cnrs.fr*

V. Nesterenko	Bogoliubov Laboratory of Theoretical Physics, Joint Institute for Nuclear Research, Dubna Moscow region, 141980, Russia *nester@theor.jinr.ru*
Y. Niu	ELI-NP, Horia Hulubei National Institute for Physics and Nuclear Engineering 30 Reactorului Street, RO-077125, Bucharest-Magurele, Romania *nyfster@gmail.com*
F. Nowacki	Université de Strasbourg, IPHC 23 rue du Loess 67037 Strasbourg, France *frederic.nowacki@iphc.cnrs.fr*
T. Otsuka	RIKEN Nishina Center, RIKEN, Wako-shi, Saitama, and Center for Nuclear Study, University of Tokyo, Hongo, Tokyo, Japan *otsuka@phys.s.u-tokyo.ac.jp*
C. Parascandolo	Istituto Nazionale di Fisica Nucleare, Sezione di Napoli, Complesso Universitario di Monte S. Angelo, Via Cintia, 80126 Napoli, Italy *Concetta.parascandolo@na.infn.it*
S. Pastore	Theoretical Division, Los Alamos National Laboratory, Los Alamos, NM 87545, USA *saori@lanl.gov*
J. Pei	School of Physics and State Key Laboratory of Nuclear Physics and Technology, Peking University Beijing 100871, China *peij@pku.edu.cn*
C. M. Petrache	Centre de Sciences Nucléaires et de Sciences de la Matière, Université Paris-Sud, CNRS/IN2P3, CNRS-IN2P3, Université Paris-Saclay, Orsay, France *petrache@csnsm.in2p3.fr*
N. Pietralla	Institut für Kernphysik, TU Darmstadt, Schlossgartenstraße 9, 64289 Darmstadt, Germany *pietralla@ikp.tu-darmstadt.de*
P. Pragati	Department of Physics, Indian Institute of Technology Roorkee Roorkee - 247667, INDIA *pragati.popli88@gmail.com*
F. Recchia	Università degli Studi di Padova and INFN-Sezione di Padova, I-35131 Padova, Italy *francesco.recchia@unipd.it*
P. Reiter	Institut für Kernphysik, Universität zu Köln Zülpicher Straße 77, D-50937 Köln, Germany *preiter@ikp.uni-koeln.de*
W. A. Richter	iThemba LABS, Somerset West, South Africa *richter@sun.ac.za*
P. Ring	Physics Department, Technical University of Munich, 85748 Garching, Germany *ring@ph.tum.de*
H. Sagawa	RIKEN Nishina Center Wako 351-0198, Japan *sagawa@ribf.riken.jp*
M. Sambataro	Istituto Nazionale di Fisica Nucleare, Sezione di Catania Via S. Sofia 64, I-95126 Catania, Italy *michelangelo.sambataro@ct.infn.it*
F. Sammarruca	Physics Department, University of Idaho Moscow, Idaho 83844-0903, USA *fsammarr@uidaho.edu*

IOP Conf. Series: Journal of Physics: Conf. Series **966** (2018) 011003 doi:10.1088/1742-6596/966/1/011003

N. Sandulescu	National Institute of Physics and Nuclear Engineering Bucharest, Romania *sandulescu@theory.nipne.ro*
S. Shlomo	Cyclotron Institute, Texas A&M University College Station, Texas 77840, USA *s-shlomo@tamu.edu*
T. Suzuki	Department of Physics, College of Humanities and Sciences, Nihon University, Setagaya-ku, Tokyo 156-8550, Japan *suzuki@phys.chs.nihon-u.ac.jp*
I. Talmi	Department of Physics, The Weizmann Institute of Science, Rehovot 76100, Israel *igal.talmi@weizmann.ac.il*
Yu. M. Tchuvil'sky	Dukhov Research Institute for Automatics 127055, Moscow, Russia *tchuvl@nucl-th.sinp.msu.ru*
J. J. Valiente Dobón	Istituto Nazionale di Fisica Nucleare, Laboratori Nazionali di Legnaro, Viale dell'Università 2, 35020 Legnaro (Pd), Italy *valiente@lnl.infn.it*
N. Van Giai	Institut de Physique Nucléaire, IN2P3-CNRS F-91406 Orsay Cedex, France nguyen@ipno.in2p3.fr
U. van Kolck	Institut de Physique Nucléaire, CNRS/IN2P3 91406 Orsay, France *vankolck@ipno.in2p3.fr*
P. Veselý	Nuclear Physics Institute, Czech Academy of Sciences, 250 68 Rez, Czech Republic *p.vesely@ujf.cas.cz*
A. Vitturi	Dipartimento di Fisica e Astronomia Università di Padova, Via Marzolo 8, 35131 Padova, Italy *vitturi@pd.infn.it*
A. A. Voinov	Flerov Laboratory of Nuclear Reaction, Joint Institute for Nuclear Research, RU-141980 Dubna, Russian Federation voinov@jinr.ru
J. Wilhelmy	Institut für Kernphysik, University of Cologne, 50937 Köln, Germany *wilhelmy@ikp.uni-koeln.de*
H. Witala	Jagiellonian University Kraków, Poland *ufwitala@cyf-kr.edu.pl*
F. Xu	School of Physics, Peking University Beijing 100871, China *frxu@pku.edu.cn*
A. Zdeb	Katedra Fizyki Teoretycznej, Uniwersytet Marii Curie-Skodowskiej w Lublinie, Poland *azdeb@kft.umcs.lublin.pl*
V. Zelevinsky	Department of Physics and Astronomy and National Superconducting Cyclotron Laboratory, Michigan State University East Lansing, Michigan 48824-1321, USA *Zelevins@nscl.msu.edu*
A. Zilges	Institut für Kernphysik, University of Cologne, 50937 Köln, Germany *zilges@ikp.uni-koeln.de*

IOP Conf. Series: Journal of Physics: Conf. Series **966** (2018) 011004 doi:10.1088/1742-6596/966/1/011004

Peer review statement

All papers published in this volume of *Journal of Physics: Conference Series* have been peer reviewed through processes administered by the proceedings Editors. Reviews were conducted by expert referees to the professional and scientific standards expected of a proceedings journal published by IOP Publishing.

IOP Conf. Series: Journal of Physics: Conf. Series **966** (2018) 012001 doi:10.1088/1742-6596/966/1/012001

Electron capture and the neutrino mass.

Amand Faessler
Institute of Theoretical Physics, University of Tuebingen, Germany,
E-mail: faessler@uni-tuebingen.de

Abstract. The electron neutrino mass can be determined by electron capture. One expects the largest influence of the neutrino mass on this decay for a small Q value of Q = 2.8 keV for ${}^{163}_{67}$Ho + e $\rightarrow$ ${}^{163}_{66}$Dy + ν. The energy of the Q value is distributed to the emitted neutrino and the excitation of the Dy atom. Thus the energy difference between the Q value and the upper end of the deexcitation spectrum is the electron neutrino mass. The electron wave functions are calculated selfconsistently by the Dirac-Hartree-Fock approach for the bound and the continuum states. To extract the neutrino mass from the spectrum is only possible, if the background is reduced relative to the present situation. The analysis presented here shows, that the determination of the electron neutrino mass by electron capture is difficult, but seems not to be impossible.

1. Introduction

The neutrino is one of the least known particles, although it plays a central role in weak interactions. We do not know the mass of the three neutrinos: electron, muon and tauon neutrinos (and antineutrinos) [1, 2, 3, 4, 5, 6, 7, 9]. We do not know their hierarchy. The nature of the neutrinos, Dirac (neutrino different from the antineutrino) or Majorana (neutrino identical with the corresponding antineutrino) is still unknown [10, 11, 12, 13, 14].

The topic of this contribution is the theoretical side of the determination of the electron neutrino mass by electron capture in ${}^{163}_{67}$Ho as developed by the Tuebingen group and coworkers [5, 15, 16, 17, 18]. For the capture

$$^{163}Ho + e \rightarrow ^{163} Dy + \nu \tag{1}$$

the Q value obtained by a Penning trap measurement by Eliseev et al. [19] (see also [20, 21, 22]) is:

$$Q = 2.833(30_{stat})(15_{sys})\ keV \tag{2}$$

This means, that electron capture in ^{163}Ho is only possible from electron orbits bound less than Q = 2.8 keV. So capture is only possible from $3s_{1/2}, M1$ with a binding energy 2.128 keV and less bound electron states. In addition the electron orbits in Ho, from which an electron can be captured must have an overlap with the nucleus to react with a proton. The energy of the Q value after capture is distributed over the excitation energy of the Dy atom (one-, two-, three-hole and higher excitations) and the emitted neutrino. The sum of the deexcitation energy can be measured by a bolometer. The maximum bolometer energy yields then the minimum neutrino

energy, which is the electron neutrino rest mass. Thus the neutrino mass can be determined by the difference of the Q-value and the upper end of the bolometer spectrum. The excitation energy of the Dy* is calculated in the sudden approximation. The excitation of the Dy* is then given by the overlap with Ho* with an electron hole in the capture orbit with the different excited configurations $|D>$ in the daughter Dy*.
According to a publication by Intemann and Polock [8] DeRujula and Lusignoli [9] use non-relativistic pure Coulomb wave functions for the electrons in Ho, which do not allow capture from $np_{1/2}$. The effect of the partial screening of the nuclear Coulomb potential by the electron cloude is considered by effective charges. Since in reference [9] the perturbation approach is used to calculate the Dy* wave functions, this requires orthonormal basis functions. Thus ref. [9] uses for each capture process, e. g. from $3s_{1/2}$, the same screening factor for all single electron orbits needed in the perturbation approach for this special capture process. Starting from these "bad" wave functions for Ho ref. [9] calculates as proposed by [8] the bound and continuum electron wave functions for Dy* with the perturbation:

$$H'(r) = +1/r - \int d^3r_1 \, |\varphi_{3s/4s}(\vec{r}_1)|^2/|\vec{r} - \vec{r}_1|. \tag{3}$$

The first term takes into account, that in Dy one has one proton less compared to Ho. The second term originates from the electron hole in the $3s_{1/2}$ or the $4s_{1/2}$ state, which are the only hole states considered in ref. [9]. We use here selfconsistent Dirac-Hartree-Fock wave functions caculated specifically for Ho* and Dy* [23, 24, 25]. The continuum electron wave functions in Yb* are determined [26] in the selfconsistent Z-1 = 65 electron potential orthogonal to all 65 bound electrons in Yb.

2. Description of Electron Capture

The spectrum of the the decay of the excited $^{163}Dy^*$ after electron capture in ^{163}Ho refs. [4, 9] and [5] assuming Lorentzian shapes for the decay resonances are:

$$\frac{d\Gamma}{dE_c} \propto \sum_{i=1,\dots N_\nu} (Q - E_c) \cdot U_{e,i}^2 \cdot \sqrt{(Q-E_c)^2 - m_{\nu,i}^2} \, *$$

$$(\sum_{f=f'} \lambda_0 B_f \frac{\Gamma_{f'}}{2\pi} \frac{1}{(E_c - E_{f'})^2 + \Gamma_{f'}^2/4} + \tag{4}$$

$$\sum_{f=f';p'<F;q'_b>F} \lambda_0 B_{f,p'<F;q'_b>F} \frac{\Gamma_{f',p'}}{2\pi} \frac{1}{(E_c - E_{f',p'})^2 + \Gamma_{f',p'}^2/4} +$$

$$\int dk_{q'} \lambda_0 B_{f,p'<F;q'_c>0} \frac{\Gamma_{f',p',q'}}{2\pi} \frac{1}{(E_c - E_{f',p',q'})^2 + \Gamma_{f',p',q'}^2/4})$$

The factor in front of the bracket is the same as for the single beta decay. It reflects the phase space of the emitted neutrino (1). The three expressions in the bracket of eq. (4) describe the decay of the 1-hole f' excitations in Dy*, of the 2-hole excitations f', p' with shake-up of p' to q' into a bound orbit and of 2-holes f', p' states with one electron p' moved to q' into the continuum (shake-off). The last term has as the others a dimensionless strength $dk_{q'} \cdot B_{f',p'<F,q'>o}$. The transformation to an integral over the energy yields the factor in front of the third term, which can be deduced from eq. (11). Different neutrino mass eigenstates $|i>$ are mixed into the electron neutrino flavor states by $U_{e,i}^2$. The Q-value is given by eq. (2)

IOP Conf. Series: Journal of Physics: Conf. Series **966** (2018) 012001 doi:10.1088/1742-6596/966/1/012001

from the ECHo collaboration [20, 21, 27, 28, 29]. The excitation energy of the final Dy* is E_c. The energy $Q - E_c$ is carried away by the neutrino. B_f, $B_{f,p'<F;q'_b>F}$ and $B_{f,p'<F;q'_c>0}$ are the overlap and exchange corrections for the 1-hole, the bound 2-hole and the shake-off 2-hole states. λ_0 contains the nuclear matrix element squared [30]. Eq. (4) yields the spectrum in arbitrary units and must be normalized at one energy to the data. We use the N1, $4s_{1/2}$ peak for the normalisation. $E_{f'}$, $E_{f',p'}$ and $E_{f',p';q'>0}$ are the 1-hole, the 2-hole shake-up and the 2-hole shake-off excitation energies in Dysprosium (see refs. [7] and [10]). $\Gamma_{f'}$, $\Gamma_{f',p'}$ and $\Gamma_{f',p';q'>0}$ are the widths of the one- and two-hole states and the two-hole states with shake-off in Dysprosium [15, 16, 17, 18].

The overlap and exchange corrections B_f, $B_{f,p'<F;q'_b>F}$ and $B_{f,p'<F;q'_c>0}$ are calculated in second quantisation with Dirac-Hartree-Fock (DHF) electron wave functions. This yields automatically the antisymmetrisation and the exchange terms. The antisymmetrized two-hole state in Dy with shake-off is for example:

$$|A'_{p',f':q'>0}>= a_1'^{\dagger} a_2'^{\dagger} ... a_{f'-1}'^{\dagger} a_{f'+1}'^{\dagger} ... a_{p'-1}'^{\dagger} a_{p'+1}'^{\dagger} ... a_Z'^{\dagger} a_{q'>0}'^{\dagger} |0> \qquad (5)$$

The probability to form a two-hole shake-off state is proportional to ($|G>$ is the DHF groundstate of Holmium):

$$P_{f',p';q'>0} = |< A'_{f',p';q'}|a_i|G>|^2 \qquad (6)$$

3. The electron wave functions in the Dy Continuum.

The selfconsistent Coulomb field for the shake-off electron is in atomic units:

$$V_{shake-off}(r) = \frac{-66}{r} + \sum_{k\ occupied\ e} g'_k \int d^3 r \cdot \frac{|\varphi_k(\vec{r'})|^2}{|\vec{r} - \vec{r'}|} \qquad (7)$$

$g'_k s$ are the number of bound electrons in the selfconsistent occupied orbits $|k> = |n, \ell, j>$ of Dy. To determine the potential for the shake-off electrons in Dy one needs the occupied selfconsistent Dirac-Hartree-Fock upper $P_k(r)$ and lower $Q_k(r)$ components of the selfconsistent electron wave functions.

$$\varphi_k(\vec{r}) * r = (\ P_k(r); Q_k(r)) \qquad (8)$$

$$V_{shake-off} \approx -\frac{66}{r} + \frac{65}{r} \cdot [1 - exp(-a \cdot r)]\ [1/(length = a.u.)];\ \ a = 3.4 \qquad (9)$$

We adjust an analytic expression (9) to the DHF potential (see figure 1). The selfconsistent Dy potential for the 66th. electron in the continuum is shown in figure 1 for Dy. For the normalization of the continuum Dirac wave function we follow here this recommendation of Perger and Karrighattam on their page 394 [32]. In the asymptotic expression (10) η is the Sommerfeld parameter, Δ the Coulomb phase shift and δ takes into account deviations from a pure Coulomb potential. The electron energy in the continuum can be due to energy conservation not larger than the Q-value of 2.8 keV minus the excitation energy of the two hole state.

$$\int_{r=0\ to\ \infty} dr \cdot 2 \cdot \sin(kr - \ell \cdot \frac{\pi}{2} - \eta \ln 2kr + \Delta + \delta) \cdot \qquad (10)$$
$$2 \cdot \sin(k'r - \ell \cdot \frac{\pi}{2} - \eta \ln 2kr + \Delta + \delta) \approx 2\pi \cdot \delta(k - k').$$

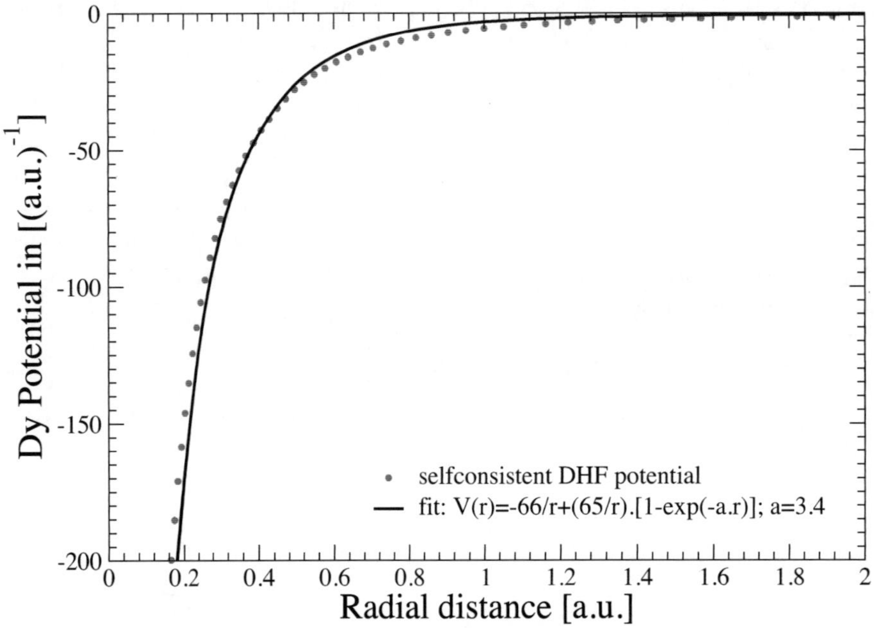

Figure 1. Selfconsistent DHF potential in Dy* with Z = 66 for a continuum electron (No 66) of 65 electrons (dimension: 1/length = atomic units) in the core and analytic approximation with a = 3.4 (See eq. 9).

The wave number is connected with the relativistic and non-relativistic energies by the equations:

$$E_{rel}^2 = c^{2-}\hbar k^2 + m^2 c^4 \;\rightarrow\; c^2 k^2 + c^4 \;\; (in\ atomic\ units); \;\; E_{n-rel} = \frac{1}{2}k^2;$$

$$k = \alpha\sqrt{E_{n-rel}(E_{n-rel} + 2\cdot c^2)}; \quad with\ c \;=\; 1/\alpha \;=\; 137.035999\ in\ [atomic\ units]. \qquad (11)$$

4. Shake-off Process.

The overlap between the bound Ho states $ns_{1/2}$ and $np_{1/2}$ with $n \geq 3$ and the continuum wave functions in Dy squared gives the probability for shake-off. We restrict this work to s-wave shake-off, which is expeted to be the largest contribution. The s-wave shake-off is given by the overlap $< n \geq 3, s_{1/2}, Ho|E,\ s, Dy >$ squared. In the spectrum with 1-hole, 2-hole and shake-off in figure 2 the shake-off contribution is hardly visible. Here and also in ref. [9] only the decay width of the 2-hole states are included. Three hole states can be neglected [17]. The electron in the continuum has an escape width, which is not included.

Figure 2 shows the logarithmic spectrum of the 1-hole, the 2-hole and the s-wave shake-off contributions. The shake-off spectrum can hardly be seen on this scale in the total spectrum.

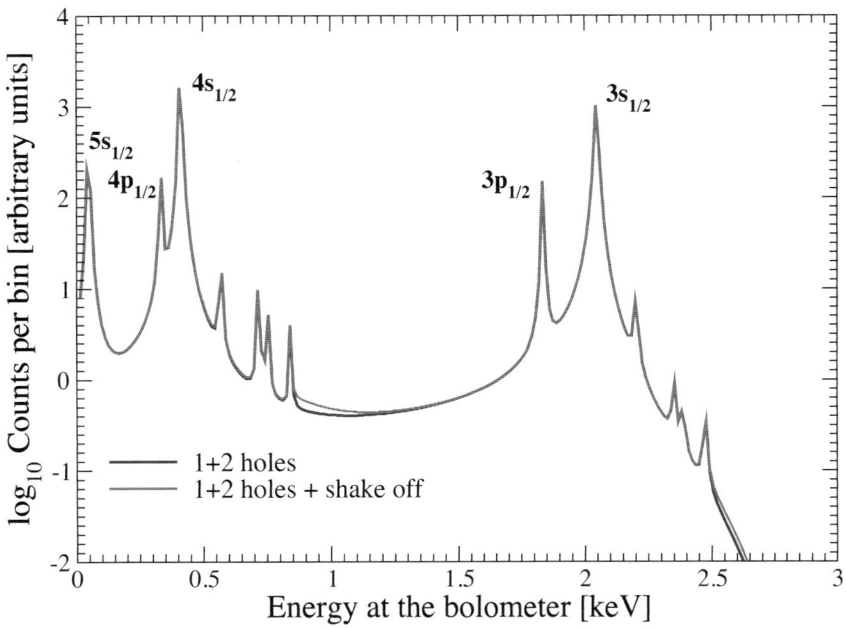

Figure 2. Theoretical results in arbitrary units of the sum of the one- and two-hole deexcitations compared to the sum of the one-, two-hole and the shake-off deexcitation as measured by the bolometer spectrum (4). The arbitrary units are adjusted to the experimental N1, $4s_{1/2}$ 1-hole peak. The nature of the one hole states are indicated. The two-hole peaks are by about two orders of magnitudes smaller than the one hole peaks. Shake-off can almost not been seen.

5. Summary

In this contribution I summarized our theoretical work on the determination of the electron neutrino mass by electron capture [5, 7, 15, 16, 17, 18] in $^{163}_{67}$Ho to $^{163}_{66}$Dy. The energy of the Q-value $= 2.833 \pm 0.05\ keV$ is distributed to the excitation of Dy and the emission of a neutrino. The recoil of ^{163}Ho can be neglected. The total deexcitation energy as X-rays and Auger electrons of Dy can be measured in a bolometer. The upper end of the spectrum requires the minimum energy of the neutrino. Thus the difference of Q minus this upper limit of the deexcitation spectrum is the rest mass of the electron neutrino.

We describe the bound states in Ho and Dy by the Dirac-Hartree-Fock approach [23, 24, 25] even including occupations in Dy due to different hole states. The s-wave continuum wave functions in Dy are determined with the Dirac equation in the selfconsistent potential [26]. (See figure: 1). The energy of the capture orbit in Ho and the excitations of Dy* including also the continuum states are limited by energy conservation to the Q-value minus the excited one-hole states in Ho* and Q minus the ecitation energies in Dy*.

Remaining discrepancies between theory and experiment could be due to configuration mixing not included here. Finally we want to stress, that the accuracy needed to extract the neutrino mass can not be obtained by theoretical calculation alone. One must fit simultaneously the neutrino mass, the Q-value, the excitation energy of the highest resonance, their width and their strength at the upper end of the spectrum to extremely accurate data. For this the present

IOP Conf. Series: Journal of Physics: Conf. Series **966** (2018) 012001 doi:10.1088/1742-6596/966/1/012001

background of the experimental spectrum has to be drastically reduced. The measurement must probably be done in an underground lab to protect against cosmic rays. This reduces the background as a first measurement in the Modane underground lab in the Fréjus tunnel has shown [33]. If the background can be reduced by one to two orders, it should be enough to adjust only the two parameters, the Q-value and the neutrinos mass, to the upper end of the spectrum for the determination of the electron neutrino mass.

Acknowledgements
I want to thank Christian Enss, Loredana Gastaldo and Fedor Šimkovic for their collaboration in earlier works [5, 15, 16, 17, 18] on this topic.

References

[1] Drexlin G, Hannen V, Mertens S and Weinheimer C 2013 *Adv. High Energy Phys.* **2013** 293986 (*Preprint* physics.ins-de/13077.0101v1)
[2] Faessler A, Huster E, Kraft O and Krahn F 1970 *Z. Physik* **238** 352
[3] Otten E W and Weinheimer C 2008 *Rep. Prog. Phys.* **71** 086201
[4] De Rujula A 2013 Two old ways to measure the electron-neutrino mass *Preprint* hep-ph/1305.4857v1; De Rujula A and Lusignoli M 1982 *Phys. Lett. B* **118** 429
[5] Faessler A, Gastaldo L and Šimkovic F 2015 *J. Phys. G* **42** 015108 (*Preprint* nucl-th/1407.6504)
[6] Robertson R G H 2015 *Phys. Rev. C* **91** 035504 (*Preprint* nucl-th/411.2906v1)
[7] Faessler A and Šimkovic F 2015 *Phys. Rev. C* **91** 045505 (*Preprint* hep-hp:/1501.04338)
[8] Intemann R L and Polock F 1967 *Phys. Rev.* **157** 41
[9] De Rujula A and Lusignoli M 2016 *JHEP* 1605 015 (*Preprint* hep-ph/1601.04990)
[10] Faessler A, Hodak R, Kovalenko S, and Šimkovic F 2013 arXiv:1304.5032v3, and *European Physical Journal Web of Conferences*
[11] Faessler A 2009 *Nucl. Phys. Proc. Suppl.* **188** 20
[12] Šimkovic F, Faessler A, Rodin V, Vogel P and Engel J 2008 *Phys. Rev. C* **77** 045503
[13] Vogel P and Engel J 2008 *Phys. Rev. C* **77** 045503
[14] Avignone F, Elliott S R and Engel J 2008 *Rev. Mod. Phys.* **80** (2008) 481
[15] Faessler A, Gastaldo L and Šimkovic F 2017 *Phys. Rev. C* **95** 045502 (*Preprint* nucl-th/1611.00325)
[16] Faessler A and Simkovic F 2016 *Phys. Scripta* **91** 043007
[17] Faessler A, Enns Ch, Gastaldo L and Šimkovic F 2015 *Phys. Rev. C* **91**, 064302
[18] Faessler A and Šimkovic F 2015 *Phys. Rev. C* **91** 045505
[19] Eliseev S, Blaum K, Block M, Chenmarek S, Dorrer H, Duellmann Ch E, Enss C, Filiamin P E, Gastaldo L, Goncharov M, Koester U, Lautenschlaeger F, Novikov Yu N, Rischka A, Schuessler R X, Schweighard L and Tuerler A 2015 *Phys. Rev. Lett.* **115** 062501
[20] Ranitzsch P C -O *et al.* 2012 *Jour. of Low Temp. Phys.* **167** 1004, and private communication
[21] Blaum K, Doerr A, Duellmann C E, Eberhardt K, Eliseev S, Enss C, Faessler A, Fleischmann A, Gastaldo L, Kempf S, Krivoruchenko M, Lahiri S, Matai M, Novikov Yu N, Ranitzsch P C, Šimkovic F, Scusc Z and Wegner M 2013 The Electron Capture ^{163}Ho Experiment ECHo *Preprint* physics.ins-det/1306.2655v1
[22] Ranitzsch P C, Hassel C, Wegner M, Kempf S, Fleischmann A, Enss C, Gastaldo L, Herlert A and Johnston K 2014 First Calorimetric Measurement of OI-line in the Electron Capture Spectrum of ^{163}Ho *Preprint* physics.ins-det/1409.007v1
[23] Grant I P 1970 *Adv. Phys.* **19** 747
[24] Desclaux J P 1975 *Comp. Phys. Com* **9** 31
[25] Ankudinov A L, Zabinsky S I and J. J. Rehr 1996 *Comp. Phys. Com.* **98** 359
[26] Salvat F, Fernandez-Varea J M and Williamson W 1995 *Comp. Phys. Com.* **90** 151
[27] Anderson J U *et al.* 1982 *Phys. Lett. B* **398** 72
[28] Gatti F *et al.* 1997 *Phys. Lett. B* **398** 415
[29] Audi G, Wapstra A H and Thibault C 2003 *Nucl. Phys. A* **279** 337
[30] Bambynek W, Behrens H, Chen M H, Crasemann B, Fitzpatrick M L, Ledingham K W D, Genz H, Mutterer M and Intemann R L 1977 *Rev. Mod. Phys.* **49** 109
[31] Williams G P 2001 *X-Ray Data Booklet* (Lawrence Berkeley National Laboratory) Table 1-1
[32] Perger W F, Vasan Karighattam 1991 *Comp. Physics Com.* **66** 392
[33] Gastaldo L private information

Recent Results from ISOLDE and HIE-ISOLDE

María J. G. Borge

Instituto de Estructura de la Materia, CSIC, Serrano113bis, 28006-Madrid, Spain
and EP-ISOLDE, CERN, Geneva, Switzerland

E-mail: mgb@cern.ch

Abstract. ISOLDE is the CERN facility dedicated to the production of rare ion beams for many different experiments in the fields of nuclear and atomic physics, materials science and life sciences. The HIE-ISOLDE, Higher Intensity and Energy upgrade has finished its stage 1 dedicated to upgrade the energy up to 5.5 MeV/u, producing the first radioactive beams with this energy in September 9th 2016. Recent results from the low energy and post-accelerated beams are given in this contribution.

1. The ISOLDE Facility

The Isotope Separator On Line, ISOLDE is the CERN facility dedicated to the production of radioactive ion beams. ISOLDE's production of radioactive nuclei began in October 1967, and half a century later, ISOLDE has developed into a versatile facility dedicated to the production of high purity rare beams for many different experiments in the fields of nuclear and atomic physics, materials science and life sciences. By a clever combination of target and ion source units pure beams of over 1000 different nuclei of 74 elements have been produced and delivered to experiments. The core of ISOLDE is its target and ion source systems, which provide a selective production of the nucleus of interest. ISOLDE produces radioactive nuclei by spallation-, fragmentation- and fission-reactions between the nuclei contained in thick targets and high energy, 1.4GeV, and high intensity, 2μA proton beam accelerated in the first circular accelerator of CERN, the Proton Synchrotron Booster (PSB). The targets are heated between 1000°C and 2000°C depending on the material to facilitate the fast release of radioactive nuclei. Target materials partly determine the radioactive nuclei production rate and their release time, thus a careful selection of "seed-material" is of paramount importance. Thus up to twenty different materials are available. Chemical selectivity is obtained by the right combination of target-ion sources giving rise to a selective production of more than 1000 isotopes of 74 different elements from He to Ra, of which germanium produced as GeS is the last addition. ISOLDE uses three types of ion sources: surface, plasma and laser, being the latter combined with the other two the one that achieves a higher degree of chemical selectivity. The Resonance Ionisation Laser Ion Source (RILIS) takes advantage of resonant excitation of atomic transitions using tunable laser radiation and allows for a high degree of isobar-free selectivity. The ISOLDE beam is then mass separated and delivered to the different experimental stations or post-accelerated. For many experiments the beam quality is of major importance. To achieve reduced emittance and energy spread an ion cooler and buncher, ISCOOL [1], was designed and installed after the high resolution separator (HRS) of ISOLDE. Using its bunching capabilities, ISCOOL permits an increase in the sensitivity for experiments using pulsed lasers such as those devoted to collinear

IOP Conf. Series: Journal of Physics: Conf. Series **966** (2018) 012002 doi:10.1088/1742-6596/966/1/012002

laser spectroscopy. In order to broaden its physics scope ISOLDE developed new ways to accelerate the singly-charged radioactive ion beams (RIB) in a universal, fast, efficient and cost-effective way. The home-made post-accelerator based on previous expertise existing in the Facility, REX-ISOLDE (Radioactive beam EXperiment at ISOLDE), is in operation since 2001. The mass separated radioactive ions are collected in an ion-trap. After cooling and bunching in the trap the ions are brought into a charge-state breeder (REXEBIS) using the technique of electron-beam ionization. The charge-bred ions passes a Nier Spectrometer, and are brought into a room temperature linear accelerator able to reach 3 MeV per nucleon (MeV/u) and A/q values up to 4.5. Although designed for low middle mass the method has proven to be universal and has provided post-accelerated beams covering the entire mass range from ^{6}He up to ^{224}Ra for reaction studies.

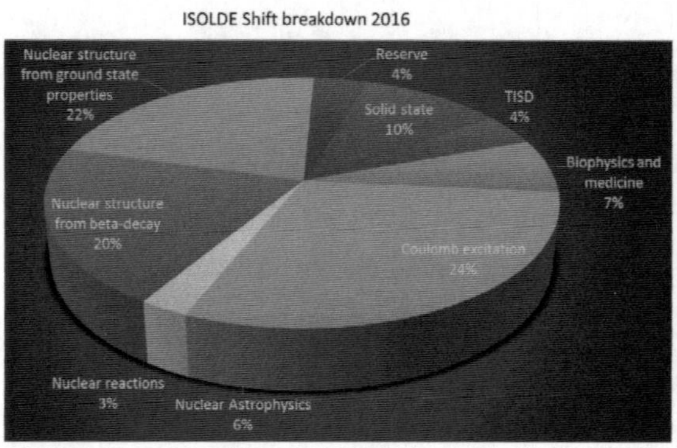

Figure 1. On the left hand side, the pie-plot shows the ISOLDE beamtime distribution in 2016 among the different physics domains. Notice that 27 % of the beam was dedicated to post-accelerated beams with energies up to 6.8 MeV/u. On the right hand side, 3D-layout of the HIE-ISOLDE post-accelerator as it looked in 2016 when stage 1 of the energy upgrade was finished. The REX-ISOLDE post-accelerator is at the right and two high-beta cryomodules are installed downstream of REX. Two beam transfer lines take the post-accelerated beam to the experiments: beamline XT01 is connected to the MINIBALL array and XT02 to the SEC scattering chamber.

2. Physics with low energy beams, Recent results

This year ISOLDE celebrates the fifty anniversary of production of radioactive beams. Since 1992 ISOLDE is connected to the PSB, that provides higher energy, increasing the cross section for spallation and fragmentation reactions than with the continuous 600 MeV proton beam of the synchrocyclotron where ISOLDE was previously located. ISOLDE is still growing and attracts presently more than 500 researchers working on 100 experiments, with a rate of about 50 experiments taking data per year. This ISOL facility has pioneered many achievements both at the level of designing new devices and of producing frontier physics. Figure 1 shows the beam distribution in 2016 with 73 % dedicated to low energy physics. In the following I present some recent highlights non covered by the contributions of G Benzoni, LM Fraile, and H. Heylen to this issue.

2.1. Beta-delayed alpha decay of ^{16}N relevant for Nuclear astrophysics

It has been known that during the helium burning process carbon is converted to oxygen through the reaction ^{12}C(α,γ)^{16}O. The astrophysical relevant center-of-mass energy for the reaction is around 0.3 MeV. At this energy the cross section is too low to be measured directly at the laboratory, and it has to be extrapolated. The present cross section estimate at this energy has an error of 20% while the desired precision is of 10 % [2]. The β-delayed alpha emission of ^{16}N has proven to be a good indirect method as it allows for the determination of the E1

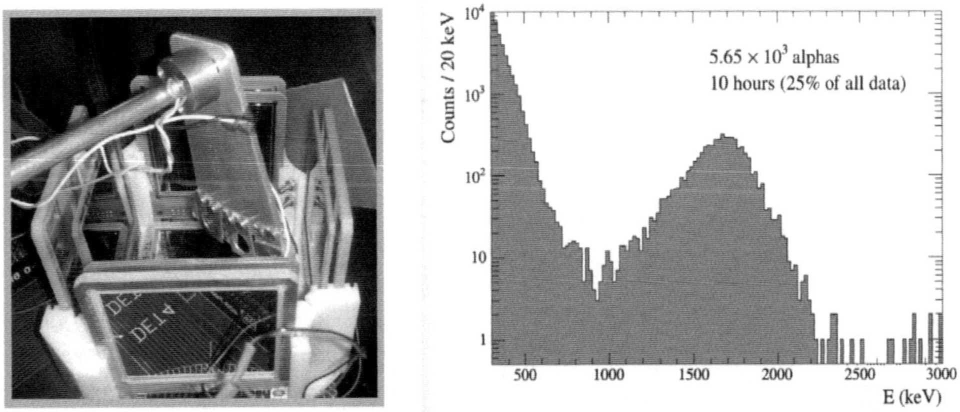

Figure 2. On the left hand side, a photo (courtesy of R. Lica) of the charged particle setup used at IDS, ISOLDE to study the $\beta\alpha$ spectrum of the ^{16}N decay. On the right hand side the $\beta\alpha$ spectrum is shown for 25% of the data. Notice the good separation between β and α in the spectrum. This separation is better for only the thinner detector of 40 μm.

component of the S-factor of this reaction. While the shape of the experimental α-spectrum is consistent among the different experiments [3, 4], this indirect method requires an improved determination of the $\beta\alpha$ branching ratio. The TUNL evaluation [5] quotes an $\beta\alpha$ branching ratio of $1.20(5)\times10^{-5}$ while recent measurement give a value of $1.49(5)\times10^{-5}$ [6]. An experiment was recently realized at the ISOLDE Decay Station (IDS) to obtain a more precise measurement of the alpha particle centroid energy and the $\beta\alpha$ branching ratio. An average of 2×10^4 ^{14}N^{16}N ions with an energy of 50 keV were stopped in a thin carbon foil surrounded by 4 telescopes with double sided Si Strip detector (DSSSD) at the front and silicon-pad detector (PAD) at the back. Another DSSSD was mounted below the carbon implantation foil. The system was surrounded by 4 HPGe clovers in the direction of the incoming beam. Figure 2 shows a photo of the very compact charged particle setup surrounded by the HPGe clovers and, on the right hand side, the charged particle spectrum for 25% of the data, where one can observe the good separation of the β and α particles obtained by the use of thin DSSSD detectors. The analysis of the data is progressing well and new results will emerge soon from this data.

2.2. Identification in ^{34}Al of the crossing of the spherical and deformed configurations at N = 20

The evolution of the N = 20 magic number has attracted a lot of attention and it has been widely studied in the past thirty years using different techniques. It has been suggested that the magicity persists along the N = 20 isotonic chain, from ^{40}Ca (Z = 20) down to ^{34}Si (Z = 14), and suddenly disappears only two protons below, in the deformed nuclide ^{32}Mg [7]. Although the crossing should occur between ^{32}Mg and ^{34}Si, the ground state of ^{33}Al (N = 20) is outside the island of inversion [8, 9]. ^{34}Al (N = 21) is a good candidate to explore this crossing as it has two β-decaying states with spin values 4^- and 1^+, with normal and intruder configurations, respectively. In addition, an abnormal crossover on the S_{2n} surface has been found between the Al and Mg chains at N = 21 [10], suggesting a structural change there.

The ^{34}Al excited structure was studied by the β-decay of ^{34}Mg produced at ISOLDE by fragmentation in an UCx target produced by the impact of a 2μA pulsed proton beam of 1.4 GeV energy. The average ^{34}Mg production rate obtained from $\beta\gamma$ measurements was of 700 ions/s. The β-decay half-life of ^{34}Mg was determined from the weighted average values extracted by gating on the most intense transitions in ^{34}Al at 364.5, 441.2, 975.0 and 1051.7 keV. The value obtained for the half-life was of 44.9(4) ms, twice as long as the previously measured value of 20(10) ms [11] obtained from βn data. Study of the $\beta\gamma$-coincidences and the

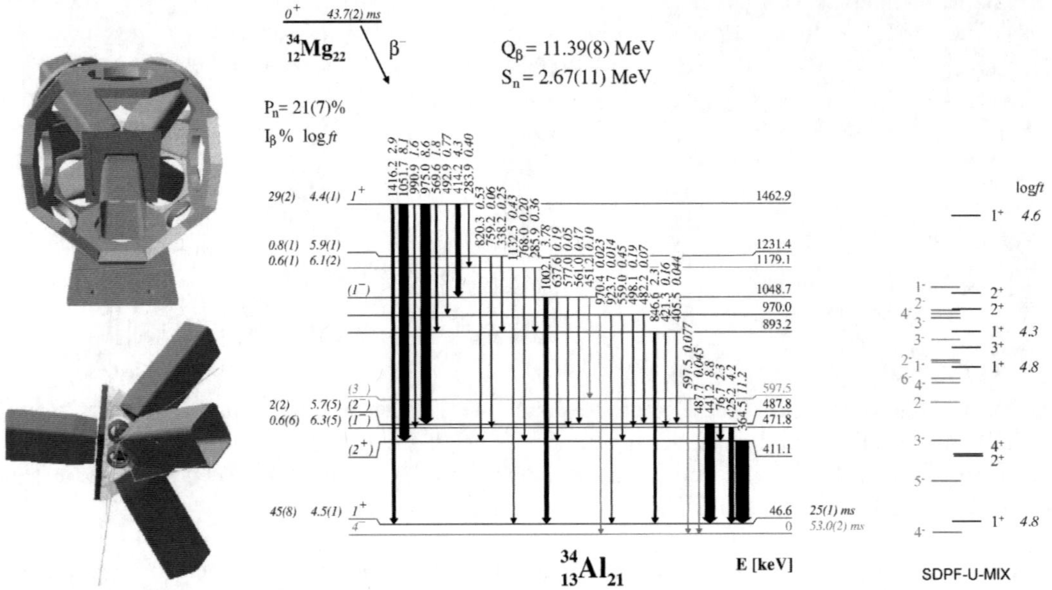

Figure 3. On the left hand side, 3D-drawings of the experimental setup at IDS used to study the decay of ^{34}Mg. On the right hand side, the experimentally deduced partial level scheme of ^{34}Al up to the second 1^+ state. In red the transitions identified that have allowed for the determination of the energy gap between the 4^- ground state and the 1^+ isomeric state. Calculations performed using the newly developed sdpf-u-mix interaction are shown at the extreme right hand part. In this plot the energy of the positive parity states are shifted down by 500 keV so that the calculated and measured energy of the first 1^+ state match.

decay pattern allowed to construct a detailed level scheme of ^{34}Al with forty two transitions identified belonging to the decay. Due to the high selectivity of the β-decay process, most of the β-feeding (29.6 %) goes to the first 1^+ state above the isomeric one at unknown excitation energy before the experiment. The summed energies of all coincident transitions as well as one direct decay consistently amount to 1416(1) keV. Due to the high statistics collected in this experiment we could identify three groups transitions decaying from three different levels and differing by 46.6 keV, the 441.2 and 487.7 keV transitions; the 970.4 and 923.7 keV; and the 1002.1 keV and (451.2.+ 597.5) keV. Assuming that the 4^- state lies 46.6 keV below the 1^+ isomeric state, we are able to build the level scheme consistently. Figure 3 shows a 3D-drawing of the setup and a partial decay scheme together with the deduced level ordering. Large scale shell-model calculations with the recently developed sdpf-u-mix interaction are compared with the new data and used to interpret the mechanisms at play at the very border of the N=20 Island of Inversion, more details in [12]. For comparison we have displayed on the right hand side of Figure 3 the positive and negative parity states shifted to fit the data. The shell model predictions overestimates by 500 keV the energy difference between the positive and negative parity states. This is not surprising as it results from the cancelation of two large energies, the energy gap between configurations and the strong correlation energy.

2.3. Study of cadmium isotopes near the double magic ^{132}Sn

The region around the doubly-magic ^{132}Sn, Z=50 and N=82, attracts a lot of interest. In this context, ISOLDE offers unique opportunities to measure ground state properties (masses, spin, electromagnetic moments) as well as to perform beta-decay and low energy reaction experiments.

For the case of the cadmium isotopes (Z=48) it has been observed an anomalous behaviour on the energy of the first 2^+ state showing a decrease when adding two neutrons to ^{124}Cd. This anomaly prompted a Coulomb excitation experiment at REX-ISOLDE of ^{122,124,126}Cd at 2.85 MeV/u. The reduced transition probabilities $B(E2; 0^+_{g.s.} \rightarrow 2^+_1)$ obtained in the safe Coulex regime and the limits for the quadrupole moments have been determined and discussed in terms of onset of collectivity in this region [13].

Going more exotic, mass measurements of the n-rich $^{129-131}$Cd isotopes were performed at the ISOLTRAP setup [14]. This device uses four traps being the last one, a Penning trap, where the mass-to-charge ratio m/q of an ion is determined by measuring its cyclotron frequency ν_c in the magnetic field of the Penning trap, m/q=B/($2\pi\nu_c$). One measures the gain in motional energy of the ions as a function of an external radio-frequency-excitation around the expected cyclotron frequency. It involves performing a frequency scan and recording the ion flight time, the so called time-of-flight ion cyclotron resonance technique (ToF-ICR) is the most precise method but it is limited by production rates and halflives. In these cases the mass measurements are realised in the Multi-Reflection-Time-of-Flight (MR-ToF) that is previously calibrated versus the Penning Trap. This device was used to determine the mass of the low produced and short-lived ($T_{1/2}$=68 ms) ^{131}Cd isotope. The new mass values showed significant deviations of more than 400 keV as compared to earlier Q-value determination. These data give evidence of a significant reduction in the N=82 shell gap for Z <50 and maybe most interesting give a more reliable description of the r-process nucleosynthesis in the region. In addition, mass separation and determination between the isomeric and ground state of n-rich cadmium isotopes have been obtained by the implementation of the phase-imaging ion- cyclotron resonance technique (PI-ICR) applied at ISOLDE for first time in 2016, see [15]. The PI-ICR method no longer includes a frequency scan so it is realised in shorter time. It constitutes a gain of factor of forty in resolving power, important to separate isomeric from ground states, and factor of four in precision with respect to the TOF-ICR technique.

The beta decay of the r-process waiting point nucleus ^{132}Cd was recently measured at IDS [16]. The beta decay strength distribution of waiting point nuclei is extremely valuable to anchor nuclear models necessary to calculate half-lives farther away from stability. Due to the expected large neutron branching ratio, the neutron time-of-flight detector VANDLE [17] was installed at IDS together with the four HPGe clover shown on the left hand side of Fig. 3. This measurement confirms previous results of the absence of bound state feeding in ^{132}In. Both the ground and the isomeric states in ^{131}In were populated in the βn process. The largest neutron transition was identified to feed the first 1^+ state in ^{131}In. This single beta transition, of $log(ft)$=4.3, dominates the decay strength and the half-life of ^{132}Cd, which is 92 ms.

3. The HIE-ISOLDE Project and first Experiments

The HIE-ISOLDE Project is a major upgrade of the existing ISOLDE facility. The first stage includes an energy increase of the REX post-accelerator up to 5.5 MeV/u, and it is already operative since the summer of 2016. A layout of the ISOLDE hall with the two experimental beamlines as it was in used in 2016 is shown on the right hand side of Fig. 1. Most of the remaining infrastructure has been installed during the winter of 2017, i.e. the third beamline, the ISOLDE Solenoidal Spectrometer for more effective transfer reaction studies and a third cryomodule that will allow to reach 7.5 MeV/u for A/q = 4.4. The increase in energy of the post-accelerated beams will finish in 2018 with the addition of a fourth cryomodule that will allow a further increase of energy to 10 MeV/u. An upgrade of the REX-Linac is also expected.

The beam intensity will be increased by two independent parts of the injectors. First the production will linearly increase with the enhanced proton intensity delivered by the new ion source and pre-injector called Linac-4 that will be implemented in the CERN second long shut down. Second the increase in energy of the PSB to 2 GeV will enhance the cross section for fragmentation and spallation reaction products by a factor of two to ten and with them the production. So when these devices will be operative for ISOLDE an increase between a factor of three, only from the Linac4, and thirty, from Linac4 + higher energy of the PSB is expected

[18]. The beam quality will be improved e.g. with the installation of a RFQ cooler. The beam purification upgrade foresees to include a new high resolution separator and a Multi-Reflection Time-of-Flight, MR-ToF, for mass diagnosis and separation. The higher energy allows for the exploration of interesting regions in the nuclear chart, increases the cross section in most cases and improves accessibility to detailed nuclear structure information.

In 2016 as much as 27 % of the yearly delivered beam time, see Figure 1, was dedicated to experiments with post-accelerated beams, 3% for reaction studies in XT02 and 24% using the MINIBALL array. Beams from ^{9}Li to ^{142}Xe were accelerated and used to produced great physics. The ^{9}Li beam at 6.8 MeV/u (A/q = 3) was used for transfer reaction studies dedicated to explore the unbound nucleus ^{10}Li using the general purpose scattering chamber SEC in XT02. The ^{66}Ni beam at 4.5 MeV/u was used to study wether the unexpected increase in the gamma strength function observed at low energy in several stable nuclei using the Oslo method is confirmed for exotic beams. They used MINIBALL together with three large $LaBr_3$ detectors in XT01. The ^{74,76,78}Zn beams were accelerated at 4.3 MeV/u in 2015 and 2016 to characterize with MINIBALL and via multi-step Coulomb excitation the evolution of the nuclear structure from very collective to more spherical structure when reaching N=50. Following the same methodology, a ^{110}Sn beam at 4.5 MeV/u was used to determine B(E2) values for the $4^+ \rightarrow 2^+$ and $2^+ \rightarrow 0^+$ transitions in order to investigate why the results obtained for light tin isotopes in previous experiments done at different energies and facilities differ greatly. Moving from the light tin to the heavier ones, a ^{132}Sn beam was accelerated at 5.5 MeV/u to also study by multi-step Coulomb excitation the reduced transition probabilities, B(E2) for the $4^+ \rightarrow 2^+$ and $2^+ \rightarrow 0^+$ transitions in this double magic nucleus. Equivalent method was applied for the ^{142}Xe beam at 4.5 MeV/u with the aim of studying the octupole correlations in this nucleus. Population up to the 8^+ excited state in ^{142}Xe was identified on the on-line analysis. More details of these very recent experiments can be found in the ISOLDE Newsletter 2017 [15].

4. Summary

The 16th of October 2017 ISOLDE will celebrate its 50 anniversary of production of radioactive beams. The success of this ISOL-Facility is due to two main factors. One is the continuous development of new radioactive ion beams with emphasis in the purity and the other the steady improvement of experimental conditions and devices reflected in a laboratory portrait recently published [19]. The facility is in continuous transformation to stay at the forefront of nuclear physics research. In this contribution I have presented very recent results obtained at ISOLDE centered in n-rich nuclei.

5. References

[1] Mane E et al. 2009 *Eur. Phys. J A* **42** 503
[2] deBoer R J et al. 2017 *Rev. Mod. Phys.l.* **89** 035007
[3] Azuma R E et al. 1994 *Phys. Rev. C* **50** 1194
[4] Buchmann L, Ruprecht G, and Ruiz C 2009 *Phys. Rev. C* **80** 045803
[5] Tilley D, Weller H, and Cheves C 1993 *Nuc. Phys* **A564** 1
[6] Refsgaard J et al. 2016 *Phys. Lett. B* **752** 296
[7] Crawford H L et al. 2016 *Phys. Rev. C* **93** 031303(R)
[8] Caurier E, Nowacki F, and Poves A 2014 *Phys. Rev. C* **90** 014302
[9] Heylen H et al. 2016 *Phys. Rev. C* **94** 034312
[10] Kwiatkowski A A et al. 2015 *Phys. Rev. C* **92** 061301(R)
[11] Langevin M et al. 1984 *Nuc. Phys* **A414** 151
[12] Lica R et al. 2017 *Phys Rev C* **95** 021301(R)
[13] Ilieva S et al. 2014 *Phys. Rev. C* **89** 014313
[14] Atanasov et al. 2016 *Phys. Rev. Lett.* **115** 232501
[15] Welker A and Karthein J 2017 *ISOLDE Newletter 2017* http://isolde.web.cern.ch/isolde-newsletter-archive
[16] Madurga M 2017 *Private communication.*
[17] Peters W A et al. 2016 *Nucl. Instrum. Methods Phys. Res. A* **836** 122
[18] Borge M J G and Riisager K 2016 *Eur. Phys. J.* **A 52** 334
[19] "Focus on Exotic Beams at ISOLDE: A Laboratory Portrait" 2017 *J. Phys. G: Nucl. Part. Phys.* **44**

IOP Conf. Series: Journal of Physics: Conf. Series **966** (2018) 012003 doi:10.1088/1742-6596/966/1/012003

Di-neutron correlations and decay dynamics of nuclei near and beyond drip line

Hiroyuki Sagawa
RIKEN Nishina Center, Wako 351-0198, Japan
Center for Mathematics and Physics, University of Aizu, Aizu-Wakamatsu, Fukushima 965-8560, Japan
E-mail: sagawa@ribf.riken.jp

Kouichi Hagino
Department of Physics, Tohoku University, Sendai 980-8578, Japan
Research Center for Electron Photon Science, Tohoku University, Sendai 982-0826, Japan
National Astronomical Observatory of Japan, 2-21-1 Osawa, Mitaka, Tokyo 181-8588, Japan

Abstract. Possible experimental probes for the strong di-neutron correlations in halo nuclei ^{6}He and ^{11}Li are discussed in the Coulomb breakup reactions. We then study a nucleus beyond the drip line ^{26}O through the direct two-neutron decay. The excited first 2^+ state of ^{26}O is also discussed. To this end, we use consistently a three-body model taking into account the coupling to the continuum. Calculated results are compared with the recent experimental data from RIBF (Radioactive Ion Beam Factory) in RIKEN.

1. Introduction

One of the most important issues in quantum many-body physics is to clarify the nature of correlations beyond the independent particle picture. In nuclear physics, the pairing correlation has been well established as a strong many-body correlation which leads to the superfluid phase characterized by phenomena such as the even-odd staggering of binding energy, the moment of inertia of rotating deformed nuclei, the difference of the excitation energy spectra between even-even and odd-even nuclei, the pair transfer reactions and also the fission barrier [1, 2, 3].

With the pairing correlation, one may naively expect that two nucleons forming a pair are located at a similar position inside a nucleus. A spatial structure of two valence neutrons was showen to be strongly clusterized in ^{210}Pb [4]. Subsequently, it was argued that two neutrons may be bound in a nucleus even though they are not bound in the vacuum [5].

The strong localization of two neutrons inside a nucleus has been referred to as the *di-neutron correlation*. It has been nicely demonstrated in Ref. [6] that an admixture of configurations of single-particle orbits with opposite parity is essential to create the strong di-neutron correlation. This implies that the pairing correlation acting only on single-particle orbits with same parity may not be sufficient in order to develop the di-neutron correlation, and the pairing model space needs to be taken sufficiently large so that both positive parity and negative parity states are included.

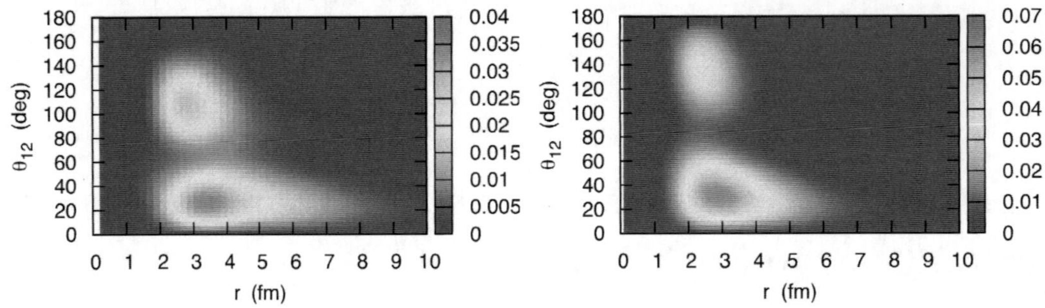

Figure 1. The two-particle densities for ^{11}Li (the left panel) and for ^{6}He (the right panel) obtained with a three-body model calculation [12]. These are plotted as a function of neutron-core distance, $r_1 = r_2 \equiv r$, and the opening angle between the valence neutrons, θ_{12}. The densities are weighted with a factor $8\pi^2 r^4 \sin\theta_{12}$.

Although the di-neutron correlation exists even in stable nuclei, it is therefore more enhanced in weakly bound nuclei because the admixtures of single-particle orbits with different parities are easier due to the couplings to the continuum spectra [7, 8]. Three-body model calculations have revealed that a strong di-neutron correlation indeed exists in weakly-bound Borromean nuclei, such as ^{11}Li and ^{6}He [9, 10, 11, 12, 13, 14]. For instance, Fig. 1 shows the two-particle density for the ^{11}Li and ^{6}He nuclei obtained with the three-body model calculation with a density dependent contact pairing interaction [12]. One can see that the densities are concentrated in the region with small opening angles, that is nothing but the di-neutron correlation. It has been shown that the di-neutron correlation exists also in heavier neutron-rich nuclei [15, 16] as well as in infinite neutron matter [17]. The di-proton correlation, which is a counter part of the di-neutron correlation, has also been shown to exist in the proton-rich Borromean nucleus, ^{17}Ne [18].

From these studies, the di-neutron correlation seems to have been theoretically established. However, it is not straightforward to probe it experimentally. In this contribution, we discuss how one can probe the di-neutron correlation. To be more specific, we shall discuss the Coulomb breakup, and the two-nucleon radioactivity as prominent probes for the correlation.

2. Electric dipole strength of Borromean nuclei and di-neutron correlations

Let us first discuss the Coulomb breakup reactions of Borromean nuclei, ^{11}Li and ^{6}He. The experimental breakup cross sections in Refs. [19, 20], especially those for the ^{11}Li nucleus, show a strong concentration in the low excitation region, reflecting the halo structure of these nuclei. Moreover, the experimental data for ^{11}Li are consistent only with the theoretical calculation which takes into account the interaction between the valence neutrons, strongly suggesting the existence of the di-neutron correlation in this nucleus (see also Ref. [21]).

The Coulomb breakup cross sections with the absorption of dipole photons are given by

$$\frac{d\sigma_\gamma}{dE_\gamma} = \frac{16\pi^3}{9\hbar c} N_\gamma(E_\gamma) \cdot \frac{dB(\mathrm{E1})}{dE_\gamma}, \tag{1}$$

where N_γ is the number of virtual photons, and

$$\frac{dB(\mathrm{E1})}{dE_\gamma} = \frac{1}{2I_i + 1} \left|\langle\psi_f||D||\psi_i\rangle\right|^2 \delta(E_f - E_i - E_\gamma), \tag{2}$$

is the $E1$ transition probability. In this equation, ψ_i and ψ_f are the wave functions for the initial and the final states, respectively, I_i is the spin of the initial state, and D_μ is the operator for the $E1$ transition. For the Borromean nuclei, assuming a three-body structure with an inert core, the $E1$ operator reads [22], $\hat{D}_\mu = \frac{e_{\rm E1}}{2}(r_1 Y_{1\mu}(\hat{\vec{r}}_1) + r_2 Y_{1\mu}(\hat{\vec{r}}_2))$, where the E1 effective charge is given by $e_{\rm E1} = 2Z_c e/(A_c+2)$, with A_c and Z_c being the mass and charge numbers for the core nucleus, and r_1 and r_2 denote the positions of particle 1 and 2 refered to the center of mass, respectively. The dipole transition of di-neutrons is entirely the recoil effect of protons in the core and the center of mass motion is exactly removed from the dipole operator $\hat{D}_\mu$. The experimental data of the transition probabilities (2) are shown in Fig. 2 with calculated results by a three-body model in ref. [21]. We can see clearly a strong enhancement of $B(E1)$ strength just above the threshold energy due to the di-neutron correlations.

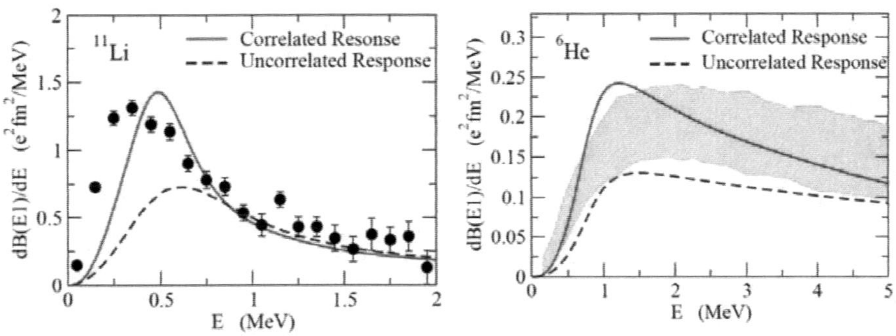

Figure 2. (Color online) The E1 transitions in ^{11}Li and ^{6}He observed by Coulomb breakup reactions. The data are taken from ref. [19] for ^{11}Li and from ref. [20] for ^{6}He, respectively.

Using Eq. (2) and the closure relation for the final state, it is easy to derive that the total E1 strength (the non-energy weighted sum rule) is proportional to the expectation value of R^2 with respect to the ground state;

$$B({\rm E1}) = \int dE_\gamma \frac{dB({\rm E1})}{dE_\gamma} = \sum_f \frac{1}{2I_i+1} |\langle\psi_f||D||\psi_i\rangle|^2 = \frac{3}{4\pi} e_{\rm E1}^2 \langle \vec{R}^2\rangle, \qquad (3)$$

where $\vec{R} = (\vec{r}_1 + \vec{r}_2)/2$, is the center of mass coordinate for the two valence neutrons. Even though the B(E1) strength distribution inevitably reflects both the correlation in the ground state and that in the final state, it is remarkable that one can extract the information which reflects *solely* the ground state properties after summing all the strength distribution. This implies that the average value of the opening angle between the valence neutrons can be directly extracted from the measured total B(E1) value once the root-mean-square distance between the valence neutrons, $\langle r_{nn}^2\rangle$, is available.

This quantity is related to the matter radius and $\langle R^2\rangle$ in the three-body model [9, 21, 23],

$$\langle r_m^2\rangle = \frac{A_c}{A}\langle r_m^2\rangle_{A_c} + \frac{2A_c}{A^2}\langle R^2\rangle + \frac{1}{2A}\langle r_{nn}^2\rangle, \qquad (4)$$

where $A = A_c + 2$ is the mass number of the whole nucleus. The matter radii $\langle r_m^2\rangle$ can be estimated from interaction cross sections. Employing the Glauber theory in the optical limit, Tanihata *et al.* have obtained $\sqrt{\langle r_m^2\rangle} = 1.57 \pm 0.04$, 2.48 ± 0.03, 2.32 ± 0.02, and 3.12 ± 0.16 fm for ^{4}He, ^{6}He, ^{9}Li, and ^{11}Li, respectively [24, 25]. Using these values, we obtain the rms neutron-neutron distance of $\sqrt{\langle r_{nn}^2\rangle} = 3.75 \pm 0.93$ and 5.50 ± 2.24 fm for ^{6}He and ^{11}Li, respectively. Combining these values with the rms core - di-neutron distance, $\sqrt{\langle R^2\rangle}$, we obtain the mean opening angle

$$\langle\theta_{nn}\rangle = 2\tan^{-1}\left(\sqrt{\langle r_{nn}^2\rangle}/2\sqrt{\langle R^2\rangle}\right) \qquad (5)$$

to be $51.56^{+11.2}_{-12.4}$ and $56.2^{+17.8}_{-21.3}$ degrees for ^{6}He and ^{11}Li, respectively [26]. These values are comparable to the result of the three-body model calculation, $\langle\theta_{nn}\rangle$=66.33 and 65.29 degree for ^{6}He and ^{11}Li, respectively [12], although the experimental values are somewhat smaller.

In the absence of the correlations, the mean opening angle is exactly $\langle\theta_{nn}\rangle$=90 degrees. The extracted values of $\langle\theta_{nn}\rangle$ are significantly smaller than this value both for ^{11}Li and ^{6}He, providing a direct proof of the existence of the di-neutron correlation in these nuclei.

3. Two-neutron decays of nuclei beyond the drip lines

Figure 3 shows the calculated decay energy spectrum of ^{26}O obtained with the ^{24}O+n+n three-body model for ^{26}O [28]. The calculations are carried out using the Green's function method as explained in Refs. [8, 22, 29, 30], together with a density-dependent contact neutron-neutron interaction, v. In this formalism, the decay energy spectrum is given by,

$$\frac{dP}{dE} = \sum_k |\langle\Psi_k|\Phi_{\rm ref}\rangle|^2\,\delta(E-E_k) = \frac{1}{\pi}\Im\langle\Phi_{\rm ref}|G(E)|\Phi_{\rm ref}\rangle, \tag{6}$$

where Ψ_k is a solution of the three-body model Hamiltonian with energy E_k and $\Phi_{\rm ref}$ is the wave function for a reference state. The reference state can be taken rather arbitrarily as long as it has an appreciable overlap with the resonance states of interest. Here, we employ the uncorrelated two-neutron state in ^{27}F for it with the $|[1d_{3/2}\otimes 1d_{3/2}]^{(I=0)}\rangle$ configuration, which is the dominant configuration in the initial state of the proton knockout reaction of ^{27}F to produce ^{26}O. In Eq. (6), G is the correlated two-particle Green's function calculated as

$$G(E) = (1+G_0(E)v)^{-1}G_0(E) = G_0(E) - G_0(E)v(1+G_0(E)v)^{-1}G_0(E), \tag{7}$$

with the uncorrelated Green's function, G_0, given by

$$G_0(E) = \lim_{\eta\to 0}\sum_{1,2}\frac{|j_1j_2\rangle\langle j_1j_2|}{\epsilon_1+\epsilon_2-E-i\eta}. \tag{8}$$

In Eq. (7), v is the two-body interaction between two neutrons.

In the figure, the uncorrelated spectrum shown by the dotted line has a peak at E = 1.498 MeV, that is twice the single-particle resonance energy of $d-$orbit, $\epsilon_{d_{3/2}}$ = 749(10) keV for ^{25}O [27]. With the pairing interaction between the valence neutrons, the calculations yield the decay energy of 18 keV of the ground state. The decay spectrum obtained by including only the $[d_{3/2}]^2$ configurations is shown by the dashed line. In this case, the peak is shifted by 0.55 MeV from the unperturbed peak at 1.498 MeV. The peak is further shifted downwards by the di-neutron correlations due to configuration mixing, and gets closer to the threshold energy.

In Ref. [27] (also in Fig. 3), a clear second peak is found at $E = 1.28^{+0.11}_{-0.08}$ MeV, which is likely attributed to the 2^+ state. In Ref. [30, 28], we have investigated the 2^+ state in the ^{26}O nucleus using the three-body model. Due to the pairing interaction between the valence neutrons, the energy of the 2^+ state is slightly shifted towards lower energies from the unperturbed energy, whereas the energy shift is much larger for the 0^+ state due to the larger overlap between the wave functions of the two neutrons. The 2^+ peak appears at 1.282 MeV, which agrees perfectly with the experimental data [28].

The angular distribution of the emitted neutrons can be calculated with the two-particle Green's function method [22, 29, 28]. The amplitude for emitting two neutrons with spin components of s_1 and s_2 and momenta $\vec{k}_1$ and $\vec{k}_2$ reads [29],

$$f_{s_1s_2}(\vec{k}_1,\vec{k}_2) = \sum_{j,l} e^{-il\pi}e^{i(\delta_1+\delta_2)}\,M_{j,l,k_1,k_2}\langle[\mathcal{Y}_{jl}(\hat{\vec{k}}_1)\mathcal{Y}_{jl}(\hat{\vec{k}}_2)]^{(00)}|\chi_{s_1}\chi_{s_2}\rangle, \tag{9}$$

IOP Conf. Series: Journal of Physics: Conf. Series **966** (2018) 012003 doi:10.1088/1742-6596/966/1/012003

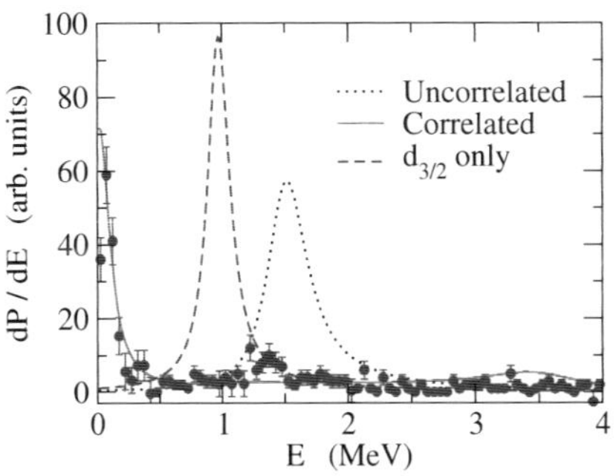

Figure 3. The decay energy spectrum for the two-neutron emission decay of ^{26}O, obtained with the ^{24}O+n+n three-body model. The dotted line is for the uncorrelated spectrum, while the solid line shows the correlated spectrum for the 0^+ state. The dashed line shows the spectrum including the pairing correlation only for the $[d_{3/2}]^2$ configuration. The experimental data are taken from Ref. [27].

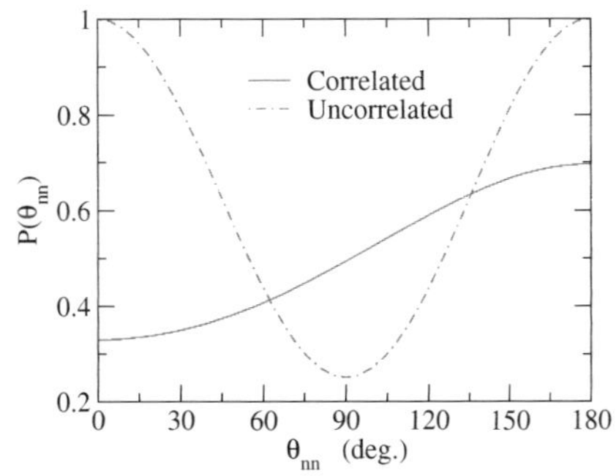

Figure 4. The angular correlation for the two emitted neutrons from the ground state decay of ^{26}O. The probability distribution for the opening angle of the momentum vectors of the emitted neutrons is shown. The solid and the dot-dashed lines denote the correlated and uncorrelated results, respectively.

where $\mathcal{Y}_{jlm}$ is the spin-spherical harmonics, χ_s is the spin wave function, and δ is the nuclear phase shift. Here, M is a decay amplitude calculated to a specific two-particle final state [22],

$$M_{j,l,k_1,k_2} = \langle (jj)^{(00)} | 1 - vG_0 + vG_0vG_0 - \cdots | \Psi_i \rangle = \langle (jj)^{(00)} | (1 + vG_0)^{-1} | \Psi_i \rangle, \quad (10)$$

where the unperturbed Green's function, G_0, is evaluated at $E = e_1 + e_2$. The angular distribution is then obtained as

$$P(\theta_{12}) = 4\pi \sum_{s_1,s_2} \int dk_1 dk_2 \, |f_{s_1 s_2}(k_1, \hat{\vec{k}}_1 = 0, k_2, \hat{\vec{k}}_2 = \theta_{12})|^2, \quad (11)$$

where we have set z-axis to be parallel to $\vec{k}_1$ and evaluated the angular distribution as a function of the opening angle, θ_{12}, of the two emitted neutrons.

Fig. 4 shows the angular correlation. The dot-dashed line shows the distribution obtained without including the nn interaction, which is symmetric around $\theta_{12} = \pi/2$. In the presence of the nn interaction, the angular distribution turns to be highly asymmetric, in which the emission of two neutrons in the opposite direction ($\theta_{12} = \pi$) is enhanced shown by the solid line.

4. Summary

We have discussed possible experimental probes for the di-neutron correlation in neutron-rich nuclei, with which two valence nucleons are localized at a similar position in the coordinate space. In particular, we have discussed the Coulomb dissociation of Borromean nuclei, and the direct two-neutron decay of the unbound ^{26}O nucleus.

For the Coulomb dissociation of Borromean nuclei, one can use the cluster sum rule (the non-energy weighted sum rule for localized systems) to deduce the mean value of the opening angle between the valence neutrons. We have demonstrated that the mean opening angle is $\langle \theta_{nn} \rangle$=74.5

$^{+11.2}_{-13.1}$ and 65.2 $^{+11.4}_{-13.0}$ degrees for ^{6}He and ^{11}Li, respectively. These values are significantly smaller than the value for the uncorrelated distribution, that is, $\langle\theta_{nn}\rangle$=90 degrees, clearly indicating the existence of the di-neutron correlation in these Borromean nuclei.

For the direct two-neutron decay, we have discussed the recent experimental data of the decay energy spectrum for the unbound ^{26}O nucleus. We have shown that the decay energy spectrum can be accounted for only with the di-neutron correlation due to the mixing of many configurations including the continuum. We have also discussed the angular correlations of the emitted two neutrons. We have argued that the di-neutron correlation enhances an emission of the two neutrons in the opposite direction (the back-to-back emission), and our three-body model calculation has revealed such feature. If the enhancement of the back-to-back emission will be observed experimentally, it will provide a direct evidence for the di-neutron correlation.

Other possible probes are proposed to detect the di-neutron correlations. Those include the two-neutron transfer reactions, the nuclear breakup reaction, the (p, d) scattering at backward angles, and the knockout reactions of Borromean nuclei. It would be extremely intriguing if the clear and direct evidence for the di-neutron correlation could be experimentally obtained in near future using the probes discussed in this manuscript.

References

[1] Bohr A and Mottelson B R 1975 *Nuclear Structure Vol. 1* (Singapore: World Scientific)
[2] Ring P and Schuck P 1980 *The Nuclear Many Body Problem* (New York: Springer-Verlag)
[3] Brink D M and Broglia R A 2005 *Nuclear Superfluidity* (Cambridge: Cambridge University Press)
[4] Bertsch G F, Broglia R A and Riedel C 1967 *Nucl. Phys.* **A91** 123
[5] Migdal A B 1973 *Soviet J. of Nucl. Phys.* **16**, 238
[6] Catara F, Insolia A, Maglione E and Vitturi A 1984 *Phys. Rev. C* **28**, 1091
[7] Hagino K, Tanihata I and Sagawa H 2013 *Exotic nuclei far from the stability line* 100 Years of Subatomic Physics ed Henley E M and Ellis S D (Singapore: World Scientific) p 231
[8] Sagawa H and Hagino K 2015 *Euro. Phys. J* **A51** 102
[9] Bertsch G F and Esbensen H 1991 *Ann. Phys.* (New York) **209** 327
[10] Zhukov et al 1993 *Phys. Rep.* **231** 151
[11] Barranco F, Bortignon P F, Broglia R A, Colo G and Vigezzi E 2001 *Eur. Phys. J.* **A11** 385
[12] Hagino K and Sagawa H 2005 *Phys. Rev. C* **72** 044321
[13] Hagino K, Sagawa H, Carbonell J and Schuck P 2007 *Phys. Rev. Lett.* **99** 022506
[14] Kikuchi Y, Kato K, Myo T, Takashina M and Ikeda K 2010 *Phys. Rev. C* **81** 044308
[15] Matsuo M, Mizuyama K and Serizawa Y 2005 *Phys. Rev. C* **71** 064326
[16] Pillet N, Sandulescu N and Schuck P 2007 *Phys. Rev. C* **76** 024310
[17] Matsuo M 2006 *Phys. Rev. C* **73** 044309
[18] Oishi T, Hagino K and Sagawa H 2010 *Phys. Rev. C* **82** 024315
[19] Nakamura T et al 2006 *Phys. Rev. Lett.* **96** 252502
[20] Aumann T et al 1999 *Phys. Rev. C* **59** 1252
[21] Esbensen H, Hagino K, Mueller P and Sagawa H 2007 *Phys. Rev. C* **76** 024302
[22] Esbensen H and Bertsch G F 1992 *Nucl. Phys.* **A542** 310
[23] Vinh Mau N and Pacheco J C 1996 *Nucl. Phys.* **A607** 163
[24] Tanihata I et al 1985 *Phys. Rev. Lett.* **55** 2676
[25] Ozawa A et al 2001 *Nucl. Phys.* **A693** 32
[26] Hagino K and Sagawa H 2007 *Phys. Rev. C* **76** 047302 (2007)
[27] Kondo Y et al 2016 *Phys. Rev. Lett.* **116** 102503
[28] Hagino K and Sagawa H 2016 *Phys. Rev. C* **93** 034330
[29] Hagino K and Sagawa H 2014 *Phys. Rev. C* **90** 027303
[30] Hagino K and Sagawa H 2014 *Phys. Rev. C* **90** 027303

IOP Conf. Series: Journal of Physics: Conf. Series **966** (2018) 012004 doi:10.1088/1742-6596/966/1/012004

Chiral Effective Field Theory for Nucleonic Matter

Francesca Sammarruca
Physics Department, University of Idaho, Moscow, ID 83844-0903, U.S.A.
E-mail: fsammarr@uidaho.edu

Abstract. Predictions for the interaction part of the symmetry energy obtained from different microscopic approaches are reviewed and discussed in comparison with updated constraints recently obtained at GSI. The discussion is then extended to the neutron skin thickness in ^{208}Pb and its relation to the density derivative of the symmetry energy. We highlight the importance of giving proper consideration to theoretical uncertainties of microscopic predictions in order to guide phenomenological analyses.

1. Introduction

The energy per particle as a function of density in infinite nuclear matter, known as the equation of state (EoS), contains rich information about the nature of the nuclear force in hadronic medium. Selected reaction observables and other nuclear properties have been found to be sensitive to the EoS, and therefore their measurements can provide useful constraints for the latter. The dependence of the EoS of isospin asymmetric matter on the neutron excess parameter brings up the *symmetry energy*, a quantity of fundamental importance for a variety of neutron-rich systems. Presently our knowledge of the symmetry energy is limited, particularly its density dependence around and above saturation density. From the experimental side, heavy-ion (HI) reactions are an established way to seek such constraints, based on the fact that the EoS is an important part of the input in transport models describing HI collisions.

Other nuclear properties have been found highly sensitive to specific aspects of the EoS of asymmetric matter. The neutron skin thickness, for instance, is sensitive to the slope of the symmetry energy, which determines to which extent neutrons are pushed outwards to form the skin. A strong correlation is also found between the pressure in the interior of a heavy nucleus and the radius of the average-mass neutron star. In summary, studies of nucleonic matter are especially timely and important, as they support on-going and future experimental effort, in both astrophysics and terrestrial laboratories.

The earlier FOPI-LAND data [1, 2], reanalysed in Ref. [3] and compared with transport model calculations, have suggested a softer-than-linear to linear term for the potential energy part of the symmetry energy, when the latter is parametrized as a power law. More recently, the directed and elliptic flows of neutrons and light charged particles in the reaction ^{197}Au + ^{197}Au at 400 MeV per nucleon were measured in the ASY-EOS experiment at GSI [4]. The updated findings confirm a moderately soft density dependence, but in Ref. [4] the authors report a more stringent constraint up to twice normal density.

Naturally, when updated constraints are made available, comparison with microscopic calculations becomes especially timely. Within that spirit, modern predictions of the symmetry energy based on chiral effective field theory (EFT) [5] are reviewed and discussed, following

IOP Conf. Series: Journal of Physics: Conf. Series **966** (2018) 012004 doi:10.1088/1742-6596/966/1/012004

closely Ref. [6]. For comparison, more "traditional" approaches are also considered, such as those based on meson-theoretic or phenomenological nucleon-nucleon (NN) potentials and three-nucleon forces (3NF). These approaches, which were particularly popular in the 1990's and are still frequently used today, follow a very different philosophy. Thus, their inclusion in the comparison will provide a realistic measure of the spreading of contemporary theoretical predictions. We will explore whether the theoretical predictions are consistent with the recent constraints. An important point we wish to raise is the relevance of including microscopic predictions when correlating the density dependence of the symmetry energy to the neutron skin thickness in ^{208}Pb.

Before moving to the next section, we recall that the parametrization used in the analysis of Ref. [4] is

$$e_{sym}(\rho) = 22\ \text{MeV}\Big(\frac{\rho}{\rho_0}\Big)^{\gamma} + 12\ \text{MeV}\Big(\frac{\rho}{\rho_0}\Big)^{2/3}, \tag{1}$$

which fixes the symmetry energy at ρ_0 to be 34 MeV. The power law coefficient, γ, is reported as 0.72 ± 0.19. The same coefficient was found to be 0.9 ± 0.4 from the FOPI-LAND data [1, 2, 3]. Other useful quantities for the discussion which follows are the so-called "L" coefficient,

$$L = 3\rho_0\Big(\frac{\partial e_{sym}}{\partial \rho}\Big)_{\rho_0}, \tag{2}$$

and the closely related symmetry pressure, $P_0 = \rho_0 L/3$.

2. Theoretical Input

2.1. The chiral EoS

Chiral EFT is presently a popular approach which starts from a low-energy realization of quantum chromodynamics [5]. In chiral EFT, one retains the basic degrees of freedom typical of low-energy nuclear physics, pions and nucleons, while fitting unresolved nuclear dynamics at short distances to the properties of two- and few-nucleon systems. More specifically, EFT is a theory in which the properties governed by low-energy physics are specified by the choice of degrees of freedom and symmetries, and can be computed systematically. Short-range physics is included through the processes of regularization and renormalization. The microscopic EoS of symmetric nuclear matter and neutron matter based on chiral EFT employed here are calculated as described in Ref. [7]. The predictions at next-to-next-to-next-to-leading order (N^{3}LO) and at next-to-next-to-leading order (N^{2}LO) are based on high-precision chiral NN potentials at the respective orders [8] together with the leading 3NF, which is treated as in Ref. [9].

Since estimating the truncation error at N^{3}LO requires the predictions at N^{2}LO, those will be shown as well. (N^{4}LO predictions are not yet available.) Note that the N^{2}LO calculation is complete, in the sense that both the two-nucleon force (2NF) and the 3NF are consistently at next-to-next-to-leading order. This is not the case, though, for the N^{3}LO calculation, where the 3NF at N^{2}LO is employed. Efforts are in progress in our group to remove this inconsistency.

2.2. Relativistic meson-theoretic potentials and the Dirac-Brueckner-Hartree-Fock approach to the equation of state

The relativistic approach to nuclear matter, particularly the Dirac-Brueckner-Hartree-Fock (DBHF) approximation, started with the observation that the DBHF theory, unlike conventional Brueckner theory, has the inherent ability to predict realistic values for the saturation energy and density of nuclear matter. The characteristic feature of the DBHF approach is the fact that important 3NF are effectively taken into account through the density dependence of the nucleon spinors. This effective 3NF, which originates from virtual excitations of nucleon-antinucleon pairs and is, for that reason, known as "Z-diagrams", provides a strong (repulsive and density-dependent) saturating effect and, thus, an important mechanism missing from conventional

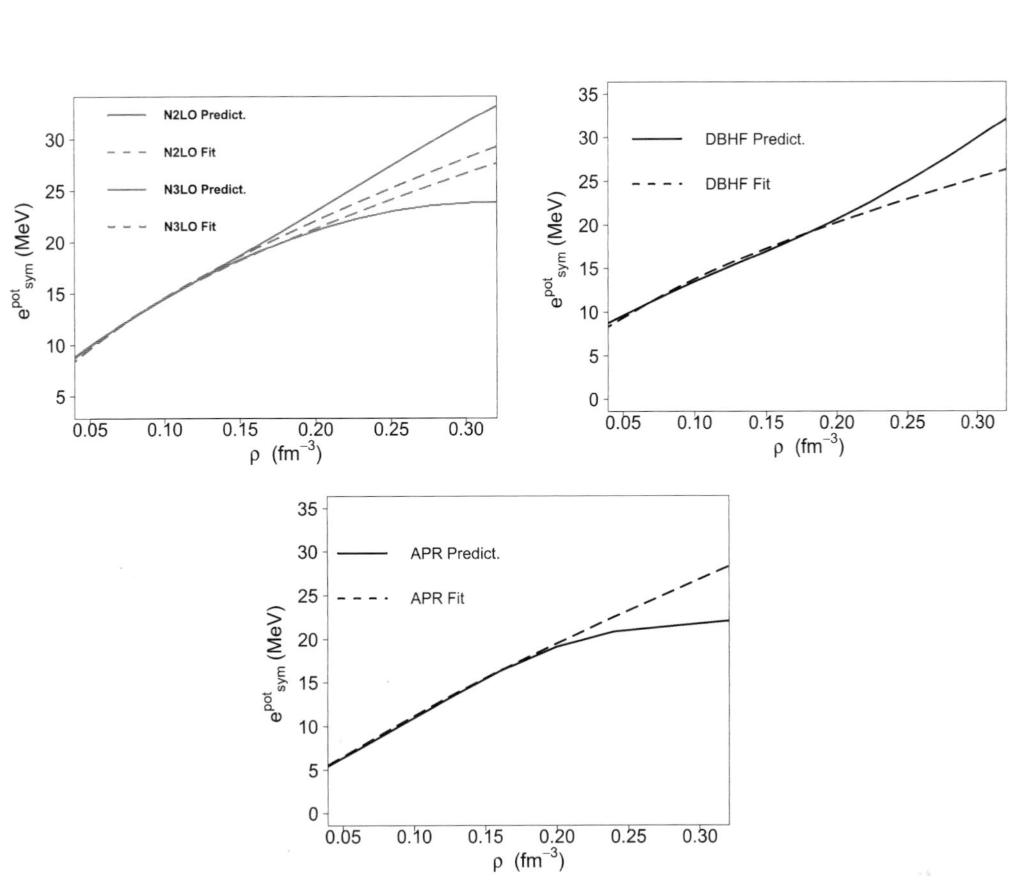

Figure 1. Frame on the upper left: microscopic predictions of the interaction part of the symmetry energy at N^2LO and N^3LO of chiral perturbation theory, and corresponding power-law fits. See inset for the definition of the various curves. Upper right: The potential part of the symmetry energy as predicted with Bonn B and the DBHF calculation compared with a power-law fit. Lower panel: the same, but for the APR model [12].

Brueckner-Hartree-Fock calculations. For a recent review, including considerations of isospin asymmetry, see Ref. [10].

2.3. Variational approaches

Alternatively, the energy per particle in nuclear matter can be obtained combining the 2NF with meson-theoretic or phenomelogical 3NF to generate the additional repulsion essential to improve saturation. Nonrelativistic calculations of symmetric and neutron matter based on variational methods [11] and phenomenological 2NF and 3NF have been used extensively. To also represent this point of view, we will include predictions [12] based on popular phenomenological 2NF and 3NF from the 90's, namely the Argonne v_{18} NN potential [13] together with the Urbana model IX [14] 3NF. These will be referred to as the "APR" model.

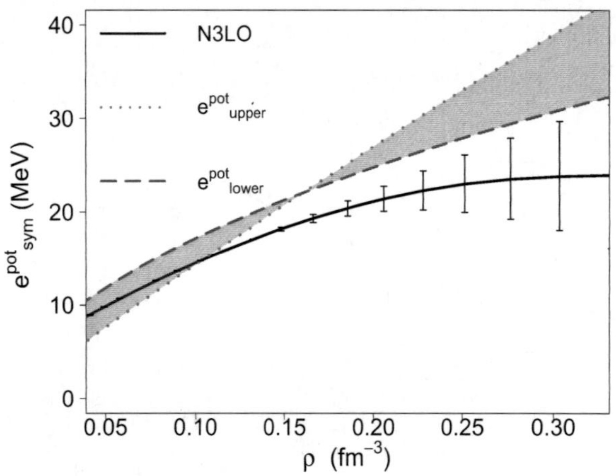

Figure 2. Microscopic predictions of the interaction part of the symmetry energy at N^3LO with EFT truncation error. The shaded area shows the empirical constraint from Ref. [4].

3. Predictions and Discussion

We calculate the symmetry energy for the theories and models mentioned in the previous section and subtract the kinetic contribution. The microscopic values of the potential energy part are then fitted with a power law for each of the models being considered, ρ_0 is the actual saturation density for that particular EoS.

On the left-hand side of Fig. 1, we display the potential energy part of the symmetry energy as obtained from microscopic calculations at N^2LO (solid green line) and at N^3LO (solid red line), together with approximations given by functions of the form

$$e_{sym}^{pot} = V_0(\rho/\rho_0)^{\gamma} \,, \tag{3}$$

dashed green for N^2LO and dashed red for N^3LO. The fit is done by searching for the single parameter γ, setting V_0 equal to the microscopically predicted value at the appropriate saturation density. The density range considered in the fit covers approximately from 0.03 to 0.33 fm^{-3}, with all points carrying the same weight.

The theoretical curves appear reasonably described by the simple *ansatz* up to their respective saturation densities and somewhat above it, whereas the constraint should be applicable up to about $2\rho_0$, which amounts to approximately 0.3 fm^{-3} by the definition of Ref. [4]. Furthermore, an analytical approximation which appears reasonable may be less than satisfactory with respect to derivatives. This point will be addressed again later in the paper.

In the middle frame of Fig. 1, one can see to which degree the interaction part of the symmetry energy can be approximated by a single-term power law for the Bonn B meson-exchange potential [15] used in a DBHF calculation, while a similar comparison is shown on the right-hand side of the same figure for the potential symmetry energy obtained from the EoS of Ref. [12].

Of course the N^3LO predictions must be seen in the context of EFT theoretical uncertainties. Therefore, in Fig. 2 the N^3LO predictions are shown with their estimated truncation error. In Fig. 2, the shaded area represents the empirical constraint. The predictions fall within the empirical constraints at the lower densities but are otherwise softer.

Obviously, the parametrization given in Eq. 1 was found to be consistent with the reaction observables measured in the GSI experiment. However, here one learns that, although a moderately soft (less than linear) dependence is preferred by microscopic models, Eq. 1 is overall not a satisfactory representation of these theoretical predictions. It would be interesting to move beyond the power-law parametrization when analyzing elliptic-flow ratios.

IOP Conf. Series: Journal of Physics: Conf. Series **966** (2018) 012004 doi:10.1088/1742-6596/966/1/012004

Table 1. Neutron skin (S) of ^{208}Pb with varying power law, γ, in the interaction symmetry energy within the range determined by the ASY-EOS analysis. The third column displays the the slope of the symmetry energy at about 2/3 of saturation density, followed by the L parameter and the symmetry pressure. For further details, see discussion in the text.

γ	S (fm)	$\frac{\partial e_{sym}}{\partial \rho}$ (MeV fm^3)	L (MeV)	P_0 (MeV/fm^3)
0.53	0.14	154	60.1	3.14
0.72	0.18	177	72.6	3.80
0.91	0.21	195	85.2	4.45

Next we extend the discussion to the neutron skin of ^{208}Pb in relation to density derivatives and pressure. Due to lack of space, the reader is referred to Ref. [6] and references therein for details on the neutron skin calculations.

In Table 1, for values of γ spanning the uncertainty of the ASY-EOS constraint, we show (second, fourth, and fifth columns, respectively), the skin thickness of ^{208}Pb, the L parameter, defined in Eq. (2), and the symmetry pressure P_0. This confirms that the skin is sensitive to the pressure in the neutron-enriched core of ^{208}Pb which pushes excess neutrons towards the low-density edges of the nucleus.

On the other hand, the average density in nuclei is less than saturation density and, therefore, typical nuclear observables (such as, for instance, those used to construct phenomenological forces) actually probe densities somewhat lower than saturation density. Therefore, Table 1 also shows how the slope of the symmetry energy at ρ=0.1 fm^{-3} varies in relation to the neutron skin (third column).

The observations collected above can be summarized as

$$S_{empirical} \approx (0.18^{+0.03}_{-0.04})\ \text{fm}; \quad \Big(\frac{\partial e_{sym}}{\partial \rho}\Big)_{\frac{2}{3}\rho_0} \approx (177^{+18}_{-23})\ \text{MeV fm}^3\ ; \quad L \approx (72.6 \pm 13)\text{MeV}\ . \tag{4}$$

Proceeding in a similar way, the spreading of the predictions from the "family" of models in Table 2 can be summarized as

$$S_{Theories} \approx (0.18 \pm 0.02)\ \text{fm}; \quad \Big(\frac{\partial e_{sym}}{\partial \rho}\Big)_{\frac{2}{3}\rho_0} \approx (154^{+22}_{-19})\ \text{MeV fm}^3\ ; \quad L \approx (58^{+14}_{-11})\text{MeV}\ . \tag{5}$$

Concentrating on the chiral predictions, the difference between the predictions at N^{2}LO and N^{3}LO is a (pessimistic) estimate of the truncation error at N^{3}LO. Proceeding as detailed in Ref. [6], the chiral predictions at N^{3}LO is estimated to be

$$S_{N3LO} \approx (0.17 \pm 0.02)\ \text{fm}; \quad \Big(\frac{\partial e_{sym}}{\partial \rho}\Big)_{\frac{2}{3}\rho_0} \approx (148 \pm 14)\ \text{MeV fm}^3\ ; \quad L \approx (48 \pm 15)\text{MeV}\ . \tag{6}$$

These observations may be more easily captured in a visual way, as presented in Figure 3. The area shaded in blue is obtained from the empirical constraint for L and the corresponding constraint for the neutron skin. Including the predictions from Table 2 generate the green area. Finally, Eq. (6) produces the pink region. Clearly, the uncertainty in the density derivative is much larger than could be inferred from the blue area, that is, from the correlation based on the empirical information.

In Ref. [4], the authors do mention that the sharp value of $e^{pot}_{sym}(\rho_0)$ is the result of choosing a power law as in Eq. (1) and that using lower values of $e^{pot}_{sym}(\rho_0)$ leads to lower values of L,

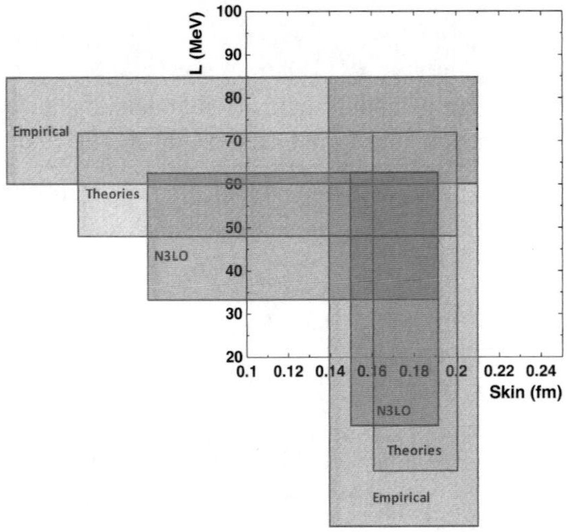

Figure 3. Relation between the L parameter and the neutron skin in ^{208}Pb. Blue shaded area: empirical constraint for L and corresponding constraint for the neutron skin. Green area: Predictions from Table 2. Equation (6) produces the pink region.

Table 2. As in Table 1, but for each of the theoretical approaches under consideration.

Theor. Approach	S (fm)	$\frac{\partial e_{sym}}{\partial \rho}$ (MeV fm^3)	L (MeV)	P_0 (MeV/fm^3)
DBHF	0.16	135	46.8	2.45
N^{3}LO	0.17	148	47.5	2.50
APR	0.18	158	65.0	3.43
N^{2}LO	0.20	176	72.0	3.79

still within acceptable error margins. However, the results from Ref. [16] (which are based on Skyrme phenomenology), are no longer met with the alternative parametrization. The results which have been reviewed here show that adhering to Eq. (1) is not recommendable from the theoretical standpoint.

4. Conclusions

Theoretical calculations of the symmetry energy, in particular from modern *ab initio* predictions, are timely and important as they complement on-going and future experimental efforts. In this paper, existing predictions for the interaction part of the symmetry energy have been explored in the light of new and more stringent constraints recently obtained for this quantity. We considered a few but fundamentally different approaches to obtain a realistic idea of the spreading in microscopic predictions.

Within the simplest assumption of a single-term power law, it is found that various predictions based on chiral EFT, relativistic meson-theory, or phenomenological forces and the variational method, can only approximately be described by a power law consistent with the constraint. The N^{3}LO prediction is generally softer than the constraint.

The focus then moves to the neutron skin thickness of ^{208}Pb. We stress the importance of experiments keeping in close touch with ab initio predictions. Caution must be excercised with regard to the possibility of constraining the density slope from the knowledge of the skin based only on correlations obtained with families of simple phenomenological interactions, as the latter procedure may underestimate the uncertainties.

Work in progress includes generating systematic order-by-order predictions of neutron star masses and radii. Our emphasis is on the role of the pressure around normal density in the formation of both the neutron skin and the radius of a typical neutron star.

Acknowledgments
This work was supported by the U.S. Department of Energy, Office of Science, Office of Basic Energy Sciences, under Award Number DE-FG02-03ER41270.

References

[1] Leifels Y *et al.* 1993 *Phys. Rev. Lett.* **71** 963.
[2] Lambrecht D *et al.* 1994 *Z. Phys.* **A 350** 115.
[3] Russotto P *et al.* 2011 *Phys. Lett.* B **697** 471.
[4] Russotto P *et al.* 2016 *Phys. Rev.* C **94** 034608.
[5] Weinberg S 1990 *Phys. Lett.* B **251** 288.
[6] Sammarruca F 2017 *Phys. Rev.* C **95** 044316.
[7] Sammarruca F, Coraggio L, Holt J, Itaco N, Machleidt R, and Marcucci L.E. 2015 *Phys. Rev.* C **91** 054311.
[8] Entem D R and Machleidt R 2003 *Phys. Rev.* C **68** 041001.
[9] Holt J W, Kaiser N, and Weise W 2009 *Phys. Rev.* C **79** 054331; 2010 *Phys. Rev.* C **81** 024002.
[10] Müther H, Sammarruca F, and Ma Z 2017, *Int. J. Mod. Phys.* E **26** 1730001, and references therein.
[11] Akmal A and Pandharipande V R 1997 *Phys. Rev.* C **56** 2261.
[12] Akmal A, Pandharipande V R, and Ravenhall D G 1998 *Phys. Rev.* C **58** 1804.
[13] Wiringa R B, Stoks V G J, and Schiavilla R 1995 *Phys. Rev.* C **51** 38.
[14] Pudliner B S, Pandharipande V R, Carlson J, and Wiringa R B 1995 *Phys. Rev. Lett.* **74** 4396.
[15] Machleidt R 1989 *Adv. Nucl. Phys.* **19** 189.
[16] Brown B A 2013 *Phys. Rev. Lett.* **111** 232502.

IOP Conf. Series: Journal of Physics: Conf. Series **966** (2018) 012005 doi:10.1088/1742-6596/966/1/012005

Nuclear-Structure Physics with MINIBALL at HIE-ISOLDE

P. Reiter[1] for the MINIBALL collaboration

[1] Institut für Kernphysik, Universität zu Köln, Zülpicher Straße 77, D-50937 Köln, Germany

E-mail: preiter@ikp.uni-koeln.de

Abstract. The MINIBALL spectrometer utilizes successfully a variety of post-accelerated radioactive ion beams provided by the new HIE-ISOLDE accelerator at CERN. In-beam γ-ray spectroscopy after Coulomb excitation (CE) or transfer reactions is performed with optimized setups of ancillary detectors for particle detection. The physics program covers a wide range of shell model investigations. Exotic heavy ion beams will enable unique studies of collective properties up to the actinide region. First data taking with HIE-ISOLDE beams started recently. The higher energies and intensities of the new post-accelerator provides a promising perspective for a new generation of MINIBALL experiments. Intriguing first results were obtained by employing beams of ^{74,76,78}Zn, ^{110,132}Sn, ^{144}Xe with beam energies in the range of 4.0 - 5.5 MeV/u for CE experiments at 'safe' energies. In all cases first results for various $B(E\lambda)$ values for these isotopes were obtained.

1. Introduction

Major strides were made within the on-going HIE-ISOLDE (High Intensity and Energy) project at the ISOLDE facility at CERN [1] by increasing the beam energy for post-accelerated radioactive beams. The previous REX linear accelerator with a maximum energy of 3 MeV/u was partially replaced and extended by an additional superconducting post accelerator. To study the properties of radioactive nuclei by different and specific nuclear reactions, the enlarged beam energy will provide essential advantages in the near future. For light-ion induced transfer reactions in inverse kinematics the wider energy range up to 10 MeV/u will allow for optimized conditions, e.g. via Q-value matching, over the whole chart of nuclei. The beam energies close to the Coulomb barrier will allow for improved CE measurements especially multiple CE with highest cross sections at 'safe' energies. Even higher beam energies will enable CE experiments at 'unsafe' energies above the barrier in order to extend the range of excitation energy. Moreover new reaction mechanisms like multi-nucleon transfer and deep inelastic scattering up to fusion will be feasible in the upcoming years. In this contribution the first in-beam γ-ray spectroscopy experiments will be introduced, which were performed at HIE-ISOLDE with a beam energies up to 5.5 MeV/u in 2016.

2. HIE-ISOLDE

In the past more than 90 different radioactive isotopes, ranging from the light ^{6}He up to the heavy actinide isotope ^{224}Rn have been efficiently post-accelerated and studied employing the REX-ISOLDE setup at beam energies up to 3.0 MeV/u [2]. Also the new HIE-ISOLDE

Figure 1. Picture of the MINIBALL setup in 2016. The radioactive ion beam delivered by HIE-ISOLDE is coming from the top, hitting the secondary target right in the middle of the MINIBALL spectrometer. Downstream of the scattering chamber additional detector systems are installed to monitor the position of the beam and its composition.

accelerator facility is based on the production of highly charged ions. For this purpose singly charged radioactive ions coming from ISOLDE are first cooled and bunched in a Penning trap (REXTRAP) [3] before they are charge bred to higher charge states, using an electron beam ion source (REXEBIS) [4]. The ions with a mass-to-charge ratio of $A/q < 4.5$ are injected into a compact linear accelerator via a mass separator. The normal conducting linear accelerator consists of a Radio Frequency Quadrupole (RFQ) accelerator which accelerates ions from 5 to 300 keV/u, a rebuncher section, an Interdigital H-type (IH) structure that boosts the energy to 1.2 MeV/u, three seven-gap spiral-resonators and a 9-gap IH resonator; it provides a maximum energy of 3.0 MeV/u [5].

The decisive new HIE-ISOLDE accelerator component for the higher beam energies are the superconducting Quarter Wave Resonators (QWRs) for post-acceleration. The QWR geometry comprises twenty high-β cavities cooled by a 4.5 K cryogenic facility for helium. The twenty cavities are grouped and installed in four cryo-modules. The accelerating cavities feature a very high voltage gradient of 6 MV/m and low heat dissipation. The transverse focusing is achieved by four superconducting solenoids housed inside the cryo-modules. Each cryo-module contains five cavities with its 101.28 MHz niobium-sputtered copper Quarter Wave Resonators and one solenoid. In 2016, the first two cryo-modules were operated successfully with the design energy of 5.5 MeV/u for A/q = 4.5.

In future, the HIE-ISOLDE upgrade will improve the experimental capabilities at ISOLDE continuously. An important step will be the increase of the beam energy from currently 5.5 MeV/u to finally 10 MeV/u by adding two more cryo-modules. The third cryo-module was successfully installed before the physics campaign in 2017 and beam energies of 7.5 MeV/u were already achieved in 2017. Installation of the last module is foreseen before the start of the 2018 campaign and will allow experiments at 10 MeV/u which will enlarge the opportunities for experiments at ISOLDE. More detailed descriptions of the complete HIE-ISOLDE project can be found in Ref. [6, 7] or in the contribution by M. J. G. Borge to this volume.

3. First MINIBALL campaign at HIE-ISOLDE

3.1. Experimental setup

The high-resolution and highly efficient MINIBALL spectrometer [8] was reinstalled around the new target position of the extended HIE-ISOLDE accelerator complex (see Figure 1). The spectrometer consists of eight triple cluster detectors, mounted on movable arms around the target chamber to allow a compact configuration. Each cryostat contains three individually large volume encapsulated six-fold segmented high-purity germanium crystals. With close distances between target and Ge-detectors the array covers a solid angle of approximately 50-60% of 4π. The photopeak efficiency of MINIBALL at 1.3 MeV is $\approx 8\%$ after cluster addback. An average energy resolution of 2.3 keV at $E = 1.33$ MeV is achieved. The azimuthal electrical segmentation of the detector increases the granularity and ensures a proper Doppler correction for in-flight γ-ray emission at higher β values well above 10% even at the very close distances between target and detector.

To ensure a proper Doppler correction for in flight γ-ray emission at increased velocities, the angular information of the γ ray, provided by the position sensitive MINIBALL detectors, is combined with the direction and velocity of the scattered beam particle that was measured in coincidence. For this purpose, the scattered beam and recoiling target nuclei are detected by silicon strip detectors which are employed in two different configurations. First, a CD-shaped double-sided silicon strip detector (DSSSD) was used for CE experiments in order to detect forward scattered beam- or target like reaction products [9]. Each quadrant comprises 16 annular strips at the front side and 24 radial strips at the back side for identification and reconstruction of the trajectories of the scattered nuclei. The distance between the scattering target and the DSSSD is about 30 mm corresponding to a forward angle range between 16.8° and 53.7° in the laboratory frame. This CD detector was mounted for the HIE-ISOLDE commissioning measurements in 2015.

A second particle detector array was employed for the experiments performed in 2016. The so-called C-REX array consists of a suit of DSSSDs: one in forward and one in backward direction, and a barrel of four position-sensitive planar detectors under backward direction [10]. The forward and backward DSSSD is cicular shaped with 16 annular strips at the front side and 24 radial strips at the back side. The distance between target and detector front side and therefore the scattering angle range was adjusted to the requirements of the kinematics of the individual experiments performed during the campaign in 2016. The angular resolution is typically better than 5°. The backward barrel detectors consist of planar Si detectors which are segmented in 16 strips perpendicular to the beam axis. The energy resolution for the protons ranges typically from 250 keV to 2 MeV, depending on their scattering angle, energy and thickness of the target.

3.2. ^{74,76,78}Zn

The first RIBs delivered by the newly installed HIE-ISOLDE accelerator to the MINIBALL spectrometer were ^{74,76}Zn ions with energies of 4.0 MeV/u as part of the last in-beam commissioning phase in 2015. CE of both isotopes on platinum and lead targets yielded straightaway the γ-decay lines from the $4_1^+ \rightarrow 2_1^+$ and $2_1^+ \rightarrow 0_{g.s.}^+$ transitions demonstrating

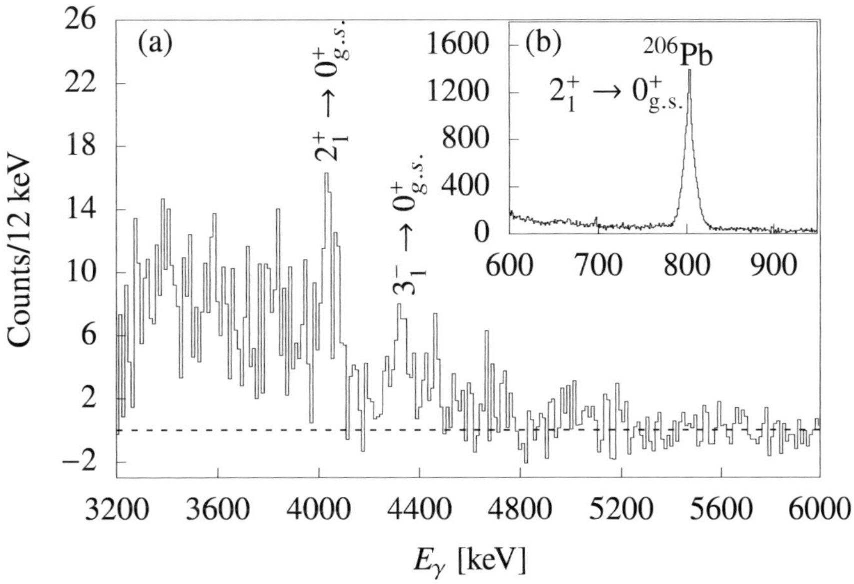

Figure 2. (a) High-energy part of the γ-ray spectrum after Doppler correction for $A = 132$ beam particles with peaks from $2_1^+ \to 0_{\text{g.s.}}^+$ and $3_1^- \to 0_{\text{g.s.}}^+$ transitions in ^{132}Sn. (b) A γ-ray spectrum, Doppler-corrected for the ^{206}Pb target excitation is shown in the inset.

the increased cross section for multiple CE. In contrast to this result a measurement at lower beam energies of 2.85 MeV/u did not show any signal from the higher lying 4_1^+ state. Another CE experiment along the chain of Zn isotopes at even higher energies of 4.3 MeV/u was performed in 2016 observing the $4_1^+ \to 2_1^+$ and $2_1^+ \to 0_{\text{g.s.}}^+$ in ^{78}Zn. Here intriguing evidence was found for excitation of even higher lying states.

These experiments explore the nature of the N=40 shell closure and the shell gap around the doubly-magic ^{78}Ni (Z=28, N=50). The nuclei around ^{78}Ni are benchmarks for latest shell-model calculations. Especially, the impact of excitations across the N=50 and Z=28 shell closures are important. A substantial set of experimental data has been obtained in this region of the nuclear chart. However, the conclusions drawn from them seem to be contradictory. The possibility of the weakening of the N=50 closure has been anticipated by [11, 12] while the contrary has been deduced by [13, 14].

Moreover, conflicting results exist between the life time of the 4_1^+ state in ^{74}Zn as deduced from Coulex measurements [14] or from direct life-time measurements using the Recoil-Distance Doppler shift method [15]. The preliminary independent results of the first HIE-ISOLDE experiment strongly support the previous findings of the Coulex measurements. In future CE studies in the neutron-rich Zn isotopes up to the N=50 shell closure are approved in order to reveal the energy of unknown excited states beside the 2_1^+ state and transition matrix element in the ^{80}Zn nucleus. These investigations are lead by E. Rapisarda (CERN), P. Van Duppen (K.U. Leuven) and M. Zielinska (CEA Saclay).

3.3. ^{110}Sn

In 2016 the in-beam campaign with MINIBALL and post-accelerated beams started in September, accelerating a beam of the semi-magic (Z=50) ^{110}Sn to 4.5 MeV/u. This first experiment addressed the determination of the transition probabilities $2_1^+ \to 0_{\text{g.s.}}^+$ that seems to differ strongly from the theoretical expectation in the light tin isotopes [16]. It is a continuation of the previous REX-ISOLDE program in the ^{100}Sn region at HIE-ISOLDE [17, 18], using the new energy $\approx$ 5 MeV/u. In a first step the γ transitions from the 2_1^+ and 4_1^+ states in ^{110}Sn were successfully detected. In future electromagnetic properties of ^{110,108,106}Sn will be addressed using CE. Even the isotope ^{104}Sn may be in reach for an in-beam study in case sufficient yield of the new RILIS will be available. These measurements are under the guidance of J. Cederkall, University of Lund.

3.4. ^{142}Xe

The second experiment in 2016 addressed the evolution of the quadrupole and octupole collectivity in ^{142}Xe. The results will shed light on the interplay between collective excitations and single particle structure when going away from the doubly-magic ^{132}Sn. The Doppler corrected γ spectrum of the ^{142}Xe beam at 4.5 MeV/u on a lead target showed a clear improvement by populating higher spin states in comparison with the equivalent spectrum obtained with the REX accelerator energies of 2.85 MeV/u. With the new energies, states up to 8^+ spin-parity are populated in multi-step CE. The experiment will aim to determine the B(E2) and B(E3) values to trace the evolution of quadrupole and octupole collectivity.

The B(E2) values connecting the ground state and the first 2_1^+ state in the even isotopes $^{138-144}$Xe have been measured at REX-ISOLDE and towards neutron-rich nuclei the trend is well described even by an empirical Grodzins-type formula. An increased dipole strength extracted from the decay of the negative parity states in ^{140,142}Xe is interpreted as indirect signature of increasing octupole correlation. However, B(E3) values are missing and subject of the new ^{144}Xe analysis. Spokesperson of this experiment is Th. Kroell, T.U. Darmstadt.

3.5. ^{132}Sn

Another important step during the 2016 campaign was to take the superconducting cavities to maximum accelerating gradient and a beam energy of 5.5 MeV/u. The vibrational first 2^+ and 3^- states of doubly-magic nucleus ^{132}Sn (Z=50, N=82) were excited via 'safe' CE after bombardment of a thin ^{206}Pb target. The optimized beam energy, the high-energy resolution and the good efficiency of the gamma spectrometer provide a favorable combination to master the demanding measurement characterized by small CE cross sections of the high-lying states with excitation energies above 4 MeV (see Figure 2). A first preliminary result for the $B(E2; 0^+ \rightarrow 2^+)$ value was obtained and also a value for the excitation strength of the 3^- state is in reach.

In the past first preliminary results for the $B(E2; 0^+ \rightarrow 2^+)$ value were obtained with an efficient scintillator array at ORNL [19, 20]. The results on excited collective states in ^{132}Sn provide crucial information on cross shell configurations that are expected to be dominated by a strong proton contribution. The results will be compared to large-scale shell model calculations, which are on its ways and new mean field calculations.

4. Summary and Outlook

The high-resolution γ-ray spectroscopy program at ISOLDE for nuclear-structure studies with reaccelerated radioactive ion beams commenced after commissioning of the first part of the HIE-ISOLDE facility. To achieve higher beam energy, better beam quality and higher secondary-beam intensity the superconducting HIE-ISOLDE linac will be in installed in stages. Two out of four superconducting cavities were put successfully into operation until spring 2016. Several CE experiments with an increased beam energy of up to 5.5 MeV/A were performed in 2016. The new facility enlarges the scope of the ongoing investigations by multi-step CE for all radioactive beams available at ISOLDE. Moreover, future transfer experiments will benefit from the higher energies and will allow an extended range for optimizing the Q-value matching condition. The final HIE-ISOLDE linac will reach an energy of 10 MeV/A. This will enable transfer reaction studies employing heavy mass beams up to masses around $A = 200$. Finally, the MINIBALL spectrometer will be augmented with new instrumentation such as the SPEDE conversion electron spectrometer [21], an upgraded version of Si detectors for the T-REX array, Anti-Compton BGO suppression shields for the HPGe triple-cluster detectors and a new digital electronics and data acquisition system in order to enlarge the detection capabilities.

Acknowledgments

The author would like to thank my colleagues P. Van Duppen (K.U. Leuven), M. Zielinska (CEA-Saclay), Th. Kroell (T.U. Darmstadt), J. Cederkall (U. Lund) for sharing preliminary and unpublished first results. The author acknowledge financial support from the German BMBF under contracts 05P12PKFNE, 05P15PKCIA, the European Unions Horizon 2020 Framework research and innovation programme under grant agreement no. 654002 (ENSAR2).

References

[1] Kugler E 2000 *Hyperfine Interact.* **129** 23–42
[2] Butler P A, Cederkall J and Reiter P 2017 *J. Phys. G: Nucl. Part. Phys.* **44** 044012
[3] Ames F *et al.* 2005 *Nucl. Instr. Meth. Phys. Res. Sect. A* **538** 17 – 32
[4] Wenander F 2002 *Nucl. Phys. A* **701** 528–536
[5] Kester O *et al.* 2003 *Nucl. Instr. Meth. Phys. Res. B* **204** 20 – 30
[6] Kadi Y *et al.* 2017 *J. Phys. G* **44** 084003
[7] Borge M J G and Jonson B 2017 *J. Phys. G: Nucl. Part. Phys.* **44** 044011
[8] Warr N *et al.* 2013 *Eur. Phys. J. A* **49** 40
[9] Ostrowski A N *et al.* 2002 *Nucl. Instr. Meth. Phys. Res. Sect. A* **480** 448 – 455
[10] Bildstein V *et al.* 2007 *Prog. Part. and Nucl. Phys.* **59** 386 – 388
[11] Perru O *et al.* 2006 *Phys. Rev. Lett.* **96**(23) 232501
[12] Stefanescu I *et al.* 2008 *Phys. Rev. Lett.* **100**(11) 112502
[13] van de Walle J *et al.* 2007 *Phys. Rev. Lett.* **99**(14) 142501
[14] van de Walle J *et al.* 2009 *Phys. Rev. C* **79**(1) 014309
[15] Louchart C *et al.* 2013 *Phys. Rev. C* **87**(5) 054302
[16] Banu A *et al.* 2005 *Phys. Rev. C* **72**(6) 061305
[17] Cederkäll J *et al.* 2007 *Phys. Rev. Lett.* **98**(17) 172501
[18] Ekström A *et al.* 2008 *Phys. Rev. Lett.* **101**(1) 012502
[19] Radford D C *et al.* 2004 *Nucl. Phys. A* **746** 83 – 89 proceedings of the Sixth International Conference on Radioactive Nuclear Beams (RNB6)
[20] Beene J *et al.* 2004 *Nucl. Phys. A* **746** 471 – 474 proceedings of the Sixth International Conference on Radioactive Nuclear Beams (RNB6)
[21] Papadakis P, Pakarinen J, Butler P A, Cox D M, Davies P, Greenlees P, Herzberg R D, Huyse M, Jenkins D G, Konki J *et al.* 2015 *Proc. Conf. Adv. Radioact. Isot. Sci.(J. Phys. Soc. Japan)*

IOP Conf. Series: Journal of Physics: Conf. Series **966** (2018) 012006 doi:10.1088/1742-6596/966/1/012006

Charge symmetry breaking in light Λ hypernuclei

Avraham Gal[1] **and Daniel Gazda**[2,3]

[1] Racah Institute of Physics, The Hebrew University, 91904 Jerusalem, Israel
[2] Department of Physics, Chalmers University of Technology, SE-412 96 Göteborg, Sweden
[3] Nuclear Physics Institute, 25068 Řež, Czech Republic

E-mail: avragal@savion.huji.ac.il

Abstract. Charge symmetry breaking (CSB) is particularly strong in the $A = 4$ mirror hypernuclei ${}^4_\Lambda$H–${}^4_\Lambda$He. Recent four-body no-core shell model calculations that confront this CSB by introducing Λ-Σ^0 mixing to leading-order chiral effective field theory hyperon-nucleon potentials are reviewed, and a shell-model approach to CSB in p-shell Λ hypernuclei is outlined.

1. Introduction

Charge symmetry of the strong interactions arises in QCD upon neglecting the few-MeV mass difference of up and down quarks. With baryon masses of order GeV, charge symmetry should break down at the level of 10^{-3} in nuclei. The lightest nuclei to exhibit charge symmetry breaking (CSB) are the A=3 mirror nuclei ^{3}H–^{3}He, where CSB contributes about 70 keV out of the 764 keV Coulomb-dominated binding-energy difference. This CSB contribution is indeed of order 10^{-3} with respect to the strong interaction contribution in realistic A=3 binding energy calculations, and is also consistent in both sign and size with the scattering-length difference $a_{pp} - a_{nn} \approx 1.7$ fm [1]. It can be explained by $\rho^0\omega$ mixing in one-boson exchange models of the NN interaction, or by considering $N\Delta$ intermediate-state mass differences in models limited to pseudoscalar meson exchanges [2]. In practice, introducing two charge dependent contact interaction terms in chiral effective field theory (χEFT) applications, one accounts quantitatively for the charge dependence of the low energy NN scattering parameters and, thereby, also for the A=3 mirror nuclei binding-energy difference [3]. CSB is manifest, of course, also in heavier nuclei.

In Λ hypernuclei, isospin invariance excludes one pion exchange (OPE) from contributing to ΛN strong-interaction matrix elements. However, it was pointed out by Dalitz and Von Hippel (DvH) that the SU(3) octet $\Lambda_{I=0}$ and $\Sigma^0_{I=1}$ hyperons are admixed in the physical Λ hyperon, thus generating a long-range OPE ΛN CSB potential $V^{\rm OPE}_{\rm CSB}$ [4]. For the mirror ${}^4_\Lambda$H–${}^4_\Lambda$He ground-state (g.s.) levels built on the ^{3}H–^{3}He g.s. cores, and using the DvH purely central wavefunction, the OPE CSB contribution amounts to $\Delta B^{J=0}_\Lambda \approx 95$ keV where $\Delta B^J_\Lambda \equiv B^J_\Lambda({}^4_\Lambda{\rm He}) - B^J_\Lambda({}^4_\Lambda{\rm H})$. This is also confirmed in our present calculations in which tensor contributions add up $\approx$100 keV. Shorter-range CSB meson-mixing contributions appear to be much smaller [5]. Remarkably, the OPE overall contribution of $\approx$200 keV to the CSB splitting of the ${}^4_\Lambda$H–${}^4_\Lambda$He mirror g.s. levels roughly agrees with the large observed g.s. CSB splitting $\Delta B^{J=0}_\Lambda$=233$\pm$92 keV shown in Fig. 1 which is of order 10^{-2} with respect to the Λ nuclear strong interaction contribution in realistic binding energy calculations of the A=4 hypernuclei. Hence, CSB in Λ hypernuclei is likely to be almost one order of magnitude stronger than in ordinary nuclei.

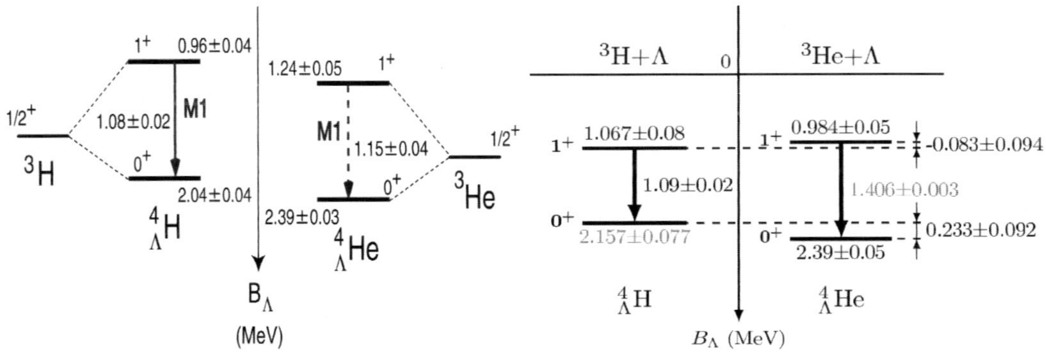

Figure 1. ${}^4_\Lambda$H-${}^4_\Lambda$He level diagram, before (left panel) and after (right panel) the recent measurements of the ${}^4_\Lambda$He excitation energy $E_\gamma(1^+_{\rm exc} \to 0^+_{\rm g.s.})$ at J-PARC [6], and of the ${}^4_\Lambda$H $0^+_{\rm g.s.}$ binding energy at MAMI [7,8], both highlighted in red in the online version. CSB splittings are shown to the very right of the ${}^4_\Lambda$He levels. Figure adapted from [8].

In addition to OPE, $\Lambda - \Sigma^0$ mixing affects also shorter range meson exchanges (e.g. ρ) that in χEFT are replaced by contact terms. Quite generally, in baryon-baryon models that include *explicitly* a charge-symmetric (CS) $\Lambda N \leftrightarrow \Sigma N$ ($\Lambda\Sigma$) coupling, the direct ΛN matrix element of $V_{\rm CSB}$ is obtained from a strong-interaction CS $\Lambda\Sigma$ coupling matrix element $\langle N\Sigma|V_{\rm CS}|N\Lambda\rangle$ by

$$\langle N\Lambda|V_{\rm CSB}|N\Lambda\rangle = -0.0297\,\tau_{Nz}\,\frac{1}{\sqrt{3}}\,\langle N\Sigma|V_{\rm CS}|N\Lambda\rangle, \tag{1}$$

where the z component of the nucleon isospin Pauli matrix $\vec{\tau}_N$ assumes the values $\tau_{Nz} = \pm 1$ for protons and neutrons, respectively, the isospin Clebsch-Gordan coefficient $1/\sqrt{3}$ accounts for the $N\Sigma^0$ amplitude in the $I_{NY} = 1/2$ $N\Sigma$ state, and the space-spin structure of this $N\Sigma$ state is taken identical to that of the $N\Lambda$ state sandwiching $V_{\rm CSB}$. The 3% CSB scale factor -0.0297 in Eq. (1) follows by evaluating the $\Lambda - \Sigma^0$ mass mixing matrix element $\langle\Sigma^0|\delta M|\Lambda\rangle$ from SU(3) mass formulae [4, 9]. The corresponding diagram for generating $\langle N\Lambda|V_{\rm CSB}|N\Lambda\rangle$ is shown in Fig. 2, demonstrating explicitly the δM CSB insertion.

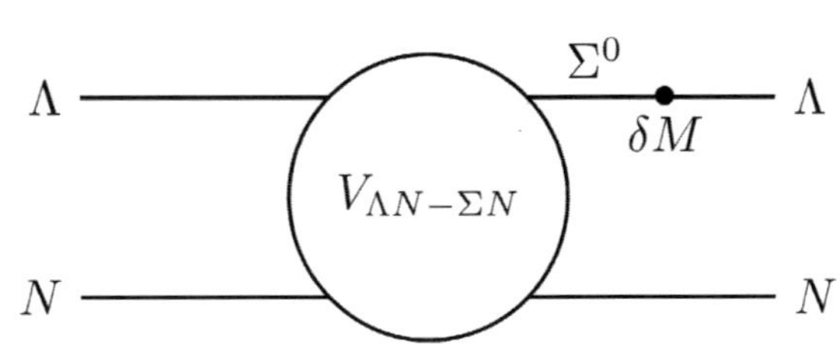

Figure 2. CSB ΛN interaction diagram describing a CS $V_{\Lambda N-\Sigma N}$ interaction followed by a CSB $\Lambda - \Sigma^0$ mass-mixing vertex.

Since the CSB ΛN matrix element in Eq. (1) is given in terms of strong-interaction CS $\Lambda\Sigma$ coupling, one wonders how strong the latter is in realistic microscopic YN interaction models. Recent four-body calculations of ${}^4_\Lambda$He levels [10], using the Bonn-Jülich leading order (LO) χEFT YN CS potential model [11], show that almost 40% of the $0^+_{\rm g.s.} \to 1^+_{\rm exc}$ excitation energy E_x arises from $\Lambda\Sigma$ coupling. This also occurs in the NSC97 models [12] as demonstrated by

Akaishi *et al* [13]. With $\Lambda\Sigma$ matrix elements of order 10 MeV, the 3% CSB scale factor in Eq. (1) suggests a CSB splitting $\Delta E_x \sim 300$ keV, in good agreement with the observed splitting $E_x(^4_\Lambda\mathrm{He})-E_x(^4_\Lambda\mathrm{H})= 320 \pm 20$ keV [6], see Fig. 1 (right) which also shows a relatively large splitting of the A=4 mirror hypernuclear g.s. levels, $\Delta B_\Lambda^{J=0}$=233±92 keV [7,8], with respect to the ≈70 keV CSB splitting in the mirror core nuclei ^{3}H and ^{3}He.

Here we review recent *ab-initio* no-core shell model (NCSM) calculaions of the A=4 Λ hypernuclei [14,15] using a LO χEFT YN CS interaction model [11] in which CSB is generated by implementing Eq. (1). We also briefly review a shell model approach [9], confronting it with some available data in the p shell.

2. LO χEFT YN interactions

N3LO NN [3] and N2LO NNN interactions [16], both with momentum cutoff $\Lambda = 500$ MeV, are used in our calculations. For hyperons, the Bonn-Jülich SU(3)-based LO YN interaction is used, plus V_{CSB} evaluated from it according to Eq. (1). At LO, V_{YN} consists of regularized pseudoscalar (PS) π, K and η meson exchanges with coupling constants constrained by SU(3)$_\mathrm{f}$, plus five central interaction contact terms simulating the short range behavior of the YN coupled channel interactions, all of which are regularized with a cutoff momentum $\Lambda \geq m_{\mathrm{PS}}$, varied from 550 to 700 MeV. Two of the five contact terms connect ΛN to ΣN in spin-singlet and triplet s-wave channels, and are of special importance for the calculation of CSB splittings. The dominant meson exchange interaction is OPE which couples the ΛN channel exclusively to the $I = \frac{1}{2}$ ΣN channel. K-meson exchange also couples these two YN channels. This V_{YN}^{LO} reproduces reasonably well, with $\chi^2/(\mathrm{d.o.f.}) \approx 1$, the scarce YN low-energy scattering data. It also reproduces the binding energy of $^3_\Lambda$H, with a calculated value $B_\Lambda(^3_\Lambda\mathrm{H})$=110±10 keV for Λ=600 MeV [17], consistent with experiment (130±50 keV [18]) and with Faddeev calculations reported by Haidenbauer *et al* [19]. Isospin conserving matrix elements of V_{YN}^{LO} are evaluated in a momentum-space particle basis accounting for mass differences within baryon iso-multiplets, while isospin breaking $(I_{NN} = 0) \leftrightarrow (I_{NN} = 1)$ and $(I_{YN} = \frac{1}{2}) \leftrightarrow (I_{YN} = \frac{3}{2})$ transitions are suppressed. The Coulomb interaction between charged baryons is included.

3. NCSM hypernuclear calculations

The NCSM approach to few-body calculations uses translationally invariant harmonic-oscillator (HO) bases expressed in terms of relative Jacobi coordinates [20] in which two-body and three-body interaction matrix elements are evaluated. Antisymmetrization is imposed with respect to nucleons, and the resulting Hamiltonian is diagonalized in a finite HO basis, admitting all HO excitation energies $N\hbar\omega$, $N \leq N_{\max}$, up to $N_{\max}$ HO quanta. This NCSM nuclear technique was extended recently to light hypernuclei [10,17]. While it was possible to obtain fully converged binding energies, with keV precision for the A=3 core nuclei ^{3}H and ^{3}He, it was not computationally feasible to perform calculations with sufficiently large $N_{\max}$ to demonstrate convergence for $^4_\Lambda$H and $^4_\Lambda$He. In these cases extrapolation to an infinite model space, $N_{\max} \to \infty$, had to be employed. For details see Ref. [15]. We note that ΔB_Λ, and to a lesser extent B_Λ, exhibit fairly weak $N_{\max}$ and ω dependence compared to the behavior of absolute energies, and the employed extrapolation scheme was found sufficiently robust. While normally using $N_{\max} \to \infty$ extrapolated values based on the last three $N_{\max}$ values, it was found that including the last four $N_{\max}$ values in the fit resulted in ΔB_Λ values that differed by $\lesssim 10$ keV.

Calculations consisting of fully converged A=3 core binding energies (8.482 MeV for ^{3}H and 7.720 MeV for ^{3}He) and ($^4_\Lambda$H, $^4_\Lambda$He) $0^+_{\mathrm{g.s.}}$ and 1^+_{exc} binding energies extrapolated to infinite model spaces from $N_{\max} = 18(14)$ for $J = 0(1)$ are reported here. The NNN interaction, was excluded from most of the hypernuclear calculations after verifying that, in spite of adding almost 80 keV to the Λ separation energies $B_\Lambda^{J=0}$ and somewhat less to $B_\Lambda^{J=1}$, its inclusion makes a difference of only a few keV for the CSB splittings ΔB_Λ^J in both the $0^+_{\mathrm{g.s.}}$ and 1^+_{exc} states.

Table 1. CS averages (in MeV) of $B^J_\Lambda(^4_\Lambda\text{H})$ and $B^J_\Lambda(^4_\Lambda\text{He})$ in four-body calculations using LO χEFT YN [11] and NLO χEFT YN [21] interaction models.

	LO (present)	LO [22]	NLO [22]	Exp. (Fig. 1)
$B^{J=0}_\Lambda$	$2.37^{+0.20}_{-0.13}$	2.5±0.1	$1.53^{+0.08}_{-0.06}$	2.27±0.09
$B^{J=1}_\Lambda$	$1.08^{+0.58}_{-0.47}$	$1.4^{+0.5}_{-0.4}$	$0.83^{+0.07}_{-0.10}$	1.03±0.09
$E_x(0^+_{\text{g.s.}}\to 1^+_{\text{exc}})$	1.29±0.38	1.05±0.25	0.71±0.04	1.25±0.02

Table 1 lists results obtained for the A=4 hypernuclear levels in the present LO-YN NCSM calculation with V_{CSB}, and in Nogga's [22] LO- and NLO-YN Faddeev-Yakubovsky calculations without V_{CSB}. To provide meaningful comparison, the 'present' column lists CS averages over mirror levels in $^4_\Lambda$H and $^4_\Lambda$He. The two LO columns are consistent with each other within the cited uncertainties, which are particularly large for J = 1, and both agree with experiment within these uncertainties. Uncertainties reflect the resulting cutoff dependence in the chosen Λ range. The NLO results are almost Λ independent, as inferred from their small uncertainties. However, NLO disagrees strongly with experiment, particularly for $J = 0$ and for the accurately determined E_x. It would be interesting in future work to modify the existing NLO χEFT version [21,22] by refitting the $\Lambda\Sigma$ contact terms to both $B^{J=0,1}_\Lambda(A{=}4)$ CS-averaged values, and then apply the CSB generating equation (1) in four-body calculations of $^4_\Lambda$H–$^4_\Lambda$He.

4. CSB in s-shell hypernuclei

Results of recent four-body NCSM calculations of the A=4 hypernuclei [14,15], using the Bonn-Jülich LO χEFT SU(3)-based YN interaction model [11] with momentum cutoff in the range Λ=550–700 MeV, are shown in Fig. 3. Plotted on the l.h.s. are the calculated $0^+_{\text{g.s.}} \to 1^+_{\text{exc}}$ excitation energies in $^4_\Lambda$H and in $^4_\Lambda$He, both of which are found to increase with Λ such that somewhere between Λ=600 and 650 MeV the γ-ray measured values of E_x are reproduced. The $\Lambda - \Sigma^0$ mixing CSB splitting ΔE_x obtained by using Eq. (1) also increases with Λ such that for Λ=600 MeV the calculated value $\Delta E_x = \Delta B^{\text{calc}}_\Lambda(0^+_{\text{g.s.}}) - \Delta B^{\text{calc}}_\Lambda(1^+_{\text{exc}}) = 330 \pm 40$ keV agrees with the measured value of $E_x(^4_\Lambda\text{He}) - E_x(^4_\Lambda\text{H}) = 320 \pm 20$ keV deduced from Fig. 1 (right).

Plotted on the r.h.s. of Fig. 3 is the $\hbar\omega$ dependence of ΔB^J_Λ, including V_{CSB} from Eq. (1) and using $N_{\max} \to \infty$ extrapolated values for each of the four possible B^J_Λ values calculated at cutoff Λ=600 MeV. Extrapolation uncertainties for ΔB^J_Λ are 10 to 20 keV. $\Delta B^{J=0}_\Lambda$ varies over the spanned $\hbar\omega$ range by a few keV, whereas $\Delta B^{J=1}_\Lambda$ varies by up to ~30 keV.

Fig. 3 demonstrates a strong (moderate) cutoff dependence of $\Delta B^{J=0}_\Lambda$ ($\Delta B^{J=1}_\Lambda$):

$$\Delta B^{J=0}_\Lambda = 177^{+119}_{-147}\ \text{keV}, \quad \Delta B^{J=1}_\Lambda = -215^{+43}_{-41}\ \text{keV}. \tag{2}$$

The opposite signs and roughly equal sizes of these ΔB^J_Λ values follow from the dominance of the $^1S_0$ contact term (CT) in the $\Lambda\Sigma$ coupling potential of the LO χEFT YN Bonn-Jülich model [11], whereas the PS SU(3)-flavor octet ($\mathbf{8}_\text{f}$) meson-exchange contributions are relatively small and of opposite sign to that of the $^1S_0$ CT contribution. This paradox is resolved by noting that regularized pieces of Dirac $\delta(\mathbf{r})$ potentials that are discarded in the classical DvH treatment survive in the LO χEFT PS meson-exchange potentials. Suppressing such a zero-range regulated piece of CSB OPE within the full LO χEFT A=4 hypernuclear wavefunctions gives [15]

$$\text{OPE(DvH)}: \quad \Delta B^{J=0}_\Lambda \approx 175 \pm 40\ \text{keV}, \quad \Delta B^{J=1}_\Lambda \approx -50 \pm 10\ \text{keV}, \tag{3}$$

with smaller momentum cutoff dependence uncertainties than in Eq. (2). Both Eqs. (2) and (3) agree within uncertainties with the CSB splittings ΔB^J_Λ marked in Fig. 1.

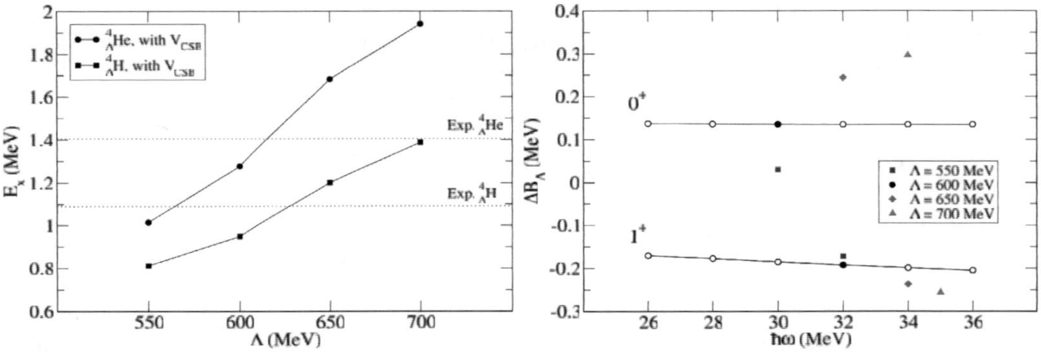

Figure 3. NCSM calculations of $^4_\Lambda$H and $^4_\Lambda$He, using CS LO χEFT YN interactions [11] and $V_{\rm CSB}$, Eq. (1), derived from these CS interactions. Left: momentum cutoff dependence of excitation energies $E_{\rm x}(0^+_{\rm g.s.}\to 1^+_{\rm exc})$. The γ-ray measured values of $E_{\rm x}$ from Fig. 1 are marked by dotted horizontal lines. Right: HO $\hbar\omega$ dependence, for Λ=600 MeV, of the separation-energy differences ΔB^J_Λ for $0^+_{\rm g.s.}$ (upper curve) and for $1^+_{\rm exc}$ (lower curve). Results for other values of Λ are shown at the respective absolute variational energy minima. Figure adapted from [15].

5. CSB in p-shell hypernuclei

Recent cluster-model work [23–25] fails to explain CSB splittings in p-shell mirror hypernuclei, apparently for disregarding the underlying CS $\Lambda\Sigma$ coupling potential. In the approach reviewed here, one introduces an effective CS $\Lambda\Sigma$ central interaction $\mathcal{V}_{\Lambda\Sigma} = \bar{V}_{\Lambda\Sigma} + \Delta_{\Lambda\Sigma}\,\vec{s}_N\cdot\vec{s}_Y$, where $\vec{s}_N$ and $\vec{s}_Y$ are the nucleon and hyperon spin-$\frac{1}{2}$ vectors. The p-shell $0p_N 0s_Y$ matrix elements $\bar{V}^{0p}_{\Lambda\Sigma}$ and $\Delta^{0p}_{\Lambda\Sigma}$, listed in the caption to Table 2, follow from the shell-model reproduction of hypernuclear γ-ray transition energies by Millener [26] and are smaller by roughly factor of two than the corresponding s-shell $0s_N 0s_Y$ matrix elements, therefore resulting in smaller Σ hypernuclear admixtures and implying that CSB contributions in the p shell are weaker with respect to those in the $A = 4$ hypernuclei also by a factor of two. To evaluate these CSB contributions, the single-nucleon expression (1) is extended by summing over p-shell nucleons [9]:

$$V_{\rm CSB} = -0.0297\,\frac{1}{\sqrt{3}}\sum_j (\bar{V}^{0p}_{\Lambda\Sigma} + \Delta^{0p}_{\Lambda\Sigma}\,\vec{s}_j\cdot\vec{s}_Y)\,\tau_{jz}. \tag{4}$$

Results of applying this effective $\Lambda\Sigma$ coupling model to several pairs of g.s. levels in p-shell hypernuclear isomultiplets are given in Table 2, abridged from Ref. [9]. All pairs except for $A = 7$ are g.s. mirror hypernuclei identified in emulsion [18] where binding energy systematic uncertainties are largely canceled out in forming the listed $\Delta B^{\rm exp}_\Lambda$ values. The B_Λ data selected for the A=7 ($^7_\Lambda$He, $^7_\Lambda$Li*, $^7_\Lambda$Be) isotriplet of lowest $\frac{1}{2}^+$ levels deserve discussion. Recall that the ^{6}Li core state of $^7_\Lambda$Li* is the 0^+ T=1 at 3.56 MeV, whereas the core state of $^7_\Lambda$Li$_{\rm g.s.}$ is the 1^+ T=0 g.s. Thus, to obtain $B_\Lambda(^7_\Lambda{\rm Li}^*)$ from $B_\Lambda(^7_\Lambda{\rm Li}_{\rm g.s.})$ one makes use of the observation of a 3.88 MeV γ-ray transition $^7_\Lambda$Li$^* \to \gamma + ^7_\Lambda$Li [28]. While emulsion $B^{\rm exp}_\Lambda$(g.s.) values [18] were used for the $^7_\Lambda$Be–$^7_\Lambda$Li* pair, more recent counter measurements that provide absolute energy calibrations relative to precise values of free-space known masses were used for the $^7_\Lambda$Li*–$^7_\Lambda$He pair [27] (FINUDA for $^7_\Lambda$Li$_{\rm g.s.}$ π^- decay [29] and JLab for $^7_\Lambda$He electroproduction [30]). Note that the value reported by FINUDA for $B_\Lambda(^7_\Lambda{\rm Li}_{\rm g.s.})$, 5.85±0.17 MeV, differs from the emulsion value of 5.58±0.05 MeV. Recent B_Λ values from JLab electroproduction experiments for $^9_\Lambda$Li [31] and $^{10}_\Lambda$Be [32] were not used for lack of similar data on their mirror partners.

Table 2. $\langle V_{CSB}\rangle$ contributions (in keV) to ΔB_Λ^{calc} in p-shell hypernuclei g.s. isomultiplets, using $\Lambda\Sigma$ coupling matrix elements $\bar{V}_{\Lambda\Sigma}^{0p}$=1.45 MeV and $\Delta_{\Lambda\Sigma}^{0p}$=3.04 MeV in Eq. (4). A similar calculation for the s-shell A=4 mirror hypernuclei [9] is included for comparison. Listed values of ΔB_Λ^{exp} are based on g.s. emulsion data [18] except for ${}^4_\Lambda$He–${}^4_\Lambda$H [8] and ${}^7_\Lambda$Li*–${}^7_\Lambda$He [27].

${}^A_\Lambda Z_>$–${}^A_\Lambda Z_<$	${}^4_\Lambda$He–${}^4_\Lambda$H	${}^7_\Lambda$Be–${}^7_\Lambda$Li*	${}^7_\Lambda$Li*–${}^7_\Lambda$He	${}^8_\Lambda$Be–${}^8_\Lambda$Li	${}^9_\Lambda$B–${}^9_\Lambda$Li	${}^{10}_\Lambda$B–${}^{10}_\Lambda$Be
I, J^π	$\frac{1}{2}, 0^+$	$1, \frac{1}{2}^+$	$1, \frac{1}{2}^+$	$\frac{1}{2}, 1^-$	$1, \frac{3}{2}^+$	$\frac{1}{2}, 1^-$
$\langle V_{CSB}\rangle$	232	50	50	119	81	17
ΔB_Λ^{calc}	226	−17	−28	+49	−54	−136
ΔB_Λ^{exp}	233±92	−100±90	−20±230	+40±60	−210±220	−220±250

The $\langle V_{CSB}\rangle$ p-shell entries listed in Table 2 were calculated with Λ-hypernuclear weak-coupling shell-model wavefunctions in terms of nuclear-core g.s. leading SU(4) supermultiplet components, except for $A = 8$ where the first excited nuclear-core level had to be admixed in. The listed $A = 7 - 10$ values of $\langle V_{CSB}\rangle$ exhibit strong SU(4) correlations, highlighted by the enhanced value of 119 keV for the SU(4) nucleon-hole configuration in ${}^8_\Lambda$Be–${}^8_\Lambda$Li with respect to the modest value of 17 keV for the SU(4) nucleon-particle configuration in ${}^{10}_\Lambda$B–${}^{10}_\Lambda$Be. This enhancement follows from the relative magnitudes of the Fermi-like interaction term $\bar{V}_{\Lambda\Sigma}^{0p}$ and its Gamow-Teller partner term $\Delta_{\Lambda\Sigma}^{0p}$. Noting that both the $A = 4$ and $A = 8$ mirror hypernuclei correspond to SU(4) nucleon-hole configuration, the roughly factor two ratio of $\langle V_{CSB}\rangle_{A=4}$=232 keV to $\langle V_{CSB}\rangle_{A=8}$=119 keV reflects the approximate factor of two for $0s_N0s_Y$ to $0p_N0s_Y$ $\Lambda\Sigma$ matrix elements discussed above. However, in distinction from the A=4 g.s. isodoublet where $\Delta B_\Lambda \approx \langle V_{CSB}\rangle$, the increasingly negative Coulomb contributions in the p-shell overcome the positive $\langle V_{CSB}\rangle$ contributions, with ΔB_Λ becoming negative definite for $A \geq 9$.

Comparing ΔB_Λ^{calc} with ΔB_Λ^{exp} in Table 2, we note the reasonable agreement reached between the $\Lambda\Sigma$ coupling model calculation and experiment for all five pairs of p-shell hypernuclei listed here. Extrapolating to heavier hypernuclei, one might naively expect negative values of ΔB_Λ^{calc}. However, this assumes that the negative Coulomb contribution remains as large upon increasing A as it is in the beginning of the p shell, which need not be the case. As nuclear cores beyond $A = 9$ become more tightly bound, the Λ hyperon is unlikely to compress these nuclear cores as much as it does in lighter hypernuclei, so that the additional Coulomb repulsion in ${}^{12}_\Lambda$C, for example, over that in ${}^{12}_\Lambda$B may not be sufficiently large to offset the attractive CSB contribution to $B_\Lambda({}^{12}_\Lambda\text{C})-B_\Lambda({}^{12}_\Lambda\text{B})$, in agreement with the value 50±110 keV suggested recently for this A=12 B_Λ(g.s.) splitting using FINUDA and JLab counter measurements [27]. In making this argument one relies on the expectation, based on SU(4) supermultiplet fragmentation patterns in the p shell, that $\langle V_{CSB}\rangle$ does not exceed ~100 keV.

Some implications of the state dependence of CSB splittings, e.g. the large difference between the calculated $\Delta B_\Lambda(0^+_{g.s.})$ and $\Delta B_\Lambda(1^+_{exc})$ in the s shell, Eqs. (2) or (3), are worth noting also in the p shell. The most spectacular one concerns the ${}^{10}_\Lambda$B g.s. doublet splitting, where adding the $\Lambda\Sigma$ coupling model CSB contribution of ≈ -27 keV to the $\approx$110 keV CS $1^-_{g.s.} \to 2^-_{exc}$ g.s. doublet excitation energy calculated in this model [26] helps bring it down well below 100 keV, which is the upper limit placed on it from past searches for a $2^-_{exc} \to 1^-_{g.s.}$ γ-ray transition [33,34].

6. Summary and Outlook

The recent J-PARC E13-experiment observation of a 1.41 MeV ${}^4_\Lambda$He($1^+_{exc} \to 0^+_{g.s.}$) γ-ray transition [6], and the recent MAMI-A1 determination of $B_\Lambda({}^4_\Lambda\text{H})$ to better than 100 keV [7,8], plus the recenly approved J-PARC E63 experiment to remeasure the $^4_\Lambda$H($1^+_{\rm exc} \to 0^+_{\rm g.s.}$) γ-ray transition, arose renewed interest in the sizable CSB already confirmed thereby in the A=4 mirror hypernuclei. It was shown in the present report how a relatively large $\Delta B_\Lambda(0^+_{\rm g.s.})$ CSB contribution of order 250 keV, in rough agreement with experiment, arises in ab-initio four-body calculations [14, 15] using χEFT YN interactions already at LO.

In p-shell hypernuclei, a $\Lambda\Sigma$ coupling shell-model approach was shown to reproduce CSB splittings of g.s. binding energies [9]. More theoretical work in this mass range, and beyond, is needed to understand further and better the salient features of $\Lambda\Sigma$ dynamics [35]. On the experimental side, the recently proposed (π^-, K^0) reaction [36] should be explored, in addition to the standard (π^+, K^+) reaction, in order to study simultaneously two members of a given Λ hypernuclear isomultiplet, for example reaching both $^{12}_\Lambda$B and $^{12}_\Lambda$C on a carbon target.

Acknowledgments

A.G. acknowledges instructive discussions with John Millener, as well as the gracious hospitality extended by the organizers of SPRING 2017 at Ischia, Italy. The research of D.G. was supported by the Grant Agency of the Czech Republic (GACR), Grant No. P203/15/04301S.

References

[1] Miller G A, Opper A K and Stephenson E J 2006 *Annu. Rev. Nucl. Part. Sci.* **56** 253
[2] Machleidt R and Müther H 2001 *Phys. Rev.* C **63** 034005
[3] Entem D R and Machleidt R 2003 *Phys. Rev.* C **68** 041001(R)
[4] Dalitz R H and Von Hippel F 1964 *Phys. Lett.* **10** 153
[5] Coon S A, Han H K, Carlson J and Gibson B F 1999 *Meson and Light Nuclei '98* ed J Adam, P Bydžovský, J Dobeš, R Mach and J Mareš (Singapore: WS) pp 407-413
[6] Yamamoto T O *et al* (J-PARC E13 Collaboration) 2015 *Phys. Rev. Lett.* **115** 222501
[7] Esser A *et al* (MAMI A1 Collaboration) 2015 *Phys. Rev. Lett.* **114** 232501
[8] Schulz F *et al* (MAMI A1 Collaboration) 2016 *Nucl. Phys.* A **954** 149
[9] Gal A 2015 *Phys. Lett.* B **744** 352
[10] Gazda D, Mareš J, Navrátil P, Roth R and Wirth R 2014 *Few-Body Syst.* **55** 857
[11] Polinder H, Haidenbauer J and Meißner U -G 2006 *Nucl. Phys.* A **779** 244
[12] Rijken Th A, Stoks V G J and Yamamoto Y 1999 *Phys. Rev.* C **59** 21
[13] Akaishi A, Harada T, Shinmura S and Myint K S 2000 *Phys. Rev. Lett.* **84** 3539
[14] Gazda D and Gal A 2016 *Phys. Rev. Lett.* **116** 122501
[15] Gazda D and Gal A 2016 *Nucl. Phys.* A **954** 161
[16] Navrátil P 2007 *Few-Body Syst.* **41** 117
[17] Wirth R, Gazda D, Navrátil P, Calci A, Langhammer J and Roth R 2014 *Phys. Rev. Lett.* **113** 192502
[18] Davis D H 2005 *Nucl. Phys.* A **754** 3c
[19] Haidenbauer J, Meißner U -G, Nogga A and Polinder H 2007 *Topics in Strangeness Nuclear Physics* ed P Bydžovský, J Mareš and A Gal (New York: Springer) *Lecture Notes in Physics* **724** pp 113-140
[20] Navrátil P, Kamuntavičius G P and Barrett B R 2000 *Phys. Rev.* C **61** 044001
[21] Haidenbauer J, Petschauer S, Kaiser N, Meißner U -G, Nogga A and Weise W 2013 *Nucl. Phys.* A **915** 24
[22] Nogga A 2013 *Nucl. Phys.* A **914** 140 and references to earlier works cited therein
[23] Hiyama E, Yamamoto Y, Motoba T and Kamimura M 2009 *Phys. Rev.* C **80** 054321
[24] Zhang Y, Hiyama E and Yamamoto Y 2012 *Nucl. Phys.* A **881** 288
[25] Hiyama E and Yamamoto Y 2012 *Prog. Theor. Phys.* **128** 105
[26] Millener D J 2012 *Nucl. Phys.* A **881** 298 and references listed therein
[27] Botta E, Bressani T and Feliciello A 2017 *Nucl. Phys.* A **960** 165
[28] Tamura H *et al* 2000 *Phys. Rev. Lett.* **84** 5963
[29] Agnello M *et al* (FINUDA Collaboration and Gal A) 2009 *Phys. Lett.* B **681** 139
[30] Gogami T *et al* (JLab HKS Collaboration) 2016 *Phys. Rev.* C **94** 021302(R)
[31] Urciuoli G M *et al* (JLab Hall A Collaboration) 2015 *Phys. Rev.* C **91** 034308
[32] Gogami T *et al* (JLab HKS Collaboration) 2016 *Phys. Rev.* C **93** 034314
[33] Chrien R E *et al* 1990 *Phys. Rev.* C **41** 1062
[34] Tamura H *et al* 2005 *Nucl. Phys.* A **754** 58c
[35] Gal A and Millener D J 2013 *Phys. Lett.* B **725** 445
[36] Agnello M, Botta E, Bressani T, Bufalino S and Feliciello A 2016 *Nucl. Phys.* A **954** 176

IOP Conf. Series: Journal of Physics: Conf. Series **966** (2018) 012007 doi:10.1088/1742-6596/966/1/012007

Odd nuclei spectroscopy within a self-consistent multiphonon approach

G. De Gregorio[1,2,3], **F. Knapp**[4], **N. Lo Iudice**[2,3], **P. Veselý**[1]

[1]Nuclear Physics Institute, Czech Academy of Sciences, 250 68 Řež, Czech Republic
[2]Dipartimento di Fisica, Università di Napoli Federico II, Naples, Italy
[3]INFN Sezione di Napoli, Naples, Italy
[4]Faculty of Mathematics and Physics, Charles University, Prague, Czech Republic

E-mail: degregorio@na.infn.it

Abstract. The extension to odd nuclei of the equation of motion phonon method (EMPM) is briefly outlined. It derives and solves iteratively a set of equations of motion which yield an orthonormal basis of states composed of an odd particle coupled to orthonormal multiphonon states, also generated within the EMPM, describing the excitations of a doubly magic core. The basis so obtained is used to solve the eigenvalue problem in the full particle-phonon space. We have applied the method to the odd neighbors of ^{16}O using an optimized chiral potential in a space spanned by particle-core basis states which include up to three-phonons. The calculations have shown that the multiphonon states enhance enormously the density of levels and compress the whole spectrum, consistently with the data. They also contribute substantially to improve the agreement with the experimental moments and transitions strengths.

1. Introduction

The correction induced by the core excitations on the single particle energies of odd nuclei is a longstanding problem, and, to a large extent, not yet solved. One of the first attempts was made during the '60s, within shell model, by including 2p-1h and 4p-3h configurations [1]. The necessity of complex configurations for describing the odd nuclei spectra was also stressed by a weak coupling model [2].

Of different nature is the particle-vibration coupling (PVC) model [3, 4]. It couples an odd particle to the collective excitations of the core commonly described in random-phase approximation (RPA). Recently, PVC calculations were performed within the framework of energy density functionals deduced from Skyrme forces [5, 6] or relativistic meson-nucleon Lagrangians [7, 8] or from the theory of finite Fermi systems [9].

Calculations using chiral forces were performed within the context of self-consistent Green's function theory [10, 11], no-core shell model (NCSM) [12] and coupled cluster theory [13, 14, 15].

Our proposal is the Equation of Motion Phonon Method (EMPM) [16, 17, 18]. It has been applied mainly to study the dipole response in neutron rich nuclei [19, 20]. Recently it was formulated in a Hartree-Fock-Bogolyubov basis and adopted to study the full spectrum as well as the dipole response of neutron-rich ^{20}O [21].

In its extension to odd nuclei [22, 23], it derives and solves iteratively a set of equations of motion to generate an orthonormal basis of states composed of a valence particle coupled to n-phonon states (n = 1, 2, ...), also generated within the EMPM, describing the multiphonon excitations of a doubly magic core. The basis is then adopted to solve the full eigenvalue problem. The formalism does not rely on approximations, and takes into fully account the Pauli principle.

2. Outline of the Equation of Motion Phonon Method for odd nuclei

Let us consider the Hamiltonian

$$H = H_0 + V. \tag{1}$$

H_0 is the one-body term

$$H_0 = \sum_r [r]^{1/2} \epsilon_r \left(a_r^\dagger \times b_r\right)^0, \tag{2}$$

where $a_r^\dagger = a^\dagger_{x_r j_r m_r}$ ($br = (-)^{j_r+m_r} a_{x_r j_r m_r}$) creates (annihilates) a particle of energy ϵ, $[r]$ stands for $[r] = 2j_r + 1$ and the symbol $\times$denotes angular momentum coupling. V is a two-body potential

$$V = \frac{1}{4} \sum_{rsqt\sigma} [\sigma]^{1/2} F^\sigma_{rsqt} \left[\left(a_r^\dagger \times b_s\right)^\sigma \times \left(a_q^\dagger \times b_t\right)^\sigma\right]^0, \tag{3}$$

where F^σ is derived from the two-body potential V^Ω through the Pandya transformation

$$F^\sigma_{rsqt} = \sum_\Omega [\omega] (-)^{(r+t-\sigma-\Omega)} W(rsqt;\sigma\Omega) V^\Omega_{rstq}. \tag{4}$$

Here $W(rsqt;\sigma\Omega)$ are Racah coefficients.

The main goal of the method is to derive for an odd system a basis of correlated particle-core states of the form

$$\mid \nu_n\rangle = \sum_{p\alpha_n} C^{\nu_n}_{p\alpha_n} \mid (p \times \alpha_n)^v\rangle = \sum_{p\alpha_n} C^{\nu_n}_{p\alpha_n} (a_p^\dagger \times \mid \alpha_n\rangle)^v \tag{5}$$

from a set of states composed of an odd particle p coupled to a a basis of correlated n-phonon core states

$$\mid \alpha_n\rangle = \sum_{\lambda\alpha_{n-1}} C^{\alpha_n}_{\lambda\alpha_{n-1}} \mid (\lambda \times \alpha_{n-1})^{\alpha_n}\rangle = \sum_{\lambda\alpha_{n-1}} C^{\alpha_n}_{\lambda\alpha_{n-1}} (O_\lambda^\dagger \times \mid \alpha_{n-1}\rangle)^{\alpha_n} \tag{6}$$

also derived within the EMPM [18]. In the above equation

$$O_\lambda^\dagger = \sum_{ph} c^\lambda_{ph} (a_p^\dagger \times b_h)^\lambda \tag{7}$$

is the TDA phonon operator, the building block of the multiphonon basis states.

In perfect analogy with the procedure adopted for even nuclei [18] we start with the equations of motion

$$\langle \alpha \parallel [b_p, H] \parallel \nu\rangle = (E_\nu - E_\alpha) X^{(\nu)}_{p\alpha}, \tag{8}$$

where E_ν are the eigenvalues to be determined and

$$X^{(\nu)}_{p\alpha} = \langle \alpha \parallel b_p \parallel \nu\rangle = \sum_{p'\alpha'} \mathcal{D}^\nu_{p\alpha p'\alpha'} C^\nu_{p'\alpha'}. \tag{9}$$

Here

$$\mathcal{D}^\nu_{p\alpha p'\alpha'} = \langle (p \times \alpha)^v \mid (p' \times \alpha')^v\rangle \tag{10}$$

is the overlap matrix which reintroduces the exchange terms among the odd particle and the n-phonon states and, thereby, re-establishes the Pauli principle. After expanding the commutator in the Eq. (8), we obtain

$$\sum_{p'\alpha'p_1\alpha_1} \Big\{(\epsilon_p + E_\alpha - E_\nu)\delta_{pp'}\delta^{(n)}_{\alpha\alpha'} + \mathcal{V}^{(\nu_n)}_{p\alpha p'\alpha'}\Big\}\mathcal{D}^{\nu}_{p'\alpha' p_1\alpha_1} C^{\nu}_{p_1\alpha_1} = 0, \tag{11}$$

where

$$\mathcal{V}^{(\nu_n)}_{p\alpha p'\alpha'} = \sum_{\sigma t q} [\sigma]^{1/2} W(\alpha\sigma v p'; \alpha' p)\langle \alpha' \mid\mid \left[a^\dagger_t \times b_q\right]^\sigma \mid\mid \alpha\rangle F^\sigma_{pp'tq} \tag{12}$$

is the particle-phonon potential. Eq. (11) represents an eigenvalue equation in the over-complete basis $\mid (p \times \alpha_n)^v\rangle$ within the n-phonon particle-core subspace. Following the procedure outlined in Refs. [16, 17], based on the Cholesky decomposition method, we extract a basis of linearly independent states $\mid (p\times\alpha_n)^v\rangle$ to obtain a non singular eigenvalue equation. Its iterative solution yields the particle-core states $\mid \nu_n\rangle$ (5) of energies E_{ν_n} for $n = 1, 2....$, which, together with the single particle states $\mid \nu_0\rangle$, form an orthonormal basis.

The eigenvalue problem in the full space spanned by $\{\mid \nu_0\rangle, \mid \nu_1\rangle, \ldots \mid\nu_n > \ldots\}$ is defined by

$$\sum_{\nu_{n'}} \Big\{(E_{\nu_n} - \mathcal{E}_\nu)\delta_{\nu_n\nu_{n'}} + \mathcal{V}^v_{\nu_n\nu_{n'}}\Big\}\mathcal{C}^{(\nu)}_{\nu_{n'}} = 0, \tag{13}$$

where $\mathcal{V}^{(v)}_{\nu_n\nu_{n'}}$ are non vanishing for $n' = n + 1$ and $n' = n + 2$. The above eigenvalue equation yields the eigenstates

$$\mid \Psi_\nu\rangle = \sum_{\nu_n} \mathcal{C}^\nu_{\nu_n} \mid \nu_n\rangle = \sum_{p\alpha_n} \mathcal{C}^\nu_{p\alpha_n} \mid (p \times \alpha_n)^v\rangle, \tag{14}$$

where ν_n are defined by the Eq. (5). The wavefunctions (14) can be used to compute the transition amplitudes of a multipole operator given by

$$\langle \Psi_{\nu'} \mid\mid \mathcal{M} \mid\mid \Psi_\nu\rangle = \sum_{nn'} \mathcal{M}^{(\nu\nu')}_{nn'}(\lambda) = \sum_{nn'} \sum_{\nu_n\nu'_{n'}} \mathcal{C}^\nu_{\nu_n}\mathcal{C}^{\nu'}_{\nu'_{n'}} \langle \nu'_{n'} \mid\mid \mathcal{M}(\lambda) \mid\mid \nu_n\rangle. \tag{15}$$

If the initial and/or final states have dominant single particle character we can safely use the truncated formula

$$\langle \Psi'_\nu \mid\mid \mathcal{M} \mid\mid \Psi_\nu\rangle \simeq \mathcal{M}^{(\nu\nu')}_{00}(\lambda) + \mathcal{M}^{(\nu\nu')}_{01}(\lambda) + \mathcal{M}^{(\nu\nu')}_{10}(\lambda), \tag{16}$$

where $\mathcal{M}^{(\nu\nu')}_{00}(\lambda)$, $\mathcal{M}^{(\nu\nu')}_{01}(\lambda)$ and $\mathcal{M}^{(\nu\nu')}_{10}(\lambda)$ are, respectively, the particle-particle, particle-phonon, and phonon-particle transition amplitudes.

3. Calculations and results

We used an Hamiltonian composed of an intrinsic kinetic operator T_{int} plus the $NNLO_{opt}$ optimized chiral potential [24] to generate the HF basis in a configuration space including up to the harmonic oscillator major shell $N_{max} = 15$. The TDA phonons are derived from a subset of HF states corresponding to $N = 12$. Their structure does not change if we use the full HF space. The spurious $J = 1^-$ TDA components induced by the center of mass (CM) motion have been removed resorting to a Gramm-Schmidt orthogonalization procedure discussed in Ref. [25].

The multiphonon particle-core basis is composed of all one-phonon particle-core states and of a restricted set of two- and three-phonon particle-core states. The inclusion of the three-phonon

IOP Conf. Series: Journal of Physics: Conf. Series **966** (2018) 012007 doi:10.1088/1742-6596/966/1/012007

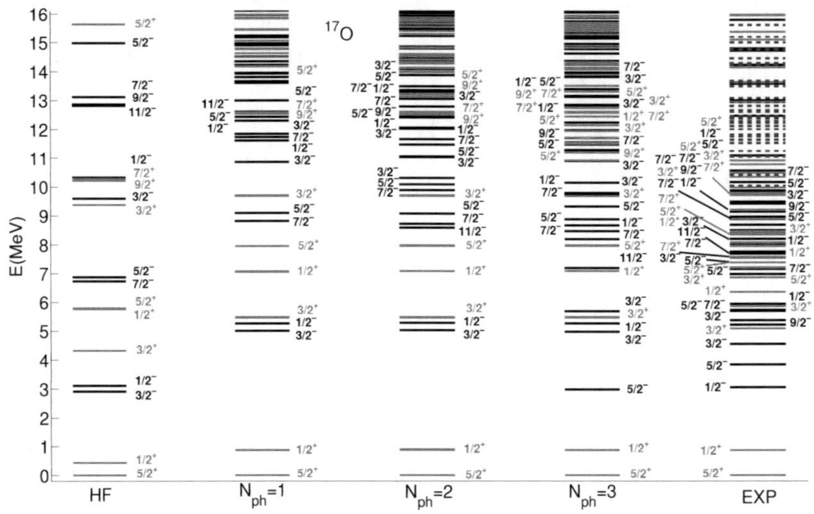

Figure 1. Theoretical versus experimental [26] spectra of ^{17}O. N_{ph} indicates the maximum phonon number. The dashed levels have unknown spin or parity or both.

Table 1. Ground state electric quadrupole Q (efm^2) and magnetic moments μ (μ_n) and $B(E2; J_i^\pi \to J_f^\pi)(e^2fm^4)$. The ground state β-decay *ft* value is also included. The experimental data are taken from [26].

		HF	EMPM	EXP
^{17}O	μ	-1.91	-1.83	-1.89
	Q	0	0.841	-2.578
	$B(E2; 5/2_1^+ \to 1/2_1^+)$	0	0.17	2.18(16)
^{17}F	μ	+4.79	+4.63	+4.72
	Q	-9.9	-7.6	± 5.8(4)
	$B(E2; 5/2_1^+ \to 1/2_1^+)$	40.71	21.89	21.64
	log *ft*	3.294	3.391	3.358(2)

particle-core states has required some approximations. We ignored the interaction $\mathcal{V}_{p\alpha_3,p'\alpha_3'}$ in the eigenvalue equation as well as the phonon-phonon potential $\mathcal{V}_{\lambda\alpha,\lambda'\alpha'}$ determining the core states $\mid \alpha_3\rangle$. We have also neglected the exchange terms between the odd particle and $\mid \alpha_3\rangle$ by putting $\mathcal{D}_{p\alpha_3,p'\alpha_3'} = \delta_{pp}\delta_{\alpha_3\alpha_3'}$.

The theoretical spectra obtained for ^{17}O and ^{17}F in different multiphonon spaces are compared to one another and with experiments in Fig.1 and 2. The one-phonon states alter appreciably the HF levels. They depress the lowest $5/2^+$ with respect to the other states thereby enhancing its distance from the other levels. In ^{17}O this energy shift is beneficial for the low-lying positive parity levels which get closer to the experimental levels of single particle nature. The negative parity states, instead, get pushed at too high energies with the exception of the $3/2^-$. In ^{17}F the energy depression promotes the inversion between the $1/2^+$ and $5/2^+$ levels. The correct ground state is thus obtained.

The action exerted by the phonon couplings is similar in both isobars. The inclusion of the two- phonon states enhances the level density only at high energy. The three phonons, by pushing few negative parity states down in energy, enrich the low energy spectra but not sufficiently to reproduce the density of the experimental low-energy level schemes.

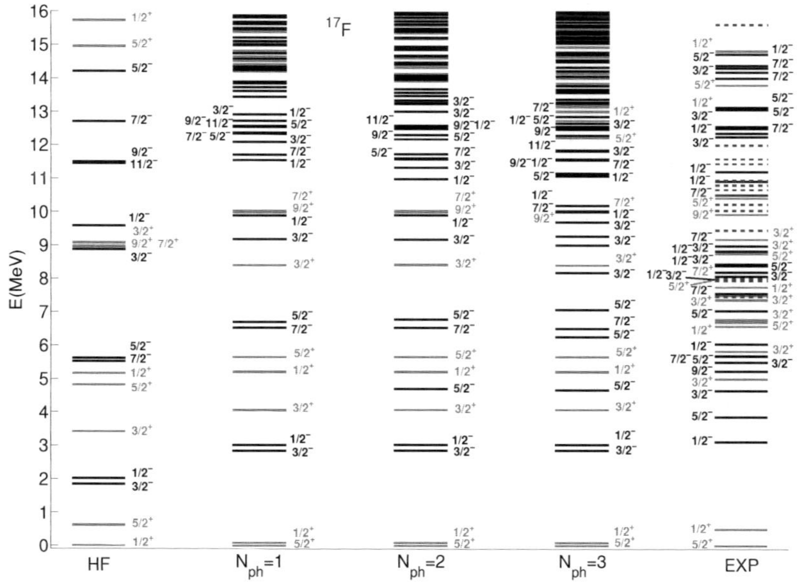

Figure 2. Theoretical versus experimental [26] spectra of ^{17}F. The notations are the same as for ^{17}O

The magnetic moments in both ^{17}O and ^{17}F are practically determined at the HF level (Tab.1). The weak quenching due to the core brings the total moments slightly more distant from the experimental values. The ground state β-decay of ^{17}F is also ruled by HF. Indeed, the ft value comes almost entirely from the transition between the HF components of the ^{17}O and ^{17}F $5/2^+$ ground states (Tab.1). The weak quenching caused by the phonon coupling brings the ft value slightly above the measured quantity.

In ^{17}O, the calculation underestimates the absolute value of the quadrupole moment by a factor three and the strength of the $E2$ transition from the ground state $5/2_1^+$ to $1/2_1^+$ by an order of magnitude (Tab.1). Since the odd particle is a neutron, and both states have a single particle nature, the contribution to the moments and transition strength comes entirely from the terms $\mathcal{M}01(E2)$ and $\mathcal{M}10(E2)$ (Eq. 16) which couple the single particle components of $5/2_1^+$ and $1/2_1^+$ to the $\lambda = 2^+$ particle-phonon pieces of $1/2_1^+$ and $5/2_1^+$, respectively. It seems, therefore, that the amplitudes of the one-phonon components and, in particular, the quadrupole phonon pieces are not sufficiently large.

In ^{17}F, the quadrupole moment, computed in HF, is $\sim$ 1.7 times the measured value. It gets considerably smaller and closer to experiments once the phonon coupling is included. The coupling is even more effective on the $5/2_1^+ \rightarrow 1/2_1^+$ $E2$ transition. Once the phonon are included, the $E2$ strength gets considerably reduced and coincides in practice with the experimental value (Tab.1). This strong effect comes from the indirect coupling among different single particle components of the ground state wave function. This mutual interference causes the quenching of quadrupole moment and transition. In ^{17}O, this interference has no effect since the odd neutron carries no charge.

4. Conclusions

The above analysis has shown that, while the two phonons have the only effect of enhancing the level density at high energy, the three phonons, through their strong coupling, push down in energy few one-phonon states of negative parity. They enrich thereby the low-energy sector but not sufficiently to reproduce the experimental level scheme. The phonon contribution to the quadrupole moments and transitions is also substantial but not not sufficient to bring those quantities close to the measured data.

This relatively weak phonon action originates from an insufficient particles-phonon and phonon-phonon admixture. In particular, the one-phonon components have too small weights in the states of dominant single particle character, lying generally at low energies, and too large amplitudes in the dominantly particle-core states. This insufficient admixing might be traced back to the large separation between Hartree-Fock levels [25]. A more compressed HF spectrum would yield more closely packed TDA phonons and, therefore, might enhance the coupling of the one-phonon components to the HF vacuum and to the other n-phonon states. This may be obtained by using a new potential which includes a three-body force.

Only four phonons, through their strong coupling to two-phonon states, would be able to push the positive parity levels down in energy and to enhance the mixing among different n-phonon components. Including four phonons may be feasible if we resort to approximations analogous to the ones we made here for three phonons.

Acknowledgements

This work was partly supported by the Czech Science Foundation (P203-13-07117S). Two of the authors (F.K. and P.V.) thank the Istituto Nazionale di Fisica Nucleare (Italy) for financial support. Highly appreciated was the access to computing and storage facilities provided by the Meta Centrum under Program No. LM2010005 and the CERITSC under the program Centre CERIT Scientific Cloud, part of the Operational Program Research and Development for Innovations, Register No. CZ.1.05/3.2.00/08.0144.

References

[1] Birkholz J and Beck F 1968 *Phys. Lett. B* **28** 18-21
[2] Engeland T and Ellis P J 1972 *Nucl. Phys. A* **181** 368-402
[3] Bohr A and Mottelson B R 1975 *Nuclear Structure* (W. A. Benjamin, New York), Vol II
[4] Mahaux C, Bortignon P F, Broglia R A, and Dasso C H 1985 *Phys. Rep.* **120**
[5] Cao L-G, Colò G, Sagawa H, and Bortignon P F 2014 *Phys. Rev. C* **89** 044314
[6] Co' G, De Donno V, Anguiano M, Bernard R N, and Lallena A M 2015 *Phys Rev. C* **92** 024314
[7] Litvinova E and Ring P 2006 *Phys Rev. C* **73** 044328
[8] Afanasjev A V and Litvinova E 2015 *Phys Rev. C* **92** 044317
[9] Gnezdilov N V, Borzov I N, Saperstein E E and Tolokonnikov S V 2014 *Phys Rev. C* **89** 034304
[10] Cipollone A, Barbieri C and Navrátil P 2013 *Phys. Rev. Lett.* **111** 062501
[11] Somá V, Barbieri C, and Duguet T 2013 *Phys Rev. C* **87** 011303(R)
[12] Barrett B R, Navrátil P, and Vary J P 2013 *Prog. Part. Nucl. Phys.* **69** 131
[13] Hagen G, Papenbrock T, and Hjorth-Jensen M 2010 *Phys. Rev. Lett.* **104** 182501
[14] Hagen G, Hjorth-Jensen M, Jansen G R, Machleidt R and Papenbrock T 2012 *Phys. Rev. Lett.* **1 08** 242501
[15] Jansen G R, Engel J, Hagen G, Navratil P, and Signoracci A 2014 *Phys. Rev. Lett.* **113** 142502
[16] Andreozzi F, Knapp F, Lo Iudice N, Porrino A and Kvasil J 2007 *Phys Rev. C* **75** 044312
[17] Andreozzi F, Knapp F, Lo Iudice N, Porrino A and Kvasil J 2008 *Phys Rev. C***78** 054308
[18] Bianco D, Knapp F, Lo Iudice N, Andreozzi F and Porrino A 2012 *Phys Rev. C***85** 014313
[19] Knapp F, Lo Iudice N, Veselý P, Andreozzi F, De Gregorio G and Porrino A 2014 *P hys Rev. C* **90** 014310
[20] Knapp F, Lo Iudice N, Veselý P, Andreozzi F, De Gregorio G and Porrino 2015, *Phys Rev. C* **92** 054315
[21] De Gregorio G, Knapp F, Lo Iudice N and Veselý P 2016 *Phys Rev. C* **93** 044314
[22] De Gregorio G, Knapp F, Lo Iudice N and Veselý P 2016 *Phys Rev. C* **94** 061301(R)
[23] De Gregorio G, Knapp F, Lo Iudice N and Veselý P 2017 *Phys Rev. C* **95** 034327
[24] Ekström A, Baardsen G, Forssén C, Hagen G, Hjorth-Jensen M, Jansen G R, Machleidt R, Nazarewicz W, Papenbrock T, Sarich J, et al. 2013 *Phys. Rev. Lett.* **110** 192502
[25] Bianco D, Knapp F, Lo Iudice N, Veselý P, Andreozzi F, De Gregorio G, and Porrino A 2014 *J. Phys. G* **41** 025109
[26] Tilley D, Weller H, and Cheves C 1993 *Nucl. Phys. A* **564** 1

IOP Conf. Series: Journal of Physics: Conf. Series **966** (2018) 012008 doi:10.1088/1742-6596/966/1/012008

A view of recent results and perspectives on nuclear structure with MAGNEX at the INFN-LNS laboratory

F Cappuzzello[1,2], C. Agodi[2], D. Carbone[2], M. Cavallaro[2]
[1] Department of Physics and Astronomy, University of Catania, Italy
[2] INFN-LNS, Catania, Italy

Abstract: An intense activity in the field of experimental, as well as theoretical, nuclear structure is ongoing at the INFN-LNS laboratory in Catania. A short overview of the main recent results is reported.

Nuclear structure studies are important research items at the INFN-LNS laboratory in Catania [1]. A characterizing feature is the use of direct quasi-elastic reactions as selective probes of specific degrees of freedom of atomic nuclei. Experimentally, heavy-ion beams are exploited to activate the interesting reactions, accelerated in a broad range of bombarding energies by the 14 MV Tandem van der Graaff and by the K800 Superconducting Cyclotron (CS). Different devices are available for the detection of the charged reaction products, as well as of neutrons and γ-rays. A particular attention is given here to the MAGNEX large acceptance magnetic spectrometer [2-3], which is a dedicated facility for nuclear structure as well as direct reaction mechanism studies. The spectrometer can detect reaction products with high mass, energy and angular resolution in a broad phase space and allows to measure at zero degree. Mass distributions, energy spectra and absolute cross section angular distributions are typically measured, giving access to important quantitative information on the selected nuclear transitions. Here we refer to results achieved in the context of old and new puzzles in nuclear structure, which we have contributed to unveil in recent years.

A campaign of experiments of heavy-ion induced one- and two- neutron transfer reactions was performed by MAGNEX with the aim, among the others, of studying the reaction mechanism and the competition between mean field versus pairing correlations in low lying nuclear states. In most of these experiments, a beam of $^{18}O^{6+}$ ions was accelerated at 84 MeV incident energy and transported to several thin enriched targets from ^{9}Be up to ^{208}Pb. Experiments at higher bombarding energies up to 270 MeV were also performed by using the Cyclotron accelerated $^{18}O^{4+}$ beams. In these experiments the MAGNEX optical axis was centered at forward angles. Neutron-neutron pairing correlations were studied through the (^{18}O,^{16}O) transfer reaction, which has allowed to give a precise determination of nuclear response to two-neutron addition operators. In particular, we have explored several nuclei from light to heavy systems [4-8], finding a clear indication of selectivity of this reaction of two-neutron + core configurations in the residual systems. The weak population of unnatural parity states and the large cross section of two-neutron transfer give a hint of the correlated two-neutron transfer as leading mechanism in the (^{18}O,^{16}O) reaction. This is a direct consequence of the existing two-neutron pair in the ^{18}O ground state wave function emphasized by the weak polarizability of ^{16}O core. An important finding was the observed signatures of the long searched Giant Pairing Vibration [9-10] in ^{14}C and ^{15}C nuclei. The population of several resonances as a distinctive feature of the measured energy spectra has also suggested to explore their decay pattern. Coupling MAGNEX with the EDEN array of NE213 liquid scintillators the neutron branching ratios of ^{15}C resonances were measured for the first time [11-13]. Exact finite range Coupled Reaction Channel (CRC) cross-section calculations were performed. The spectroscopic amplitudes were determined by shell-model and Interacting Boson Model (IBM) calculation. In the analyses a quantitative determination of pairing correlation in the wave function is achieved for many of the populated states, giving access to deeper understanding of two-neutron clustering inside nuclear matter. A specific theoretical approach, called Microscopic Cluster Model was also developed to better emphasize this aspect of nuclear structure [14].

Another explored research item is the study of the effect of nuclear structure in the heavy-ion elastic scattering. The coupling of the elastic channel with the other reaction channels was the main topics investigated in a series of experiments at the MAGNEX facility. Above the Coulomb barrier, nuclear internal degrees of freedom are more and more involved in the elastic scattering, which thus gain sensitivity on the details of the nuclear structure of the involved systems. A striking feature is that state of the art Coupled Channel (CC) calculations predict new kind of rainbow-like structures [15]. Experimentally these effects are stronger at large scattering angles, where elastic cross sections are sensibly smaller than inelastic and the channel coupling mechanism can be a dominant source of the elastic flux. However, in this angular region the elastic cross sections become extremely small ($d\sigma/d\omega$ ~ 100 nb/sr and less) and experiments are consequently quite challenging. Three experimental campaigns have been set-up to study ^{16}O + ^{27}Al scattering at 100 MeV and 280 MeV and ^{16}O + ^{60}Ni at 280 MeV. In all the experiments the cross section was accurately measured down to about 100 nb/sr and less [16-17]. The data shows a clear deviation from optical model results and demand challenging CC calculations with state of art optical potentials and nuclear structure models to reproduce the experimental data [18-19]. In particular, it was found that the coupling of the elastic channel with low-lying and even high-lying collective states as giant resonances is essential to reproduce the observed deviation of the experimental cross section angular distributions from Fraunhofer scattering.

Finally, we mention here the intense activity at the LNS for the exploration of nuclear response to isospin by charge exchange reactions. This study started with a systematic exploration of the (^{7}Li, ^{7}Be) charge-exchange reaction at about 8 MeV/u was performed in light neutron rich nuclei, showing that at such incident energy this process proceeds with a considerable predominance of the direct one-step mechanism, thus being a useful probe for spectroscopic studies [20-24]. In particular, the ^{19}F(^{7}Li, ^{7}Be)^{19}O reaction at 52 MeV incident energy was studied with MAGNEX. More recently the attention has been focused on heavier ion induced reactions such as the (^{18}O,^{18}F) for their connection with the NUMEN project (see below). First results from the ^{40}Ca(^{18}O,^{18}F)^{40}K, in comparison with state of art (p,n) and (^{3}He,t) data give encouraging indication that this probe can be useful for nuclear structure investigations. Charge exchange reactions have been recently extended to second order by the exploration of (^{18}O,^{18}Ne) and (^{20}Ne,^{20}O) Double Charge Exchange (DCE) reactions. Despite the challenging scenario, characterized by very low cross sections (of the order of nb) and the necessity to measure at very forward angles, including zero degrees, these studies are very important for us, due to the possible connections to neutrinoless double beta decay studies.

The NUMEN project [25], which represents a very ambitious perspective for the INFN-LNS laboratory in nuclear science, has started as a natural extension of the charge exchange program at the MAGNEX facility, aims at producing experimentally driven information useful for double beta decay research [26]. NUMEN [27-28] proposes to use heavy-ion induced DCE (HI-DCE) reactions as tools to access quantitative information, relevant for $0\nu\beta\beta$ decay NME. These reactions are characterized by the transfer of two charge units, leaving the mass number unchanged, and can proceed by a sequential nucleon transfer mechanism or by meson exchange. Despite $0\nu\beta\beta$ decays and HI-DCE reactions are mediated by different interactions, they present a number of similarities. Among those, the key aspects are that initial and final nuclear states are the same and the transition operators in both cases present a superposition of isospin, spin-isospin and rank-two tensor components with a relevant available momentum (100 MeV/c or so). In a pioneering experiment, performed at the INFN-LNS laboratory, we studied the DCE reaction ^{40}Ca(^{18}O,^{18}Ne)^{40}Ar at 270 MeV, with the aim to measure the cross section at zero degree [26]. The key elements in the experiment were the high resolution CS beams and the MAGNEX spectrometer. In the "pilot experiment" we have shown that high resolution and statistically significant experimental data can be measured for DCE processes and that precious information towards NME determination could be at our reach. To move towards nuclei candidates for 0νββ decay important experimental limits need to be overcome. As a consequence, the present limits of beam power (~100 W) for the CS accelerator and acceptable rate for the MAGNEX focal plane detector (few kHz) must be sensibly overcome. For a systematic study of the many "hot" cases of ββ decays an upgraded set-up, able to work with at least two orders of magnitude more luminosity than the present, is thus necessary. This goal can be achieved by a substantial change in the technologies implemented in the beam extraction [29], in the control of the beam induced radioactivity, in the detection of the ejectiles [30-33] and in the power dissipation of the thin targets [34]. In addition, the project demands for an enhancement of the maximum accepted magnetic rigidity, preserving the geometry and field uniformity of the magnetic field [35-38] in order to keep the high-precision of the present trajectory reconstruction.

Finally, NUMEN is fostering the development of a specific theory program to allow an accurate extraction of nuclear structure information from the measured cross sections. Relying on the use of the Distorted Wave Born Approximation for the cross section, the theory is focused on the development of microscopic models for DCE reactions, employing several approaches (Quasi-particle Random Phase Approximation, IBM, shell model, among others) for inputs connected to nuclear structure quantities. We are also investigating the possible link between the theoretical description of the 0νββ decay and DCE reactions.

References

[1] https://www.lns.infn.it/it/
[2] Cappuzzello F *et al.* 2016 *Eur. Phys. J.* A **52**:167
[3] Cunsolo A *et al.* 2007 *Eur. Phys. Jour.-Special Topics* **150** 343-346
[4] Cappuzzello F *et al.* 2012 *Phys. Lett. B* **711** 347
[5] Cavallaro M *et al.* 2013 *Phys. Rev. C* **88**, 054601
[6] Carbone D *et al.* 2011 J. *Phys. Conf. Ser.* **312** 082016
[7] Carbone D *et al.* 2014 *Phys. Rev. C* **90**, 064621
[8] Ermamatov M J *et al.* 2016 *Phys. Rev. C* **94**, 024610
[9] Cappuzzello F *et al.*, 2015 *Nature Comm.* **6** 6743
[10] Carbone D 2015 *Eur. Phys. J. Plus* **130**: 143
[11] Laurent H *et al.* 1993 *Nucl. Instr. and Meth.* **326** 517
[12] Cavallaro M *et al.* 2013 *Nucl. Instr. and Meth.* A **700** 65–69
[13] Cavallaro M *et al.* 2016 *Phys. Rev. C* **93** 064323
[14] Carbone D *et al.* 2017 *Phys. Rev. C* **95**, 034603
[15] Pereira D *et al.* 2012 *Phys. Lett. B* **710** 426
[16] Cavallaro M *et al.* 2011 *Nucl. Instr. and Meth. A* **648** 46–51
[17] Cappuzzello F *et al.* 2014 *Nucl. Instr. and Meth A* **763** 314–319
[18] Oliveira J R B *et al.* 2013 *J. Phys. G: Nucl. Part. Phys.* **40** 105101
[19] Cappuzzello F *et al.* 2016 Eur. Phys. J. A **52**: 169
[20] Cappuzzello F *et al.* 2004 *Nuclear Physics A* **739** 30
[21] Nociforo C *et al.* 2006 *Eur. Phys. Journ. A* **27**, s01, 283
[22] Orrigo S E A *et al.* 2006 *Phys. Lett. B* **633** 499
[23] Cappuzzello F *et al.* 2004 *Europhys. Lett.* **65** 766
[24] Cavallaro M 2011 *Nuovo Cimento C* **34** 1

[25] https://web.infn.it/NUMEN/index.php/it/
[26] Cappuzzello F *et al.* 2015 *Eur. Phys. J. A* **51**: 145
[27] Cappuzzello F *et al.* 2015 *Journal of Physics Conference Series* **630** 012018
[28] Agodi C *et al.* 2015 *Nuclear and Particle Physics Proceedings* **265** 28-30
[29] Calabretta L *et al.* 2017 *Modern Physics Letters A* **32** 17
[30] Cortesi M *et al.* 2017 *Review of Scientific Instruments* **88** 013303
[31] Muoio A *et al.* 2016 *EPJ Web of Conferences* **117** 10006
[32] Carbone D *et al.* 2016 *Results in Physics* **6** 863–865
[33] De Geronimo G *et al.* 2013 *IEEE Transactions on Nuclear Science* **60** 2314-2321
[34] Iazzi F et al. 2017 *WIT Transactions on Engineering Sciences* **116**, Materials and Contact Characterisation VIII 61-70.
[35] Lazzaro A *et al.* 2008 *Nucl. Instr. and Methods A* **591** 394
[36] Lazzaro A *et al.* 2008 *Nucl. Instr. and Methods A* **585** 136
[37] Lazzaro A *et al.* 2007 *Nucl. Instr. and Methods A* **570** 192
[38] Lazzaro A *et al.* 2009 *Nucl. Instr. and Methods A* **602** 494

IOP Conf. Series: Journal of Physics: Conf. Series **966** (2018) 012009 doi:10.1088/1742-6596/966/1/012009

Nuclear Structure for Double Beta Decay

Mihai Horoi

Department of Physics, Central Michigan University, Mount Pleasant, Michigan 48859, USA

E-mail: mihai.horoi@cmich.edu

Abstract. Neutrinoless double-beta decay, if observed, would signal physics beyond the Standard Model that would be discovered at energies significantly lower than those at which the relevant degrees of freedom can be excited. Accurate nuclear structure calculation of the nuclear matrix elements (NME) necessary to analyze the decay rates could be helpful to narrow down the list of competing mechanisms, and to better identify the more exotic properties of the neutrinos. In this paper we present and analyze the status of the NME shell model calculations, and their relevance for discriminating the possible competing mechanisms that may contribute to the neutrinoless double-beta decay process.

1. Introduction

Should the neutrinoless double-beta transitions occur, then the lepton number conservation is violated by two units, and the black-box theorems [1, 2, 3, 4] indicate that the light left-handed neutrinos are Majorana fermions. As such, through black-box theorems alone, it is not possible to disentangle the dominant mechanism contributing to this process. Most of the theoretical effort dedicated to this subject consists of calculations of leptonic phase-space factors and nuclear matrix elements that are computed via several nuclear structure methods and within specific models. One of the most popular models is the left-right symmetric model [5, 6, 7, 8, 9], which is currently investigated at the Large Hadron Collider (LHC) [10]. In two recent papers [11, 12] we have discussed ways to identify some of the possible contributions to the decay rate by studying the angular distribution and the energy distribution of the two outgoing electrons that could be measured. However, there are still many other possible contributions to this process that one cannot yet dismiss. For these reasons, a more general beyond standard model (BSM) effective field theory would be preferable, as it would not be limited to relying on specific models, but rather considering the most general BSM effective field theoretical approach that describes this process. An important outcome of such a theory is the evaluation of the energy scales up to which the effective field operators are not broken, together with limits for the effective low-energy couplings.

The analysis of the $0\nu\beta\beta$ decay process is generally done at three levels. At the lowest level the weak interaction of the quarks and leptons is considered, and the BSM physics is treated within a low-energy effective field theory approach. At the next level the hadronization process to nucleons and exchanging pion is considered. The nucleons are treated in the impulse approximation leading to free space $0\nu\beta\beta$ transition operators. At the third level the nucleon dynamics inside the nuclei is treated using nonperturbative nuclear wave functions, which are further used to obtain nuclear matrix elements (NME) needed to calculate the $0\nu\beta\beta$ observables, such as half-lives and two-electron angular and energy distributions [11].

To accomplish this goal we need reliable NME. Although there seem to exist many NME results to choose from, most of the literature is dedicated calculations for the light left-handed Majorana neutrino exchange. Ref. [13] provides tables and plots that compare the latest results for the light left-handed neutrino exchange and for the heavy right-handed neutrino exchange. We calculate the NME using shell model techniques, which are consistent with previous calculations [14, 15, 16, 17, 18, 19, 20, 21, 22, 23, 24, 13]. The reason for choosing shell model NME is our belief that these are better suited and more reliable for $0\nu\beta\beta$ calculations, as they take into account all the correlations around the Fermi surface, respect all symmetries, and take into account consistently the effects of the missing single particle space via many-body perturbation theory (the effects were shown to be small, about 20%, for ^{82}Se [25]). Furthermore, we have tested the shell model methods and the effective Hamiltonians used by comparing the calculated spectroscopic observables to the experimental data, as presented in Ref. [15, 26, 13]. We do not consider any quenching for the bare $0\nu\beta\beta$ operator in these calculations. Such a choice is different from that for the simple Gamow-Teller operator used in the single beta and $2\nu\beta\beta$ decays where a quenching factor of about 0.7 is necessary [27]. For the PSF we use an effective theory based on the formalism of Ref. [28], but fine-tuned as to take into account the effects of a Coulomb field distorting finite-size proton distribution in the final nucleus. To our knowledge, some of the NME presented here were calculated for the first time using shell model techniques.

2. $0\nu\beta\beta$ decay effective field theory formalism

Throughout the literature, the main mechanism considered to be responsible for the neutrinoless double beta decay is the so called "mass mechanism" that assumes that the neutrinos are Majorana fermions, and requires that the light left-handed neutrinos have mass. However, the possibility that right-handed currents could contribute to the neutrinoless double-beta decay ($0\nu\beta\beta$) has been already considered for some time [29, 28]. Recently, $0\nu\beta\beta$ studies [30, 9] have adopted the left-right symmetric model [31, 7] for the inclusion of right-handed currents. In addition, the R-parity violating ($\not{R}_p$) supersymmetric (SUSY) model can also contribute to the neutrinoless double beta decay process [32, 33, 34]. In the framework of the left-right symmetric model and R-parity violating SUSY model, the $0\nu\beta\beta$ half-life can be written as a sum of products of PSF, BSM LNV parameters, and their corresponding NME [11]:

$$\left[T^{0\nu}_{1/2}\right]^{-1} = G_{01} g_A^4 \left|\eta_{0\nu} M_{0\nu} + \left(\eta^L_{N_R} + \eta^R_{N_R}\right) M_{0N} + \eta_{\tilde{q}} M_{\tilde{q}} + \eta_{\lambda'} M_{\lambda'} + \eta_\lambda X_\lambda + \eta_\eta X_\eta\right|^2 . \quad (1)$$

Here, G_{01} is a phase space factor that can be calculated with good precision for most cases [35, 36, 37, 38], g_A is the axial vector coupling constant, $\eta_{0\nu} = \frac{\langle m_{\beta\beta}\rangle}{m_e}$, with $\langle m_{\beta\beta}\rangle$ representing the effective Majorana neutrino mass, and m_e the electron mass. $\eta^L_{N_R}$, $\eta^R_{N_R}$ are the heavy neutrino parameters with left-handed and right-handed currents, respectively [20, 9], $\eta_{\tilde{q}}$, $\eta_{\lambda'}$ are $\not{R}_p$ SUSY LNV parameters [39], η_λ, and η_η are parameters for the so-called "$\lambda-$" and "$\eta-$mechanism", respectively [9]. $M_{0\nu}$, M_{0N}, are the light and the heavy neutrino exchange NME, $M_{\tilde{q}}$, $M_{\lambda'}$ are the $\not{R}_p$ SUSY NME, and X_λ and X_η denote combinations of NME and other PSF ($G_{02} - G_{09}$) corresponding to the $\lambda-$mechanism involving right-handed leptonic and right-handed hadronic currents, and the $\eta-$mechanism with right-handed leptonic and left-handed hadronic currents, respectively [11].

A more general approach is based on the effective field theory extension of the Standard Model. The analysis based on the BSM contributions to the effective field theory is more desirable, because it does not rely on specific models, and their parameters could be extracted/constrained by the existing $0\nu\beta\beta$ data, and by data from LHC and other experiments. In fact, the models considered above always lead to a subset of terms in the low-energy ($\sim$ 200

IOP Conf. Series: Journal of Physics: Conf. Series **966** (2018) 012009 doi:10.1088/1742-6596/966/1/012009

MeV) effective field theory Lagrangian. We consider all the terms in the Lagrangian allowed by the symmetries. Some of the couplings will correspond to the model couplings in Eq. (1), but they might have a wider meaning. Others are new, not corresponding to specific models.

We treat the long-range component of the $0\nu\beta\beta$ diagrams [40] as two point-like vertices at the Fermi scale, which exchange a light neutrino. In this case, the dimension 6 Lagrangian can be expressed in terms of effective couplings [41]:

$$\mathcal{L}_6 = \frac{G_F}{\sqrt{2}} \left[j^{\mu}_{V-A} J^{\dagger}_{V-A,\mu} + \sum_{\alpha,\beta}^{*} \epsilon^{\beta}_{\alpha} j_{\beta} J^{\dagger}_{\alpha} \right], \tag{2}$$

where $J^{\dagger}_{\alpha} = \bar{u}\mathcal{O}_{\alpha}d$ and $j_{\beta} = \bar{e}\mathcal{O}_{\beta}\nu$ are hadronic and leptonic Lorentz currents, respectively. The definitions of the $\mathcal{O}_{\alpha,\beta}$ operators are given in Eq. (3) of Ref. [41]. The LNV parameters are $\epsilon^{\beta}_{\alpha} = \{\epsilon^{V+A}_{V-A}, \epsilon^{V+A}_{V+A}, \epsilon^{S+P}_{S\pm P}, \epsilon^{TR}_{TL}, \epsilon^{TR}_{TR}\}$. The "*" symbol indicates that the term with $\alpha = \beta = (V-A)$ is explicitly taken out of the sum. However, the first term in Eq. (2) still entails BSM physics through the dimension-5 operator responsible for the Majorana neutrino mass [42, 11]. Here $G_F = 1.1663787 \times 10^{-5}$ GeV^{-2} denotes the Fermi coupling constant.

In the short-range part of the $0\nu\beta\beta$ diagrams [40] we consider the interaction to be point-like. The general Lorentz-invariant Lagrangian in terms of effective couplings [33] can be written as:

$$\mathcal{L}_9 = \frac{G_F^2}{2m_p} \left[\varepsilon_1 JJj + \varepsilon_2 J^{\mu\nu} J_{\mu\nu} j + \varepsilon_3 J^{\mu} J_{\mu} j + \varepsilon_4 J^{\mu} J_{\mu\nu} j^{\nu} + \varepsilon_5 J^{\mu} J j_{\mu} \right], \tag{3}$$

with the hadronic currents of defined chirality $J = \bar{u}(1 \pm \gamma_5)d$, $J^{\mu} = \bar{u}\gamma^{\mu}(1 \pm \gamma_5)d$, $J^{\mu\nu} = \bar{u}\frac{i}{2}[\gamma^{\mu}, \gamma^{\nu}](1 \pm \gamma_5)d$, leptonic currents $j = \bar{e}(1 \pm \gamma_5)e^C$, $j^{\mu} = \bar{e}\gamma^{\mu}(1 \pm \gamma_5)e^C$, and $\varepsilon^{\beta}_{\alpha} = \varepsilon^{xyz}_{\alpha} = \{\varepsilon_1, \varepsilon_2, \varepsilon_3^{LLz(RRz)}, \varepsilon_3^{LRz(RLz)}, \varepsilon_4, \varepsilon_5\}$. These parameters have dependence on the chirality of the hadronic and the leptonic currents involved, with $xyz = L/R, L/L, L/R$. In the case of ε_3, one can distinguish between different chiralities, thus we express them separately as $\varepsilon_3^{LLz(RRz)}$ and $\varepsilon_3^{LRz(RLz)}$.

The contribution of the light left-handed neutrino exchange and that of long-range diagram to the $0\nu\beta\beta$ decay amplitude is proportional to the time-ordered product of two effective $\mathcal{L}_6$ Lagrangians [41],

$$T(\mathcal{L}_6^{(1)} \mathcal{L}_6^{(2)}) = \frac{G_F^2}{2} T \left[j_{V-A} J^{\dagger}_{V-A} j_{V-A} J^{\dagger}_{V-A} + \epsilon^{\beta}_{\alpha} j_{\beta} J^{\dagger}_{\alpha} j_{V-A} J^{\dagger}_{V-A} + \epsilon^{\beta}_{\alpha} \epsilon^{\delta}_{\gamma} j_{\beta} J^{\dagger}_{\alpha} j_{\delta} J^{\dagger}_{\gamma} \right], \tag{4}$$

while the contribution of the short-range diagram is proportional to $\mathcal{L}_9$.

However, when calculating the $0\nu\beta\beta$ half-life it is necessary to identify the contributions corresponding to different hadronization prescriptions. In addition to these contributions that were also considered in Ref. [41], we also include the long range diagrams [40] that involve pion(s) exchange. These diagrams were considered before as contributing to the $0\nu\beta\beta$ decay rate, but in the context of $\not{R}_p$ SUSY mechanism.

After hadronization, the extra terms in the Lagrangian require the knowledge of 23 individual NME [32, 41, 43, 44, 42, 39]. We can write the half-life in a factorized compact form

$$\left[T^{0\nu}_{1/2}\right]^{-1} = g_A^4 \left[\sum_i |\mathcal{E}_i|^2 \mathcal{M}_i^2 + \mathrm{Re}\left[\sum_{i \neq j} \mathcal{E}_i \mathcal{E}_j \mathcal{M}_{ij} \right] \right]. \tag{5}$$

Here, the $\mathcal{E}_i$ contain the neutrino physics parameters, with $\mathcal{E}_1 = \eta_{0\nu}$ representing the exchange of light left-handed neutrinos, $\mathcal{E}_{2-7} = \{\epsilon^{V+A}_{V-A}, \epsilon^{V+A}_{V+A}, \epsilon^{S+P}_{S\pm P}, \epsilon^{TR}_{TL}, \epsilon^{TR}_{TR}, \eta_{\pi\nu}\}$ are the long-range

LNV parameters, and $\mathcal{E}_{8-15} = \{\varepsilon_1,\ \varepsilon_2,\ \varepsilon_3^{LLz(RRz)},\ \varepsilon_3^{LRz(RLz)},\ \varepsilon_4,\ \varepsilon_6,\ \eta_{1\pi},\ \eta_{2\pi}\}$ denote the short-range LNV parameters at the quark level involved in the pion-exchange diagrams [40]. The rational of including the $\eta_{\pi\nu}$ in the same class with the LNV entering the quark-level long range diagrams is that Ref. [34] indicate that $\eta_{\pi\nu}$ is proportional to ϵ_{TR}^{TR}. In the same vein, Ref. [45] indicates that ε_1 and ε_2 are proportional to a combination of $\eta_{1\pi}$ and $\eta_{2\pi}$. Therefore the $\eta_{1\pi}$ and $\eta_{2\pi}$ were included in the list LNV couplings associated with quark-level short-range diagrams. Denoted with $\mathcal{M}_i^2$ are combinations of NME [43, 44, 41, 39] and integrated PSF [38] (denoted as $G_{01} - G_{09}$). Our values of the PSF are taken from Ref [38]. In some cases the interference terms $\mathcal{E}_i\mathcal{E}_j\mathcal{M}_{ij}$ are small [46] and can be neglected, but not all of them. In Ref. [11] we analyzed a subset of terms contributing to the half-life formula, Eq. (1) originating from the left-right symmetric model. In that restrictive case we showed that one can disentangle different contributions to the $0\nu\beta\beta$ decay process using two-electron angular and energy distributions as well as half-lives of two selected isotopes. Obviously, this number of observables is not enough to extract all coupling appearing in the effective field theory Lagrangian. However, they can be used to constrain these couplings, thus adding to the information extracted from the Large Hadron Collider and other related experiments.

In the following, we present the individual NME that enter $\mathcal{M}_i^2$ and are needed to analyze the outcome of Eq. (5). These NME are written as products of two-body transition densities (TBTD) and two-body matrix elements (TBME), with summation over all the nucleon states. Their numerical values when calculated within the shell model approach are presented in Table 1 for the light left-handed Majorana neutrino exchange, in Table 2 for the long-range part, and in Table 3 for the short-range component. The general expressions for the NME are (see Refs. [28, 20, 11]):

$$M_\alpha = \sum_{j_p j_{p'} j_n j_{n'} J^\pi} TBTD\left(j_p j_{p'}, j_n j_{n'}; J^\pi\right) \times \left\langle j_p j_{p'}; J^\pi \left\| \tau_{-1}\tau_{-2}\mathcal{O}_{12}^{\gamma,\phi,\theta,P,R} \right\| j_n j_{n'}; J^\pi \right\rangle .$$

We group the operators that share similar structure into five families.

$$\begin{aligned}
\text{Gamow-Teller operator} : \mathcal{O}_{12}^{\gamma} &= \vec{\sigma}_1 \cdot \vec{\sigma}_2 H_\gamma(r), \\
\text{Fermi operator} : \mathcal{O}_{12}^{\phi} &= H_\phi(r), \\
\text{Tensor operator} : \mathcal{O}_{12}^{\theta} &= \left[3(\vec{\sigma}_1 \cdot \hat{\mathbf{r}})(\vec{\sigma}_2 \cdot \hat{\mathbf{r}}) - \vec{\sigma}_1 \cdot \vec{\sigma}_2\right] H_\theta(r), \\
\text{P operator} : \mathcal{O}_{12}^{P} &= (\vec{\sigma}_1 - \vec{\sigma}_2)\, H_P(r), \\
\text{R operator} : \mathcal{O}_{12}^{R} &= \vec{\sigma}_1 \cdot \vec{\sigma}_2 H_R(r).
\end{aligned}$$

Here, $\gamma = GT,\ GT\omega,\ GTq,\ GTN,\ GT',\ GT'',\ GT\pi\nu,\ GT1\pi,\ GT2\pi$, $\phi = F,\ F\omega,\ Fq,\ FN,\ F'$, and $\theta = T,\ Tq,\ T',\ T'',\ T\pi\nu,\ T1\pi,\ T2\pi$. Detailed expression for all these NME are found in the Appendix of Ref. [40].

3. Results

The NME presented in this section (Eq. (6)) are calculated using shell model approaches. To take into account the two-nucleon short-range correlation (SRC) we multiply the relative wavefunctions by $f(r) = 1 - ce^{-ar^2}(1 - br^2)$; in the CD-Bonn parametrization used in this work $a = 1.52\ \text{fm}^{-2}$, $b = 1.88\ \text{fm}^{-2}$, and $c = 0.46\ \text{fm}^{-2}$ [47]. This method is described in greater detail in Refs. [14, 15, 16, 17, 19, 20, 21, 22, 23, 24, 13]

Table 1 presents the M_{GT}, M_F, and M_T NME involved in the standard mass mechansim with left-handed currents of Eq. (5). For these NME, an optimal closure energy $\langle \bar{E} \rangle$ was used for each effective Hamiltonian [23]: $\langle \bar{E} \rangle = 0.5\text{MeV}$ for ^{48}Ca [19] and the GXPF1A Hamiltonian

IOP Conf. Series: Journal of Physics: Conf. Series **966** (2018) 012009 doi:10.1088/1742-6596/966/1/012009

[48], $\langle \bar{E} \rangle = 3.4$MeV for ^{76}Ge [26] and ^{82}Se [23] calculated with the JUN45 Hamiltonian [49], and $\langle \bar{E} \rangle = 3.5$MeV for ^{130}Te [24] and ^{136}Xe [17] calculated with the SVD Hamiltonian [50].

Table 1. NME values for the exchange of light left-handed Majorana neutrinos corresponding to the "mass mechanism" .

	^{48}Ca	^{76}Ge	^{82}Se	^{130}Te	^{136}Xe
M_{GT}	0.807	3.206	3.005	1.662	1.505
M_F	−0.233	−0.674	−0.632	−0.438	−0.400
M_T	0.080	0.011	0.012	−0.007	−0.008

The long-range NME M_α (with $\alpha = GTq$, Fq, Tq, $GT\omega$, $F\omega$, P, R, GT', F', T', GT'', T'') and $M_{GT\pi\nu}$ and $M_{T\pi\nu}$ that appear in Eq. (5) are presented in Table 2.

Table 2. NME for the long-range part of the $0\nu\beta\beta$ decay rate.

	^{48}Ca	^{76}Ge	^{82}Se	^{130}Te	^{136}Xe
M_{GTq}	0.709	3.228	3.034	1.587	1.440
$M_{GT\omega}$	0.930	3.501	3.287	1.855	1.682
$M_{GT'}$	0.841	2.699	2.567	2.120	1.935
$M_{GT''}$	3.581	11.982	11.490	12.210	11.202
$M_{GT\pi\nu}$	86.2	331.6	313.3	184.2	167.7
M_{Fq}	−0.121	−0.383	−0.362	−0.249	−0.230
$M_{F\omega}$	−0.232	−0.659	−0.618	−0.427	−0.391
$M_{F'}$	−0.258	−0.812	−0.772	−0.635	−0.581
M_{Tq}	−0.173	−0.059	−0.058	−0.013	−0.012
$M_{T'}$	0.337	0.015	0.025	−0.077	−0.085
$M_{T''}$	2.231	0.028	0.118	−0.773	−0.861
$M_{T\pi\nu}$	21.3	7.3	6.9	1.2	1.1
M_P	0.395	−2.466	−2.332	−1.729	−1.617
M_R	1.014	3.284	3.127	2.562	2.341

Shown in Table 3 are the short-range NME M_{GTN}, M_{FN}, and M_α (with $\alpha = GT1\pi$, $T1\pi$, $GT2\pi$, $T2\pi$) that appear in Eq. (5).

Table 3. The short-range NME.

	^{48}Ca	^{76}Ge	^{82}Se	^{130}Te	^{136}Xe
M_{GTN}	58.5	162.3	150.1	107.6	96.6
$M_{GT1\pi}$	−1.354	−3.559	−3.282	−2.421	−2.171
$M_{GT2\pi}$	−0.676	−1.983	−1.854	−1.257	−1.136
M_{FN}	−22.9	−62.6	−58.1	−41.0	−36.9
$M_{T1\pi}$	−0.590	−0.010	−0.027	0.106	0.115
$M_{T2\pi}$	−0.227	−0.010	−0.015	0.038	0.040

4. Conclusions

In this paper, the shell model values for 23 individual nuclear matrix elements are presented. These NME could be used to disentangle the 15 lepton number violating parameters that may contribute to the $0\nu\beta\beta$ decay rate, and for obtaining the energy and angular distributions of the outgoing electrons. To our knowledge the $M_{GT'}$, $M_{GT''}$, $M_{F'}$, $M_{T'}$, and $M_{T''}$ NME were calculated for the first time using shell model techniques.

Acknowledgments
Support from the NUCLEI SciDAC Collaboration under U.S. Department of Energy Grant No. DE-SC0008529, the U.S. Department of Energy Grant No. DE-SC0015376, and the U.S. NSF Grant No. PHY-1404442 is acknowledged.

References
[1] Schechter J and Valle J W F 1982 *Phys. Rev. D* **25** 2951
[2] Nieves J 1984 *Phys. Lett. B* **147** 375
[3] Takasugi E 1984 *Phys. Lett. B* **149** 372
[4] Hirsch M, Kovalenko S and Schmidt I 2006 *Phys. Lett. B* **642** 106
[5] Pati J and Salam A 1974 *Phys. Rev. D* **10** 275
[6] Mohapatra R and Pati J 1975 *Phys. Rev. D* **11** 2558
[7] Senjanovic G and Mohapatra R N 1975 *Phys. Rev. D* **12** 1502
[8] Keung W Y and Senjanovic G 1983 *Phys. Rev. Lett.* **50** 1427
[9] Barry J and Rodejohann W 2013 *J. High Energy Phys.* 153
[10] Khachatryan V, Sirunyan A M, Tumasyan A, Adam W, Bergauer T, Dragicevic M, Er J, Fabjan C, Friedl M, Fruhwirth R *et al.* (CMS-Collaboration) 2014 *Eur. Phys. J. C* **74** 3149
[11] Horoi M and Neacsu A 2016 *Phys. Rev. D* **93**(11) 113014 (*Preprint* `arXiv:1511.00670 [hep-ph]`)
[12] Neacsu A and Horoi M 2016 *Advances in High Energy Physics* **2016**
[13] Horoi M and Neacsu A 2016 *Phys. Rev. C* **93** 024308
[14] Horoi M, Stoica S and Brown B A 2007 *Phys. Rev. C* **75** 034303
[15] Horoi M and Stoica S 2010 *Phys. Rev. C* **81** 024321
[16] Neacsu A, Stoica S and Horoi M 2012 *Phys. Rev. C* **86** 067304
[17] Horoi M and Brown B A 2013 *Phys. Rev. Lett.* **110** 222502
[18] Neacsu A and Stoica S 2014 *J. Phys. G* **41** 015201
[19] Sen'kov R A and Horoi M 2013 *Phys. Rev. C* **88** 064312
[20] Horoi M 2013 *Phys. Rev. C* **87** 014320
[21] Brown B A, Horoi M and Sen'kov R A 2014 *Phys. Rev. Lett.* **113** 262501
[22] Sen'kov R A and Horoi M 2014 *Phys. Rev. C* **90** 051301(R)
[23] Sen'kov R A, Horoi M and Brown B A 2014 *Phys. Rev. C* **89** 054304
[24] Neacsu A and Horoi M 2015 *Phys. Rev. C* **91** 024309
[25] Holt J D and Engel J 2013 *Phys. Rev. C* **87** 064315
[26] Sen'kov R A and Horoi M 2016 *Phys. Rev. C* **93**(4) 044334
[27] Brown B A, Fang D L and Horoi M 2015 *Phys. Rev. C* **92**(4) 041301
[28] Doi M, Kotani T and Takasugi E 1985 *Prog. Theor. Phys. Suppl.* **83** 1
[29] Doi M, Kotani T, Nishiura H and Takasugi E 1983 *Progr. Theor. Exp. Phys.* **69** 602
[30] Rodejohann W 2012 *J. Phys. G* **39** 124008
[31] Mohapatra R N and Pati J C 1975 *Phys. Rev. D* **11** 566
[32] Hirsch M, KlapdorKleingrothaus H and Kovalenko S 1996 *Phys. Rev. D* **53** 1329
[33] Kolb S, Hirsch M and Klapdor-Kleingrothaus H V 1997 *Phys. Rev. D* **56**(7) 4161–4165
[34] Faessler A, Gutsche T, Kovalenko S and Šimkovic F 2008 *Phys. Rev. D* **77**(11) 113012
[35] Suhonen J and Civitarese O 1998 *Phys. Rep.* **300** 123
[36] Kotila J and Iachello F 2012 *Phys. Rev. C* **85** 034316
[37] Stoica S and Mirea M 2013 *Phys. Rev. C* **88** 037303
[38] Horoi M and Neacsu A 2016 *Adv. High Energy Phys.* **2016** 7486712
[39] Vergados J D, Ejiri H and Simkovic F 2012 *Rep. Prog. Phys.* **75** 106301
[40] Horoi M and Neacsu A 2017 (*Preprint* `1706.05391`)
[41] Deppisch F F, Hirsch M and Pas H 2012 *J. Phys. G* **39** 124007
[42] Deppisch F F, Harz J, Huang W C, Hirsch M and Päs H 2015 *Phys. Rev. D* **92**(3) 036005
[43] Pas H, Hirsch M, Klapdor-Kleingrothaus H V and Kovalenko S G 1999 *Phys. Lett. B* **453** 194–198
[44] Pas H, Hirsch M, Klapdor-Kleingrothaus H V and Kovalenko S G 2001 *Phys. Lett. B* **498** 35–39 (*Preprint* `hep-ph/0008182`)
[45] Wodecki A, Kamiński W A and Šimkovic F 1999 *Phys. Rev. D* **60**(11) 115007
[46] Ahmed F, Neacsu A and Horoi M 2017 *Physics Letters B* **769** 299–304
[47] Simkovic F, Faessler A, Muether H, Rodin V and Stauf M 2009 *Phys. Rev. C* **79** 055501
[48] Honma M, Otsuka T, Brown B A and Mizusaki T 2005 *Eur. Phys. J. A* **25 Suppl. 1** 499
[49] Honma M, Otsuka T, Mizusaki T and Hjorth-Jensen M 2009 *Phys. Rev. C* **80** 064323
[50] Qi C and Xu Z X 2012 *Phys. Rev. C* **86** 044323

IOP Conf. Series: Journal of Physics: Conf. Series **966** (2018) 012010 doi:10.1088/1742-6596/966/1/012010

^{8}B + ^{208}Pb Elastic Scattering at Coulomb Barrier Energies

M. La Commara[1,2], M. Mazzocco[3,4], A. Boiano[2], C. Boiano[5], C. Manea[4], C. Parascandolo[2], D. Pierroutsakou[2], C. Signorini[3,4], E. Strano[4], D. Torresi[4], H. Yamaguchi[6], D. Kahl[6], P. Di Meo[2], J. Grebosz[7], N. Imai[6], Y. Hirayama[8], H. Ishiyama[8], N. Iwasa[9], S.C. Jeong[8], H.M. Jia[10], Y.H. Kim[8], S. Kimura[8], S. Kubono[11], C.J. Lin[10], H. Miyatake[8], M. Mukai[8], T. Nakao[12], M. Nicoletto[4], Y. Sakaguchi[6], A.M. Sánchez-Benítez[13], F. Soramel[3,4], T. Teranishi[14], Y. Wakabayashi[11], Y.X. Watanabe[8], L. Yang[10], Y.Y. Yang[10].

[1]Department of Physics "E.Pancini", University "Federico II", Napoli, Italy
[2]INFN-Sezione di Napoli, Napoli, Italy
[3]Department of Physics and Astronomy, University of Padova and INFN, Padova, Italy
[4]INFN-Sezione di Padova, Padova, Italy
[5]INFN-Sezione di Milano, Milano, Italy
[6]CNS - The University of Tokyo, Wako-shi, Saitama, Japan
[7]IFJ PAN, Krakow, Poland
[8]KEK, Tsukuba, Ibaraki, Japan
[9]Department of Physics, Sendai, Miyagi, Japan
[10]China Institute of Atomic Energy, Beijing, China
[11]RIKEN Nishina Center, Wako, Saitama, Japan
[12]JAEA, Tokai, Ibaraki, Japan
[13]Department of Applied Physics, University of Huelva, Huelva, Spain
[14]Department of Physics, Kyushu University, Hakozaki, Fukuoka, Japan

E-mail: Marco.LaCommara@na.infn.it

Abstract. The scattering process of weakly-bound nuclei at Coulomb barrier energies provides deep insights on the reaction dynamics induced by exotic nuclei. Within this framework, we measured for the first time the scattering process of the short-lived Radioactive Ion Beam (RIB) ^{8}B (S_p = 0.1375 MeV) from a ^{208}Pb target at 50 MeV beam energy. The ^{8}B RIB was produced by means of the in-flight facility CRIB (RIKEN, Japan) with an average intensity on target of 10 kHz and a purity about 25%. Elastically scattering ions were detected in the angular range $\theta_{c.m.}$ = 10°-160° by means of the detector array EXPADES. A preliminary optical model analysis indicates a total reaction cross section of about 1 b, a value, once reduced, 2-3 times larger than those obtained for the reactions induced by the stable weakly-bound projectiles ^{6,7}Li on a ^{208}Pb target in the energy range around the Coulomb barrier.

1. Introduction

Nuclei far from $\beta-$stability valley, as ^{6}He, ^{11}Be, ^{8}B, are very interesting because of their peculiarities as far as their structure and their way to interact with other nuclei are concerned, indeed during the last decades many theoretical and experimental efforts have been devoted to the investigation of halo nuclei in order to shed light on their unusual and intriguing features [1]. Nuclear collisions involving stable and unstable weakly-bound nuclei have attracted great interest nowadays. However, despite great efforts in experiments and theories over the past decades, the precise description on the influence of breakup channel on fusions is still elusive, due to the limited knowledge on the breakup process, especially for the unstable reaction systems. More knowledge in this framework can be obtained starting from the investigation of simple processes, as elastic scattering. It is well accepted by the nuclear physicists community that the presence of a proton (neutron) halo or skin may heavily influence the behavior of an exotic nucleus in terms of nuclear reaction mechanism as well as can strongly influence its structure [1] [2]. Elastic scattering, a simple process, is a very appropriate probe to study the size and surface diffuseness of exotic nuclei by comparing similarities and differences in reactions induced by weakly bound and tightly bound nuclei. Anyway, elastic scattering data for proton-halo isotopes around the Coulomb barrier are still not enough to fully understand the dynamics of this process. ^{8}B is a proton-drip-line $\beta-$decaying isotope, conveying pretty much interest because of the role it plays in the production of high-energy neutrinos in the sun and its quite unusual structure, being considered by now a one-proton halo nucleus [3]. It is a very interesting isotope since it has a very low proton separation energy, $S_p = 0.1375$ MeV, which makes it a very lightly bound nucleus being an ideal candidate for nuclear-reaction mechanism investigations with exotic nuclei. Furthermore, the studies on ^{8}B reaction-mechanism processes are rather exiguous due to the difficulty of producing the ^{8}B beam, being it produced only in very few facilities and anyhow with low intensities [4] [5] [6]. In this paper we report new experimental data for the elastic scattering of ^{8}B on a ^{208}Pb-target, at 50 MeV, which means just at the Coulomb barrier, showing very preliminary results.

2. Experimental setup and data analysis

The radioactive beam ^{8}B at 50 MeV (6,25 A MeV) has been produced via the reaction ^{3}He(^{6}Li,^{8}B)n ($Q_{value} = -1.97$ MeV) at the facility CRIB [7], situated at the RIKEN laboratory in Japan. The primary beam of ^{6}Li had an intensity of 3eμA, and a cryogenic ^{3}He$-$target with pressure of 1 bar and a typical temperature of 90 K was used. The secondary beam had an intensity (on target) of 10^4 pps and a purity of 20% consistent with the expected one, being the main contaminants ^{7}Be and ^{3}He, besides the ^{6}Li scattered beam. Anyhow all the contaminants had different times of flight through the separator and then it was possible to separate events coming from the various components of the beam by using proper gates in the ToF spectra. Charged products arising from the reaction of the ^{8}B beam impinging on a 2.2 mg/cm^2 ^{208}Pb target (evaporated on a 1.5 μm thick mylar backing foil) were detected with the detector array EXPADES [8] utilized in a configuration consisting of 6 $\Delta E - E_{res}$ double-sided silicon-strip detector telescopes, placed at 113 mm from the center of the ^{208}Pb target in a slightly asymmetric configuration such to ensure the whole angular-coverage ranging from 16° to 164° in the laboratory frame. In Fig.1 a very preliminary evaluation of the angular distribution for the elastic scattering process in the system ^{8}B+^{208}Pb is shown, where each of the four plotted points corresponds to the angular-range mean placement of the EXPADES detectors. The experimental data have been normalized with a Monte-Carlo simulation accounting for the detector-geometry and the elastic-scattering of the process, whereas the simulated events have been generated according to a pure Rutherford cross-section. The continuous line represents the result of an optical model best-fit analysis (FRESCO code). In this way we could extract a $-$preliminary$-$ evaluation of the reaction cross-section for the system ^{8}B+^{208}Pb at 50 MeV beam energy.

In Fig.2 the reaction X-sections for the systems ^{6,7}Li, ^{7}Be and ^{8}B +^{208}Pb in the energy range around the Coulomb barrier are shown [10], [11] and [12]. The data for the different systems have been reduced as reported in ref. [8] and multiplied by a normalization factor of the ^{7}Be+^{208}Pb

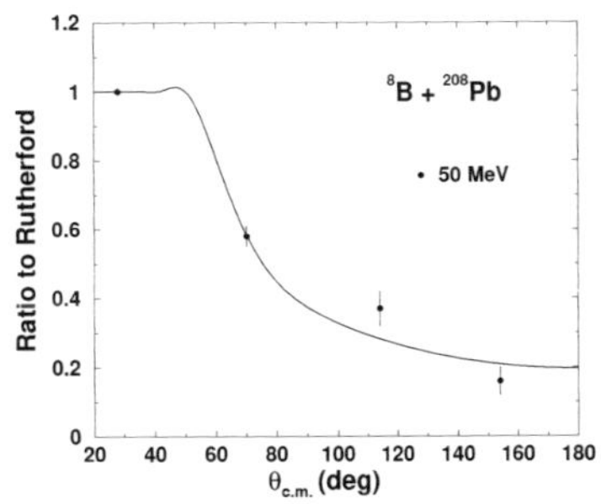

Figure 1. Preliminary elastic scattering angular distribution for the system ^{8}B+^{208}Pb at 50 MeV beam energy. The continuous line is the result of the optical model analysis of the collected data

reaction, in order to compare them with each other. It seems, even if the data analysis is in the first stages of its route, that the reaction induced by the one-proton halo nucleus ^{8}B (S_p =0.1375 MeV) on the ^{208}Pb target has a total reaction X-section much larger than those measured for reactions induced by other weakly-bound isotopes on the same target.

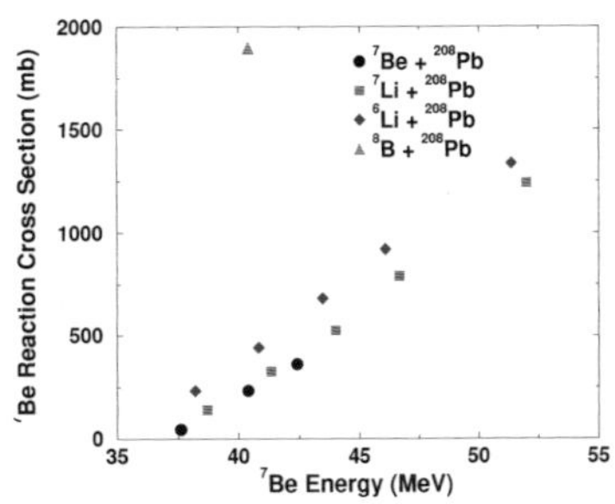

Figure 2. Reaction X-sections for the systems ^{6,7}Li, ^{7}Be and ^{8}B +^{208}Pb (see [9], [10] and [11]). The data were reduced according to the procedure recommended in [8] and then normalized to the system ^{7}Be +^{208}Pb

3. Discussion and outlook

This is to some extent consistent with the outcome for reactions induced by the same projectiles on ^{58}Ni [3]. If this will be confirmed, the full data analysis outcome (elastic scattering data, possible break-up data, etc.) will hopefully allow us to disentangle which mechanism, direct processes such as break-up and/or transfer or fusion, is responsible for accounting such large X-section value.

References

[1] Keeley N *et al.* 2007 *Prog. Part. Phys.* **59** 579
[2] Gomes P R S *et al.* 2005 *Phys. Rev.* **C71** 034608
[3] Aguilera E F *et al.* 2009 *Phys. Rev.* **C79** 021601(R)
[4] Aguilera E F and Kolata J J 2012 *Phys. Rev.* **C85** 014603
[5] Pakou A *et al.* 2013 *Phys. Rev.* **C87** 014619
[6] Yang Y Y *et al.* 2013 *Phys. Rev.* **C87** 044613
[7] Yanagisawa Y *et al.* 2005 *Nucl. Instr. and Meth. A* **539** 74
[8] Pierroutsakou D *et al.* 2016 *Nucl. Instr. and Meth. A* **834** 46

IOP Conf. Series: Journal of Physics: Conf. Series **966** (2018) 012010 doi:10.1088/1742-6596/966/1/012010

[9] Mazzocco M *et al.* 2017 Fusion17 proceedings, to be published.
[10] Keeley N *et al.* 1994 *Nucl. Phys. A* **571** 326
[11] Mazzocco M *et al.* 2016 *Eur. Phys. Journal*, Web of Conferences **117** 06006

IOP Conf. Series: Journal of Physics: Conf. Series **966** (2018) 012011 doi:10.1088/1742-6596/966/1/012011

Consistent, high-quality two-nucleon potentials up to fifth order of the chiral expansion

R. Machleidt

Department of Physics, University of Idaho, Moscow, Idaho 83844, USA

E-mail: machleid@uidaho.edu

Abstract. We present NN potentials through five orders of chiral effective field theory ranging from leading order (LO) to next-to-next-to-next-to-next-to-leading order (N^4LO). The construction may be perceived as consistent in the sense that the same power counting scheme as well as the same cutoff procedures are applied in all orders. Moreover, the long-range parts of these potentials are fixed by the very accurate πN low-energy constants (LECs) as determined in the Roy-Steiner equations analysis by Hoferichter, Ruiz de Elvira and coworkers. In fact, the uncertainties of these LECs are so small that a variation within the errors leads to effects that are essentially negligible, reducing the error budget of predictions considerably. The NN potentials are fit to the world NN data below pion-production threshold of the year of 2016. The potential of the highest order (N^4LO) reproduces the world NN data with the outstanding χ^2/datum of 1.15, which is the highest precision ever accomplished for any chiral NN potential to date. The NN potentials presented may serve as a solid basis for systematic *ab initio* calculations of nuclear structure and reactions that allow for a comprehensive error analysis. In particular, the consistent order by order development of the potentials will make possible a reliable determination of the truncation error at each order. Our family of potentials is non-local and, generally, of soft character. This feature is reflected in the fact that the predictions for the triton binding energy (from two-body forces only) converges to about 8.1 MeV at the highest orders. This leaves room for three-nucleon-force contributions of moderate size.

1. Introduction

The quest for a practically feasible, and yet fundamental, theory of hadronic interactions at low energy (where QCD is non-perturbative) has spanned several decades. At the present time, there exists a general consensus that chiral effective field theory (chiral EFT) may provide the best answer to the quest. By its nature, chiral EFT is a model-independent approach with firm roots in QCD, due to the fact that interactions are subjected to the constraints of the broken chiral symmetry of low-energy QCD. Moreover, the approach is systematic in the sense that the various contributions to a particular dynamical process can be arranged as an expansion in terms of a suitable "parameter". The latter is chosen to be the ratio of a typical external momentum (soft scale) to the chiral symmetry breaking scale ($\approx$ 1 GeV, hard scale). Recent comprehensive reviews on the subject can be found in Refs. [1, 2].

In its early stages, chiral perturbation theory (ChPT) was applied mostly to $\pi\pi$ and πN dynamics, because, due to the Goldstone-boson nature of the pion, these are the most natural scenarios for a perturbative expansion to exist. Subsequently, chiral EFT has been applied in nucleonic systems with quantitative NN potentials (at N^3LO) making their appearance in the early 2000's [3, 4].

IOP Conf. Series: Journal of Physics: Conf. Series **966** (2018) 012011 doi:10.1088/1742-6596/966/1/012011

Since then, a wealth of applications of N^3LO NN potentials together with chiral three-nucleon forces (3NFs) have been reported. These investigations include few-nucleon reactions, structure of light- and medium-mass nuclei, and infinite matter [5]. Although satisfactory predictions have been obtained in many cases, persistent problems continue to pose serious challenges, such as the well-known 'A_y puzzle' of nucleon-deuteron scattering. Naturally, one would invoke 3NFs as the most likely mechanism to solve this problem. Unfortunately, the chiral 3NF at NNLO does only very little to improve the situation with nucleon-deuteron scattering, while inclusion of the N^3LO 3NF produces an effect in the wrong direction [6]. The next step is then to proceed systematically in the expansion, namely to look at N^4LO (or fifth order). This order is interesting for diverse reasons. From studies of some of the 3NF topologies at N^4LO, we know that a complete set of isospin-spin-momentum 3NF structures (a total of 20) are present at this order [7] and that contributions can be of substantial size. Even more promising, at this order a new set of 3NF contact interactions appears, which has recently been derived by the Pisa group [8]. Contact terms are relatively easy to work with and, most importantly, come with free coefficients and, thus, provide larger flexibility and a great likelihood to solve persistent problems such as the A_y puzzle as well as other issues (like, the "radius problem" and the overbinding of intermediate-mass nuclei).

A principle of all EFTs is that, for meaningful predictions, it is necessary to include *all* contributions that appear at the order at which the calculation is conducted. Thus, when nuclear structure problems require for their solution the inclusion of 3NFs at N^4LO, then also the two-nucleon force involved in the calculation has to be of order N^4LO. This is one reason why in Ref. [9] we derived the N^4LO two-pion exchange (2PE) and three-pion exchange (3PE) contributions to the NN interaction and tested them in peripheral partial waves. In this contribution, we will present complete N^4LO NN potentials that also include the lower partial waves which receive contributions from contact interactions.

In Ref. [9], we also demonstrated that the next-to-next-to-leading order (NNLO), the N^3LO, and the N^4LO contributions to the NN interaction are all of about the same size, thus, not showing much of a trend towards convergence. Therefore, in Ref. [10] we calculated the N^5LO (sixth order) contribution which, indeed, turned out to be small. The latter result may be perceived as an indication of convergence showing up at N^5LO. This adds to the significance of order N^4LO.

Besides the above, we are faced with another set of convergence issues: The convergence of the predictions for the properties of nuclear few- and many-body systems, in which also chiral many-body forces are involved. To investigate these issues, one needs (besides those many-body forces) NN potentials at all orders of chiral EFT, ranging from leading order (LO) to N^4LO, and constructed consistently, i. e., using the same power-counting scheme, consistent low-energy constants (LECs), etc..

For that reason, we present in this contribution NN potentials through five orders from LO to N^4LO, constructed with the above-stated consistencies and with a reproduction of the NN data of the maximum quality possible at the respective orders [11]. These potentials will allow for systematic investigations of nuclear few- and many-body systems with clear implications for convergence and uncertainty quantifications (truncation errors).

2. The pion-nucleon low-energy constants

Chiral symmetry establishes a link between the dynamics in the πN-system and the NN-system through common low-energy constants. Therefore, consistency requires that we use the LECs for subleading πN-couplings as determined in analysis of low-energy πN-scattering. Over the years, there have been many such determinations of questionable reliability. Fortunately, that has changed recently with the analysis by Hoferichter and Ruiz de Elvira *et al.* [12], in which the Roy-Steiner (RS) equations are applied. The RS equations are a set of coupled partial-wave

Table 1. The πN LECs as determined in the Roy-Steiner-equation analysis of πN scattering conducted in Ref. [12]. The given orders of the chiral expansion refer to the NN system. Note that the orders, at which the LECs are extracted from the πN system, are always lower by one order as compared of the NN system in which the LECs are applied. The c_i, $\bar{d}_i$, and $\bar{e}_i$ are the LECs of the second, third, and fourth order πN Lagrangian and are in units of GeV^{-1}, GeV^{-2}, and GeV^{-3}, respectively. The uncertainties in the last digits are given in parentheses after the values.

	NNLO	N^{3}LO	N^{4}LO
c_1	–0.74(2)	–1.07(2)	–1.10(3)
c_2	—	3.20(3)	3.57(4)
c_3	–3.61(5)	–5.32(5)	–5.54(6)
c_4	2.44(3)	3.56(3)	4.17(4)
$\bar{d}_1 + \bar{d}_2$	—	1.04(6)	6.18(8)
$\bar{d}_3$	—	–0.48(2)	–8.91(9)
$\bar{d}_5$	—	0.14(5)	0.86(5)
$\bar{d}_{14} - \bar{d}_{15}$	—	–1.90(6)	–12.18(12)
$\bar{e}_{14}$	—	—	1.18(4)
$\bar{e}_{17}$	—	—	–0.18(6)

dispersion relations constraint by analyticity, unitarity, and crossing symmetry. In the work of Ref. [12], they are used to extract the LECs from the subthreshold point in πN scattering instead of the physical region. This is the preferred method for LECs to be applied in chiral potentials where, e. g., a one-loop πN amplitude leads to a two-loop contribution in NN. Such diagrams are best evaluated by means of Cutkosky rules. The πN amplitude that enters the dispersion integrals is weighted much closer to subthreshold kinematics than to the threshold point. The LECs determined in Ref. [12] carry very small uncertainties (cf. Table 1) for, essentially, two reasons: first, because of the constraints built into the RS equations; second, because of the use of the high-accuracy πN scattering lengths extracted from pionic atoms. In fact, the uncertainties are so small that they are negligible for our purposes. This makes the variation of the πN LECs in NN potential construction obsolete and reduces the error budget in applications of these potentials. For the potentials constructed in this contribution, the central values of Table 1 are applied.

3. NN scattering and the deuteron

We have constructed NN potentials through five orders of the chiral expansion, ranging from LO (Q^0) to N^{4}LO (Q^5). In each order, we consider three cutoffs, namely, $\Lambda = 450$, 500, and 550 MeV. Since we take charge-dependence into account, each NN potential comes in three versions: pp, np, and nn. We will present now the results from these potentials for NN scattering and the deuteron.

The χ^2/datum for the reproduction of the NN data at various orders of chiral EFT are shown in Table 2 for different energy intervals below 290 MeV laboratory energy ($T_{\rm lab}$). The bottom line of Table 2 summarizes the essential results. For the close to 5000 pp plus np data below 290 MeV (pion-production threshold), the χ^2/datum is 51.4 at NLO and 6.3 at NNLO. Note that the number of NN contacts terms is the same for both orders. The improvement is entirely due to an improved description of the 2PE contribution, which is responsible for the crucial intermediate-range attraction of the nuclear force. At NLO, only the uncorrelated 2PE

IOP Conf. Series: Journal of Physics: Conf. Series **966** (2018) 012011 doi:10.1088/1742-6596/966/1/012011

Table 2. χ^2/datum for the fit of the 2016 NN data base by NN potentials at various orders of chiral EFT ($\Lambda = 500$ MeV in all cases).

T_{lab} bin (MeV)	No. of data	LO	NLO	NNLO	N^{3}LO	N^{4}LO
			proton-proton			
0–100	795	520	18.9	2.28	1.18	1.09
0–190	1206	430	43.6	4.64	1.69	1.12
0–290	2132	360	70.8	7.60	2.09	1.21
			neutron-proton			
0–100	1180	114	7.2	1.38	0.93	0.94
0–190	1697	96	23.1	2.29	1.10	1.06
0–290	2721	94	36.7	5.28	1.27	1.10
			pp* plus *np			
0–100	1975	283	11.9	1.74	1.03	1.00
0–190	2903	235	31.6	3.27	1.35	1.08
0–290	4853	206	51.5	6.30	1.63	1.15

is taken into account which is insufficient. From the classic meson-theory of nuclear forces [13], it is well known that π-π correlations and nucleon resonances need to be taken into account for a realistic model of 2PE that provides a sufficient amount of intermediate attraction to properly bind nucleons in nuclei. In the chiral theory, these contributions are encoded in the subleading πN vertexes with LECs denoted by c_i. These enter at NNLO and are the reason for the substantial improvements we encounter at that order. This is the best proof that, starting at NNLO, the chiral approach to nuclear forces is getting the physics right.

To continue on the bottom line of Table 2, after NNLO, the χ^2/datum then further improves to 1.63 at N^{3}LO and, finally, reaches the almost perfect value of 1.15 at N^{4}LO—a fantastic convergence.

Corresponding np phase shifts are displayed in Fig. 1, which reflect what the χ^2 have already proven, namely, an excellent convergence when going from NNLO to N^{3}LO and, finally, to N^{4}LO. However, at LO and NLO there are large discrepancies between the predictions and the empirical phase shifts as to be expected from the corresponding χ^2 values. This fact renders applications of the LO and NLO nuclear force useless for any realistic calculation (but they could be used to demonstrate truncation errors).

The evolution of the deuteron properties from LO to N^{4}LO of chiral EFT are shown in Table 3. In all cases, we fit the deuteron binding energy to its empirical value of 2.224575 MeV using the non-derivative $^3S_1$ contact. All other deuteron properties are predictions. Already at NNLO, the deuteron has converged to its empirical properties and stays there through the higher orders.

At the bottom of Table 3, we also show the predictions for the triton binding as obtained in 34-channel charge-dependent Faddeev calculations using only 2NFs. The results show smooth and steady convergence, order by order, towards a value around 8.1 MeV that is reached at the highest orders shown. This contribution from the 2NF will require only a moderate 3NF. The relatively low deuteron D-state probabilities ($\approx 4.1\%$ at N^{3}LO and N^{4}LO) and the concomitant generous triton binding energy predictions are a reflection of the fact that our NN potentials are soft (which is, at least in part, due to their non-local character).

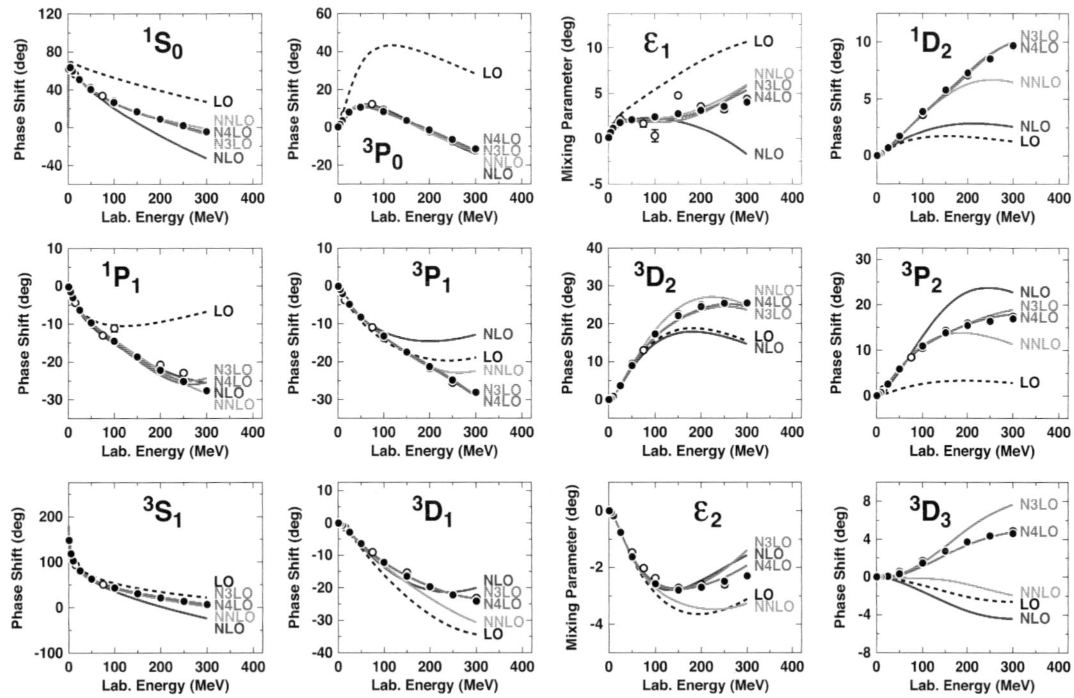

Figure 1. (Color online). Chiral expansion of neutron-proton scattering as represented by the phase shifts in S, P, and D waves and mixing parameters ϵ_1 and ϵ_2. Five orders ranging from LO to N^{4}LO are shown as denoted. A cutoff $\Lambda = 500$ MeV is applied in all cases. The filled and open circles represent the results from phase shift analyses.

4. Summary and Conclusions

We have presented chiral NN potentials through five orders of chiral EFT ranging from LO to N^{4}LO. The construction may be perceived as consistent, because the same power counting scheme as well as the same cutoff procedures are applied in all orders. Moreover, the long-range part of these potentials are fixed by the very accurate πN LECs as determined in the Roy-Steiner equations analysis of Ref. [12]. In fact, the uncertainties of these LECs are so small that a variation within the errors leads to effects that are essentially negligible at the current level of precision. Another aspect that has to do with precision is that, at least at the highest order (N^{4}LO), the NN data below pion-production threshold are reproduced with the outstanding χ^2/datum of 1.15. This is the highest precision ever accomplished with any chiral NN potential to date.

The NN potentials presented in this contribution may serve as a solid basis for systematic *ab initio* calculations of nuclear structure and reactions that allow for a comprehensive error analysis. In particular, the order by order development of the potentials will make possible a reliable determination of the truncation error at each order.

Our family of potentials is non-local and, generally, of soft character. This feature is reflected in the fact that the predictions for the triton binding energy (from two-body forces only) converges to about 8.1 MeV at the highest orders. This leaves room for moderate three-nucleon forces.

These features of our potentials are in contrast to other families of chiral NN potentials of semi-local or local character that have recently entered the market [14, 15]. Such potentials are

Table 3. Two- and three-nucleon bound-state properties as predicted by NN potentials at various orders of chiral EFT ($\Lambda = 500$ MeV in all cases). (Deuteron: Binding energy B_d, asymptotic S state A_S, asymptotic D/S state η, structure radius $r_{\rm str}$, quadrupole moment Q, D-state probability P_D; the predicted $r_{\rm str}$ and Q are without meson-exchange current contributions and relativistic corrections. Triton: Binding energy B_t.) B_d is fitted, all other quantities are predictions.

	LO	NLO	NNLO	N^{3}LO	N^{4}LO	Empirical
Deuteron						
B_d (MeV)	2.224575	2.224575	2.224575	2.224575	2.224575	2.224575(9)
A_S (fm$^{-1/2}$)	0.8526	0.8828	0.8844	0.8853	0.8852	0.8846(9)
η	0.0302	0.0262	0.0257	0.0257	0.0258	0.0256(4)
$r_{\rm str}$ (fm)	1.911	1.971	1.968	1.970	1.973	1.97507(78)
Q (fm^2)	0.310	0.273	0.273	0.271	0.273	0.2859(3)
P_D (%)	7.29	3.40	4.49	4.15	4.10	—
Triton						
B_t (MeV)	11.09	8.31	8.21	8.09	8.08	8.48

less soft and, consequently, require stronger three-body force contributions.

The availability of families of chiral NN potentials of different character offers the opportunity for interesting systematic studies that may ultimately shed light on issues, like, the "radius problem", the overbinding of intermediate-mass nuclei, and others.

Note that the differences between the above-mentioned families of potentials are in the off-shell character, which is not an observable. Thus, any off-shell behavior of a NN potential is legitimate. There is no wrong off-shell character. However, some off-shell behaviors may lead in a more efficient way to realistic results than others. That is of interest to the many-body practitioner. We are now in a position to systematically investigate this issue for chiral forces.

Acknowledgments

This work was supported in part by the U.S. Department of Energy under Grant No. DE-FG02-03ER41270.

References

[1] Machleidt R and Entem D R 2011 *Phys. Rep.* **503** 1
[2] Epelbaum E, Hammer H-W and Meißner U-G 2009 *Rev. Mod. Phys.* **81** 1773
[3] Entem D R and Machleidt R 2003 *Phys. Rev.* C **68** 041001
[4] Epelbaum E, Glöckle W and Meißner U-G 2005 *Nucl. Phys.* **A747** 362
[5] Sammarruca F, Coraggio L, Holt J W, Itaco N, Machleidt R and Marcucci L E 2015 *Phys. Rev.* C **91** 054311
[6] Golak J *et al* 2014 *Eur. Phys. J.* A **50** 177
[7] Epelbaum E, Gasparyan A M, Krebs H and Schat C 2015 *Eur. Phys. J.* A **51** 26
[8] Girlanda L, Kievsky A and Viviani M 2011 *Phys. Rev.* C **84** 014001
[9] Entem D R, Kaiser N, Machleidt R and Nosyk Y 2015 *Phys. Rev.* C **91** 014002
[10] Entem D R, Kaiser N, Machleidt R and Nosyk Y 2015 *Phys. Rev.* C **92** 064001
[11] Entem D R, Machleidt R and Nosyk Y 2017 *Phys. Rev.* C **96** 024004
[12] Hoferichter M, Ruiz de Elvira J, Kubis B and Meißner U-G 2015 *Phys. Rev. Lett.* **115** 192301
[13] Machleidt R, Holinde K and Elster Ch 1987 *Phys. Rep.* **149** 1
[14] Epelbaum E, Krebs H and Meißner U-G 2015 *Eur. Phys. J.* A **51** 53
[15] Piarulli M *et al* 2016 *Phys. Rev.* C **94** 054007

IOP Conf. Series: Journal of Physics: Conf. Series **966** (2018) 012012 doi:10.1088/1742-6596/966/1/012012

EXOGAM at the ILL: the EXILL campaign

G. de France [1], A. Blanc [2], F. Drouet [3], M. Jentschel [2], U. Köster [2], P. Mutti [2], J.M. Régis [4], G. Simpson [3], T. Soldner [2], O. Stezowski [5], C.A. Ur [6,7], W. Urban [2,8] and A. Vancrayenest [3]

[1] Grand Accélérateur National d'Ions Lourds (GANIL), CEA/DSM-CNRS/IN2P3, Bd Henri Becquerel, BP55027, F-14076 Caen Cedex 5, France
[2] Institut Laue-Langevin, BP 156, 6, rue Jules Horowitz, 38042 Grenoble Cedex 9, France
[3] LPSC, Université Joseph Fourier Grenoble 1, CNRS/IN2P3, Institut National Polytechnique de Grenoble, F-38026 Grenoble Cedex, France
[4] Institut für Kernphysik der Universitt zu Köln, Zülpicher Str. 77, 50937 Köln, Germany
[5] Institut de Physique Nucléaire de Lyon CNRS/IN2P3, Université Claude Bernard, 4 rue Enrico Fermi, F-69622 Villeurbanne Cedex, France
[6] Dipartimento di Fisica, Università di Padova, Via Marzolo 8, I-35131 Padova, Italy
[7] ELI-NP, National, IFIN-HH30, Reactorului, 077125 Bucharest-Magurele, Romania
[8] Faculty of Physics, University of Warsaw, ul. Hoza 69, 00-681 Warsaw, Poland

E-mail: gilles.defrance@ganil.fr

Abstract. A combination of germanium detectors has been installed at the PF1B neutron guide of the ILL to perform the prompt spectroscopy of neutron-rich nuclei produced in the neutron-capture induced-fission of ^{235}U and ^{241}Pu. Radiative capture reactions on rare targets have also been performed. $LaBr_3$ detectors from the FATIMA collaboration have also been installed in complement with the EXOGAM clovers to measure lifetimes of low-lying excited states. The measured characteristics indicate very good performances of the overall setup. Some recent results will be discussed.

1. Introduction

It is known since long that fission produces neutron-rich nuclei in a very effective way. One of the best mechanism is to induce fission of actinide targets using cold neutrons from a neutron guide. In this regime, the neutrons bring just enough energy into the system to produce fission, whilst preserving the neutron-richness of the fragments. The average spin generated in this reaction is around 6-7$\hbar$ [1] and states with spins as high as 20$\hbar$ can be observed. Several possible actinide targets have been envisaged and it came out that of particular interest in this context is the use of ^{235}U and ^{241}Pu fissile targets which give access to nuclei that are too weakly produced using spontaneous fission sources or another production mean to perform their prompt spectroscopy. In addition to the spectroscopy of nuclei produced in such a cold fission process, neutron capture reactions using rare targets have also been performed. In this contribution we will describe the installation of the EXOGAM array, usually standing at GANIL, at the Institut Laue Langevin (ILL), Grenoble (hence the name EXILL for EXOGAM at the ILL), and the experimental campaign which took place in 2012 and 2013.

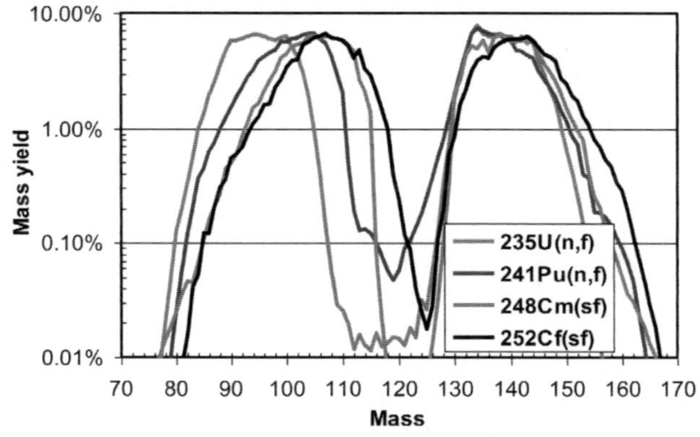

Figure 1. Mass yield distributions measured for the spontaneous fission of ^{252}Cf and ^{248}Cm compared to the thermal neutron induced fission of ^{235}U and ^{241}Pu (data from [2])

2. Physics case

The basic idea of EXILL was to install a large number of efficient germanium detectors around a fissile target irradiated by a well collimated cold neutron beam. Large arrays like Euroball or Gammasphere were already used to study fission fragments produced in the spontaneous fission of ^{252}Cf and ^{248}Cm sources but it was the first time that such a large array was installed in a reactor facility, around a fissile target irradiated by cold neutrons. We used ^{235}U and ^{241}Pu targets giving access to many nuclei where little was known, especially in the regions north-east of ^{78}Ni and beyond ^{132}Sn. This is what is shown in Figure 1 which compares the mass yield distribution from the thermal neutron induced fission of ^{235}U and ^{241}Pu and the yield from the spontaneous fission of ^{252}Cf and ^{248}Cm. The production of nuclei in the mass A$\sim$90 region is an order of magnitude larger using a ^{235}U target compared to the two spontaneous fission sources. The same is true for nuclei in the mass A$\sim$125 region using a ^{241}Pu target. These two mass regions were the focus of the EXILL physics case.

One main subject was the so-called Shell Model (SM) nuclei with just a few particles or holes outside a double shell closure like ^{132}Sn. They allow very sensitive tests of the interactions that are used in this kind of calculations. One way to address this key question is to measure spectroscopic properties. Correct predictions of nuclear structure are important as nuclei here lie on, or next to, the astrophysical rapid neutron capture process. Properties, such as neutron binding energies and low-lying decaying states can change the speed and direction of the predicted r-process path. The evolution of the N=50 gap in exotic nuclei and the question of whether ^{78}Ni is a doubly magic nucleus is highly debated. From the structure of nuclei with N=49-50 close to ^{78}Ni, one can see how well SM interactions can reproduce excited states and to what level contributions from core excitations need to be included. In particular, we are missing the single-particle energies and the two body matrix elements for the residual proton-neutron interactions. Low lying excited states in nuclei located in the vicinity of doubly closed shell nuclei are usually dominated by the coupling of single-particle excitations to core excitations. The nature of these couplings might be responsible of basic phenomena such as the quenching of spectroscopic factors or anharmonicities in vibrational spectra. Such phenomena have been investigated in particular in the region of ^{132}Sn with the EXILL setup. The reasons for the rapid change from a spherical to a deformed shape in the A$\sim$100 region, with the addition of just a few neutrons, remained a mystery for a long time. In recent years, studies point to the role played by deformation-driving, and resisting, Nilsson orbitals. However, such arguments

IOP Conf. Series: Journal of Physics: Conf. Series **966** (2018) 012012 doi:10.1088/1742-6596/966/1/012012

neglect any role played by the protons in such phenomena and the neutron-proton interaction is thought to be largely responsible for the onset of collectivity when moving away from closed shells. Studies of the intermediate spin states of the Kr and Rb nuclei with N$\sim$59 will help to clarify the role played by the protons.

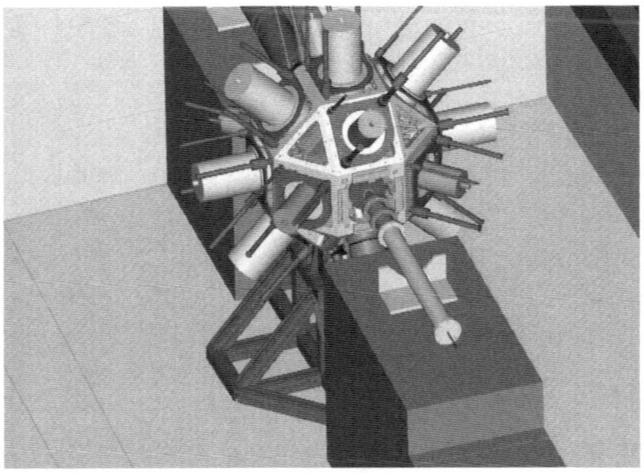

Figure 2. Configuration 1: 8 EXOGAM large Clovers, 6 coaxial detectors from GASP and two clovers from the ILL.

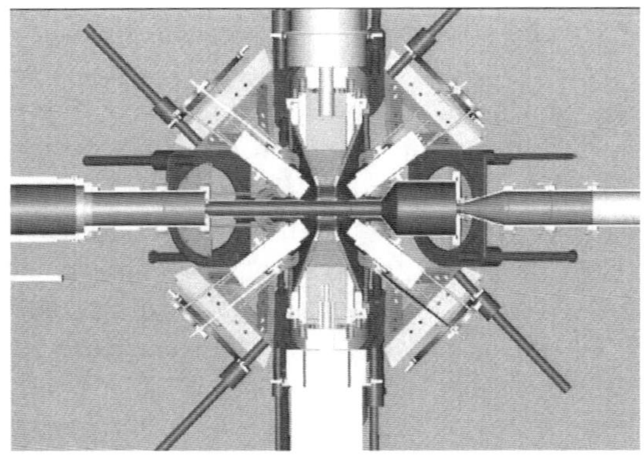

Figure 3. Configuration 2: 8 EXOGAM Clovers and 16 $LaBr_3$ for lifetime measurements.

The generation of angular momentum in the fission mechanism itself remains not understood. The angular correlations of γ-rays emitted by complementary fragments will give insights on how angular momentum up to $20\hbar$ can be generated from an initial system with zero angular momentum. Of high interest to better understand fission is also the mass split and the prompt neutron multiplicity. Precise fission yields and prompt γ-ray spectra from ^{235}U(n,f) and ^{241}Pu(n,f) are also crucial for calculating the heating in nuclear reactors and have thus strong impact on the design of the future reactors and fuel elements. Finally, the high photopeak efficiency of the array allowed the study of samples that have low neutron capture cross sections of tens of millibarn or where only small milligram quantities of enriched material are available. An important property of capture reaction is its non selectivity which ensures the feeding of all available states in a certain energy and spin range for complete spectroscopy. Therefore, studies on collective models, symmetries and statistical properties of nuclei close to stability become possible.

3. Setups

Two setups were envisaged (see Figures 2 and 3 and reference [3] for a complete description). The first one consisted in 8 EXOGAM large clovers [4], 6 large coaxial detectors from GASP [5] and the 2 clovers from the ILL. In the second setup, the GASP and ILL detectors were replaced by 16 $LaBr_3$ crystals from the FATIMA collaboration [6]. The EXOGAM and GASP detectors were mounted with their anti-Compton shield. Both configurations were instrumented with a full digital triggerless data acquisition system based on 14 bit 100MHz CAEN digitizers. The two setups were installed around the target point of the PF1B line of the ILL. The ILL reactor is the worlds brightest continuous neutron source with an in-pile flux up to 1.5×10^{15} neutrons per s.cm^2. Neutrons are guided first via the H113 ballistic neutron guide [7] toward the experimental area. The neutron beam collimation is then realized by using a series of lithium and boron collimators mounted upstream the target. In order to produce around 10^4 fissions per second in a 400 μg thick, 1cm^2, ^{235}U target a flux of 10^8 neutrons per s.cm^2 is required.

Details of the collimation can be found in [3]. The campaign covered two reactor cycles (93 days of effective beam time for experiments) organized as follows: 22 (14) days for spectroscopy studies with ^{235}U (^{241}Pu); 13 (10) days for fast timing measurements using ^{235}U (^{241}Pu) targets; during the remaining time several (n,γ) reactions were performed using ^{46,48}Ca, ^{48}Ti, ^{68,70}Zn, ^{77}Se, ^{95}Mo, ^{96}Zr, ^{143}Nd, ^{155,157}Gd, ^{161}Dy, ^{167}Er, ^{194,195}Pt and ^{209}Bi: 24 (10) for spectroscopy (fast timing) studies.

In the absence of any fragment identification, it is crucial to acquire high-fold events (3 or larger) to possibly associate the observed γ-rays to a given fragment. This technique relies on the knowledge of the transitions deexciting the first excited states in the two fragments and the measurements of the evolution of the unknown γ-rays intensity gated by known transitions in a series of isotopes. ^{235}U and ^{241}Pu targets have been used leading for a given nucleus to two distinct complementary fragments, which greatly helps to confirm the assignment. The method relies therefore primarily on the efficient detection of high-fold data, which is achieved using a large number of efficient HPGe detectors. Such a technique is now well established and has been proven to work very well to analyze fission data [8, 9]. The total photopeak efficiency for the 16 Ge detectors setup was about 6% at E_γ=1.3 MeV. The granularity is also crucial: even though 2/3 of this efficiency is coming from the EXOGAM clover, the gain in the number of γ-γ-γ coincidences obtained by adding the GASP and the ILL detectors is a factor of 4.2. As an example we can compare the spectra obtained after setting two gates to select ^{92}Rb in ^{248}Cm fission data obtained with EUROGAM2 (10 days experiment) and our EXILL dataset. A gain of 30 in statistics is obtained. The second setup using the $LaBr_3$ detectors was dedicated to lifetime measurements. In this configuration, the EXOGAM detectors are used to select the fragment of interest and the $LaBr_3$ crystals give the lifetimes via the electronics fast timing technique using $LaBr_3$-$LaBr_3$ coincidences [10, 11, 12, 13]. This method requires therefore Ge-$LaBr_3$-$LaBr_3$ or Ge-Ge-$LaBr_3$-$LaBr_3$ coincidences. Recently, a novel method to analyze such data has been developed [14] and lead to the determination of lifetimes with an accuracy of 5 ps between 40 keV and 6.7 MeV.

4. Example of results

4.1. ^{133}Sb: Coupling collective and single-particle states in the vicinity of a doubly closed shell nucleus

We have studied particle-core excitations in the one-valence proton nucleus ^{133}Sb. This study used the full complementarity of the various runs within the EXILL campaign: spectroscopy and lifetime measurements using ^{235}U and ^{241}Pu targets in both cases. In the spectroscopy part we search for excited states above the $21/2^+$ isomer using the capability of EXILL to build prompt-delayed and prompt-prompt coincidences (see Figure 4). With the first condition, several transitions were found in coincidence with known, delayed transitions in ^{133}Sb while with the prompt-prompt coincidences, ordering of these new transitions above the isomer was possible. We also used fission events from ^{235}U and ^{241}Pu to ensure that the newly observed transitions were observed in both cases hence not associated to a complementary fragment of ^{133}Sb.

We then concentrated on lifetime measurements of the $15/2^+$ and $13/2^+$ states located just below the isomer. Using Ge-$LaBr_3$-$LaBr_3$ coincidences, the Ge gate to select properly the decay path, the two $LaBr_3$ ones to measure the lifetime, and the centroid shift technique [12] we could extract the lifetimes and deduce the reduced transition probabilities: $B(M1 : 15/2^+ \rightarrow 13/2^+) > 0.24$ W.u and $B(M1 : 13/2^+ \rightarrow 11/2^+) = 0.0042(15)$ W.u. This difference of two orders of magnitude clearly indicates a dramatic change in the structure of the two levels. In order to get insight into this puzzle a new model called the Hybrid Configuration Mixing Model was developed [15, 16] with the aim to couple collective and single-particle excitations in a fully self-consistent manner. This model suggests that the low spin states are built from a proton

IOP Conf. Series: Journal of Physics: Conf. Series **966** (2018) 012012 doi:10.1088/1742-6596/966/1/012012

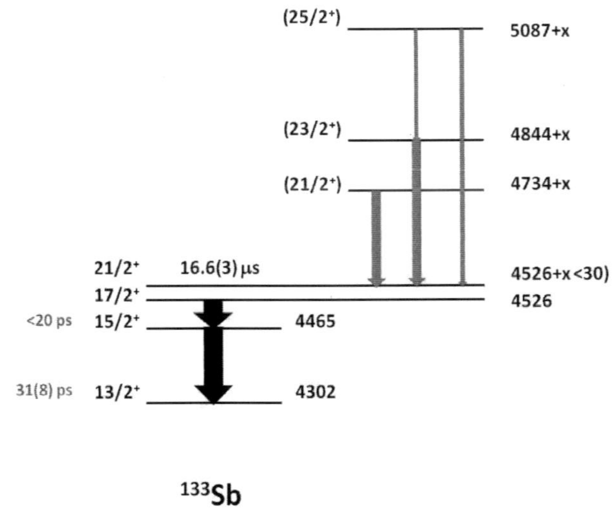

Figure 4. Partial level scheme of ^{133}Sb showing only states discussed in this paper with the newly observed γ-rays above the 21/2 isomer in red and the new lifetimes in blue.

$g_{7/2}$ coupled to the 2^+ ^{132}Sn core excitation while the higher lying levels are arising from the valence proton coupled to single-neutron $h^{-1}_{11/2}f_{7/2}$ excitations. As a result the model gives 0.021 W.u and 0.001 W.u for the $15/2^+ \rightarrow 13/2^+$ and $13/2^+ \rightarrow 11/2^+$ transitions respectively, in fair agreement with data, which is not the case for shell model calculations [17].

4.2. Symmetry studies

^{196}Pt has been proposed a long time back as a good candidate to test the validity of the SO(6) limit of the sd Interacting Boson Model ([18, 19]). During the EXILL campaign, the Cologne group led by J Jolie proposed to use the fast timing technique for this purpose. In the SO(6) limit, the E2 transition operator is a generator of SO(6) and strictly speaking only transitions between states with $\Delta\sigma$=0 are allowed (σ being one of the quantum number associated to the irreducible representations of SO(6)). Therefore, E2 transitions between different SO(6) representations ($\Delta\sigma \neq 0$) are forbidden. The states of interest have low spin and high excitation energy and are not populated in heavy-ion induced reactions. In contrary they might be well populated in (n,γ) reactions. This was realized during the EXILL campaign using a ^{195}Pt target and using Ge-LaBr$_3$-LaBr$_3$ coincidences to deduce the lifetimes of the third 0^+ state at 1402 keV and the first 2^+ state at 356 keV (which lifetime was already known at 49.2(2) ps). The measurements yielded the values of 50(5) ps for the 2^+_1 state and an upper limit of 12 ps for the 0^+_3 state. Using also the lower limit from Börner et al [19], it was established that 0.56 W.u $< B(E2; 0^+_3 \rightarrow 2^+_1) <$ 5 W.u and 0.05 W.u $< B(E2; 0^+_3 \rightarrow 2^+_2) <$ 0.41 W.u. The absence of collective B(E2) values demonstrates for the first time the validity of the SO(6) symmetry [20].

4.3. ^{195}Pt: the link between nuclear structure and medical physics

The structure of ^{195}Pt was also studied and more specifically the population of the $21/2^+$ isomer using the radiative capture ^{194}Pt(n,γ) reaction. This nucleus has been recognized since long as a good candidate to probe multi-j supersymmetry ([21, 22]). In addition to this fundamental interest, ^{195}Pt is also a promising candidate for nuclear medicine since it combines therapy and diagnosis capabilities. Indeed its $T_{1/2}$=4.01 days isomer is compatible with such a usage (long enough for transportation and short enough for biology). This isomer decays via a highly converted M4 transition which provides an interesting possibility for treatment, with a very high linear energy transfer deposited within a few nm^3, capable to induce double strand DNA breaks. In addition, the 99 keV gamma-ray and 65-78 keV X-rays emitted in the cascade from the isomer allow imaging using Single Photon Emission Computed Tomography and gamma cameras. During the EXILL campaign we searched for new doorway states that would significantly feed the isomer and decay to the ground state. In this study several new states have been identified in the decay from the $1/2^+$ neutron capturing state at 6.1 MeV. To envisage an abundant production of the isomer for nuclear medicine, a method to populate specifically the doorway state of interest has to be applied. For this, another experiment was realized at the HIGS facility (TUNL) where we tried to use a gamma beam at energies tuned around the newly identified states in order to measure the population of the isomer of interest. This work is still under analysis and is part of the thesis work of D. Wilmsen.

5. Conclusion

The EXILL campaign ran smoothly for two reactor cycles (~100 days) at the ILL, harvesting results from 73 proposals and gathering together 18 different laboratories. Today more than 30 papers have been published in refereed journals, showing the success of this adventure. The installation of a large and efficient gamma-ray array at a neutron beam facility is proven to be very effective in the study of neutron rich nuclei from cold neutron induced fission as well as in performing the detailed spectroscopy of close-to-stability nuclei in radiative capture reactions. Recent and new results have been discussed in the various aspects of the campaign, aiming at highlighting the diversity of the physics cases.

Acknowledgements

The EXILL campaign would not have been possible without the support of several services at the ILL and the LPSC. We are grateful to the EXOGAM collaboration for the loan of the detectors, to GANIL for assistance during installation and dismantling, and to the INFN Legnaro laboratory for the loan of the GASP detectors.

References

[1] Hotchkis M A C *et al.* 1990 *Phys. Rev. Lett.* **64** 3123
[2] JEFF-3.1.1 database, http://www.oecd-nea.org/dbdata/jeff/
[3] Jentschel M *et al.* 2017 JINST **12**
[4] Simpson J and the EXOGAM collaboration 2000 *Acta Physica Hungarica, New Series, Heavy Ion Physics* **11** 159.
[5] Bazzacco D and the GASP Collaboration 1992 *Proc. Int. Conf. Nuclear Structure at High Angular Momentum, Ottawa, AECL-10613* 376
[6] http://nuclear.fis.ucm.es/fasttiming/
[7] Abele H *et al.* 2006 *Nucl. Instr. and Meth. A* **562** 407
[8] Hotchkis M A C *et al.* 1991 *Nucl. Phys. A* **530** 111
[9] I. Ahmad and W.R. Phillips 1995 *Rep. Prog. Phys.* **58** 1415
[10] Löbner K E G *The electromagnetic interaction in nuclear spectroscopy, (ed. W. D. Hamilton, North-Holland New York 1975)* 173
[11] Mach H, Gill R L, Moszyński M 1989 *Nucl. Instr. and Meth. A* **280** 49
[12] Régis J M *et al.* 2010 *Nucl. Instr. and Meth.* **A 622** 83
[13] Régis J M *et al.* 2014 *Nucl. Instr. and Meth.* **A 763** 210
[14] Régis J M *et al.* 2012 *Nucl. Instr. and Meth.* **A 684** 36
[15] Bocchi G et al 2016 *Phys. Lett. B* **760**, 273
[16] Colò G et al 2017 *Phys. Rev. C* **95**, 034303
[17] Urban W et al 2000 *Phys. Rev. C* **62**, 027301; 2009 *Phys. Rev. C* **79**, 037304
[18] Cizewski J A, Casten R F, Smith G J, Stelts M L, Krane W R, Brner H G, Davidson W F 1978 *Phys. Rev. Lett.* **40** 167
[19] Börner H G, Jolie J, Robinson S J, Casten R F, Cizewski J A 1990 *Phys. Rev. C* **42** R2271
[20] Jolie J *et al.* 2015 *Nucl. Phys. A* **934** 1
[21] Warner D D, Casten R F, Stelts M L, Borner H G and Barreau G 1982 *Phys. Rev. C* **26** 1921
[22] Casten R F, Colvin G G and Schreckenbach K 1987 *J. Phys. G* **13** 221

IOP Conf. Series: Journal of Physics: Conf. Series **966** (2018) 012013 doi:10.1088/1742-6596/966/1/012013

Effect of short-range correlations on the single proton $3s_{1/2}$ wave function in ^{206}Pb

S. Shlomo[1,2], I. Talmi[2], M. R. Anders[1] and G. Bonasera[1]
1Cyclotron Institute, Texas A&M University, College Station, Texas 77843, USA
2Department of Particle Physics and Astrophysics, The Weizmann Institute of Science, Rehovot, Israel

E-mail: s-shlomo@tamu.edu; igal.talmi@weizmann.ac.il; manders14@tamu.edu; gbonasera@comp.tamu.edu

Abstract. We consider the experimental data for difference, $\Delta\rho_c(r)$, between the charge density distributions of the isotones ^{206}Pb – ^{205}Tl, deduced by analysis of elastic electron scattering measurements and corresponds to the shell model $3s_{1/2}$ proton orbit. We investigate the effects of two-body short-range correlations. This is done by: (a) Determining the corresponding single particle potential (mean-field), employing a novel method, directly from the single particle proton density and its first and second derivatives. We also carried out least-square fits to parametrized single particle potentials; (b) Determining the short-range correlations effect by employing the Jastrow correlated many-body wave function to derive a correlation factor for the single particle density distribution. The $3s_{1/2}$ wave functions of the determined potentials reproduce fairly well the experimental data within the quoted errors. The calculated charge density difference, $\Delta\rho_c(r)$, obtained with the inclusion of the short-range correlation effect does not reproduce the experimental data.

1. Introduction

The shell model has been very successful in explaining many features of nuclei [1,2]. The relation between the shell model wave functions and the real nuclear ones is rather complicated. Important information about it may be deduced from the measured charge density difference, $\Delta\rho_c(r)$, between charge density distributions of the isotones ^{206}Pb – ^{205}Tl [3]. The experimental data of $\Delta\rho_c(r)$ show a clear maximum at the center of ^{206}Pb with two additional maxima. This seems to be the shape obtained from a $3s_{1/2}$ single proton wave-function, in agreement with the simple shell model. It was pointed out in the literature that commonly used central potentials, such as the Woods-Saxon potential, lead to a $3s_{1/2}$ charge density in disagreement with experimental data. In particular, the central density obtained from the Woods-Saxon potential is too large by about 30%. This difference between data and the Woods-Saxon results was considered earlier in the literature resulting with the statement that it is accounted for by the effect of two-body short range correlations (SRC) on the shell model wave functions [3,4]. However, it was also stated in Ref, [5] that "there is no *significant* discrepancy between the experimental data and the Hartree-Fock calculations using reasonable effective interactions." In this work we take a closer look at this problem and ask: (i) Is there a single particle potential V(r) whose $3s_{1/2}$ wave function reproduces the experimental data? To answer this question we have developed a novel method, using the single particle Schrodinger equation for a wave function $\psi(\vec{r})$ with eigenenergy E, to determine the central potential $V(\vec{r})$ directly from the measured single particle proton density of

$3s_{1/2}$ wave function and its first and second derivatives, assuming known for all $\vec{r}$. We apply the method to the experimental data of the charge distribution of the proton $3s_{1/2}$ orbit given by the charge density difference, $\Delta\rho_c(r)$ [3]. We have also carried out fits to parametrized single particle potentials. (ii) Is the effect of short-range correlations (SRC) on the nuclear density explains the experimental data? To assess the effect of SRC we have used the Jastrow many-body correlated wave function, with a two-body correlation factor, and derived a simple and a good approximate method for calculating the effect of SRC on the single particle density distribution.

We point out that the resulting single particle potential, if found, will provide a stringent limit on the effects of short correlation on the expected values of long-range operators, an important test for the shell model. The potential can also be used as an additional experimental constraint in determining a modern energy density functional (EDF) for providing more reliable predictions of properties of nuclei and nuclear matter [6,7]. In the next section we provide a short presentation of the novel methods used in the calculations and in section 3 we present some results with the conclusion given in section 4. Some results were presented in Refs. [8,9,10].

2. Formalism

2.1 Deducing the potential from the single-particle density

From the single particle Schrodinger equation,

$$-\frac{\hbar^2}{2m}\Delta\psi + V\psi = E\psi\,, \tag{1}$$

where $V(\vec{r})$ is a real local and non-singular potential, follows that for a given single particle wave function $\psi(\vec{r})$, known for all $\vec{r}$, and given eigenvalue E, the corresponding single particle potential V is uniquely determined by

$$V(\vec{r}) = E + \frac{\hbar^2}{2m}S(\vec{r}), \quad S(\vec{r}) = \frac{\Delta\psi(\vec{r})}{\psi(\vec{r})}. \tag{2}$$

For a non-singular V, $\Delta\psi(\vec{r}) = 0$, when $\psi(\vec{r}) = 0$. The general relation for $[\psi(\vec{r})]^b$, where b is a positive integer, is given in Refs. [9,10]. Here we limit our consideration to the spherically symmetric case where,

$$\psi_{nlj}(\vec{r}) = \frac{R_{nlj}(r)}{r}Y_{lj}\,. \tag{3}$$

Here, $R_{nlj}(r)$ is the radial wave function for the orbit with principal number n, orbital angular momentum l and total angular momentum j and Y_{lj} is the eigenfunction of the angular momenta l and j. In the following we limit the discussion to the proton $3s_{1/2}$ orbit, in which the angular momentum $l =$ 0 and therefore no centrifugal and spin-orbit potentials appear in $V(\vec{r})$. Substituting (3) in (1) we obtain the Schrodinger equation for the radial wave function $R_{nlj}(r)$. Using the Schrodinger Equation for $R_{nlj}(r)$ we find that,

$$V_{cen}(r) = E + \frac{\hbar^2}{2m}S(r) - \frac{1}{2}(1-\tau_z)V_{coul}(r), \quad S(r) = \frac{1}{R_{nlj}(r)}\frac{d^2R_{nlj}}{dr^2}\,. \tag{4}$$

where $V_{cen}(r)$ and $\frac{1}{2}(1-\tau_z)V_{coul}(r)$, are the central and coulomb potentials, respectively. Here, τ_z=1 for a neutron and -1 for a proton.

Using the relation

$$R^2_{nlj}(r) = r^2\rho_{nlj}(r)\,, \tag{5}$$

where $\rho_{nlj}(r)$ is obtained by integrating the single particle density distribution over the angles and spin variables, it is possible to extract the wave function $R_{nlj}(r)$ from the experimental data for the charge density of the $3s_{1/2}$ proton orbit and deduce the corresponding single particle potential using Eq. (4). This leads to having some numerical complications. Therefore, we develop below a method to determine the potential directly from $\rho_{nlj}(r)$ and its first and second derivatives. Using Eq. (4) for the relation between $S(r)$ and $R_{nlj}(r)$, we obtain the simple relation between $S(r)$ and R^2_{nlj},

$$S(r) = \frac{1}{2R^2_{nlj}}\left[\frac{d^2(R^2_{nlj})}{dr^2} - \frac{1}{2}\frac{1}{R^2_{nlj}}\left[\frac{d(R^2_{nlj})}{dr}\right]^2\right]. \tag{6}$$

When $R_{nlj}^2 = 0$, $dR_{nlj}^2/dr = 0$. From Eq. (4), for non-singular potential, the term on the right hand side (r.h.s) of Eq. (6) in the large square brackets also vanishes. From Eqs. (5) and (6) we obtain the relation,

$$S(r) = \frac{1}{2\rho_{nlj}}\left[\frac{d^2\rho_{nlj}}{dr^2} + \frac{2}{r}\frac{d\rho_{nlj}}{dr} - \frac{1}{2\rho_{nlj}}\left(\frac{d\rho_{nlj}}{dr}\right)^2\right]. \quad (7)$$

When $\rho_{nlj} = 0$, $d\rho_{nlj}/dr = 0$. From Eq. (4), for non-singular potential, the term in the square brackets of (7) vanishes and $\frac{d^2\rho_{nlj}}{dr^2}$ also vanish.

We add that a commonly used central nuclear potential is the Woods Saxon (WS) potential,

$$V(r) = V_0/[1 + exp((r - R_1)/a_0)]\,, \quad (8)$$

where, V_0, R_1 and a_0 are the depth, half radius and diffuseness parameters, respectively. For the Coulomb potential we adopt the form obtained from a uniform charge distribution of radius R_{ch},

$$V_{coul}(r) = Ze^2\begin{cases}(3 - r^2/R_{ch}^2)/2R_{ch} & r \le R_{ch} \\ 1/r & r > R_{ch}\end{cases}, \quad (9)$$

where R_{ch} is deduced from the experimental value of the charge root mean square radius.We emphasize that in elastic electron-nucleus scattering measurements the charge density distribution, $\rho_{ch}(\vec{r})$ is deduced but in theoretical models the point proton density distribution, $\rho_p(\vec{r})$ is calculated. They are related by

$$\rho_{ch}(\vec{r}) = \int \rho_p(\vec{r'})\,\rho_{pfs}(\vec{r} - \vec{r'})d^3\vec{r'}\,, \quad (10)$$

where $\rho_{pfs}(\vec{r})$ is the charge density distribution of the proton. The experimental elastic electron scattering data on a free proton can be well reproduced by the expression

$$\rho_{pfs}(r) = \frac{1}{8\pi a^3}e^{-r/a}\,, \quad (11)$$

where $a^2 = \frac{1}{12}r_{pfs}^2$ with $r_{pfs} = 0.85$ fm being the corresponding charge root mean square (rms) radius [11]. The Fourier transform of the charge density $\rho_{ch}(\vec{r})$, determined by the convolution relation of Eq. (10) is given by

$$F_{ch}(q) = F_{pfs}(q)F_p(q)\,, \quad (12)$$

where $F_{ch}(q)$, $F_{pfs}(q)$ and $F_p(q)$, are the Fourier transfoms of $\rho_{ch}(r)$, $\rho_{pfs}(r)$ and $\rho_p(r)$, respectively. Eq. (12) can be used to determine the form factor $F_p(q)$. Then $\rho_p(r)$ can be obtained from $F_p(q)$ by the inverse Fourier transform and compared with theoretical predictions.

2.2 Effect of short-range correlations on the single particle density

Recently [12], we have derived and employed a simple and a good approximate method for determining the effect of short-range correlations (SRC) on the single particle densities of nucleons in nuclei. The method is obtained assuming: (i) the Jastrow ground state many body wave-function in the independent pair approximation [13,14], and (ii) the relation between the exchange terms and the direct terms of the two-body matrix element of the correlation function is similar to that of the contact interaction $V_0\delta(\vec{r}_i - \vec{r}_j)$. The Jastrow ground state wave-function has the form,

$$\Psi(\vec{r}_1,\cdots\vec{r}_A) = N_{Corr}\prod_{1\le i<j}^{A} f(|\vec{r}_i - \vec{r}_j|)\,\Phi(\vec{r}_1,\cdots\vec{r}_A), \quad (13)$$

where A is the number of nucleons, N_{Corr} is the normalization constant, $\Phi(\vec{r}_1,\cdots\vec{r}_A)$ is the independent particle model (IPM) wave-function for the ground state, given by the Slater determinant,

$$\Phi(\vec{r}_1,\cdots\vec{r}_A) = \frac{1}{\sqrt{A!}}det|\phi_i(\vec{r}_j)|\,, \quad (14)$$

of the single particle wave-functions ϕ_i with $\vec{r}_j$ stands for the set of $\vec{r}$, σ and τ, of space, spin and isospin coordinates, respectively. Note that for open shell nucleus, $\Phi(\vec{r}_1,\cdots\vec{r}_A)$ in (13) is a linear combination of slater determinants. The function $f_{ij} = f(|\vec{r}_i - \vec{r}_j|)$ in (13) is the SRC function taken in the form

$$f_{ij} = f(|\vec{r}_i - \vec{r}_j|) = 1 - e^{-\alpha|\vec{r}_i - \vec{r}_j|^2} = 1 - h_{ij}, \quad (15)$$

where the parameter α is deduced from the experimental data on the charge form factor. The effect of SRC on the proton single particle density distribution is determined by evaluating the matrix element of the density operator using the expansion

$$\prod_{1\le i<j}^{A} f_{ij}^2 = \prod_{1\le i<j}^{A}(1-g_{ij}) = 1-\sum_{1\le i<j}^{A} g_{ij} + \frac{1}{2}\sum_{\substack{i<j,k<l\\(ij)\neq(kl)}}^{A} g_{ij}g_{kl} - \cdots, \quad (16)$$

where, $g = 2h - h^2$. We adopt in our calculations the approximation of Ref. [14] with the ratio $d/r_0 \sim 0.5$, where d being the "healing distance" and r_0^3 is the average volume per particle. In the calculations we use the shell model harmonic oscillator wave functions.

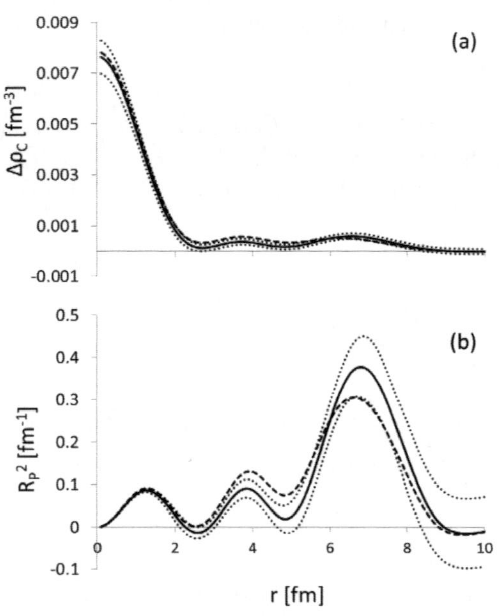

Fig.1. (a) Experimental data for the charge density difference ^{206}Pb – ^{205}Tl. The solid, dashed and dotted lines show the data, data with effect of rearrangement effect and the experimental errors, respectively. (b) Same as (a) for the $R_p^2(r)$ obtained from the experimental data.

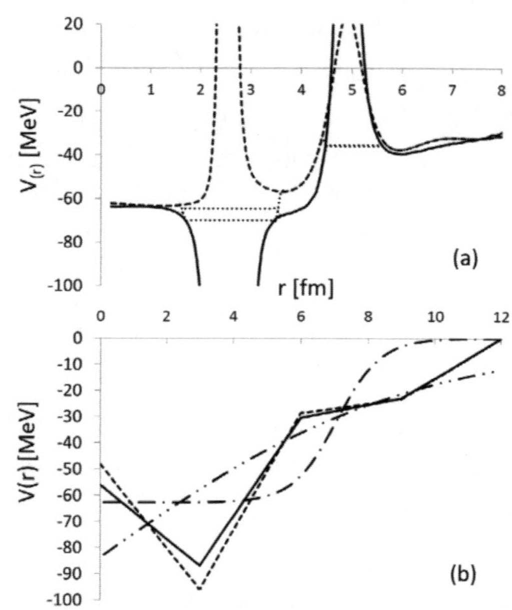

Fig.2. (a) The solid and dashed lines are the potentials deduced from Eq. (6) using Figure 1b. (b) The solid and dashed lines are the fitted broken potentials using Figure 1b. The dashed-dotted line is a fit to Woods-Saxon potential. The dashed-dotted-dotted line is the standard Woods Saxon.

3. Results

We present in Figure 1a (solid line) the experimental data [3] for the charge density difference,

$$\Delta\rho_c(r) = \rho_c(r;\, ^{206}\text{Pb}) - \rho_c(r;\, ^{205}\text{Tl}), \quad (17)$$

between the isotones ^{206}Pb – ^{205}Tl. The dotted lines indicate the experimental uncertainty. The two nodes associated with the proton $3s_{1/2}$ orbit are clearly seen in the figure. To assess the possible rearrangement effect (from ^{205}Tl to ^{206}Pb) on the charge rms radius of the 81 protons core in ^{206}Pb, we adopt the scaling approximation,

$$\Delta\rho_{Rc}(r) = \rho_c(r;\, ^{206}\text{Pb}) - \alpha^3\rho_c(\alpha r;\, ^{205}\text{Tl}), \quad (18)$$

assuming an increase by 0.005 fm, similar to the change between nuclei in this region [15]. We obtain the scaling parameter α = 5.4792/(5.4792 + 0.005) = 0.9990, the ratio between the charge rms radius of ^{205}Tl to that of the 81 core protons in ^{206}Pb. The results for $\Delta\rho_{Rc}(r)$ is shown in Figure1a (dashed line).

IOP Conf. Series: Journal of Physics: Conf. Series **966** (2018) 012013 doi:10.1088/1742-6596/966/1/012013

We determined the values of $R_p^2(r) = 4\pi r^2 \Delta\rho_p(r)$ and $R_{Rp}^2(r) = 4\pi r^2 \Delta\rho_{Rp}(r)$, deduced from the results of Figure 1a using Eqs. (10) and (12), shown as solid and dashed lines in Figure 1b, respectively The magnitude of the difference between $R_p^2(r)$ and $R_{Rp}^2(r)$ is similar to that of the experimental uncertainty. The corresponding potentials, for $R_p^2(r)$ and $R_{Rp}^2(r)$ of Figure 1b, obtained by employing Eqs. (4) - (6), are shown in Figure 2a by the solid and dashed lines, respectively. The dotted lines (constant potentials) are extracted from fits of the corresponding wave-function to Ψ= Csin(kr+ϕ), in the vicinity of the minima. The Coulomb potential of Eq. (9), with R_{ch} = 7.1 fm, was adopted in the calculations. Note the large uncertainties in the extracted potential in the vicinity of the nodes (at 2.6 and 4.9 fm). We have therefore considered several nuclear central potentials with parameters obtained by fits to the corresponding experimental data. In Figure 2b we show the resulting potentials fitted to $R_p^2(r)$ of Figure 1b. The solid line is a fit to broken lines potential to the solid line of Figure 1b, resulting with a good agreement with data and χ^2/N = 1.15. A similar fit to the dashed line of Figure 1b leads to the dashed line of Figure 2b with χ^2/N = 1.81. The dashed-double dotted line is a fit of the solid line of Figure 1b to a parametrized Woods-Saxon potential (9), resulting with χ^2/N = 3.28. The dashed-dotted line is the conventional Woods-Saxon potential (9), resulting with χ^2/N = 8.85 when comparing to the experimental data.

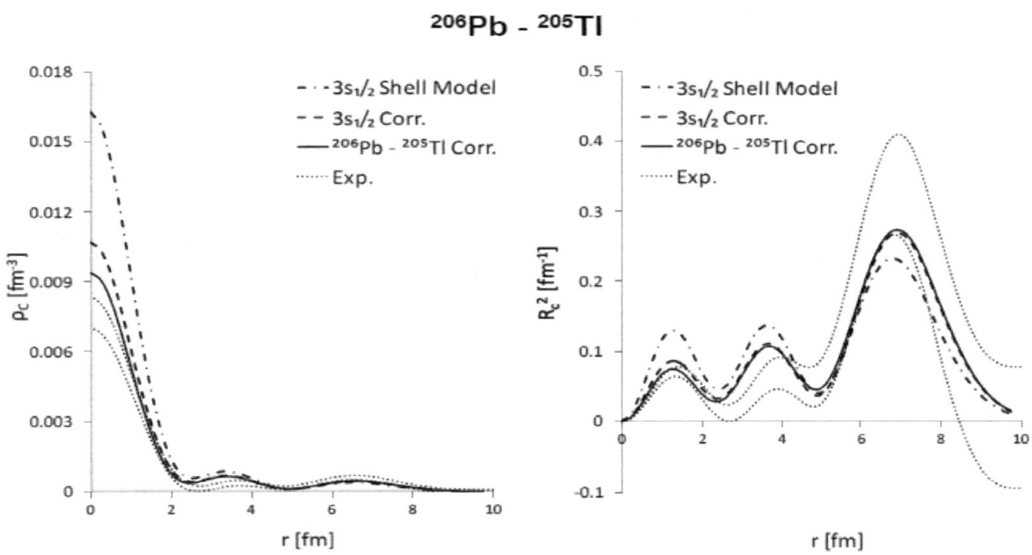

Fig. 3. Comparison between experiment and theory for the charge density of the proton $3s_{1/2}$ orbit using the Jastrow correlated wave function.

Fig. 4. Same as Figure 3 for $R_c^2(r)$.

Using the Jastrow wave function, Eq. (13), we determined the effect of SRC on the charge density distributions of ^{206}Pb and ^{205}Tl. In the calculations we adopted the shell model harmonic oscillator wave function with the size parameter $\nu = 1/5.9$ fm^{-2} and the parameter $\alpha = 1.96$ fm^{-2}. In Figs. 3 and 4 we show the results for the $3s_{1/2}$ proton charge density and the square of the radial wave-function, respectively, and compare with experimental data. The dotted lines show the experimental results for the difference between ^{206}Pb and ^{205}Tl charge distributions and the solid line is the corresponding calculated results. The dashed and dashed dotted lines are for the harmonic oscillator $3s_{1/2}$ proton orbit in ^{206}Pb with and without the effect of correlations, respectively. It is seen from Figures 3 and 4 that although the effect of SRC is to reduce the $3s_{1/2}$ single proton charge density at r = 0 by about 30%, the calculated density disagrees with the experimental data by more than a factor of 2, particularly in the region of r = 2 – 4 fm.

4. Conclusions

We have considered the experimental data for the charge density difference between the isotones ^{206}Pb – ^{205}Tl, deduced by the analysis of elastic electron scattering measurements and corresponds to the shell model $3s_{1/2}$ proton orbit. We have investigated the effects of two-body short-range correlations. This was done by: (a) Determining the corresponding single particle potential (mean-field), employing a novel method, directly from the single particle proton density distribution and its first and second derivatives. We also carried out least-square fits to parametrized single particle potentials. The $3s_{1/2}$ wave functions of the determined potentials reproduce fairly well the experimental data within the quoted errors.; (b) Determining the effect of short-range effect correlations by employing the Jastrow correlated many-body wave function to derive a correlation factor for the density distributions of single particle orbits. The calculated $3s_{1/2}$ density obtained with the inclusion of the effect of short-range correlations does not reproduce the experimental data. Our conclusion does not contradict the existence of SRC effects in the real nuclear wave function, as demonstrated by the high momentum tail in the experimental data for the charge form factor. Our results indicate that the shell-model wave function may yield charge density distributions which are in agreement with experimental data. More accurate experimental data, with uncertainty smaller by a factor of two or more, may answer the question how well can the data be reproduced by a $3s_{1/2}$ single particle wave function obtain from mean-field calculations. It will also help to determine more exactly the deviation between data and our SRC calculations. Of course, more accurate calculations of the effect of SRC are needed.

Acknowledgments

This work was supported in part by US Department of Energy under Grant No. DOE-FG03-93ER40773. S. Shlomo thanks the Weizmann Institute for the Weston Visiting Professorship Award and the nice hospitality extended to him.

References

[1] Mayer MR 1995 *Phys. Rev.* **75** 1969
[2] de Shalit A and Talmi I 1963 *Nuclear Shell Model* (New York, NY: Academic Press)
[3] Frois B et al 1986 *Nucl. Phys.* **A396** 409c
[4] Pandhripande VR, Sick I, and deWitt Huberts PKA 1997 *Rev. Mod. Phys.* **69** 981
[5] Bennour L, Heenen PH, Bonche P, Dobaczewski J, and Flocard H 1989, *Phys. Rev.* C **40** 2834
[6] Agrawal BK, Shlomo S and Au VK 2005, *Phys. Rev. C* **72** 014310-1-13
[7] Shlomo S 2013 *The Universe Evolution: Astrophysical and Nuclear Aspects* ed L Blokhintsev and Strakovsky (Nova Science Publishers)
[8] Anders MR, Shlomo S and Talmi I 2015 *Phys. Rev. C* **92** 034318
[9] Shlomo S, Anders MR and Talmi I 2015 *J. Phys. Conf. Series* **633** 012028
[10] Shlomo S and Anders MR 2017 *Bulletin of the Russian Academy of Science: Physics* **81** 1230
[11] de Vries H, de Jager CW and de Vries C 1987 *At. Data Nucl. Data Tables* **36** 495
[12] Bonasera G, *et al*, to be published
[13] Jastrow R 1955 *Phys. Rev.* **98** 1479
[14] de Shalit A and Weisskopf VF 1958 *Ann. Phys.* **5** 282
[15] Euteneuer H, Friedrich J and Voegler N 1978 *Nucl. Phys.* **A298** 452

IOP Conf. Series: Journal of Physics: Conf. Series **966** (2018) 012014 doi:10.1088/1742-6596/966/1/012014

Nuclear physics from an expansion around the unitarity limit

U van Kolck[1,2]

[1] Institut de Physique Nucléaire, CNRS/IN2P3, Univ. Paris-Sud, Université Paris-Saclay, 91406 Orsay, France

[2] Department of Physics, University of Arizona, Tucson, AZ 85721, USA

E-mail: vankolck@ipno.in2p3.fr

Abstract. Many features of the structure of nuclei can be understood in the unitarity limit, where the two-nucleon S waves have bound states at zero energy. In this limit, the only dimensionful parameter, which is needed for proper renormalization of the relevant effective field theory, is set by the triton binding energy. While the complexity of some many-body systems may stem from a profusion of distinct scales, this one three-body scale is sufficient to generate rich structures already in few-body systems due to the anomalous breaking of continuous to discrete scale invariance. I discuss how the spectra of light nuclei arise from a controlled, perturbative expansion around the unitarity limit. I also present some implications of discrete scale invariance for nuclear matter.

1. Introduction

What is essential in the physics of nuclei? Somehow a perfect plate of *rigatoni alla genovese* at La Colombaia on the beach in Ischia seemed to me the ideal occasion to ponder this question.

The traditional answer is: a nuclear potential obtained by fitting nucleon-nucleon (NN) scattering (up to some arbitrarily chosen energy, such as that of pion production) and bound-state data nearly exactly and then appending some three-body forces (3BFs) as needed to fix the most glaring discrepancies in the description of three- and/or four-nucleon systems. In the last quarter of a century, the field-theoretical roots of this program have been revived by the use of effective field theory (EFT), particularly in the form of Chiral EFT, the generalization of Chiral Perturbation Theory to systems with $A \geq 2$ nucleons. The approach has taken hold in the community of focusing on "chiral potentials" and building them up according to increasing A. The essential features of the interactions get lost in the details of the high "orders" (NLO, N^{2}LO, N^{3}LO, N^{4}LO, and counting...) deemed necessary to follow this program. In most reports we do not even find leading-order (LO) results, because they are supposed to be so poor.

This was not the original goal of the EFT program. EFT concerns the S matrix —the potential is only an intermediate step— and is organized by energy, not particle number. LO is supposed to provide an overall description of most low-energy observables —to represent the essential physics. One would like to have a rough but basically correct picture of few- and perhaps many-nucleon observables already at LO, with details obtained in perturbation theory. Apart from the difference in the magnitude of the expansion parameter, this is what we do in, say, non-relativistic QED, where the Coulomb interaction forms the basis on top of which finer interactions are accounted for with distorted-wave perturbation theory.

IOP Conf. Series: Journal of Physics: Conf. Series **966** (2018) 012014 doi:10.1088/1742-6596/966/1/012014

Here I discuss a recent proposal [1] for how to accomplish this through Pionless EFT [2]. I will argue that a good starting point is the limit where the $A = 2$ system is at unitarity and a single momentum scale $\Lambda_\star$ appears through the 3BF that ensures renormalizability for $A \geq 3$ systems. This scale arises from the anomalous breaking of scale symmetry down to a residual discrete scale invariance (DSI), which then determines the essential features of spectra and reactions. The emergence of complex spectra from such simple interactions is striking, and in the context of cold-atom physics has provided impetus to a vigorous experimental program. We now propose that it applies to nuclei as well. Our emphasis shifts from an accurate description of the *NN* system to an approximate description of three-, four- and hopefully more-nucleon bound states.

2. Why two-body unitarity?

For non-relativistic systems of identical particles of mass m, the $A = 2$ T matrix $T_2(k)$ at momentum k much smaller than the inverse of the potential range R is given by the effective-range expansion (ERE) [3]. When there is a shallow S-wave bound state with binding energy $B_2 \ll (mR^2)^{-1}$, the scattering length is large, $|a_2| \approx (mB_2)^{-1/2} \gg R$, while the effective range, shape parameter and higher ERE parameters have natural size, $|r_2| \sim |P_2| \sim \ldots \sim R$. The unitarity limit, in which $a_2^{-1} = r_2 = P_2 = \ldots = 0$, is an idealization of what I call the "unitarity window",

$$T_2(a_2^{-1} \ll k \ll R^{-1}) = \frac{4\pi}{m}(ik)^{-1}\left[1 + \mathcal{O}\left(kR, (ka_2)^{-1}, \alpha m k^{-1}\right)\right], \quad (1)$$

where I allowed for electromagnetic effects $\alpha m \lesssim a_2^{-1}$, with α the fine-structure constant. In EFT, the expansion of T_2 in kR is rephrased as a more general expansion of *all* amplitudes in powers of $M_{\rm lo}/M_{\rm hi} < 1$, the ratio between the typical momentum of interest and the theory's breakdown scale. The $(ka_2)^{-1}$ expansion is an additional expansion around the unitarity limit in powers of $\aleph/M_{\rm lo} < 1$, where $\aleph$ denotes a fine-tuned scale associated with the small two-body binding momentum.

$A \geq 3$ ground states, whose average binding momentum is identified with $M_{\rm lo}$, fall within the unitarity window, $\aleph \ll Q_{A\geq 3} \ll M_{\rm hi}$. An estimate of the typical binding momentum is

$$Q_A \sim \sqrt{2mB_A/A}, \quad (2)$$

since in this case each particle contributes an equal amount $-Q_A^2/(2m)$ to the total energy, while Q_2 reproduces the pole location in the complex relative-momentum plane. For nuclear systems, where the range of the interaction is given by the pion mass, $R \sim m_\pi^{-1} \simeq 1.4$ fm, we find $Q_A/m_\pi = 0.3, 0.5, 0.8, 0.7, 0.7, \ldots, 0.9, \ldots$ for $A = 2, 3, 4, 5, 6, \ldots, 56, \ldots$. An expansion in $M_{\rm lo}/M_{\rm hi} \sim 0.7$ should not be quickly convergent but is useful if we can calculate to many orders. (In fact, we have some indirect evidence that the expansion parameter is smaller than this.) The unitarity expansion [1] is likely no worse, since $\aleph_1/M_{\rm lo} \sim 0.45$ if we take Q_2, or alternatively the isospin-singlet inverse scattering length, as representative of the $^3S_1$ scale $\aleph_1$. A more limited, $^1S_0$ unitarity expansion [4, 5] converges much faster because $\aleph_0/M_{\rm lo} \sim 0.05$, using the middle component ($I_3 = 0$) of the isospin-triplet inverse scattering length as representative of the $^1S_0$ scale $\aleph_0$. Isospin splitting does not spoil this expansion because the difference between $I_3 = +1$ and $I_3 = 0$ inverse scattering lengths is $\sim \alpha m \sim \aleph_0$, while that between $I_3 = -1$ and $I_3 = 0$ is $\sim m_d - m_u < \aleph_0$, where m_d (m_u) is the down (up) quark mass.

We do not have a precise explanation for why the *NN* system is close to unitarity. A possibility [6] is that the physical quark masses happen to be near the critical values where the $^1S_0$ virtual state (deuteron) goes (un)bound, similarly to cold atoms around a Feshbach resonance produced by an external magnetic field. Whatever the reason, it looks like nuclear systems are in the sweet spot of the unitarity window. The estimates above are admittedly very crude; a more convincing case is made by explicit calculations, which I describe in the following.

3. Pionless EFT

EFT is the framework to generate the most general S matrix consistent with assumed symmetries. In the case of QCD, we are concerned with Lorentz, parity, time-reversal and baryon-number symmetries —breaking of all of which can be accounted for when/if desired— and color $SU(3)_c$ and electromagnetic $U(1)_{em}$ gauge invariance. At the energies of interest, the relevant degrees of freedom are nucleons whose annihilation is encoded in a color-singlet, isospin-doublet, two-component spinor field ψ, for which Lorentz invariance can be implemented through a k/m_N expansion. The most general action is

$$S = \int \frac{dt\, d^3r}{2m} \left\{ \psi^\dagger \left(2im\frac{\partial}{\partial t} + \nabla^2 \right) \psi - 4\pi \sum_{I=0,1} C_{0I}\, \psi^\dagger \psi^\dagger P_I \psi\psi - \frac{(4\pi)^2}{3} D_0\, \psi^\dagger\psi^\dagger\psi^\dagger\psi\psi\psi + \ldots \right\}, \tag{3}$$

where P_I stands for a projector onto isospin $I = 0, 1$, and C_{0I}, D_0, *etc.* are "low-energy constants" (LECs). The "…" include terms with more derivatives, fields, and powers of $e = \sqrt{4\pi\alpha}$ and $m_d - m_u$. The mild constraints of QCD symmetries in this energy regime imply that essentially the same EFT applies also to very different systems, such as atoms at distances larger than the van der Waals length l_{vdW} defined from the coordinate-space potential $V(r) = -l_{vdW}^4/(2mr^6)$. For ^{4}He atoms, for example, ψ is a bosonic field representing the annihilation of an atom and there is only one C_{0I} interaction, with $P_I = 1$.

The action involves singular interactions which need to be regularized as in any other quantum field theory. The momentum regulator parameter Λ arbitrarily splits short-range physics between loops and LECs. Renormalization is the procedure that fixes the cutoff dependence of the LECs in such a way that observables are independent of the arbitrary regularization, up to terms that go as inverse powers of Λ. At any given order, this residual cutoff dependence can be made as small as higher-order terms by taking $\Lambda \gtrsim M_{hi}$. Failure to eliminate non-negative powers of Λ for one choice of regulator implies model dependence on the form of the regulator function and signals the omission of important interactions at that order.

The EFT's non-relativistic character simplifies its solution tremendously. Unless they are present in initial or final states, antiparticles do not appear explicitly, and the A-body system can be solved without worrying about systems with more bodies.

4. $A = 2$

The order-by-order solution of the two-body system is straightforward [7]. At LO it consists of the sum of all ladder diagrams with the LO components of C_{0I} interactions. Renormalization at unitarity requires the latter to be $C_{0I}^{(0)}(\Lambda) = -(\theta_0\Lambda)^{-1}$, where $\theta_0 = \mathcal{O}(1)$ is a number that depends on the specific form of the regulator. The LO amplitude is the first term in Eq. (1). The first two terms in the action (3) are invariant under continuous scale transformations,

$$r \to \alpha r, \qquad t/m \to \alpha^2 t/m, \qquad \Lambda \to \alpha^{-1}\Lambda, \qquad \psi \to \alpha^{-3/2}\psi, \tag{4}$$

with a parameter $\alpha > 0$. The system is also conformally invariant [8] and Wigner $SU(4)_W$-symmetric [9].

Higher-order terms in Eq. (1) arise from perturbative insertions of higher-order interactions. At NLO, which consists of one insertion of NLO interactions, scale invariance is broken explicitly by $(ka_2)^{-1}$, kr_2 and αmk^{-1} effects. The scattering lengths enter through the NLO components of C_{0I}, $C_{0I}^{(1)}$, in the combination $(a_2\Lambda)^{-1}$. The effective ranges appear through two-derivative, four-field interactions in the "…" of Eq. (3). Electromagnetic effects arise from (long-range) Coulomb photon exchange stemming from the gauge-covariant derivatives also not written explicitly in Eq. (3), and from an isospin-breaking contact interaction needed for renormalization, which accounts for the splitting between $I_3 = +1$ and $I_3 = 0$ isotriplet inverse scattering lengths. At this order the predictions, apart from the improved energy dependence of $T_2(k)$, are the equality of $I_3 = -1$ and $I_3 = 0$ isotriplet scattering lengths and of all isotriplet effective ranges, which is reasonably satisfied experimentally. Higher orders include isospin breaking from $m_d - m_u$. For more details, see Refs. [4, 1, 5].

5. $A = 3$

The three-body problem in Pionless EFT also requires an exact solution with the LO two-body interactions. The EFT is still renormalizable, but only with the D_0 interaction in Eq. (3) [10, 11, 12]. With a sharp-cutoff regulator at unitarity,

$$D_0^{(0)}(\Lambda) \propto \frac{1}{\Lambda^4} \frac{\sin\Big(s_0 \ln(\Lambda/\Lambda_\star) - \arctan s_0^{-1}\Big)}{\sin\Big(s_0 \ln(\Lambda/\Lambda_\star) + \arctan s_0^{-1}\Big)}, \tag{5}$$

where $s_0 \simeq 1.00624$ and $\Lambda_\star$ is a physical parameter, which is however only defined up to a factor $\exp(n\pi/s_0) \simeq (22.7)^n$, with n an integer. This is a renormalization-group (RG) limit cycle.

The appearance at LO of a single dimensionful parameter $\Lambda_\star$ breaks continuous scale invariance down to a discrete subgroup with $\alpha \to \alpha_n \equiv \exp(n\pi/s_0)$, as can be verified directly from Eqs. (3) and (5). DSI manifests itself in a geometric tower of bound "Efimov states" [13], with the tower position determined by $\Lambda_\star$. The tower is truncated from below, with states more bound than the physical ground state falling outside the regime of validity of the EFT. In the nuclear case we can fix $\Lambda_\star$ from the triton binding energy B_t (equal to the helion energy $B_h^{(0)}$ at this order). Three-body observables are correlated so that variation in $\Lambda_\star$ leads to a curve on the plane of two observables, such as the empirical "Phillips line" [14] relating B_t and the spin-doublet neutron-deuteron (*nd*) scattering length. The 3BF is $SU(4)_W$-symmetric [12, 15].

At NLO, the amplitude is renormalized by a shift $D_0^{(1)}$ in the 3BF LEC, which can be determined by demanding that B_t does not change. Isospin-breaking *NN* interactions split helion from triton and we can predict [1]

$$B_h^{(1)} - B_t = -(0.92 \pm 0.18) \text{ MeV}, \tag{6}$$

to be compared with the experimental value, -0.764 MeV. Subleading corrections affect the shallow Efimov states, which are sensitive to the finite values of the scattering lengths. Indeed, a virtual state in *nd* scattering becomes the first triton excited state as Q_2 is decreased [16].

6. $A = 4$

For the various regulators that have been studied, the four-body system seems to be renormalizable in Pionless EFT without a four-body force in LO [17, 18, 19, 20, 21, 22, 23]. As a consequence, DSI should be approximately valid. Indeed, for bosons two towers of bound states have been identified [19], where each Efimov state spawns two four-body states: one $\simeq 1.002$ times deeper than the trimer-particle threshold, the other $\simeq 4.6$ times deeper [24]. The correlation with the three-body tower manifests itself in a line in the space of $A = 3, 4$ observables produced by changing $\Lambda_\star$, such as the empirical "Tjon line" [25] on the plane of ground-state energies.

There has been no $A = 4$ calculation where the full NLO interaction is accounted for in perturbation theory. Existing calculations [20, 26] treat NLO interactions nonperturbatively, which precludes a study of renormalizability.

For nucleons at unitarity [1], we have considered LO and an "incomplete NLO" where only $(ka_2)^{-1}$ corrections were included in first-order perturbation theory. We indeed observe convergence with the cutoff in both cases. At LO the alpha-particle binding energy $B_\alpha^{(0)}$ is consistent with that for four bosons, while at incomplete NLO

$$B_\alpha^{(1,\mathrm{inc})} = 29.5 \pm 8.7 \text{ MeV}, \tag{7}$$

to be compared with the experimental value, 28.4 MeV. We also observed an excited state close to the triton-nucleon threshold. Thus, an expansion around unitarity captures the basic elements of the four-nucleon system.

7. $A > 4$

We have recently shown that no higher-body force appears at LO, either: for $A = 5, 6$ bosons, ground-state energies converge with the cutoff [22], and the same seems true for nucleons [27], as expected on the basis of the further suppression of many-body contact forces coming from the derivatives required by the exclusion principle. This has far-reaching consequences. First, at unitarity the same parameter $\Lambda_\star$, which can be traded for the three-body binding energy B_3, determines the spectrum of all systems within the range of applicability of Pionless EFT, generating "generalized Tjon lines" [22]. For ground states [28, 29, 22], for example,

$$\frac{B_A}{A} = \kappa_A \frac{B_3}{3}, \tag{8}$$

where κ_A is a set of universal numbers, which however depend on whether the particles are bosons or fermions. Second the spectrum should display DSI in the form of geometric towers, just as for $A = 4$, which indeed has been demonstrated explicitly for bosons [28, 30, 31, 32].

Moreover, using potential-model results for $A = 2, 3$ ^{4}He atoms in order to determine the LECs, LO EFT works well for $A = 4, 5, 6$ [22]. This suggests that small ^{4}He-atom clusters are within the range of applicability of the EFT, and are thus universal: they depend essentially on two parameters, which one can take as the energies of the dimer and one trimer state. This is in spite of the fact that binding energies grow rapidly, $\kappa_A \simeq 3(A-2)^2/A$ [22]. The growth tapers off, and a liquid-drop extrapolation of potential-model results gives saturation with $\lim_{A\to\infty} \kappa_A \simeq 180$ [33]. It seems that large ^{4}He-atom clusters are not universal [34]. In contrast, at unitarity bosonic clusters saturate with $\lim_{A\to\infty} \kappa_A \simeq 90$ [35]. The corresponding distance remains larger than that associated with the next-deeper trimer, which is beyond the validity of the EFT. In this sense, unitary bosonic clusters are universal.

Since for nucleons binding energies grow much more slowly, we have speculated [1] that also larger nuclear systems can be described in a first approximation in terms of $\Lambda_\star$ alone. If that is true we might expect DSI to be relevant for nuclear matter as well, in which case at LO [36]

$$\lim_{A\to\infty} \frac{B_A(\rho)}{A} = -\left(\frac{3\pi^2}{2}\right)^{2/3} \frac{\rho^{2/3}}{m}\, \Upsilon\left(\frac{s_0}{3} \ln\left(\frac{3\pi^2\rho}{2\Lambda_\star^3}\right)\right), \tag{9}$$

where ρ is the particle density and Υ is a (real) periodic function with period π, which can be expanded in a Fourier series in terms of real coefficients $\gamma_{n\geq 0}$ and phases $\phi_{n\geq 1}$,

$$\Upsilon\left(x + l\pi\right) = \Upsilon\left(x\right) = \frac{3\gamma_0}{10} - \sum_{n=1}^{\infty} \gamma_n \sin\left(2nx + \phi_n\right). \tag{10}$$

The minimum of Eq. (9) describes a quadratic curve under variation of $\Lambda_\star$, just as the empirical "Coester line" [37]. A truncation of the Fourier series in Eq. (10) can describe empirical properties of nuclear matter around the minimum. As before, the two-body scattering lengths and effective ranges enter at NLO.

IOP Conf. Series: Journal of Physics: Conf. Series **966** (2018) 012014 doi:10.1088/1742-6596/966/1/012014

8. Conclusion

The answer to the opening question, if we are right, is: a single-parameter three-body force needed for renormalization-group invariance around the unitarity limit. Evidence is that this works for $A = 3, 4$ systems. There is a long way to go, though, to show that this framework works also for larger nuclei. I am already looking forward to reporting new results at the next Ischia meeting!

Acknowledgments

I thank Angela Gargano, Nunzio Itaco and especially Luigi Coraggio for the great hospitality at Ischia. Many thanks also to Harald Grießhammer, Hans-Werner Hammer and Sebastian König for the collaboration that led to the main ideas summarized above. This material is based upon work supported in part by the U.S. Department of Energy, Office of Science, Office of Nuclear Physics, under award DE-FG02-04ER41338 and by the European Union Research and Innovation program Horizon 2020 under grant No. 654002.

References

[1] König S, Grießhammer H W, Hammer H-W and van Kolck U 2017 *Phys. Rev. Lett.* **118** 202501
[2] Bedaque P F and van Kolck U 2002 *Ann. Rev. Nucl. Part. Sci.* **52** 339
[3] Bethe H A 1949 *Phys. Rev.* **76** 38
[4] König S, Grießhammer H W, Hammer H-W and van Kolck U 2016 *J. Phys.* G **43** 055106
[5] König S 2017 *J. Phys.* G **44** 064007
[6] Beane S R, Bedaque P F, Savage M J and van Kolck U 2002 *Nucl. Phys.* A **700** 377
[7] van Kolck U 1999 *Nucl. Phys.* A **645** 273
[8] Mehen T, Stewart I W and Wise M B 2000 *Phys. Lett.* B **474** 145
[9] Mehen T, Stewart I W and Wise M B 1999 *Phys. Rev. Lett.* **83** 931
[10] Bedaque P F, Hammer H-W and van Kolck U 1999 *Phys. Rev. Lett.* **82** 463
[11] Bedaque P F, Hammer H-W and van Kolck U 1999 *Nucl. Phys.* A **646** 444
[12] Bedaque P F, Hammer H-W and van Kolck U 2000 *Nucl. Phys.* A **676** 357
[13] Efimov V 1970 *Phys. Lett.* **33B** 563
[14] Phillips A C 1968 *Nucl. Phys.* A **107** 209
[15] Vanasse J and Phillips D R 2017 *Few-Body Syst.* **58** 26
[16] Adhikari S K and Tomio L 1982 *Phys. Rev.* C **26** 83
[17] Platter L, Hammer H-W and Meißner U-G 2004 *Phys. Rev.* A **70** 052101
[18] Platter L, Hammer H-W and Meißner U-G 2005 *Phys. Lett.* B **607** 254
[19] Hammer H-W and Platter L 2007 *Eur. Phys. J.* A **32** 113
[20] Kirscher J, Grießhammer H W, Shukla D and Hofmann H M 2010 *Eur. Phys. J.* A **44** 239
[21] Kirscher J, Barnea N, Gazit D, Pederiva F and van Kolck U 2015 *Phys. Rev.* C **92** 054002
[22] Bazak B, Eliyahu M and van Kolck U 2016 *Phys. Rev.* A **94** 052502
[23] Contessi L, Lovato A, Pederiva F, Roggero A, Kirscher J and van Kolck U 2017 *Phys. Lett.* B **772** 839
[24] Deltuva A 2010 *Phys. Rev.* A **82** 040701
[25] Tjon J A 1975 *Phys. Lett.* B **56** 217
[26] Lensky V, Birse M C and Walet N R 2016 *Phys. Rev.* C **94** 034003
[27] Stetcu I, Barrett B R and van Kolck U 2007 *Phys. Lett.* B **653** 358
[28] von Stecher J 2010 *J. Phys.* B **43** 101002
[29] Nicholson A N 2012 *Phys. Rev. Lett.* **109** 073003
[30] Gattobigio M, Kievsky A and Viviani M 2011 *Phys. Rev.* A **84** 052503
[31] von Stecher J 2011 *Phys. Rev. Lett.* **107** 200402
[32] Gattobigio M, Kievsky A and Viviani M 2012 *Phys. Rev.* A **86** 042513
[33] Pandharipande V R, Zabolitzky J G, Pieper S C, Wiringa R B and Helmbrecht U 1983 *Phys. Rev. Lett.* **50** 1676
[34] Kievsky A, Polls A, Juliá Díaz B J and Timofeyuk N K 2017 *Phys. Rev.* A **96** (2017) 040501
[35] Carlson J, Gandolfi S, van Kolck U and Vitiello S A 2017 Ground-state properties of unitary bosons: from clusters to matter *Preprint* arXiv:1707.08546 [cond-mat.quant-gas]
[36] van Kolck U 2017 *Few Body Syst.* **58** 112
[37] Coester F, Cohen S, Day B and Vincent C M 1970 *Phys. Rev.* C **1** 769

IOP Conf. Series: Journal of Physics: Conf. Series **966** (2018) 012015 doi:10.1088/1742-6596/966/1/012015

Recent Applications of Self-Consistent Green's Function Theory to Nuclei

Carlo Barbieri, Francesco Raimondi and Christopher McIlroy
Department of Physics, University of Surrey, Guildford GU2 7XH, United Kingdom
E-mail: C.Barbieri@surrey.ac.uk

Abstract. We discuss recent *ab initio* calculations based on self-consistent Green's function theory. It is found that a simple extension of the formalism to account for two-nucleon scattering outside the model space allows to calculate non-soft interactions. With this, it is possible to make predictions for Lattice QCD potentials, obtained so far at pion masses of $m_\pi = 0.47$ GeV/c^2. More traditional calculations that use saturating chiral EFT forces yield a good description of nuclear responses and nucleon knockout spectroscopy.

Ab initio nuclear theory has seen remarkable advances in the last 15 years. These resulted from the combination of improved theories of realistic two-nucleon (NN) and three-nucleon forces (3NFs) and of advances in many-body computations that have reached nuclear masses of the order of A$\sim$100 [1].

The self consistent Green's function (SCGF) approach is based on the one-body propagator, $g(\omega)$, that describes the evolution of particle and hole states in (i.e., nucleon attached or removed to/from) the exact ground state [2]. The propagator is the solution of the Dyson equation:

$$g_{\alpha\beta}(\omega) = g^{(0)}_{\alpha\beta}(\omega) + \sum_{\gamma,\delta} g^{(0)}_{\alpha\gamma}(\omega)\left[\Sigma^{(\infty)}_{\gamma\delta} + \tilde{\Sigma}_{\gamma\delta}(\omega)\right] g_{\delta\beta}(\omega) , \tag{1}$$

where the indices $\alpha, \beta, \ldots$ label the states of a single-particle model space basis and $g^{(0)}(\omega)$ is the uncorrelated propagator, which correspond to a mean field (MF) reference state. The central quantity is the irreducible self-energy, $\Sigma^\star(\omega) \equiv \Sigma^{(\infty)} + \tilde{\Sigma}(\omega)$, which acts as an optical potential for nucleons inside the correlated medium. One separates an enegy-independent term $\Sigma^{(\infty)}$ that describes the average MF potential felt by all particles. The dynamic part $\tilde{\Sigma}(\omega)$ contains the effect of correlations beyond MF. We calculate this in the third-order algebraic diagrammatic construction [ADC(3)] approximation, which is a *non perturbative* resummation of $2p1h$ and $2h1p$ configurations [3]. We generally construct the self-energy based on a reference MF state, which is derived to best approximate $g(\omega)$ and, therefore, solve Eq. (1) iteratively. The interested reader is referred to Refs. [3, 4, 5] for details of the formalism and of our implementation.

Once the propagator is known, it is easy to extract information on the ground state energy, nucleon-nucleus scattering, single-particle spectroscopy, response to external probes, and so on [3]. In the following we describe a few very recent applications based on nuclear forces obtained either directly from Lattice QCD calculations or from chiral effective field theory (EFT).

1. Taming the hard-core of forces from Lattice QCD

The HAL QCD collaboration has devised an approach to extract n-nucleon forces from QCD simulation of $3n$ quarks in a space-time lattice [6, 7, 8]. This method generates systematically consistent two-,

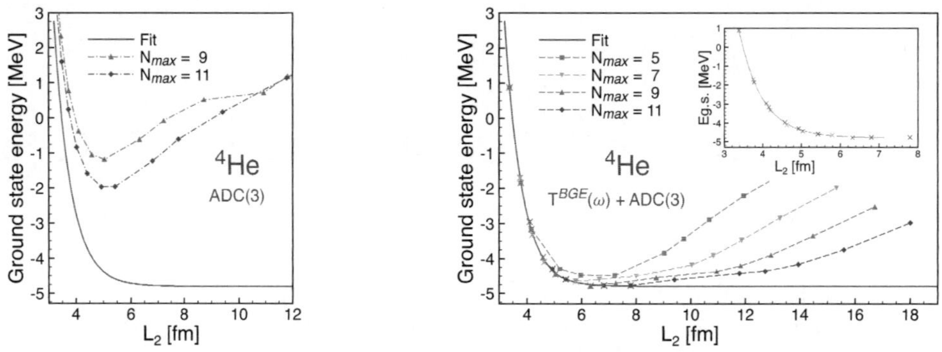

Figure 1. Calculated ground state energies of ^{4}He for the HAL469 potential as a function of the effective radius L_2. *Left*: Solution for the bare interaction at N_{max}=9 and 11 and varying oscillator frequencies without ladders from the excluded space. *Right*: Full calculation, including all ladder diagrams from outside the model space. Different colors and broken lines are a guide to the eye connecting results of the same N_{max}. The data points included in the fit are marked with crosses and are also shown separately in the inset. For *both panels*, the full black line is the result of the IR extrapolation, with the inclusion of T^{BGE} ladders.

three- and many-nucleon interactions that are faithful to the few-body data and scattering phase shifts by construction. An earlier set of potentials was derived in the flavour SU(3) limit with different masses of the (pion) pseudo-scalar meson. Among these the force with lightest value of M_{PS} = 469 MeV/c^2 shows saturation on nuclear matter [9]. We refer to this as the HAL469$_{SU(3)}$ interaction. Note that more advanced interactions at near the physical pion mass are currently being computed [10].

For forces with a strong short-range repulsion, like the HAL QCD interactions, usual truncations of the oscillator space (of up to 12 shells in this case) are not sufficient and a resummation of ladder diagrams outside the model space is required. We do this by solving the Bethe-Goldstone Equation (BGE) in the excluded space according to Refs. [11, 12] and add the corresponding diagrams to the mean field (MF) term of the self-energy, which becomes energy dependent [4]:

$$\Sigma^{(\infty)}_{\alpha\beta}(\omega) = \sum_{\gamma\delta} \int \frac{d\omega'}{2\pi i} T^{BGE}_{\alpha\gamma,\beta\delta}(\omega + \omega')\, g_{\delta\gamma}(\omega')\, e^{i\omega'\eta} \tag{2}$$

where $T^{BGE}_{\alpha\gamma,\beta\delta}(\omega)$ are the elements of the scattering t-matrix in the excluded space. A static effective interaction is then extracted that we use to calculate the ADC(3) self-energy within the included model space. To do this, we solve the Hartree-Fock (HF) equations with the MF potential of Eq. (2):

$$\sum_{\beta} \left\{ \langle\alpha|\frac{p^2}{2m}|\beta\rangle + \Sigma^{(\infty)}_{\alpha\beta}(\omega = \varepsilon^{HF}_r) \right\} \psi^r_\beta = \varepsilon^{HF}_r\, \psi^r_\alpha \,, \tag{3}$$

where latin indices label HF states. We then define a static interaction in this HF basis similarly to Refs. [4, 13]:

$$V_{rs,pq} = \frac{1}{2}\left[T^{BGE}_{rs,pq}(\varepsilon^{HF}_r + \varepsilon^{HF}_s) + T^{BGE}_{rs,pq}(\varepsilon^{HF}_p + \varepsilon^{HF}_q) \right] . \tag{4}$$

It should be noted that the BGE used to generate $T^{BGE}(\omega)$ resums scattering states where at least one nucleon is outside the whole model space, while full ADC(3) correlations are computed for all nucleons inside the space. Hence, $T^{BGE}(\omega)$ does not suffer from ambiguities with the choice of the single-particle spectrum at the Fermi surface encountered with the usual G-matrix used in Brueckner HF calculations.

The combination of the ladder diagrams contained in the ADC(3) expansion and those in the $T^{BGE}(\omega)$ accounts for the complete diagonalization of short-distance degrees of freedom. One can then study the

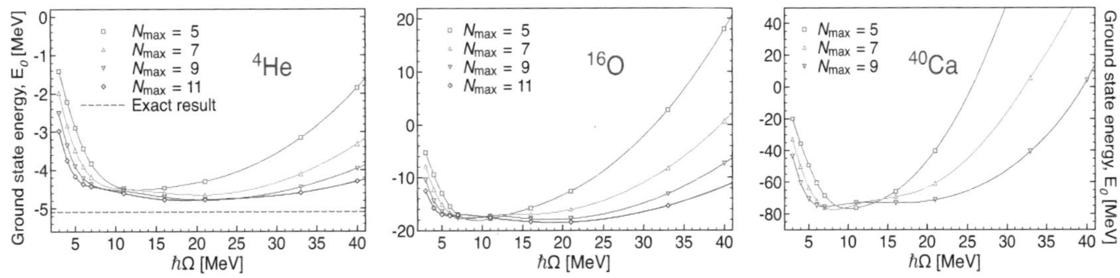

Figure 2. Ground state energy of ^{4}He, ^{16}O and ^{40}Ca as a function of the harmonic oscillator frequency, $\hbar\Omega$, and the model space size, N_{max}. All results are for the HAL469 potential from full self-consistent calculations in the $T^{BGE}(\omega)$ plus ADC(3) approach.

infrared (IR) convergence of total binding energies to a complete basis set. Ref. [14] established that a harmonic oscillator (HO) model space, of frequency $\hbar\Omega$ and truncated to the first $N_{\text{max}} + 1$ shells, behaves as a hard wall spherical box of radius

$$L_2 = \sqrt{2(N_{\text{max}} + 3/2 + 2)}\, b\,, \tag{5}$$

where $b \equiv \sqrt{\hbar c^2/m_N\Omega}$ is the oscillator length ($\hbar$=c=1). Given an interaction that is independent of the model space, if the ultraviolet (UV) degrees of freedom are diagonalised exactly then the calculated ground state energies are expected to converge exponentially when increasing the effective radius L_2:

$$E_0^A[N_{\text{max}}, \hbar\Omega] = E_\infty \;+\; C\, e^{-2k_\infty L_2}. \tag{6}$$

For the bare HAL469 interaction, if we use the SCGF without ladder diagrams from outside the model space, the extrapolation according to Eq. (6) will fail because the short-distance repulsion requires extremely large model spaces (N_{max} >> 20) to reach UV convergence, see the left panel of Figure 1. The results using the interaction (4) and including $T^{BGE}(\omega)$ are displayed in the right panel and show good IR convergence. The total energy is extracted from a nonlinear least-squares fit to Eq. (6).

The one-body propagators of ^{4}He, ^{16}O and ^{40}Ca have been calculated using $T^{BGE}(\omega)$ to remove UV modes. We employed spherical harmonic oscillator spaces of different frequencies, $\hbar\Omega$, and sizes up to N_{max}=max$\{2n + \ell\}$ =11 (and $N_{\text{max}} \leq 9$ for ^{40}Ca) and we used the Koltun sum rule to extract the intrinsic ground state energies from $g(\omega)$. Fig. 2 shows that the complete resummation of ladder diagrams outside the model space results in a somewhat flat behaviour of the total energies for $\hbar\Omega \approx$5-20 MeV. While some oscillations w.r.t. $\hbar\Omega$ are still present, the IR convergence pattern (shown in Fig. 1 for ^{4}He) is very clean for all three isotopes. This gives confidence that the short-range repulsion of HAL469 is relatively mild and that it is accounted for accurately. From calculations up to $\hbar\Omega$=50 MeV and the IR extrapolation, we estimate a converged binding energy of 4.80(3) MeV for ^{4}He, where the error corresponds to the uncertainties in the extrapolation. This is to be benchmarked on the *exact* result of -5.09 MeV, which is known from Stochastic Variational calculations [15]. Since the method is size extensive, we assume a 10% error due to many-body truncations for all isotopes, as a conservative estimate.

We obtain -17.9(0.3)(1.8) MeV for the ground state energy of ^{16}O, where the first error is from the IR extrapolation. Hence, ^{16}O is unstable with respect to 4-α break up, by $\approx$2.5 MeV, although allowing an error in our binding energies of more than 10% could make it slightly bound. This is in contrast to the experiment, at the physical quark masses, where the 4-α breakup requires 14.4 MeV. On the other hand, ^{40}Ca we calculate -75.4(6.7)(7.5) MeV and this is stable with respect to α breakup by $\approx$24 MeV.

IOP Conf. Series: Journal of Physics: Conf. Series **966** (2018) 012015 doi:10.1088/1742-6596/966/1/012015

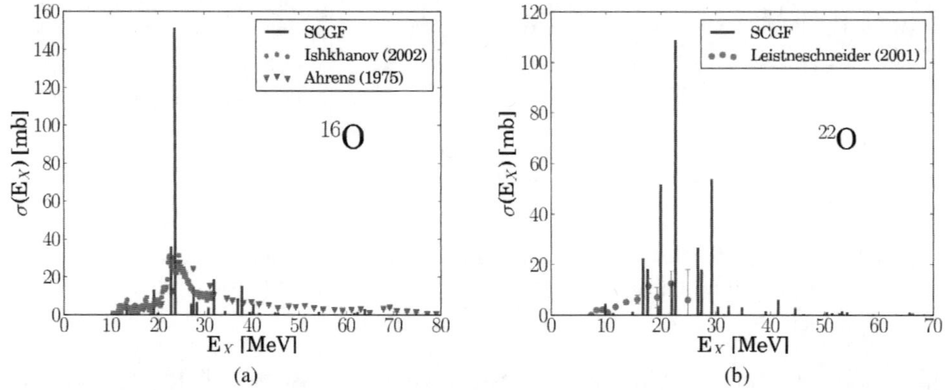

Figure 3. Computed total cross section for ^{16}O (a) and ^{22}O (b) compared to experimental data obtained from photoabsorption [16, 17] and electromagnetic excitation [18] experiments, respectively. The size of the HO basis in the calculation is $N_{\rm max} = 13$ (i.e., 14 major shells) and frequency $\hbar\Omega = 20$ MeV. The interaction used is NNLO$_{\rm sat}$.

2. Isovector Dipole Nuclear Response

In this Section we focus on the ^{16}O, ^{22}O, ^{40}Ca and ^{48}Ca nuclear response produced by an isovector dipole electric field, E1. The corresponding operator, corrected for the center-of-mass displacement, is:

$$\hat{Q}^{T=1}_{1m} = \frac{N}{N+Z}\sum_{p=1}^{Z} r_p Y_{1m} - \frac{Z}{N+Z}\sum_{n=1}^{N} r_n Y_{1m}\,, \tag{7}$$

which probes the excitation spectrum with multipolarity and parity J^π=1$^-$. We calculate the response function $R(E)$ of Eq. (7) by solving the usual random phase approximation (RPA) in the particle-hole channel and starting from the same MF propagator that is used as reference state to obtain the self-energy in previous sections. As mentioned in the introduction, this includes effects of correlations that go beyond the usual HF mean field and, in particular, those responsible for reproducing the correct centroid of giant resonances. We then compute the total photoabsorption cross section as

$$\sigma(E) = 4\pi^2\alpha\, E\, R(E) \tag{8}$$

and the dipole polarizability, which is the accumulated E1 strength weighted by the inverse of the energy:

$$\alpha_{\rm D} = 2\alpha \int_{E_{thr}}^{+\infty} dE\, \frac{R(E)}{E}\,. \tag{9}$$

In the above relations, α denotes the fine-structure constant.

Our results for the total cross sections for ^{16}O, ^{22}O, ^{40}Ca and ^{48}Ca are compared with photoabsorption and Coulomb excitation experiments in Figs. 3 and 4. The computed spectrum is collected into peaks since we diagonalize the RPA matrix in a finite model space. We performed calculations using the

Nucleus	SCGF	CC-LIT	Exp
^{16}O	0.50	0.57(1)	0.585(9)
^{22}O	0.72	0.86(4)	0.43(4)
^{40}Ca	1.79	1.87	1.87(3)
^{48}Ca	2.08	2.45	2.07(22)

Table 1. Isovector dipole polarizabilities $\alpha_{\rm D}$ for ^{16}O, ^{22}O, ^{40}Ca and ^{48}Ca, computed in the SCGF approach (second column), the CC-LIT approach (third column) and obtained by integrating the experimental data (fourth column). The errors in CC-LIT are estimated form the dependence on the oscillator frequency $\hbar\Omega$ of model space [19, 20].

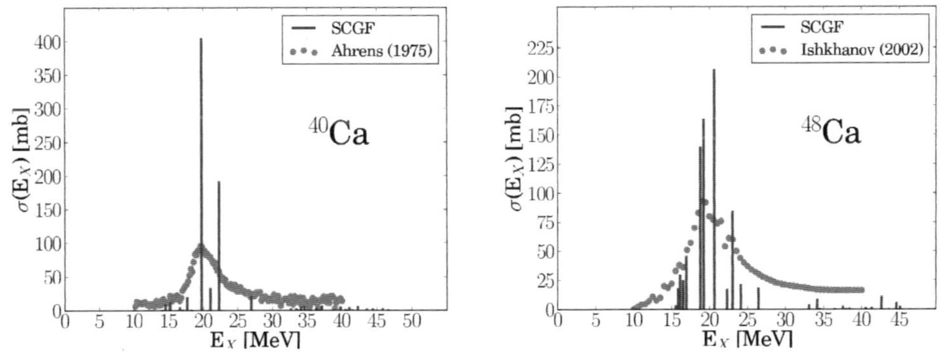

(a) (b)

Figure 4. Same as Fig. 3 but for ^{40}Ca (a) and ^{48}Ca (b). Experimental data are taken from Ahrens *et al* [17] and Ishkhanov *et al* [16] respectively.

NNLO$_{\rm sat}$ chiral interaction that reproduces radii in this region and has good saturation properties [21]. For all isotopes considered, the position of the giant dipole resonance is fairly well reproduced.

By moving from the N=Z nucleus ^{16}O to the neutron-rich ^{22}O, the strength of the response is distributed towards lower energies, with the first RPA peak shifted to energy below 10 MeV. In the corresponding experimental curve, there appears a soft dipole mode of excitation, with a weaker strength compared to the giant resonance at higher energies.

In Table 1, dipole polarizabilities $\alpha_{\rm D}$ of Eq. (9) are compared with values obtained from the integrated experimental spectrum and from a Coupled Cluster combined with the Lorentz Integral Transform (CC-LIT) approach calculation [19, 20]. For the ^{16}O, where the experimental dipole polarizability is fully integrated over the entire spectrum, our computed $\alpha_{\rm D}$ lacks 15% of the strength with respect to both the experimental value and result of the CC-LIT approach. The total strength in the SCGF calculation is better recovered for higher mass ^{40}Ca and ^{48}Ca, and even within the experimental error for ^{48}Ca. Correcting the discrepancies with CC-LIT may require improvements of the many-body truncation to go beyond the RPA and we are currently working to implement such extensions.

3. Spectroscopic factors

The fragmentation of the single-particle spectral function is mostly determined by long-range correlations (LRC) and it is constrained by general properties at the energy surface, such as the density of states and gaps [22]. As an example, Fig. 5 demonstrates the dependence of spectroscopic factors (SFs) for the dominant quasiparticle peaks in ^{56}Ni on the particle-hole gap ΔE_{ph} (which has an experimental value of 6.1 MeV). This is calculated from a chiral N3LO two-nucleon interaction only but with a modification of its monopole strength in the pf shell. The monopole correction controls the separation among the $p_{3/2}$ and the $f_{7/2}$ orbits. Hence, by varying its strength one can control the predictions for ΔE_{ph}. Calculations for spaces of different sizes and HO frequencies are not converged w.r.t. the model space but all lines lay on top of each other, showing that there exists a clear correlation and that SFs can be strongly constrained by observable quantities at low energy.

Since the correlation of SFs with ΔE_{ph} is strong, we calculated the oxygen chain using NNLOsat Hamiltonian of Ref. [21], which has the advantage of predicting gaps accurately. The spectroscopic factors obtained in the ADC(3) scheme are shown in the right panel of Fig. 5 and are sensibly smaller than previous results from older Hamiltonians that had a too dilute spectrum at the Fermi surface [23]. Remarkably, the NNLOsat results are very close (almost equal) to past FRPA calculations of ^{16}O where the quasiparticle energies were phenomenologically constrained to their experimental values [24]. These are also in good agreement with $(p,2p)$ measurements from R^{3}B at GSI [25].

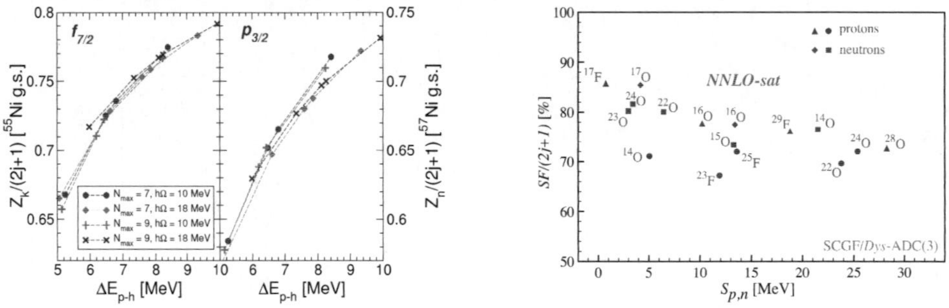

Figure 5. *Left.* Dependence of spectroscopic factors of the first quasiparticle and quasihole peaks in ^{56}Ni as a function of the particle-hole gap ΔE_{ph} [4]. The calculated quenching is strongly correlated to the gap even when full convergence with respect to the model space is still not achieved. Note that $\Delta E_{ph} \equiv E_0^{A+1} - E_0^{A-1}$ is an experimentally observable quantity. Here, it is varied over a range of values by tuning the monopoles of the NN interaction. *Right.* Calculated spectroscopic factors for protons and neutrons around Oxygen isotopes obtained from NNLOsat. Each point refer to the separation of a nucleon from the isotope indicated nearby to the ground state of the daughter nucleus. Likewise, stronger quenching is found for those isotopes with smaller ΔE_{ph} gap.

Acknowledgments

We thank the HAL QCD collaboration for providing the HAL469$_{SU(3)}$ interaction. This work was supported in part by the United Kingdom Science and Technology Facilities Council (STFC) under Grants No. ST/L005743/1 and ST/L005816/1 and by the Royal Society International Exchanges Grant No. IE150305. Calculations were performed at the DiRAC Complexity system at the University of Leicester (BIS National E- infrastructure capital Grant No. ST/K000373/1 and STFC Grant No. ST/K0003259/1).

References

[1] Hebeler K, Holt J, Menéndez J and Schwenk A 2015 *Annual Review of Nuclear and Particle Science* **65** 457
[2] Dickhoff W and Barbieri C 2004 *Progress in Particle and Nuclear Physics* **52** 377
[3] Barbieri C and Carbone A 2017 *Self-Consistent Green's Function Approaches* (*Lect. Notes Phys.* vol 936) (Cham: Springer International Publishing) chap 11, pp 571–644
[4] Barbieri C and Hjorth-Jensen M 2009 *Phys. Rev. C* **79** 064313
[5] Somà V, Barbieri C and Duguet T 2014 *Phys. Rev. C* **89** 024323
[6] Ishii N, Aoki S and Hatsuda T 2007 *Phys. Rev. Lett.* **99** 022001
[7] Aoki S *et al.* (HAL QCD Collaboration) 2012 *PTEP* **2012** 01A105
[8] Doi T *et al.* (HAL QCD Collaboration) 2012 **127** 723
[9] Inoue T *et al.* (HAL QCD Collaboration) 2013 *Phys. Rev. Lett.* **111** 112503
[10] Doi T *et al.* 2016 *PoS LATTICE* **2015** 086 (*Preprint* `arXiv:1512.01610`)
[11] Hjorth-Jensen M, Kuo T T and Osnes E 1995 *Physics Reports* **261** 125
[12] Engeland T, Hjorth-Jensen M and Jansen G R CENS, a Computational Environment for Nuclear Structure URL `https://github.com/ManyBodyPhysics/CENS/tree/master/MBPT/`
[13] Gad K and Müther H 2002 *Phys. Rev. C* **66** 044301
[14] More S N, Ekström A, Furnstahl R J, Hagen G and Papenbrock T 2013 *Phys. Rev. C* **87** 044326
[15] Nemura H 2014 *Int. Jour. Mod. Phys. E* **23** 1461006
[16] Ishkhanov B S *et al.* 2002 Cross sections of photon absorption by nuclei with nucleon numbers 12 - 65, Tech. Rep. MSU-INP-2002-27/711 *(Institute of Nuclear Physics, Moscow State University)* URL `http://www.ntse-2013.khb.ru/Proc/JPVary.pdf`
[17] Ahrens J *et al.* 1975 *Nuclear Physics A* **251** 479
[18] Leistenschneider A *et al.* 2001 *Phys. Rev. Lett.* **86** 5442
[19] Miorelli M, Bacca S, Barnea N, Hagen G, Jansen G R, Orlandini G and Papenbrock T 2016 *Phys. Rev. C* **94** 034317
[20] Birkhan J *et al.* 2017 *Phys. Rev. Lett.* **118** 252501
[21] Ekström A *et al.* 2015 *Phys. Rev. C* **91** 051301
[22] Barbieri C 2009 *Phys. Rev. Lett.* **103** 202502
[23] Cipollone A, Barbieri C and Navrátil P 2015 *Phys. Rev. C* **92** 014306
[24] Barbieri C and Dickhoff W H 2002 *Phys. Rev. C* **65** 064313
[25] Aumann T 2017 See contribution to these proceedings.

IOP Conf. Series: Journal of Physics: Conf. Series **966** (2018) 012016 doi:10.1088/1742-6596/966/1/012016

Nuclear structure studies performed using the (^{18}O,^{16}O) two-neutron transfer reactions

D. Carbone[1], C. Agodi[1], F. Cappuzzello[1,3], M. Cavallaro[1], J. L. Ferreira[2], A. Foti[3,4], A. Gargano[5], S.M. Lenzi[6,7], R. Linares[2], J. Lubian[2], G. Santagati[1]

[1]INFN Laboratori Nazionali del Sud, Catania, Italy
[2]Instituto de Fisica, Universidade Federal Fluminense, Niteroi, RJ, Brazil
[3]Dipartimento di Fisica e Astronomia, Università degli Studi di Catania, Italy
[4]INFN Sezione di Catania, Catania, Italy
[5]INFN Sezione di Napoli, Napoli, Italy
[6]Dipartimento di Fisica e Astronomia, Università di Padova, Padova, Italy
[7]INFN Sezione di Padova, Padova, Italy

carboned@lns.infn.it

Abstract. Excitation energy spectra and absolute cross section angular distributions were measured for the ^{13}C(^{18}O,^{16}O)^{15}C two-neutron transfer reaction at 84 MeV incident energy. This reaction selectively populates two-neutron configurations in the states of the residual nucleus. Exact finite-range coupled reaction channel calculations are used to analyse the data. Two approaches are discussed: the extreme cluster and the newly introduced microscopic cluster. The latter makes use of spectroscopic amplitudes in the centre of mass reference frame, derived from shell-model calculations using the Moshinsky transformation brackets. The results describe well the experimental cross section and highlight cluster configurations in the involved wave functions.

1. Introduction

Direct two-nucleon transfer reactions play an important role in the study of specific features of the atomic nucleus and indeed they were extensively explored [1] [2] [3] [4] [5] to study for example pairing correlations. Among these, heavy-ion direct transfer reactions at energies close to the Coulomb barrier are useful tools to obtain precise spectroscopic information.

In this context, a wide range of systems were explored by heavy-ion induced one- and two-neutron transfer reactions at INFN-LNS (Italy) by the (^{18}O,^{17}O) and (^{18}O,^{16}O) reactions. The MAGNEX spectrometer [6] was used to detect the ejectiles. The large acceptance and high resolution of it allowed to obtain high quality energy spectra up to the region above the two-neutron separation energy in the residual nucleus [7] [8] [9]. New phenomena were observed, e.g. the dominance of the direct one-step transfer of the two neutrons [10] and the presence of the first experimental signature of the Giant Pairing Vibration [11] [12] at high excitation energy in the ^{14}C and ^{15}C spectra. The analysis of the broad structures in the ^{15}C spectrum at high excitation energy was presented in ref. [13] and the neutron decay of these structures was investigated in ref. [14], exploiting the use of the EDEN array coupled to the

IOP Conf. Series: Journal of Physics: Conf. Series **966** (2018) 012016 doi:10.1088/1742-6596/966/1/012016

spectrometer [15] [16]. Moreover, it was demonstrated in several works that the (^{18}O,^{16}O) two-neutron transfer reaction can be used for quantitative spectroscopic studies of pair configurations in nuclear states [17] [18] [19].
A complete theoretical treatment of the transfer process should contain a description of three terms: the one-step channel with the inclusion of all the possible inelastic excitations of the target, projectile, ejectile or residual nucleus; the sequential channel, with the inclusion of intermediate partitions and the non-orthogonal term deriving from the limited model space used in actual calculations. However, in the state-of-the-art theories a complete model is still not available. As long as heavy-ions are concerned, it is necessary to include explicitly the inelastic excitations of the involved nuclei using the coupled-channels approach [20] [21] [22]. In the lack of a complete two-step Coupled Reaction Channel (CRC) theory, the one-step and two-step calculations can be performed separately and their amplitudes summed by considering the relative phase as an additional parameter.
In ref. [17] the experimental absolute cross sections of the one- and two- neutron transfer reactions induced by an ^{18}O beam on a ^{12}C target were reproduced without any scaling factor by means of Exact Finite Range (EFR) CRC calculations. Two approaches were used: the extreme cluster model and the independent coordinate scheme. The description of these approaches can be found in refs. [23] and [24], respectively. The same framework was recently applied to describe the absolute cross section angular distributions of the states populated by the ^{13}C(^{18}O,^{16}O) reaction [19]. A new approach consisting in a fully microscopic cluster calculation is introduced, performed by using the two-neutron spectroscopic amplitudes calculated in the shell-model framework. This new approach can be more extensively used with respect to the extreme cluster model to evaluate the presence of cluster components in the involved wave functions.

2. Experimental results

The experiment was performed at the INFN-LNS laboratory using an ^{18}O^{6+} beam at 84 MeV incident energy on a thin ^{13}C target. The same measurement was also done on ^{12}C target to estimate the background coming from the ^{12}C impurities in the ^{13}C target. The MAGNEX spectrometer was used to detect the ejectiles [6]. The particle identification and the data reduction technique are the same described in detail in refs. [12] [25]. An example of the obtained energy spectra for the ^{15}C nucleus is shown in Fig. 1 (left panel), in which the ^{14}C background spectrum coming from the ^{12}C impurities in the ^{13}C target is superimposed, after normalization. The strongest bound and resonant states appearing in the spectrum of the ^{15}C nucleus are almost the same strongly populated in the (t,p) reactions on ^{13}C [26]. In particular, below the one-neutron separation energy (S_n = 1.218 MeV) the only two ^{15}C bound states are identified, i.e. the ground (J^{π} = 1/2$^+$) and the 5/2$^+$ state at E_x = 0.74 MeV, that are characterized by a dominant single particle configuration [27]. In the region between S_n and the two-neutron separation energy (S_{2n} = 9.39 MeV), narrow resonances at E_x = 3.103 (1/2$^-$), 4.22 (5/2$^-$), 4.66 (3/2$^-$), 6.84 (9/2$^-$, 7/2$^-$), 7.35 (9/2$^-$, 7/2$^-$) MeV are populated. All of these states are indicated to consist mainly of $2p$-$1h$ configurations (with respect to the ^{14}C$_{g.s.}$ vacuum state) [26]. Resonances with a single-particle configuration of a ^{14}C$_{g.s.}$ + 1n [28] are very weakly populated in the present reaction.
Examples of the obtained absolute cross section angular distributions for the ground state and the resonance at 4.22 MeV are shown in Fig. 1 (right panel).

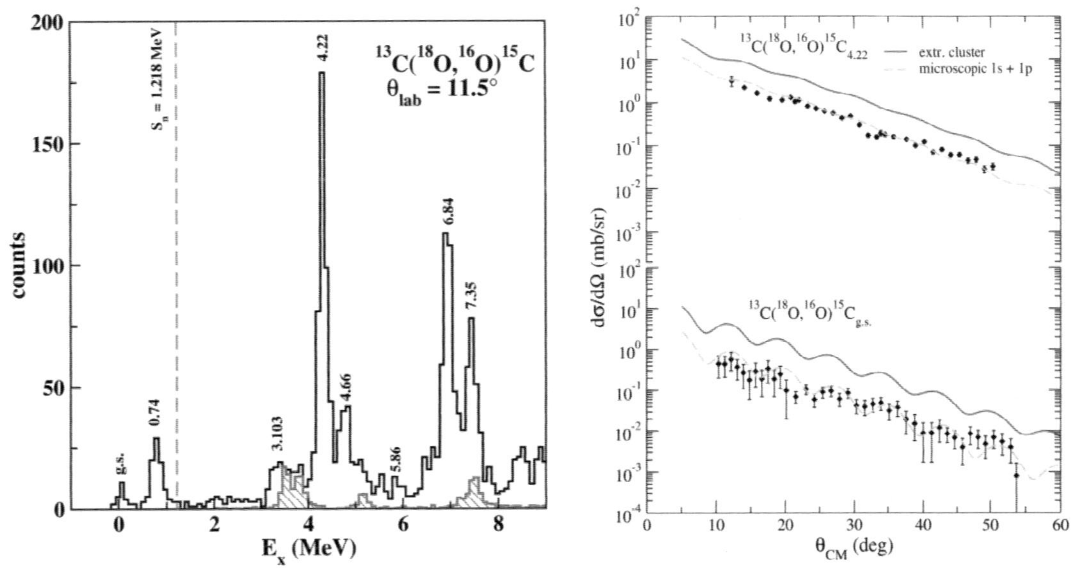

Fig. 1. (left panel) Excitation energy spectrum of the $^{13}C(^{18}O,^{16}O)^{15}C$ reaction for $10° < \theta_{lab} < 11°$. The red-hatched area corresponds to the background that comes from ^{12}C impurities in the target. (right panel) Experimental cross section angular distributions for the g.s. and the state at 4.22 MeV in ^{15}C. Theoretical calculations: extreme cluster calculations (red line) and microscopic cluster model calculations including 1s + 1p waves (green dashed line) .

3. Theoretical analysis

Exact finite range coupled reaction channel (CRC) and two-step distorted wave Born approximation (DWBA) calculations for two-neutron transfer reactions were performed to describe the cross section, using the FRESCO code [29], with the same ingredients of ref. [17]. A direct, simultaneous transfer of the two particles was considered. The two-particle wave functions for the one-step two-neutron transfer mechanism were obtained considering two different schemes: i) the extreme cluster model, in which the relative motion between the two transferred neutrons is frozen and separated from the core and the two neutrons are coupled antiparallel to an intrinsic angular momentum $S = 0$, with spectroscopic amplitudes for both target and projectile set to 1.0 and ii) the *microscopic cluster model*, which will be introduced in Section 3.1. The ^{12}C is assumed as a closed core and the model space includes the $1p_{1/2}$, $1d_{5/2}$, $2s_{1/2}$ orbits. The effective phenomenological *zbm* interaction [30] was used.

When a cluster model is used, the parameters relevant for the definition of the wave function are the principal quantum number N and the orbital angular momentum L relative to the core. These parameters are obtained from the conservation of the total number of quanta in the transformation of the wave function of two independent neutrons in orbits (n_i, l_i) $(i = 1, 2)$ into a cluster with internal state (n, l) [31]: $2(n_1 - 1) + l_1 + 2(n_2 - 1) + l_2 = 2(N - 1) + L + 2(n - 1) + l$. In the extreme cluster model hypothesis, we consider $n = 1$ and $l = 0$ so that the cluster is in a $1s$ internal state. The spectroscopic amplitudes for both target and projectile were set to 1.0. The resulting differential cross sections using the extreme cluster hypothesis for the ground and the state at 4.22 MeV are shown in Fig. 1 (right panel). Using this extreme approximation the calculations gives a cross section larger than the experimental one. The main reason for this overestimation might lie in the approximation that the two neutrons are coupled to the total spin $S = 0$ with 100% of probability.

3.1. The microscopic cluster model

A natural way to go beyond the assumptions of the extreme cluster model is to introduce both parallel and antiparallel couplings for the two-neutrons. Realistic spectroscopic amplitudes are required for all the possible combinations of single-particle configurations in this enlarged space that can be derived from shell-model calculations. To achieve this goal, we made use of transformations from individual (*j*-*j* coupling) to relative and centre of mass coordinates (*LS* coupling) for the harmonic-oscillator wave functions of the two-particle system [19], using the Moshinsky brackets [31]. This approach is what we call *microscopic cluster calculations*.

As a first step, we performed microscopic cluster calculations considering that the cluster relative motion state is represented exclusively by $n = 1$ and $l = 0$ quantum numbers, i.e. the cluster is in the 1*s* intrinsic state. The obtained cross sections are much lower than data for all the states. Thus, we included also the 1*p* ($n = 1$ and $l = 1$) cluster relative motion states and the results are shown in Fig. 1 (right panel). We see that the results of 1*s* + 1*p* microscopic cluster calculations are in rather good agreement with the experimental angular distributions for both the ground and the state at 4.22 MeV.

To conclude, the new microscopic cluster model allows to describe rather well the experimental cross section for the states populated by the $^{13}C(^{18}O,^{16}O)^{15}C$ reaction, thus demonstrating the importance of a two-neutron correlation in the nuclear wave function in the two-neutron transfer mechanism. A dominance of the 1*s* and 1*p* waves in the two-neutron cluster internal wave functions is found.

Acknowledgments
This project has received funding from the European Research Council (ERC) under the European Union's Horizon 2020 research and innovation programme (grant agreement No 714625).

References

[1] Mordechai S et al. 1978 *Nucl. Phys.* A **301** 463.
[2] Guazzoni P et al. 2011 *Phys. Rev.* C **83** 044614.
[3] Guazzoni P et al. 2004 *Phys. Rev.* C **69** 024619.
[4] Potel G et al. 2011 *Phys. Rev. Lett.* **107** 092501.
[5] Potel G, Barranco F, Vigezzi E, and Broglia R A 2010 *Phys. Rev. Lett.* **105** 172502.
[6] Cappuzzello F, Agodi C, Carbone D, and Cavallaro M 2016 *Eur. Phys. J.* A **52** 167.
[7] Carbone D et al. 2014 *Phys. Rev.* C **90** 064621.
[8] Nicolosi D et al. 2013 *Acta Phys. Pol.* B **44** 657.
[9] Carbone D et al. 2014 *Acta Phys. Pol.* B **45** 431.
[10] Carbone D et al. 2011 *J. Phys.: Conf. Ser.* **312** 082016.
[11] Cappuzzello F et al. 2015 *Nat. Commun.* **6** 6743. doi:10.1038/ncomms7743.
[12] Carbone D 2015 *Eur. Phys. J. Plus* **130** 143.
[13] Cappuzzello F et al. 2012 *Phys. Lett.* B **711** 347.
[14] Cavallaro M et al., *Phys. Rev.* C 93 (2016) 064323.
[15] Laurent H et al. 1993 *Nucl. Instr. and Meth.* A **326** 517.
[16] Cavallaro M et al. 2013 *Nucl. Instr. and Meth.* A **700** 65.
[17] Cavallaro M et al. 2013 *Phys. Rev.* C **88** 054601.
[18] Ermamatov M J et al. 2016 *Phys. Rev.* C **94** 024610.
[19] Carbone D et al. 2017 *Phys. Rev.* C **95** 034603.

[20] Bond P D, Korner H J, Lemaire M C, Pisano D J, and Thorn C E1977 *Phys. Rev.* C **16** 177.
[21] Lemaire M C and Low K S 1977 *Phys. Rev.* C **16** 183.
[22] Pereira D et al. 2012 *Phys. Lett.* B **710** 426.
[23] Satchler G R 1983 *Direct Nuclear Reactions* Oxford University press, Oxford.
[24] Thompson I J 1988 *Comp. Phys. Rep.* **7** 167.
[25] Cappuzzello F et al. 2014 *Nucl. Instr. and Meth.* A **763** 314.
[26] Truong S and Fortune H T 1983 *Phys. Rev.* C **28** 977.
[27] Murillo G, Sen S, and Darden S E 1994 *Nucl. Phys.* A **579** 125.
[28] Goss J D et al. 1973 *Phys. Rev.* C **8** 514.
[29] Thompson J I, http://www.fresco.org.uk/.
[30] Zuker A P, Buck B, and McGrory J B 1968 *Phys. Rev. Lett.* **21** 39.
[31] Moshinsky M 1959 *Nucl. Phys.* **13** 104.

Nuclear matrix elements for the $\beta\beta$ decay of ^{76}Ge

B. A. Brown[1], **D. L. Fang**[2], **and M. Horoi**[3]

[1] Department of Physics and Astronomy and National Superconducting Cyclotron Laboratory Michigan State University, East Lansing, Michigan 48824-1321, USA
[2] College of Physics, Jilin University, Changchun 130012, China
[3] Department of Physics, Central Michigan University, Mount Pleasant, Michigan 48859, USA

E-mail: brown@nscl.msu.edu

Abstract. The nuclear matrix elements for two-neutrino double-beta ($2\nu\beta\beta$) and zero-neutrino double-beta ($0\nu\beta\beta$) decay of ^{76}Ge are evaluated in terms of the configuration interaction (CI), quasiparticle random phase approximation (QRPA) and interacting boson model (IBM) methods. We determine nuclear matrix elements with error bars that take into account the truncation of the model space used for CI.

Many properties of the active neutrinos are measured, but it is not yet established whether they are Dirac or Majorana type particles and their absolute masses are not known. Left-right symmetric extensions to the standard model provide an explanation for the non-zero masses of the left-handed light neutrinos and also predict the existence of right-handed heavy neutrinos [1]. Neutrinoless double-beta ($0\nu\beta\beta$) decay of nuclei provides unique information and constraints on these neutrino properties The $\beta\beta$ decay process and the associated nuclear matrix elements (NME) have been investigated by using several approaches including the quasiparticle random phase approximation (QRPA), the configuration interaction (CI) model, and the interacting boson model (IBM). A recent review of the theoretical methods and results is given in [2].

Assuming contributions from the light left-handed (ν) neutrino-exchange mechanism and the heavy right-handed (N) neutrino-exchange mechanism, the decay rate of a neutrinoless double-beta decay process can be written as [3], [4]

$$\left[T^{0\nu}_{1/2}\right]^{-1} = G^{0\nu}\left(\left|M^{0\nu}\right|^2 |\eta_\nu|^2 + \left|M^{0N}\right|^2 |\eta_N|^2\right),$$

where $G^{0\nu}$ is the phase space factor [5], [6], M are the nuclear matrix elements (NME), and η are combinations of the neutrino masses [4], [3].

Since the experimental decay rate is proportional to the square of the calculated nuclear matrix elements, it is important to calculate these matrix elements with good accuracy to be able to determine the absolute scale of neutrino masses. However, the theoretical methods used give results that differ from one another by factors of up to 2-3. It is important to understand the nuclear structure aspects of these matrix elements and why the models give differing results.

In this talk we discuss the NME for the $\beta\beta$ decay of ^{76}Ge obtained with the CI, QRPA and IBM-2 methods [7]. We will show that all of these methods have deficiencies. Some of the deficiencies can be addressed with many-body perturbation theory (MBPT) approaches, and connections to other observables.

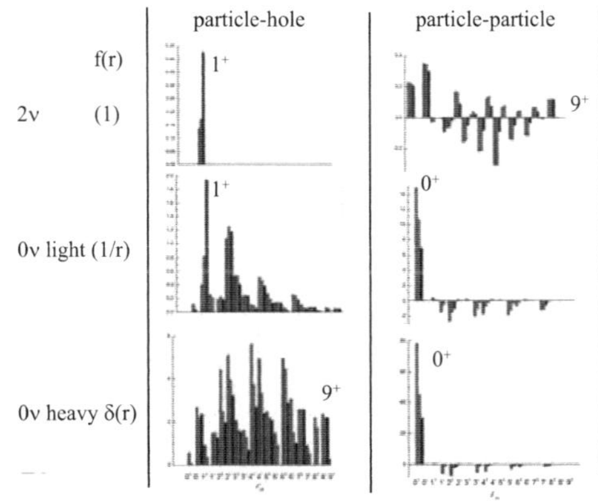

Figure 1. (Colour online) The NME obtained in the QRPA calculations expressed in terms of their contributions from the J_{ph} states in ^{76}As on the left-hand side and the J_{pp} states in ^{74}Ge on the right-hand side. The color code is blue for the results obtained in $jj44$, black for the results obtained in fpg and red for the results obtained in the 21-orbit model spaces.

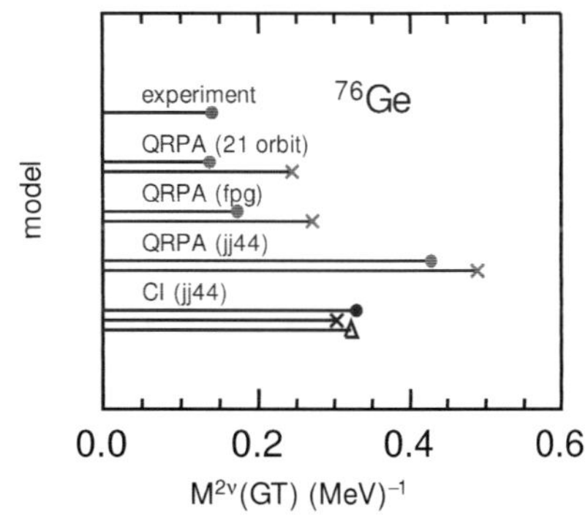

Figure 2. (Colour online) Nuclear matrix elements for $2\nu\beta\beta$ decay of ^{76}Ge. The top point in green is the experimental value [8]. The QRPA results are shown for $g_{pp}^{T=0} = 0.673$ (red dots) and $g_{pp}^{T=0} = 0.643$ (red crosses). The CI results are shown for the JUN45 (dot), jj44bpn (cross) and gcn28:50 (triangle) Hamiltonians.

The nuclear matrix elements can be presented as a sum of Gamow-Teller (M_{GT}), Fermi (M_F), and Tensor (M_T) matrix elements (see, for example, Refs. [9], [10]), $M = M_{GT} - (g_V/g_A)^2 M_F + M_T$, where g_V and g_A are the vector and axial constants, correspondingly. We use $g_V = 1$ and $g_A = 1.27$. The M_α are matrix elements of scalar two-body potentials. The Gamow-Teller has the form $V_{GT}(r, A, \mu)\ \sigma_1 \cdot \sigma_2\ \tau_1^- \tau_2^-$ and the Fermi has the form $V_F(r, A, \mu)\ \tau_1^- \tau_2^-$, where τ^- are the isospin lowering operators. The neutrino potentials depend on the relative distance between the two decaying nucleons, r, the mass number A, and the closure energy μ [11]. The radial forms are given explicitly in [9]. For the heavy-neutrino exchange, the potential does not depend on μ. For the light neutrino matrix element the closure approximation is good to within 10% [12].

The operators for M_{GT} are given to a good approximation by $f(r)\ \sigma_1 \cdot \sigma_2\ \tau_1^- \tau_2^-$, where $f(r)^{2\nu} = 1$ (in closure), $f(r)^{0\nu} = a/r$ and $f(r)^{0N} = b\,\delta(r)$ where the constants a and b depend on A, μ and the SRC. The results discussed below follow from the expansions of the many-body matrix elements for these three operators in terms of the particle-hole (ph) in ^{76}As or particle-particle (pp) intermediate states in ^{74}Ge [13].

The expansion of the NME over intermediate J values obtained from the QRPA calculations is shown in Fig. 1. This figure shows the different types of correlations for the three operators. The 2ν decay is completely determined by the 1^+ states of the ph channel and its expansion over pp is complicated. The heavy neutrino is dominated by the 0^+ states of the pp channel and its expansion over ph states is complicated. The light neutrino NME is some where between these and but looks simplest in the pp channel.

The 2ν tensor NME is zero and the Fermi NME is zero since isospin is conserved. For 0ν and $0N$ the Fermi and tensor parts are both relatively small, and we define a correction factor for these given by $R_{GT} = M/M_{GT}$, where M contains all three terms of Eq. 2. The CI calculations give $R_{GT}^{0\nu} = 1.10(3)$. Larger values of 1.23 for QRPA [14] and 1.33 for IBM-2 [15] were obtained with the older calculations. But more recently, it was found that the 2ν Fermi matrix element was not zero because isospin was being treated incorrectly in QRPA [16] and IBM-2 [17]. After this was corrected the new $M_F^{2\nu}$ values are now zero in all methods. The new results for $R_{GT}^{0\nu}$ are 1.10 [16] and 1.19 [18] for QRPA, and 1.04 [17] for IBM-2. Taking these results into account we adopt a correction factor from the tensor plus Fermi contributions of $R_{GT}^{0\nu} = 1.12(7)$. The ratio for the heavy neutrino is 1.20 for CI, 1.26 for QRPA [18] and 1.00 for IBM-2 [17]. The adopted correction factor is $R_{GT}^{0N} = 1.13(13)$.

In the following we first focus on results for M_{GT}. At the end, the total matrix element M will be obtained from M_{GT} via a product of correction factors R given by $M = [M_{GT}(\mathrm{CI})][R_V][R_S][R_{GT}]$. R_{GT} is defined above. We start with the use of short-range correlations (SRC) [10] based on the CD-Bonn potential [19]. At the end we will give a value and error for the correction to this, R_S, based on a range of assumptions about the SRC. R_V represents the correction coming from a "vertical" expansion of the CI model space that includes the effect of orbitals outside those in $jj44$. R_V is the main focus of attention of this talk.

The model space for CI and IBM-2 is $jj44$ that consists of the four valence orbitals $0f_{5/2}$, $1p_{3/2}$, $1p_{1/2}$ and $0g_{9/2}$ for protons and neutrons. The model space for QRPA are the 21 orbitals with oscillator quanta $N \leq 5$ where $N = 2n + \ell$ for protons and neutrons. The QRPA results are also given when the evaluation of the NME are restricted to $jj44$ and to fpg ($jj44$ plus $0f_{7/2}$ and $0g_{7/2}$). In addition to our own CI calculations with the JUN45 [20] and jj4bpn [21] Hamiltonians, we will show results from the gcn28:50 Hamiltonian [22] for 2ν [23], 0ν [24] and $0N$ [25]. The method and parameters used for the QRPA calculations are similar to those used in [16]. For the particle-particle channel in order to restore the isospin symmetry, we follow the formalism introduced in [26], [16], by separately fitting the $T = 0$ and $T = 1$ parts of the interaction.

Results for the $2\nu\beta\beta$ NME are shown in Fig. 2. This NME is completely determined by $J_{ph}^{\pi} = 1^+$ intermediate states in ^{76}As. In CI the summation over intermediate including the energy denominator (Eq. 2 in [23]) is obtained with the strength-function method [27]. The IBM-2 result is not shown because it uses an approximation for the NME based on the closure result for the operator $\sigma_1 \cdot \sigma_2\ \tau_1^- \tau_2^-$ together with average closure energies from other methods (Eq. 16 in Ref. [17]). Experiment is reduced by a factor of about $R_V^{2\nu} = 0.45$ compared to CI. $R_V^{2\nu} = M^{2\nu}/M^{2\nu}(\mathrm{CI})$ denotes the correction beyond the $jj44$ model space, due to a "vertical" expansion that includes correlations from orbitals below and above the $jj44$ model space. The QRPA results for $jj44$ and pfg show that part of this reduction is due to the missing spin-orbit partners in the $jj44$ model space. The particle-hole correlations are dominated by a strong repulsive interaction in the 1^+ channel. Relative to the non-interacting single-particle distribution, Gamow-Teller strength is reduced in low-lying states and shifted into the giant Gamow-Teller resonance. As shown by the QRPA results for $jj44$ and fpg, both spin-orbit partners are important for the reduction. A similar behavior was observed for CI in the case of ^{136}Xe [28].

Beyond QRPA, it is known that two-particle two-hole (2p-2h) admixtures into the model space wavefunctions are important for Gamow-Teller beta decay. The experimental Gamow-Teller strength is observed to be reduced by a factor of $R_V' = 0.5 - 0.6$ relative to the CI calculations in the *sd* [29] and *pf* [30] model spaces. Also the strength extracted from charge-exchange reactions for the total Gamow-Teller strength up to about 25 MeV in excitation energy is reduced by this factor relative to QRPA [31] and the $3(N-Z)$ Ikeda sum rule [32]. Arima et al. [33] and Towner [34] have explained this reduction using MBPT in terms of 2p-2h admixtures into the model-space wavefunctions. Earlier calculations claimed that the reduction in GT strength was due to Δ excitations [35] in the nucleus. However, calculations with a realistic $N\Delta\pi$ interaction vertex have shown that the influence of Δ (and other mesonic-exchange currents) is small [33], [34]. These results are compared to the empirical *sd* results in Fig. 13 of [29]. In order to conserve the Ikeda sum rule, the reduction in low-lying B(GT) strength is associated

IOP Conf. Series: Journal of Physics: Conf. Series **966** (2018) 012017 doi:10.1088/1742-6596/966/1/012017

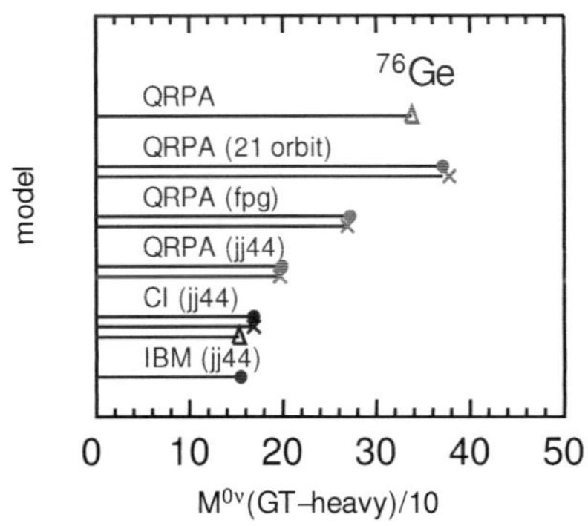

Figure 3. (Colour online) The $0N$ NME for heavy neutrino decay of ^{76}Ge. See caption to Fig. 2. The QRPA point with the triangle is from Ref. [18].

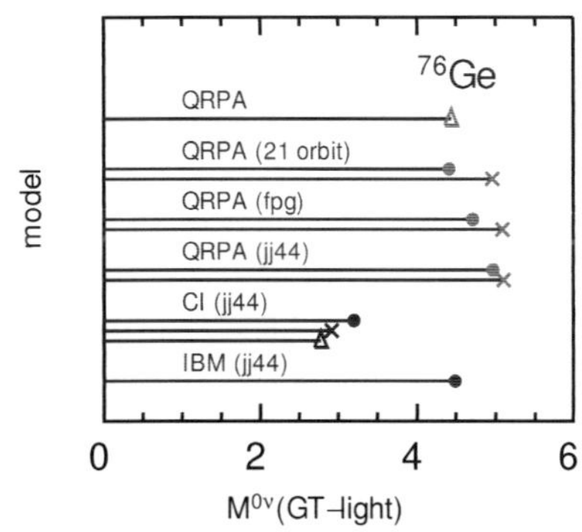

Figure 4. (Colour online) The 0ν NME for the light-neutrino decay of ^{76}Ge. See caption2 to Figs. 2 and 3.

with a spreading of strength to high excitation energy [36] that gets removed from the 2ν NME due to the energy denominator in the summation over intermediate states. In summary, relative to CI in the $jj44$ model space, reductions due to a spin-orbit complete model space, together with 2p-2h admixtures are required for the $2\nu\beta\beta$ NME. The observed factor of $R_V = 0.45$ is consistent with expectations.

The results for $0N$ (heavy neutrino) are shown in Fig. 3. In addition to our own QRPA results, we show the QRPA result from [18]. The J_{pp} intermediate states are dominated by the 0^+ ground state of ^{74}Ge (see Ref. [13] for details on the analysis). In QRPA the NME increases by a factor of $R_V^{0N} = 1.9$ as the number of orbitals included in the sums increases from $jj44$ to full (21 orbitals). This is due to the strong pairing (particle-particle) part of the Hamiltonians and the resulting increase in the number of coherent pairs contributing to the $0N$ NME. The pairing also gives rise to the odd-even staggering of the nuclear binding energies quantified by the pairing energies D [37], [38]. For the germanium isotopes the experimental pairing energies are a factor of 1.45 larger than that obtained with the first-order expectation value of the CD-Bonn Hamiltonian. Based on the average of this result and the increase observed in QRPA, we will use $R_V^{0N} = 1.65(25)$.

The results for $0\nu\beta\beta$ (light neutrino) are shown in Fig. 4. The largest term in the 0ν NME is from the $J_{pp}^{\pi} = 0^+$ ground state of ^{74}Ge [13]. In QRPA the NME is nearly constant as the number of orbitals included in the sums increase. Qualitatively this is due to a competition between the reduction from the particle-hole channel observed for 2ν and the enhancement due to the particle-particle channel observed for $0N$. The connection of the 0ν matrix elements with pairing has been previously discussed [39]. The new point of our analysis is that the increase expected from pairing coming from MBPT beyond the $jj44$ model space is cancelled by the reduction from the ph-type correlations.

Contributions from states with $J_{pp} > 0$ cancel part of the NME from $J_{pp} = 0^+$. Within $jj44$ the reduction is dominated by the $J_{pp} = 2^+$ states [13]. For the 0ν NME within $jj44$, one finds $R_{pp}^{0\nu} = \{M_{GT}^{0\nu}/[M_{GT}^{0\nu}(J_{pp} = 0^+)]\} = 0.53$ for CI [13], 0.90 for IBM-2 [15] and 0.72 for QRPA. The reason for these differences may be due to the truncation within $jj44$ made by IBM-2 and QRPA. For the $0N$ NME this ratio is $R_{pp}^{0N} = 0.89$ in CI [13]; the cancellation from higher J_{pp} is much less, and the result is dominated by the $J_{pp} = 0^+$ contribution and its connection to pairing is discussed above. In the $jj44$ model space the agreement between the $0N$ NME (Fig.

3) for CI, QRPA and IBM-2 is much better than that for 0ν (Fig. 4) since the cancellation from higher J_{pp} terms is small.

Holt and Engel [40] considered the effect of 2p-2h admixtures beyond the $jj44$ model space by treating the effective transition operator in MBPT. They found a 20% increase of the 0ν NME for ^{76}Ge. Part of these MBPT contributions go beyond QRPA. At present this is the best estimate for the correction beyond CI in the $jj44$ model space. We will use $R_V^{0\nu} = 1.2(2)$ with a generously large value of 20% for its uncertainty.

The results shown above are based on the CD-Bonn SRC. This is the weakest of several SRC that have been used [10]. The strongest is the AV18 SRC, and the UCOM [41] SRC is about half way between. For our final result we use the average of CD-Bonn and AV18 with an error that encompasses both. The result is that the $0N$ NME are multiplied by $R_S^{0N} = 0.80(20)$ and the 0ν NME are multiplied by $R_S^{0\nu} = 0.97(3)$, where R_S is the SRC correction relative to the CD-Bonn starting point.

Finally, we combine all of the factors discussed above in the form $M = [M_{GT}(\text{CI})][R_V][R_S][R_{GT}]$. Based on the experimental value [8] for 2ν the NME is, $M^{2\nu} = 0.140(5) = [0.31(3)][0.45][1][1]$. The second term is the empirical correction for R_V due to mixing beyond the $jj44$ model space. The error in the CI NME reflects the spread obtained with the three different Hamiltonians used (Fig. 2). For $0N$,$M^{0N} = [155(10)][1.65(25)][0.80(20)][1.13(13)] = 232(80)$, where the CI value is from Fig. 3. The error for $0N$ is dominated by the SRC correction. Finally For 0ν, $M^{0\nu} = [3.0(3)][1.2(2)][0.97(3)][1.12(7)] = 3.9(8)$, where the CI value is from Fig. 4. The error for 0ν is dominated by an estimated uncertainty of 20% in the correction beyond $jj44$. Comparison to previous values must take into account the isospin correction for QRPA and IBM discussed above, and the choice of SRC (in our R_S factor). The range is from 2.8 for CI [24] to 4.7 for IBM-2 [17] and 5.3 for QRPA [18]. Our result is in between these, but it is not an average since we have made comments on the deficiencies of all of these models. Using Eq. 1 with the experimental limit of the half-life ($T_{1/2}^{0\nu} > 3 \times 10^{25}$ yr [42]), and the phase space factor from [5], we obtain $\mid \eta_\nu \mid m_e c^2 < 0.3$ eV.

Sometimes the 2ν correction factor (0.45 in this case) is expressed in terms of an effective g_A value ($g'_A = 0.85$ in this case). Since the factor $(g_A)^4$ appears inside the phase-space factor of Eq. 1, one might think that the decay rate for 0ν and $0N$ could be reduced by a factor of $(g'_A/1.27)^4 = 0.20$ [17], [43]. However, this g'_A is only for a specific operator associated with a specific observable ($2\nu\beta\beta$ decay) relative to a specific model (CI in $jj44$ in this case). The operators involved in 0ν and $0N$ decay are different (short ranged), and corrections beyond CI cannot be expressed in terms of an overall change in g_A. It is better to express the renormalizations in terms of factors such as R_V that are operator and model space dependent.

The model-space truncation contributions to R_{pp} should be understood. The error for the R_{GT} correction could be reduced if reasons for the variations within the models is understood. The error for the R_V correction could be reduced if the MBPT results such as those in [40] should be expanded to include the renormalization of the separate effects in the ph and pp channels in order to compare to the results found previously relative to the $jj44$ model space. This includes the reduction in Gamow-Teller beta decay strength [33], [34], and the enhancements of the pairing strength seen in the D values. The basic division between CI and its MBPT corrections from all other orbitals can be checked by no-core and ab-initio CI in lighter nuclei where they are tractable. Other methods such as in-medium SRG [44] and coupled cluster [45] can be used in place of MBPT, and at this level the division between short-range renormalization, R_S, and long-range renormalization, R_V, might be merged. The CI results for the $A = 76$ region can be further checked against spectroscopic observables (occupations number are in good agreement with CI [24]) including two-nucleon transfer. Future results should be presented in terms of changes relative to various contributions we have discussed, and evaluations for other cases of interest [46] should be made.

Acknowledgements We acknowledge suppot from NSF Grant No. PHY-1404442 and the NUCLEI SciDAC Collaboration under U.S. Department of Energy Grant No. DE-SC0008529.

References

[1] Pati J C and Salam A 1974 *Phys. Rev. D* **10** 275; Mohapatra R N and Pati J C 1975 *Phys. Rev. D* **11** 2558; Keung W-Y and Senjanovic G 1983 *Phys. Rev. Lett.* **50** 1427
[2] Engel J and Menendez J 2017 *Rep. Prog. Phys.* **80** 046301; and references therin
[3] Vergados J D Ejiri H and Simkovic F 2012 *Rep. Prog. Phys.* **75** 106301
[4] Horoi M 2013 *Phys. Rev. C* **87** 014320
[5] Kotila J and Iachello F 2012 *Phys. Rev. C* **85** 034316
[6] Stoica S and Mirea M 2013 *Phys. Rev. C* **88** 037303
[7] Brown B A Fang D L and Horoi M 2015 *Phys. Rev. C* **92** 041301
[8] Barabash A S 2010 *Phys. Rev. C* **81** 035501
[9] Senkov R A and Horoi M 2013 *Phys. Rev. C* **88** 064312
[10] Horoi M and Stoica S (2010) *Phys. Rev. C* **81** 024321
[11] Senkov R A Horoi M and Brown B A 2014 *Phys. Rev. C* **89** 054304
[12] Senkov R A and Horoi M 2014 *Phys. Rev. C* **90** 051301(R)
[13] Brown B A Horoi M and Senkov R A 2014 *Phys. Rev. Lett.* **113** 262501
[14] Yousef M S Rodin V Faessler A and Simkovic F 2009 *Phys. Rev. C* **79** 014314
[15] Barea J and Iachello F 2009 *Phys. Rev. C* **79** 044301
[16] Simkovic F Rodin F V Faessler A and Vogel P 2013 *Phys. Rev. C* **87** 045501 (2013)
[17] Barea J Kotila J J and Iachello F 2015 *Phys. Rev. C* **91** 034304
[18] Hyvarinen J and Suhonen J 2015 *Phys. Rev.* **91** 024613
[19] Machleidt R 2001 *Phys. Rev. C* **63** 024001
[20] Honma M Otsuka T Mizusaki T and Hjorth-Jensen M 2009 *Phys. Rev. C* **80** 064323
[21] Cheal B *et al.* 2010 *Phys. Rev. Lett.* **104** 252502
[22] Gniady A Caurier E and Nowacki F (unpublished)
[23] Caurier E Nowacki F and Poves A 2012 *Phys. Lett. B* **711** 62
[24] Menendez J Poves A Caurier E and Nowacki F 2009 *Phys. Rev. C* **80** 048501
[25] Blennow Fernandez-Martinez M E Lopez-Pavon J and Menendez J 2010 *JHEP07* 096
[26] Rodin V and Faessler A 2011 *Phys. Rev. C* **84** 014322 .
[27] Horoi M 2010 *AIP Conf. Proc.* **1304** 106
[28] Horoi M and Brown B A 2013 *Phys. Rev. Lett.* **110** 222502
[29] Brown B A and Wildenthal B H 1988 *Annu. Rev. Nucl. Part. Sci.* **38** 29 (
[30] Martinez-Pinedo G Poves A Caurier E and Zuker A P 1996 *Phys. Rev. C* **53** R2602 .
[31] Litvinova E Brown B A Fang D L Marketin T and Zegers R G T 2014 *Phys. Lett B* **730** 307
[32] Anderson B D Chittrakarn T C Baldwin A R Lebo C Madey R Watson J W Brown B A and Foster C C 1985 *Phys. Rev. C* **31** 1161, and references therein
[33] Arima A Schimizu K Bentz W and Hyuga H 1987 *Adv. Nucl. Phys.* **18** 1
[34] Towner I S 1987 *Phys. Rep.* **155** 263
[35] Rho M 1974 *Nucl. Phys. A* **231** 493
[36] Bertsch G F and Hamamoto I 1982 *Phys. Rev. C* **26** 1323; Bertsch G F and Esbensen H 1987 *Rep. Prog. Phys.* **50** 607
[37] Brown B A 2013 *Phys. Rev. Lett.* **111** 162502
[38] Brown B A 2013 *Journal of Physics: Conference Series* **445** 012010
[39] Caurier Menendez E J Nowacki F and Poves A 2008 *Phys. Rev. Lett.* **100** 052503
[40] Holt J D and Engel J 2013 *Phys. Rev. C* **87** 064315
[41] Roth Hergert R H Papakonstantinou P Neff T and Feldmeier H 2005 *Phys. Rev. C* **72** 034002
[42] Agostini M *et al.* 2013 *Phys. Rev. Lett.* **111** 12250
[43] Menendez J Gazit D and Schwenk A 2011 *Phys. Rev. Lett.* **107** 062501
[44] Bogner S K Hergert H Holt J D Schwenk A Binder S Calci A Langhammer J and Roth R 2014 *Phys. Rev. Lett.* **113** 142501
[45] Jansen G R Engel J Hagen G Navratil P and Signoracci A 2014 *Phys. Rev. Lett.* **113** 142502
[46] http://science.energy.gov/~/media/np/nsac/pdf/docs/ 2014/NLDBD_Report_2014_Final.pdf

IOP Conf. Series: Journal of Physics: Conf. Series **966** (2018) 012018 doi:10.1088/1742-6596/966/1/012018

Experiments with brilliant gamma beams at ELI-NP: A glimpse in the future

Dimiter L. Balabanski

Extreme Light Infrastructure – Nuclear Physics, Horia Hulubei National Institute for R&D in Physics and Nuclear Engineering, Reactorului Str. 30, 077125 Bucharest – Magurele, Romania

dimiter.balabanski@eli-np.ro

Abstract. The emerging experimental program with brilliant gamma beams at the Extreme Light Infrastructure – Nuclear Physics facility (ELI-NP), which is under construction in Magurele, Romania is presented with emphasis on the prepared day-one experiments. Experiments at ELI-NP will cover nuclear resonance fluorescence (NRF) measurements, studies of large-amplitude motions in nuclei, photofission and photonuclear reactions of astrophysics interest, and measurements of photonuclear reaction cross sections. The physics cases of the flagship experiments at ELI-NP are discussed, as well as the related instruments which are under construction for their realization.

1. Introduction

The Extreme Light Infrastructure (ELI) Pan-European facility initiative represents a major step forward in quest for extreme electromagnetic fields and related research. The Extreme Light Infrastructure – Nuclear Physics (ELI-NP), which is under construction in Magurele, Romania, is one of the three pillars of the ELI, and aims at utilization of extreme electromagnetic fields for nuclear physics and quantum electrodynamics research and applications. At ELI-NP, high-power laser pulses together with brilliant gamma beams are the main research tools. It is one of the three Pan-European nuclear physics laboratories, which are at present under construction in the EU as landmark facilities within the ESFRI scheme [1]. The ELI-NP facility will be commissioned in 2018 and will become operational as a user facility in 2019.

The ELI-NP gamma-beam system (GBS) [2] will be superior to the laboratories which are operational at present [3,4] in terms of beam intensity and bandwidth. It aims at providing almost fully polarized (> 95%), narrow bandwidth ($\leq 0.5\%$) γ-ray beams in the energy range between 200 keV and 19.5 MeV with a spectral density of 10^4 photons/(s·eV). The quasi-monochromatic γ beams will be produced via Compton backscattering of laser pulses off electron bunches accelerated to relativistic energies. The 100-Hz laser pulses will be delivered by a Yb:YAG laser. The electron bunches will be accelerated by a warm linac. The high-brilliance and narrow bandwidth of the gamma beam is due to the high spectral luminosity of the photon-electron collision, $L = 2.5\cdot10^{35}$ cm$^{-2}\cdot$s^{-1} [5], which is achieved by tight focusing of the electron beam at the interaction point, *e.g.* the focal spot sizes of the laser and electron beams are ≤ 30 μm. The electron beam emittance is of upmost importance for the production of a brilliant quasi-monochromatic gamma beam by the ELI-NP GBS.

The ELI-NP GBS research program, which is under discussion within the nuclear photonics community, covers nuclear resonance fluorescence (NRF) measurements, studies of large-amplitude motions in nuclei, photofission and photonuclear reactions of astrophysics interest, and measurements of photonuclear reaction cross sections. Recently, the ELI-NP GBS research program was published in a series of technical design reports (TDR) [6-9] and reviewed in a number of papers [10-12]. The targeted experiments, which take advantage of the expected parameters of the ELI-NP gamma beams, cannot be done at present elsewhere and focus at key topics of photonuclear research. Given the fact that the ELI-NP project is realized within the EU Structural Funds, it is of importance to demonstrate early results. Thus, the day-one experimental program at the ELI-NP GBS is designed following few basic principles, *e.g.* start experiment preparations early enough and pick up the lowest-hanging fruits first, prioritize and identify flagship topics that are most attractive, yet doable on the defined short time scale, which is needed for generating further energy in the project. Here, the current ideas for day-one experiments, related to NRF and gamma-ray experiments above the particle evaporation threshold (GANT), photonuclear reactions of astrophysics interest and photofission measurements, are discussed. The expected performance of the instruments for their realization is presented, too.

2. NRF experiments

The NRF technique is an outstanding tool for the investigation of low-lying dipole excitations in atomic nuclei. In the experiment photons are scattered off bound nuclear states. For a recent review see Ref. [13]. The high-brilliance, narrow-bandwidth, pencil-size, polarized ELI-NP beams will result in unique experiments addressing the availability, sensitivity and precision experimental frontiers of NRF measurements. There are experiments that can performed already in the commissioning phase of ELI-NP. A most appealing case, which can be done already at low energies, is a photo-activation measurement of ^{180m}Ta, *i.e.*, a measurement of the subsequent γ decay as a function of the gamma-beam energy. There are indications from a bremsstrahlung experiment about the doorway states, through which the 9^- isomer is de-excited [14,15]. At ELI-NP this experiment can be done with the needed sensitivity and precision. Other ideas for day-one NRF experiments include studies of actinides that have 10,000-year lifetimes. NRF experiments at present-day facilities require large-mass targets, of the order of grams, while at ELI-MP milligram targets will be used. Thus, the availability frontier will be push further down at ELI-NP. On a longer term, parity-doublet studies [6,16] would be of a high impact, addressing both the sensitivity and precision frontiers.

For the realization of this experimental program, the ELIADE detector array [6] is constructed with highest priority. It consists of eight segmented HPGe Clover detectors with a digital readout. Both features are of importance for dealing with the time structure of the gamma beam, which has 10-Hz macropulses, each of them consisting of a train of 32 micropulses separated by 16 ns. The ELIADE array is designed in such a way that polarization and angular correlation measurements will be possible. The detector efficiency is about 6%, which will make possible $\gamma\gamma$ coincidence measurements. An artistic view of the ELIADE array is presented in Fig. 1, which reveals the positioning of the Clover detectors in two rings at 90^0 and 135^0 with respect to the gamma beam direction and in a horizontal and a vertical plane. Recently, in-beam tests of the detectors were done, using a triggerless data acquisition system (DAQ), and the expected performance has been achieved [17].

3. GANT experiments

Experiments above the neutron evaporation threshold include gamma-induced photodisintegration and studies of collective modes in nuclei, as described in the corresponding TDR [7]. The designed GANT research program is complimentary to what is possible to be done at other facilities round the world and aims at addressing hot problems in a number of research fields, such as photodisintegration cross sections for low-abundance nuclei relevant to the *p*-process nuclear synthesis, nuclear structure of GDR from the studies of the γ and neutron decays, studies of neutron-capture reactions on *s*-process branching points, where γ-decay of longer lived isotopes competes with neutron capture, and measurements of the cross sections of photonuclear reactions. Some studies can be done already during the commissioning of the GBS, *e.g.* ^{9}Be(γ,n) reaction at γ-beam energies of 1.65 to 3.20 MeV. The cross section for this reaction has been measured several times. However, there is considerable discrepancy between the reported results [18,19]. An independent measurement could resolve this discrepancy in early ELI-NP operation.

The γ decay of GDR in ^{208}Pb is a case for a day-one experiment because very detailed measurement on the absorption cross sections with fine structures is available for this case [20], while only few data with large uncertainties exist for the γ-decay branch [21]. On a longer term, detailed studies of E1 strength in atomic nuclei will be possible, which will provide reliable estimates of nuclear polarizability, an observable, which some theories relate to the symmetry term of the nuclear equation of state (EoS) [22-24]. In addition, detailed studies of the gamma-decay of the GDR both to the ground and excited states will be possible, which will set constraints on different theoretical models. It is worth noting that these experiments will have similar resolution to what is achieved with the Grand Raiden spectrometer at RCNP, Osaka in (p,p′) experiments [25]. Thus, the ELI-NP experimental program will provide high-resolution complimentary measurements of the distribution of the E1 and M1 strength in nuclei using a γ probe.

It has been demonstrated that the neutron-capture rates can be determined by inverse photodisintegration and photoexcitation studies [26,27]. Such measurements be pursued for many of the critical *s*-process branching point nuclei, which will help to determine the neutron flux and the temperature conditions at the *s*-process sites. An important aspect of this research program, is the resolution of large discrepancies of the partial (γ,*xn*) cross sections, which are reported [28-30].

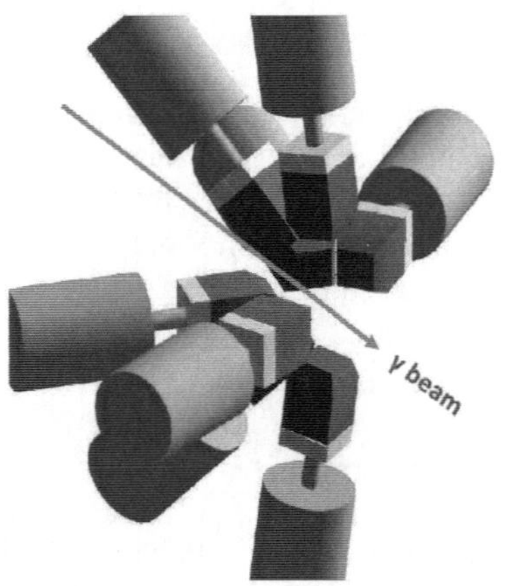

Figure 1. An artistic view of the ELIADE array, revealing the positioning of the HPGe Clover detectors with respect to the direction of the γ beam.

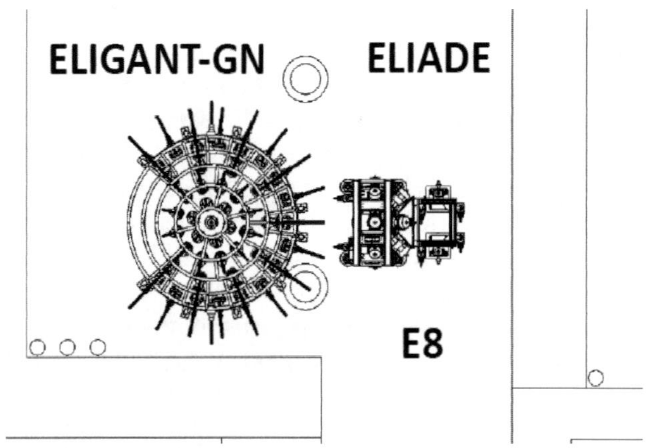

Figure 2. Schematic view of the placement of the ELIADE and ELIGANT-GN detector arrays in the E8 experimental hall of the ELI-NP laboratory complex. The γ beam is transported from right to left along the axis of the detector arrays.

Two main experimental instruments are been built to pursue the foreseen research program. The first one, named ELIGANT-TN, is based on ^{3}He tubes embedded in polyethylene for measurements of thermal neutrons. The ELIGANT-TN array is designed to have flat efficiency of about 40% over a large range of energies of the incident neutrons [31]. The second instrument, named ELIGANT-GN, is a 4π array of scintillators for measurements of γ rays and fast neutrons in different energy ranges. It consists of 34 $LaBr_3$:Ce and $CeBr_3$ 3″ x 3″ scintillators of cylindrical shape, up to 60 BC501A liquid scintillators and up to 30 GS20 ^{6}Li glass scintillators [32]. The ELIGANT-GN array will allow simultaneous studies of the (γ,γ′) and (γ,n) de-excitation of the GDR. The γ-ray efficiency will be sufficient for γγ coincidence measurements, too. All detectors, which are needed for the construction of the arrays have been delivered and the tests demonstrated that the expected performance is achieved. At present the implementation team works on the construction of the detector assembly. A schematic view of the placement of the ELIADE and the ELIGANT-GN arrays in the E8 experimental hall of the ELI-NP laboratory building is shown in Fig. 2.

4. Charged particle detection experiments

Photonuclear reactions with detection of charged particles are another topic of interest within the ELI-NP research program. The physics cases under study are related to nuclear reactions of astrophysics relevance. Two instruments are under construction for these studies, a 4π silicon strip detector (SSD) array, named ELISSA, and a time-projection chamber with electronic readout, ELITPC, which are described in the corresponding TDR [9]. Two flagship experiments are considered, *e.g.* studies of the ^{7}Li(γ,α)^{3}H and the ^{16}O(γ,α)^{12}C reactions. The measurement of the ^{7}Li photodisintegration is closely related to the understanding of the Big Bang nucleosynthesis (BBN), while the study of the ^{16}O(γ,α)^{12}C reaction cross section will look for an answer of the decades old problem of understanding of the ^{4}He burning in massive stars. The first experiment will be carried out with the ELISSA detector array, where one can measure effectively t and α particles, the products of the ^{7}Li photodisintegration. A pilot measurement [33] was performed at the HIγS facility with the ORRUBA array [34] from ORNL. This study will continue with better resolution and at lower γ-beam energies at ELI-NP. The ELISSA SSD detectors have undergone laboratory tests and the expected performance have been reached [35].

Since both, electromagnetic and nuclear interactions are time reversal, following the principle of detailed balance, the cross-section of an (α,γ) process can be determined from the measurement of the time inverse (γ,α) reaction. An advantage of measuring the photo-dissociation of ^{16}O is a factor of 50–100 gain in cross-section. Such an experiment requires intense γ-ray beams of energies of 10 MeV and less. The proposed measurements of the ^{16}O(γ,α) photo-dissociation reaction with the gas-filled active target ELITPC detector, which is under construction at ELI-NP [9], are a continuation of the studies performed at the HIγS facility [36]. A schematic drawing of the ELITPC detector is presented in Fig. 3. The figure reveals the electrode case structure and the positioning of the GEM detectors [37] and the segmented anode of the electronic readout. The ELITPC will use GET electronics [38] for data processing of the 1024-channel readout.

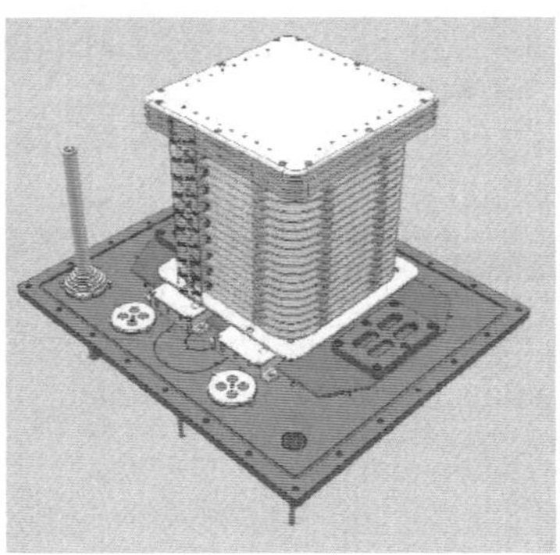

Figure 3. A 3D model view of the ELITPC detector, revealing the electrode cage structure and the positioning of the GEM detectors and the segmented anode.

ELITPC experiments at the ELI-NP GBS will benefit from the γ-beam brilliance, the higher reaction rates which will be achieved due to the greater thickness of the extended gas target and, simultaneously, the low-energy charged particles produced can be detected with sufficient angular and energy resolution. In addition, more complex decay events can be reconstructed, and background events can be detected, and rejected from the analysis. Quick and easy changes of the gas target material will allow for studies with different chemical elements, *i.e.* studies of different reactions of interest. The experiment will profit from the usage of highly isotopically enriched gases. The design of the ELITPC is in an advanced stage. A prototype, called mini-TPC, which has a 256-channel readout, has been successfully tested in-beam, using an α beam, and the track reconstruction of scattered α-particles was demonstrated [39].

5. Photofission experiments

The goal of these studies is the exploration of the unique photofission reactions enabled by the GBS at ELI-NP, which are described in the corresponding TDR [8]. The flagship experiment is the investigation of fission transmission resonances in the second and third potential minima in the actinides, thus mapping the fission barriers as a function of the photon-beam energy in the range of about 5 to 7 MeV [40]. The experiments aim at the characterization of these transmission resonances through angular distributions, masses and charges of the fission fragments. A pilot experiment on ^{238}U has been carried out at the HIγS facility [41], but the available beams there lack resolution to resolve the resonance states. The advantages of the photofission studies at ELI-NP are the high flux, $1.5 \cdot 10^8$ γ/s, and the small bandwidth $\Delta E/E = 0.3\%$, *e.g.* 15 keV at 5 MeV. This will allow resolving the fewer 1^- and 2^+ transmission resonances in the second and third minimum that will be preferentially populated in photofission in contrast to the many more resonances that are populated with hadronic probes. This will lead to an accurate mapping of the potential energy surface (PES) of the actinide nuclei.

Related to GBS photofission studies, the research programme includes also measurement of kinetic energy, angular, mass and charge distribution of fission fragments, measurements of absolute photofission cross-sections, studies of rare photofission events, such as triple fission, highly asymmetric fission, clusterization phenomena, the predicted cold valleys of fission potential, etc [8].

Two instruments are being developed for photofission studies at the ELI-NP GBS, the ELI-BIC and ELITHGEM detector arrays [8]. Their design is in an advanced phase. The ELI-BIC array consists of four double-sided Frisch-grid Bragg ionization chambers (DSBIC). Each ionization chamber in the ELI-BIC array will be coupled to eight ΔE-E detectors covering a solid angle of π, for the study of light charge particles, mostly long range α-particles from ternary fission events. A ΔE-E detector consists of a gas detector coupled with a double sided Si strip detector. The ELITHGEM array consists of thick gas electron multipliers (THGEM), which allows measurement of angular distributions of fission fragments. The ELI-BIC and ELITHGEM detectors have undergone different tests and it has been demonstrated that the expected performance has been reached [42,43].

6. Summary

The progress of the preparation of the day-one research program at the ELI-NP GBS has been reported. All instruments, which are required to take data at the start of the facility, are in advanced stage of construction and will be commissioned during the commissioning phase of the GBS. The detectors and the electronics, which are needed for construction of the arrays, have been delivered and the tests demonstrate that the expected performance have been achieved. Cases of flagship experiments were defined, and certain low-hanging fruits which should be measured first at the start of the facility have been identified.

In short, the ELI-NP physics team will be ready in time with all the instrumentation needed for taking data at the ELI-NP GBS, and will be prepared to provide services to the international nuclear physics community when ELI-NP becomes operational as an open access facility in 2019.

Acknowledgments

Work supported by the Extreme Light Infrastructure Nuclear Physics (ELI-NP) Phase II, a project co-financed by the Romanian Government and the European Union through the European Regional Development Fund – the Competitiveness Operational Programme (1/07.07.2016, COP, ID 1334).

References

[1] *ESFRI Strategy Report on Research Infrastructures: Roadmap 2016*, ISBN: 978-0-9574402-4-1, www.esfri.eu/roadmap-2016
[2] Adriani O *et al.* 2014 arXiv:1407.3669 [physics.acc-ph]
[3] NewSUBARU synchrotron radiation facility, University of Hyogo, http://www.lasti.u-hyogo.ac.jp/NS-en/
[4] High Intensity Gamma-Ray Source (HIGS), Triangle Universities Nuclear Laboratory,

http://www.tunl.duke.edu/facilities/
[5] Bacci A *et al.* 2013 *J. Appl. Phys.* **113** 194508
[6] Ur C A *et al.* 2016 *Rom. Rep. Phys.* **68** S483
[7] Camera F *et al.* 2016 *Rom. Rep. Phys.* **68** S539
[8] Balabanski D L *et al.* 2016 *Rom. Rep. Phys.* **68** S621
[9] Tesileanu O *et al.* 2016 *Rom. Rep. Phys.* **68** S699
[10] Filipescu D *et al.* 2015 *Eur. Phys. J. A* **51** 185
[11] Gales S *et al.* 2016 *Phys. Scr.* **91** 093004
[12] Balabanski D L *et al.* 2017 *Europhys. Lett.* **117** 28001
[13] Kneissl U, Pietralla N, Zilges A, 2006 *J. Phys. G* **32** R217
[14] Belic D *et al.* 2002 *Phys. Rev. C* **65** 035801
[15] Belic D *et al.* 1999 *Phys. Rev. Lett.* **83** 5242
[16] Beller J *et al.* 2015 *Phys. Lett. B* **741** 128
[17] Capponi L *et al.* 2017 *Nucl. Instr. Meth. Phys. Res. A* (in preparation)
[18] Unsunomiya H *et al.* 2015 *Phys. Rev. C* **92** 064323
[19] Arnold C W *et al.* 2012 *Phys. Rev. C* **85** 044605
[20] Tamii A *et al.* 2011 *Phys. Rev. Lett.* **107** 062502
[21] Beene J R *et al.* 1990 *Phys. Rev. C* **41** 920
[22] Reinhardt P-G, Nazarewicz W 2010 *Phys. Rev. C* **81** 051303
[23] Piekarewicz J 2011 *Phys. Rev. C* **83** 034319
[24] Roca-Maza X *et al.* 2011 *Phys. Rev. Lett.* **106** 252501
[25] Tamii A 2009 *Nucl. Instr. Meth. Phys. Res. A* **605** 326
[26] Raut R *et al.* 2013 *Phys. Rev. Lett.* **111** 112501
[27] Sauerwein A *et al.*, 2014 *Phys. Rev. C* **89** 035803
[28] Varlamov A V *et al.* 1999 *Atlas of Giant Dipole Resonances, Preprint* INDC (NDS) 394
[29] Berman B, Fultz S 1975 *Rev. Mod. Phys.* **47** 713
[30] Dietrich S, Berman B 1988 *At. Data Nucl. Data Tables* **38** 199
[31] Unsunomiya H *et al.* 2017 *Nucl. Instr. Meth. Phys. Res. A* **871** 135
[32] Krzysiek M *et al.* 2017 *Nucl. Instr. Meth. Phys. Res. A* (in preparation)
[33] Matei C *et al.* 2017 *Phys. Rev. C* (in preparation)
[34] Pain S D *et al.* 2007 *Nucl. Instr. and Meth. in Phys. Res. B* **261** 1122
[35] Chesnevskaya S *et al.* 2017 *JINST* (in preparation)
[36] Zimmerman W R *et al.* 2012 *Phys. Rev. Lett.* **110** 152502
[37] Shalem S K *et al.* 2006 *Nucl. Instr. Meth. Phys. Res. A* **558** 468
[38] Pollacco E *et al.* 2012 *Phys. Procedia* **37** 1799
[39] Bihałowicz J S 2015 *AIP Conf. Proc.* **1645** 301
[40] Krasznahorkay A 2011 *Handbook of Nuclear Chemistry* vol 1, ed A Vertes *et al.* (Dordrecht: Springer) p 281
[41] Csige L *et al.* 2013 *Phys. Rev. C* **87** 044321
[42] Choudhury D *et al.* 2017 *Acta Phys. Pol. B* **48** 559
[43] Balabanski D L *et al.* 2017 *EPJ Web Conf.* (in print)

IOP Conf. Series: Journal of Physics: Conf. Series **966** (2018) 012019 doi:10.1088/1742-6596/966/1/012019

Electromagnetic reactions from coupled-cluster theory

Sonia Bacca[1,2,3], **Mirko Miorelli**[2,4], **Gaute Hagen**[5,6]

[1] Institut für Kernphysik, Johannes Gutenberg-Universität Mainz, 55128 Mainz, Germany
[2] TRIUMF, 4004 Wesbrook Mall, Vancouver, BC V6T 2A3, Canada
[3] Department of Physics and Astronomy, University of Manitoba, Winnipeg, MB, Canada R3T 2N2
[4] Department of Physics and Astronomy, University of British Columbia, Vancouver, BC, V6T 1Z4, Canada
[5] Physics Division, Oak Ridge National Laboratory, Oak Ridge, TN 37831, USA
[6] Department of Physics and Astronomy, University of Tennessee, Knoxville, TN 37996, USA

E-mail: s.bacca@uni-mainz.de, miorelli@triumf.ca, hageng@ornl.gov

Abstract. We review recent results for electromagnetic reactions and related sum rules in light and medium-mass nuclei obtained from coupled-cluster theory. In particular, we highlight our recent computations of the photodisintegration cross section of ^{40}Ca and of the electric dipole polarizability for oxygen and calcium isotopes. We also provide new results for the Coulomb sum rule for ^{4}He and ^{16}O. For ^{4}He we perform a thorough comparison of coupled-cluster theory with exact hyperspherical harmonics.

1. Introduction

Electromagnetic reactions have traditionally been a golden tool for studying nuclear dynamics. They lead, for instance, to the discovery of giant dipole resonances and their interpretation in terms of collective modes [1, 2]. While a complete body of data has been collected over the years for stable nuclei, only recently have we begun to elucidate them in terms of first principle calculations. This was achieved thanks to the introduction of a new ab initio computational tool, obtained from a coupled cluster theory formulation of the Lorentz integral transform method [3], which we will review below.

The key ingredient to study reactions induced by external electromagnetic probes, such as photons or electrons, is the nuclear response function. The latter is defined as

$$R(\omega, q) = \sum_f \left| \langle \psi_f | \hat{\Theta}(q) | \psi_0 \rangle \right|^2 \delta\left(E_f - E_0 - \omega\right), \tag{1}$$

where $\hat{\Theta}(q)$ is the excitation operator, which will in general depend on the momentum-transfer q and whose form will be specific to the considered probe. The nuclear response function is a dynamical observable which requires knowledge on the whole spectrum of the nucleus, being $|\psi_0\rangle$ and $|\psi_f\rangle$ the ground- and excited-states, respectively. The difficulty of studying Eq. (1) lies in the fact that one needs, in principle, to calculate all the excited states $|\psi_f\rangle$, including all open channels in the continuum. While this is possible for two- and three-body problems, starting at four nucleons it can only be done with restrictions in energy and/or with approximations. To overcome this complication, one can employ the Lorentz integral transform (LIT) [4] and reduce the problem to the solution of a bound state equation. In this approach, rather than directly calculating $R(\omega, q)$, one computes the following transform

$$L(q, \omega_0, \Gamma) = \int_{\omega_{\rm th}}^{\infty} d\omega \frac{R(\omega, q)}{(\omega - \omega_0)^2 + \Gamma^2}\,. \tag{2}$$

Here $\omega_{\rm th}$ is the threshold energy and $\Gamma > 0$ is the Lorentzian width, which plays the role of a resolution parameter. By using the closure relation one finds

$$L(q,z) = \langle\psi_0|\hat{\Theta}^\dagger(q)\frac{1}{\hat{H}-z^*}\frac{1}{\hat{H}-z}\hat{\Theta}(q)|\psi_0\rangle = \langle\widetilde{\psi}|\widetilde{\psi}\rangle\,, \tag{3}$$

where we introduced the complex energy $z = E_0 + \omega_0 + i\Gamma$ and $\hat{H}$ is the nuclear Hamiltonian, which will be taken from chiral effective field theory [5, 6]. The LIT of the response function, $L(q,z)$, can be computed directly by solving the Schrödinger-like equation

$$(\hat{H} - z)|\widetilde{\psi}\rangle = \hat{\Theta}(q)|\psi_0\rangle \tag{4}$$

for different values of z. The solution $|\widetilde{\psi}\rangle$ has bound-state-like asymptotic, thus $L(q,z)$ can be calculated even for $A > 3$ with any good bound-state method, such as hyperspherical harmonics expansions [7]. Results for $R(q,\omega)$ are obtained from an inversion of the LIT and are independent on the choice of Γ [4].

Recently, we have merged the advantage of the LIT method with the mild computational scaling that characterizes coupled-cluster theory [8] for increasing mass number, obtaining a new computational tool [3]. In coupled-cluster theory, as originally introduced by Coester and Kümmel [9], the exact many-body wave function is written in the exponential *ansatz*

$$|\Psi_0\rangle = \exp{(T)}|\Phi_0\rangle\,, \tag{5}$$

where $|\Phi_0\rangle$ is a Slater determinant. The operator T, typically expanded in n-particle-n-hole excitations (or clusters) $T = T_1 + T_2 + \cdots + T_A$, is responsible for introducing correlations.

The LIT method in the coupled-cluster language requires the solution of a bound-state equation

$$(\overline{H} - z)|\tilde{\Psi}_R(z)\rangle = \overline{\Theta}(q)|0_R\rangle\,, \tag{6}$$

where the right ground-state $|0_R\rangle \equiv |\Phi_0\rangle$, while $\overline{H}$ and $\overline{\Theta}$ are the similarity transformed Hamiltonian and excitation operators, respectively

$$\overline{H} = \exp(-T)\hat{H}\exp(T)\,, \tag{7}$$
$$\overline{\Theta} = \exp(-T)\hat{\Theta}\exp(T)\,. \tag{8}$$

The solution of Eq. (6) is found as linear superposition of particle-hole excitation on top of the reference Slater determinant as

$$|\widetilde{\Psi}_R(z)\rangle = \mathcal{R}(z)|\Phi_0\rangle = (\mathcal{R}_1 + \mathcal{R}_2 + \cdots + \mathcal{R}_A)|\Phi_0\rangle\,.$$

In Ref. [3] we have adopted the most commonly used approximation, i.e., the single and doubles scheme (CCSD), where $T = T_1 + T_2$ and $\mathcal{R} = \mathcal{R}_1 + \mathcal{R}_2$ and have successfully benchmarked this method with exact hyperspherical harmonics on ^{4}He, showing that the CCSD approximation differs only by 1-2% from the exact result.

Because the computational cost of the coupled-cluster method scales mildly with respect to the mass number, this theory paves the way for many future investigations of electroweak reactions in medium-mass nuclei, as it provides an *ab initio* approach where the continuum is properly taken into account.

2. The photodisintegration cross section

The first observables where we used the power of coupled-cluster theory to address medium-mass nuclei has been the photodisintegration cross section. In the unretarded dipole approximation, the photodisintegration cross section is

$$\sigma_\gamma(\omega) = 4\pi^2\alpha\omega R(\omega)\,, \tag{9}$$

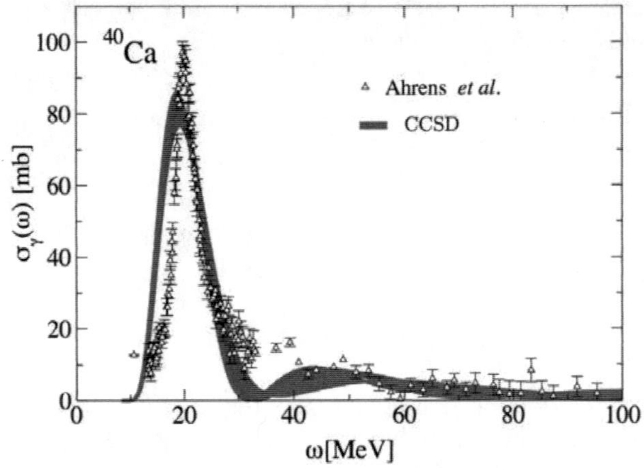

Figure 1. ^{40}Ca photodisintegration cross section at the CCSD level [10] compared to experimental data by Ahrens *et al.* [12]. Result are obtained with a two-nucleon chiral force [11]. Figure adapted from Ref. [10].

where $R(\omega)$ is the dipole response function, with $\omega = q$ for real photons. The excitation operator is the translationally invariant dipole operator

$$\hat{\Theta} = \sum_{k}^{A} (\mathbf{r}_k - \mathbf{R}_{\rm cm}) \left(\frac{1 + \tau_k^3}{2} \right), \tag{10}$$

where $\mathbf{r}_k$ and $\mathbf{R}_{\rm cm}$ are the coordinates of the k-th particle and the center-of-mass, respectively, while $(1 + \tau_k^3)/2$ defines the projection operator on the Z protons.

By solving Eq. (6) with the dipole operator in Eq. (10) we have been able, e.g., to calculate the photodisintegration cross section of ^{40}Ca [10] using a realistic two-body chiral potential [11]. Results are shown in Fig. 1. The width of the blue curve is obtained from various inversions of the integral transform and constitutes a lower band of the total theoretical error-bar associated to this observable. A very good description of this quantity with respect to the experimental data from Ref. [12] is obtained already at this level of approximation in coupled-cluster theory. This result highlights the relevance of the newly introduced method, which allowed (i) to depart from the classical few-body systems and reach medium mass nuclei with only little more computational expense and (ii) to obtain, for the first time, a microscopic description of experimental data in this mass range.

3. The electric dipole polarizability

We now focus our attention on the electric dipole polarizability. The latter is defined as an inverse energy weighted sum rule of the dipole response function

$$\alpha_D = 2\alpha \int_{\omega_{ex}}^{\infty} d\omega \, \frac{R(\omega)}{\omega} \,. \tag{11}$$

It is well known from the nuclear droplet and hydrodynamic models [13], as well as from density functional theory [14], that the electric dipole polarizability is correlated with the nuclear charge radius r_{ch}. Thus, it is interesting to see if this correlation is also observed in first principle calculations and the developed LIT-CC technology allows for such a study.

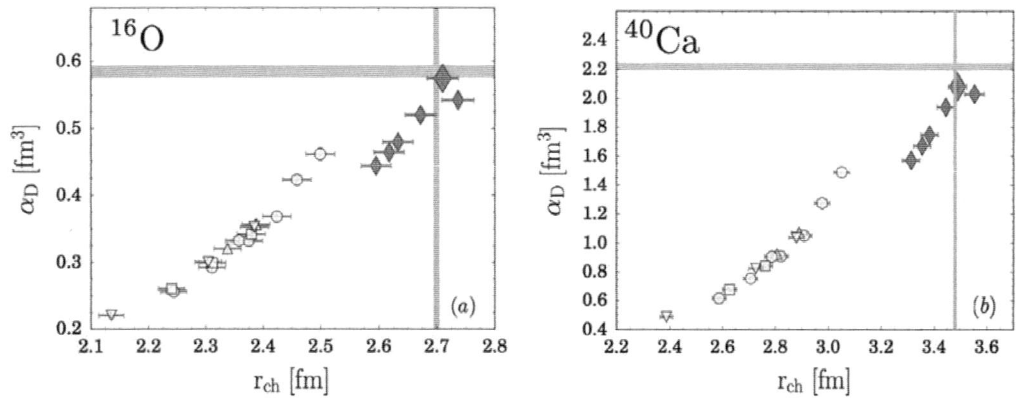

Figure 2. α_D versus r_{ch} in ^{16}O (a) and ^{40}Ca (b). Empty symbols refer to calculations with nucleon-nucleon potentials only. The red (full) diamonds refer to calculations that include three-body forces. Details on the potentials used are found in Ref. [15]. The green bands denote the experimental values for these two quantities. Figure adapted from Ref. [15].

In order to study correlations, one needs a considerable number of interactions, such that results span a wide range of values both for the polarizability and the charge radius. For this purpose, we take some of the few realistic nucleon-nucleon interaction and evolve them with similarity transformations. Fig. 2 shows the correlation of the electric dipole polarizability α_D with the charge radius in ^{16}O (a) and in ^{40}Ca (b). The empty symbols refer to calculations with two-body forces only. They clearly show a strong correlation between the two observables. On the other hand, one also observes that the use of two-body interactions only leads to a strong underestimation of the experimental values of both quantities, denoted by the green bands. When we include three-nucleon forces, by using a variety of Hamiltonians from chiral effective field theory [16, 17], we obtain the solid diamonds, which still exhibiting a strong correlation but cluster much closer to the experimental value. This fact indicates that three nucleon forces are important to obtain a correct description of both observables. It needs to be noted though, that we have made a specific choice of Hamiltonians, preferring those that provide a reasonable description of the charge radius. It is well known that other Hamiltonians from chiral effective field theory suffer from and under-estimation of the radii, which, in the light of Fig. 2, would also lead to an underestimation of the electric dipole polarizability. In particular, the largest diamond data are obtained with the NNLO$_{sat}$ [17] potential, which has been fit to reproduce also the ^{16}O charge radius. It is interesting to note that this constraint enables a correct prediction also for ^{40}Ca.

By studying similar correlations in ^{48}Ca and combining them with experimental information from electron scattering, we also were able to provide narrow constraints on the neutron skin thickness and on the electric dipole polarizability [18, 19].

4. The Coulomb sum rule

Finally, we turn our attention to the Coulomb sum rule. With the objective in mind to eventually tackle neutrino-nucleus interactions relevant for long-baseline neutrino experiments, we initiated a study of electron-scattering reactions within coupled-cluster theory. Following the work done with the Green's function Monte Carlo method in Ref. [20], we first investigate the Coulomb sum rule.

Omitting for simplicity the role of the nucleon form factor, the longitudinal Coulomb sum rule is

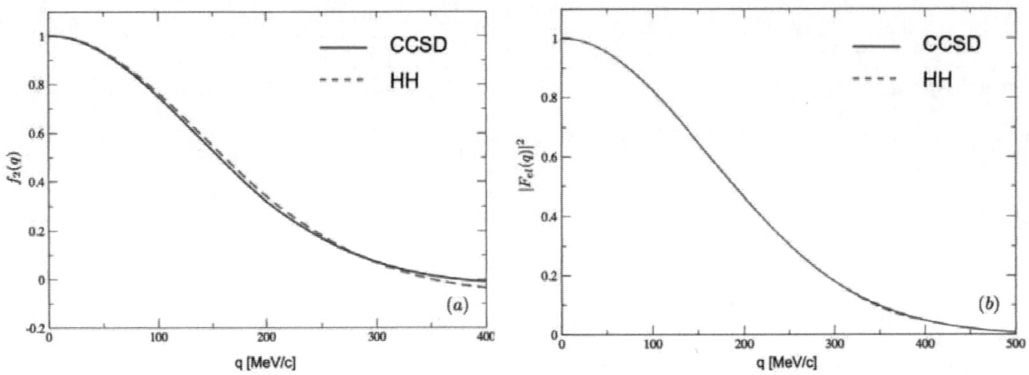

Figure 3. Comparison of CCSD and exact hyperspherical harmonics (HH) computations of $f_2(q)$, panel (a) and $|F_{el}(q)|^2$, panel (b) as a function of the momentum transfer q using a chiral two-body force [11] for the nucleus of ^{4}He.

defined as as total strength of the inelastic longitudinal response function as

$$CSR(q) = \int_{\omega_{th}^+}^{\infty} d\omega R_L(\omega, q)\,, \tag{12}$$

where

$$R_L(\omega, q) = \sum_{f\neq 0} |\langle \Psi_f|\hat{\rho}(q)|\Psi_0\rangle|^2 \delta\left(E_f + \frac{q^2}{2M} - E_0 - \omega\right), \tag{13}$$

with $\hat{\rho}(q)$ being the charge operator and $\frac{\mathbf{q}^2}{2M}$ a recoil term. One possible approach to compute the Coulomb sum rule is to perform a multipole expansion of the charge operator and then solve Eq. (4) for every multipole on a grid of q, as was done in Ref. [21]. However, the Coulomb sum rule can be rewritten also as expectation value on the ground state which is much easier to calculate

$$CSR(q) = Z + \langle\psi_0|\sum_{i\neq j} e^{i\mathbf{q}\cdot(\mathbf{r}_i - \mathbf{r}_j)}|\psi_0\rangle - |F_{el}(q)|^2 Z^2 = Z + Z(Z-1) f_2(q) - |F_{el}(q)|^2 Z^2\,, \tag{14}$$

where $f_2(q)$ is the Fourier transform of the proton-proton correlation density and $F_{el}(q)$ is the elastic form factor [22]. To first compare the CCSD approximation with exact calculation we compare computations of $f_2(q)$ and $|F_{el}(q)|^2$ with exact hyperspherical harmonics, as shown in Fig. 4 vs. the momentum transfer. Calculations are performed with a chiral two-body force [11] and show a very nice agreement between the two methods, proving the CCSD approximation to be very reliable.

In Fig. 4 we finally compare the Coulomb sum rule calculated from coupled-cluster theory for ^{4}He and ^{16}O with experimental data. Measurements of the longitudinal response functions at low momentum transfers (q = 200 and 250 MeV/c) are taken from Buki *et al.* [23], while intermediate momentum transfer values are collected in Ref. [24]. Unfortunately, no data exist for ^{16}O, so we show data for ^{12}C taken from [25]. Similar results were also obtained in Refs. [25, 26]. Fig. 4 shows that our calculations agree quite well with the experimental data. Thus, extensions to heavier nuclei and to neutrino reactions are promising.

5. Conclusions

We reviewed some of our recent results on electromagnetic reactions and related sum rules obtained from coupled cluster theory and the Lorentz integral transform. The method has allowed us to surpass previous

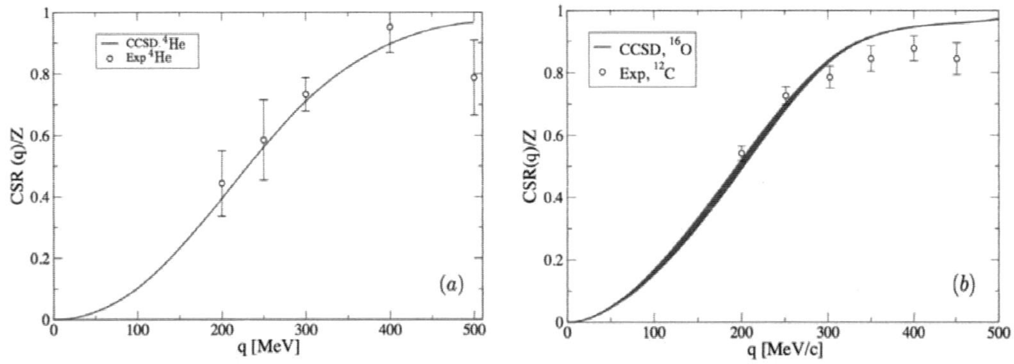

Figure 4. Coulomb sum rule for ^{4}He (a) and ^{16}O (b) from coupled-cluster theory using a chiral two-body force [11], compared to experimental results from [23, 24] for ^{4}He (a), and from Ref. [25] for ^{12}C (b).

mass limits and paves the way for ab initio studies of neutrino-nucleus cross section. We addressed first electron scattering observables, such as the Coulomb sum rule, and performed a comparison to exact calculations and to data. This is essential for any theory that is going to be used to model neutrino-nucleus interactions.

Acknowledgments

This work was supported in parts by the Natural Sciences and Engineering Research Council (NSERC), the National Research Council of Canada, the PRISMA Cluster of Excellence and the Sonderforschungbereich CRC 1044.

References

[1] Goldhaber M and Teller E 1948 *Phys. Rev.* **74** 1046
[2] Steinwedel H and Jensen J H D 1950 Z. *Naturforsch* **5A** 413
[3] Bacca S *et al.* 2013 *Phys. Rev. Lett.* **111** 122502
[4] Efros V D *et al* 2007 *J. Phys. G: Nuclear and Particle Physics* **34** R459
[5] Epelbaum E, Hammer H-W and Meißner U-G 2009 *Rev. Mod. Phys.* **81** 1773
[6] Machleidt R and Entem D R 2011 *Phys. Rep.* **503** 1
[7] Barnea N, Leidemann W and Orlandini G 2001 *Nucl. Phys. A* **693** 565
[8] Hagen G, Papenbrock T, Hjorth-Jensen M and Dean D J 2014 *Rep. Prog. Phys.* **77** 096302
[9] Coester F and Kümmel H 1960 *Nucl. Phys.* **17** 477
[10] Bacca S, Barnea N, Hagen G, Miorelli M, Orlandini G and Papenbrock T 2014 *Phys. Rev. C* **90** 064619
[11] Entem D R and Machleidt R 2003 *Phys. Rev. C* **68** 041001
[12] Ahrens J *et al.* 1975 *Nucl. Phys. A* **251**, 479
[13] Lipparini E and Stringari S 1989 *Phys. Rep.* **175** 104
[14] Roca-Maza X *et al.* 2015 *Phys. Rev. C* **82** 064304
[15] Miorelli M, Bacca S, Barnea N, Hagen G, Jansen G R, Orlandini G, Papenbrock T 2016 *Phys. Rev. C* **94** 034317
[16] Hebeler K and Bogner S K, Furnstahl R J, Nogga A and Schwenk A 2011 *Phys. Rev. C* **83** 031301
[17] Ekström A *et al.* 2015 *Phys. Rev. C* **91** 051301
[18] Hagen G *et al.* 2016 *Nature Physics* **12**, 186
[19] Birkhan J *et al.* 2017 *Phys. Rev. Lett.* **118** 252501
[20] A. Lovato *et al.* 2013 *Phys. Rev. Lett.* **111** 092501
[21] Bacca S, Barnea N, Leidemann W and Orlandini G 2009 *Phys. Rev. C* **80** 064001
[22] Orlandini G and Traini M 1991 *Rep. Prog. Phys.* **54** 257
[23] Buki A *et al.* 2006 *Phys. Lett. B* **641** 156
[24] Carlson J, Jourdan J, Schiavilla R and Sick I 2002 *Phys. Rev. C* **65** 024002
[25] Mihaila B and Heisenberg H J 2000 *Phys. Rev. Lett.* **84** 1403
[26] Lonardoni D *et al.* 2017 *Phys. Rev. C* **96** 024326

How sharp is the transition into the N=20 island of inversion for the Mg isotopes ?

B. Fernández-Domínguez[1,2,3], B. Pietras[1], W.N. Catford[4], N.A. Orr[2], M. Petri[1,5,6], M. Chartier[1], S. Paschalis[1,6], N. Patterson[4], J.S. Thomas[4], M. Caamaño[3], T. Otsuka[7], A. Poves[8], N. Tsunoda[7], N.L. Achouri[2], J-C. Angélique[2], N.I. Ashwood[9], A. Banu[10,1], B. Bastin[2], R. Borcea[11], J. Brown[1], F. Delaunay[2], S. Franchoo[12], M. Freer[9], L. Gaudefroy[13], S. Heil[5], M. Labiche[14], B. Laurent[2], R.C. Lemmon[14], A.O. Macchiavelli[15], F. Negoita[11], E.S. Paul[1], C. Rodríguez-Tajes[13], P. Roussel-Chomaz[13], M. Staniou[11], M. Taylor[6,2], L. Trache[10,11], G.L. Wilson[4]

[1]Department of Physics, University of Liverpool, Liverpool, L69 7ZE, UK
[2]LPC-Caen, IN2P3/CNRS, ENSICAEN, UNICAEN et Normandie Université, 14050 Caen Cedex, France
[3]Universidade de Santiago de Compostela, 15754 Santiago de Compostela, Spain
[4]Department of Physics, University of Surrey, Guildford GU2 5XH, UK
[5]Institut für Kernphysik, Technische Universität Darmstadt, 64289 Darmstadt, Germany
[6]Department of Physics, University of York, Heslington, York YO10 5DD, UK
[7]CNS, University of Tokyo, 7-3-1 Hongo, Bunkyo-ku, Tokyo, Japan
[8]Departamento de Fisica Teórica and IFT-UAM/CSIC, Universidad Autónoma de Madrid, E-28049 Madrid, Spain
[9]School of Physics and Astronomy, University of Birmingham, Birmingham, B15 2TT, UK
[10]Cyclotron Institute, Texas A&M University, College Station, TX-77843, USA
[11]IFIN-HH Bucharest-Magurele, RO-077125, Romania
[12]IPN-Orsay, 91406 Orsay, France
[13]GANIL, CEA/DRF–CNRS/IN2P3, BP 55027, 14076 Caen Cedex 05, France
[14]Nuclear Structure Group, CCLRC Daresbury Laboratory, Daresbury, Warrington WA4 4AD, UK
[15]Nuclear Science Division, Lawrence Berkeley National Laboratory, Berkeley, California 94720, USA

E-mail: beatriz.fernandez.dominguez@usc.es

Abstract.
The N=20 island of inversion is an excellent playground for testing shell model calculations. The Mg chain is a region of shell evolution still far from being well understood. In this paper we present preliminary results of a single-neutron knockout experiment from ^{31}Mg performed at GANIL to study the structure of ^{31}Mg and of the core ^{30}Mg. The level scheme and longitudinal momentum distributions were mesured and spectroscopic factors were deduced. Negative parity states arise at low energy and the spectroscopic factor for the isomeric 0^+_2 in ^{30}Mg was determined to be smaller than foreseen in the standard picture. The preliminary

[1] Present address: Department of Physics and Astronomy, James Madison University, Harrisonburg, VA 22807, USA
[2] Present address: Division of cancer sciences, University of Manchester, Manchester, M13 9PL, UK

IOP Conf. Series: Journal of Physics: Conf. Series **966** (2018) 012020 doi:10.1088/1742-6596/966/1/012020

experimental results are compared to state-of the art shell model calculations revealing opposed interpretations.

1. Introduction

The N=20 island of inversion is a region of sudden changes in shell structure where intruder configurations become dominant in the ground state wave function of several neutron-rich nuclei with atomic numbers in between Z=10 and Z=12. Over the past years, large-scale shell model calculations have been very succesful in describing the main properties of the nuclei around this region [1] and predicting new islands of inversions [2]. However, just recently a newly developed interaction, EEdf1 [3], that reproduces as well most of the electromagnetic properties of the nuclei around the N=20 region, predicts an increase of particles in the pf shell and holes in the sd shell over the Z = N = 20 shell gap. In particular along the Mg chain, nuclei at the border of the island of inversion are ideal to investigate the enhancement of $np - nh$ excitations across the N=20 shell gap. Energy levels are important but overlap functions are key inputs in the calculations. Therefore, a single-neutron knockout experiment from ^{31}Mg was performed at GANIL to study the structure of ^{31}Mg and of the core ^{30}Mg.

2. Experimental details

The experiment was performed at the GANIL coupled cyclotron facility. A primary beam of ^{36}S incident on a ^{12}C production target was used to produce a cocktail beam, analysed and purified using the SISSI device and the ALPHA spectromer [4]. The secondary beam of ^{31}Mg at $\sim$ 55 MeV/u impinged on a 171 mg/cm^2 ^{12}C target placed at the entrance of the SPEG spectrometer [5]. An array of 8 segmented EXOGAM clovers of high purity Ge detectors [6] surrounded the target. The array was arranged in a configuration with two rings at 45° and 135° polar angles with four clovers detectors each. The photopeak efficiency of the array was measured to be 3.3% at 1.3 MeV and the energy resolution, after Doppler correction, was 2.7%.

3. Results

The add-back reconstructed and doppler corrected γ-ray spectrum detected in coincidence with ^{30}Mg is shown in Fig. 1. Transitions previously reported [7, 8, 9, 10] can be clearly seen at: 799, 954, 985, 1482, 1816, 1898, 1975 and 3534 keV. A weak transition not reported before at 1660 keV is also visible. Furthermore, a peak at $\sim$ 300 keV corresponding to the known isomeric 306 γ-ray, coming from the 1789 keV level and decaying into the 1482 keV 2_1^+ state, is also observed. This state was reported to have a half-life of 3.9(5) ns based on fast timing measurements $\beta\gamma\gamma(t)$ [11]. The line shape shown by the forward energy spectrum in Fig. 2 reflects the isomeric nature of the state and agrees well with the reported value of the lifetime. The number of counts below each transition N_γ, was obtained in a statistical fit of the γ-ray spectrum to GEANT4 simulations. The γ-ray intensities for each transition, I_γ, are then calculated from the ratio of the efficiency corrected photopeak counts $N_\gamma^{'}$, and the number of residual knockout cores. Branching ratio for the direct population of each level, b, has been obtained after feeding correction, assuming the level scheme proposed in Fig. 4.

For the isomeric state, the branching ratio was deduced only from the E2 radiative contribution since the branching ratio between the E0 and E2 transitions is shown to be small $\sim$ 1.4 10^{-2}, as deduced from the partial lifetime of the E0 decay ($\tau(E0)$=396 ns) measured by Schwerdtfeger et al. [12]. Besides, the non-radiative contribution of the E2 transition was found to be negligeable (internal conversion coefficient α_{tot}=9.1 10^{-4}). The exclusive cross section for

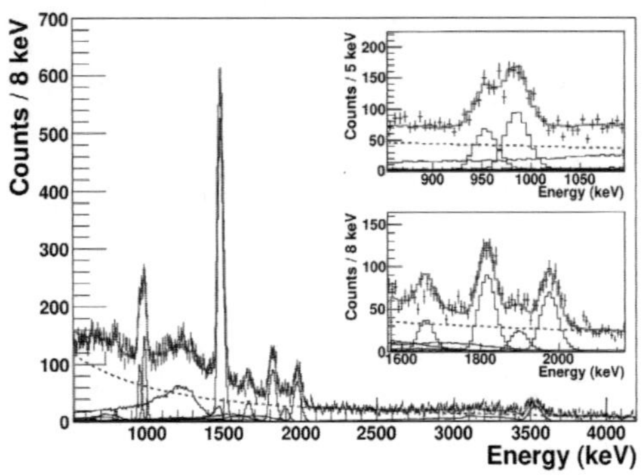

Figure 1. The γ-ray energy spectrum ($E\gamma$ > 500 keV) compared to Geant4 simulated line shapes for each transition (black lines). The total fit (red line) includes also a continuum background of an exponential function (blue dashed line), where the slope and scale parameters are left free to vary in the fit.

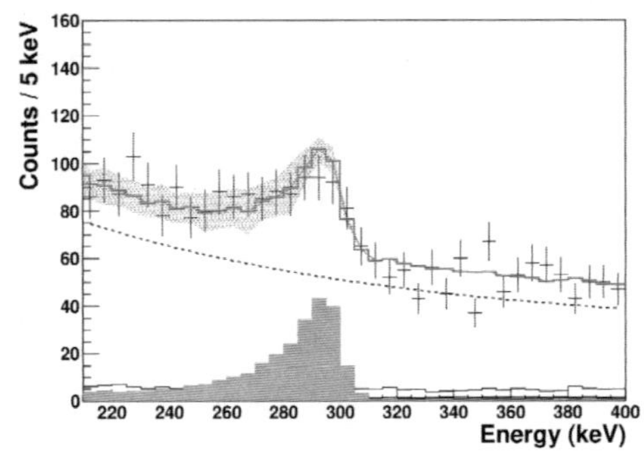

Figure 2. The γ-ray energy spectrum ($E\gamma$ < 500 keV) reconstructed using the post-target velocity (β_{post} = 0.303) and using only the EXOGAM clovers in the forward array owing to a higher sensitivity to the lineshape. The grey histogram shows a simulated isomeric transition at $\sim$ 300 keV with a lifetime of 3.9(5) ns

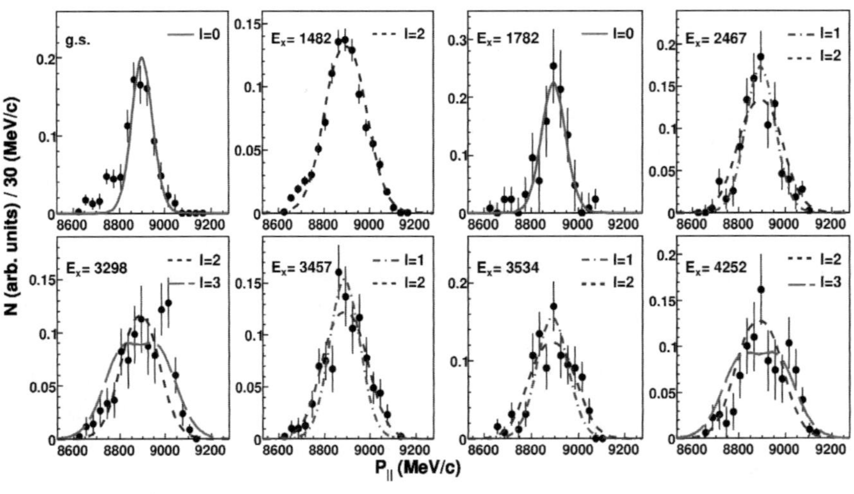

Figure 3. Experimental momentum distributions (solid marker) compared to eikonal-model calculations with ℓ= 0, 1, 2 and 3 (red solid, blue short-dashed, green dot-dashed, and brown long-dashed lines, respectively) for the states in ^{30}Mg. The theoretical lineshapes have been convoluted with the experimental resolution and are normalized to the best fit of data.

each state is given by the product of the branching ratio and the inclusive cross section (σ_{inc} = 90(12) mb).

The exclusive longitudinal momentum distributions for the ground state and for the states at: 1482, 1782, 2467, 3298, 3457, 3534 and 4252 keV are presented in Fig. 3. The ground state momentum distribution was obtained by subtracting away the exclusive momentum distributions for the excited states shown in Fig. 3. The eikonal model calculations [13, 14] have been folded with the experimental filter including the beam spread, energy loss and angular straggling, and the resolution of the spectrometer and are compared in Fig. 3 to the experimental results.

The experimental shape of the ground state distribution is reproduced with a s-wave, whereas the first excited state at 1482-keV is in agreement with a l=2 value. Despite the low statistics of the momentum distribution gated on the isomeric transition at 300-keV, the narrow shape is consistent with an assignment of l=0. The momentum distribution for the state at 2467-keV was obtained applying a narrow gate on the high-energy side of the 985-keV photopeak and the shape is compatible with l=1 since $\chi^2_\nu(l=1) = 1.0$ whereas $\chi^2_\nu(l=2) = 1.6$. For the state at 3298-keV, the distribution is broad and in best agreement with a l=3 value ($\chi^2_\nu(l=3) = 1.9$ and $\chi^2_\nu(l=2) = 2.8$). The distribution gated on the 1975-keV transition is best described by a l=2 value ($\chi^2_\nu(l=2) = 0.8$ and $\chi^2_\nu(l=1) = 2.2$) and that to the state at 3534-keV is consistent with l=1 and l=2 ($\chi^2_\nu(l=1) = 1.9$ and $\chi^2_\nu(l=2) = 2.0$). The level at 4252 keV exhibits a broad distribution compatible with a l=3 orbital ($\chi^2_\nu(l=3) = 1.4$ and $\chi^2_\nu(l=2) = 1.6$). Spectroscopic factors were deduced from the ratio between the exclusive cross section for each state and the corresponding single-particle cross sections computed using the eikonal formalism [13, 14]. Tentative assignments of the spin-parity of the levels were based on the rules of gamma decay (see text). The results discussed above are summarized in Fig. 4.

4. Discussion

Figure 4 shows the preliminary experimental results compared to state-of-the-art shell model predictions using both microscopically derived interactions from chiral effective field theory, EEdf1 [3], and phenomenological theories with the newly available SDPF-U-MIX interaction [1]. No quenching [16] has been applied to the predicted spectroscopic factors of both theories.

Knocking a neutron out the $2s_{1/2}$ orbital in ^{31}Mg leads to the population of $0^+_{1,2}$ states in ^{30}Mg. These correspond to the measured ground and isomeric state at 1782(5). Given the upper limit obtained for $C^2S(0^+_1)$=0.42(8), the experimental result is in agreement with both theories but with completely different descriptions. While the SDPF-U-MIX interaction describes the ground-state wave function of ^{30}Mg as dominated by $0p-0h$, the newly developped, ab initio-type, EEdf1 interaction, produces much enhanced *np-nh* excitations, indicating that this nucleus belongs to the island of inversion. However, the moderately low $C^2S(0^+_2)$ obtained here suggests a more complex structure rather than the conventional interpretation of this state as being purely 2p-2h. The deduced spectroscopic factors for both $0^+_{1,2}$ are in good agreement with the results from the EEdf1 interaction and only the $C^2S(0^+_1)$ agrees well with the SDPF-U-MIX.

Removing a neutron from the $1d_{3/2}$ orbital in ^{31}Mg populates spins and parity states of $(1,2)^+$ in ^{30}Mg. The first 2^+_1 state at 1482 keV in ^{30}Mg shows a relatively large strength of 0.44(13) in agreeement with the EEdf1 interaction. Assuming some population to the 2^+_1 state could come from dynamical excitation of the core, the resulting spectroscopic factor is thus an upper limit and also consistent with the SDPF-U-MIX interaction. The next 2^+_2 is placed at 3457-keV since 1^+ states are expected at higher energy. Fusion-evaporation measurements employing a $^{14}C(^{18}O,2p)$ reaction of Deacon et al. [8], have assigned this state to be a 4^+ based on an angular ratio $R_{ang} < 1$ for the 1975-keV γ-ray, however in a direct knockout process from ^{31}Mg, the neutron had to be removed from the $1g_{9/2}$ orbital and the shape of the momentum distribution is clearly not compatible with $l = 4$. Therefore a 2^+_2 assignment is made for this state, in line with the results from [10]. The spectroscopic factor is rather large, 0.48(7), and no equivalent strength for positive parity states is found in shell model calculations at these energies. The large spectroscopic factor found here evidences a substantial overlap with the ground state wavefunction of ^{31}Mg pointing to the intruder nature of the 2^+_2 state. This result is supported by the large β and γ-branching ratios observed in [10].

Negative parity states can also be populated when knocking a neutron out the fp-shell. The state at 2467-keV shows a l=1 shape, thus spin and parities of $(1,2)^-$ are possible. However,

IOP Conf. Series: Journal of Physics: Conf. Series **966** (2018) 012020 doi:10.1088/1742-6596/966/1/012020

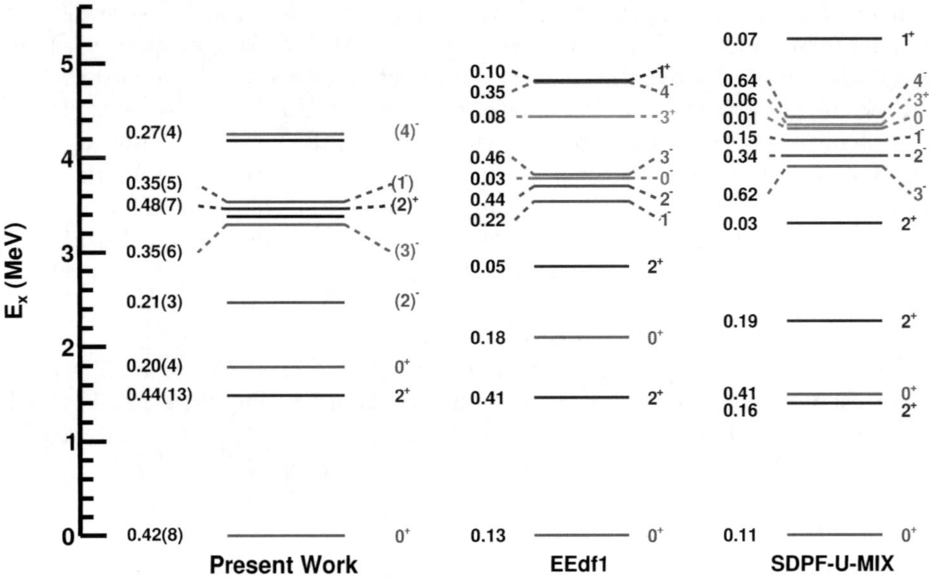

Figure 4. Experimental level scheme of ^{30}Mg (first column) compared to shell model calculations using the EEdf1 and SDPF-U-MIX interactions (second and third column, respectively). Spin and parity assignments are displayed on the right and the spectroscopic factors on the left of the levels.

based on the branching scheme, this state must surely be the 2^- because otherwise it would decay to the ground state by an E1 rather than to the 2^+. This is unexpected according to shell model calculations that predict these negative parity states at much higher excitation energy. Despite the experimental C^2S of 0.20(3) being in agreeement with the 1^- state predicted by the EEdf1 and close to the value obtain by the SDPF-U-MIX, the low position in energy for this negative parity state remains a challenge for current theories. An assignment of 2^+ was originally suggested by H. Mach [11], later adopted by [8] and by β-decay measurements from ^{30}Na [10], although with a remarkably weak feeding and large logft value for this later measurement. It is also worth to notice here, that according to our spin and parity assingment, the corresponding E1 transition is still in agreement with the upper limit for the lifetime of this state measured in [11] The momentum distribution for the state at 3534-keV is compatible with $l = 1, 2$ which yields a spin-parity assignment of $(1, 2)^+$ or $(1, 2)^-$. Preference to the l=1 assignment is made based on the direct decay to the ground state. A tentative assignment of 1^- is suggested for this state.

Some occupancy of the $f_{7/2}$ orbital in ^{31}Mg results in $(3,4)^-$ states in ^{30}Mg. The distributions of the 3298-keV and the 4252-keV states are both consistent with $l = 3$ shapes and therefore spins and parities of 3^- and 4^- are assigned for these states, respectively. The order is deduced from the typical Weisskopf estimates, that favour a E1 transition from the 3^- to the 2^+ state over a E3 for an anssignement of 4^- for that level. The upper state with $l = 3$ is a 4^- in agreement with a M1 transition to the 3^- state. A remarkably good agreement is found with the EEdf1 interaction while the predictions from the SDPF-U-MIX interaction overstimate the experimental results. The difference might come from the approximately 30% $3\hbar\omega$ configurations present in the predictions with the EEdf1. Shimoda et al. [10] however, suggested a 2^+

assignment for the state at 3298-keV, (found at 3.302 MeV). Based on the measured logft > 6.5 in the β-decay experiment, this transition could still be first-forbidden in agreement with our assignment of 3^-. We also note that a tentative assignment of 4^+ has been made in a 2-neutron removal measurement [15].

We have studied single-neutron removal from ^{31}Mg using the SPEG spectrometer and the EXOGAM array for gamma-ray detection. Inclusive and exclusive cross sections and momentum distributions for the populated states in ^{30}Mg have been measured. The two 0^+ states have been measured simultaneously allowing us to extract information on the mixing in ^{30}Mg. The low spectroscopic factor of the 0_2^+ suggests a different structure from the well established 2p2h configuration of the ground state of ^{31}Mg. The large spectroscopic factor found for the 2_2^+ state points to the intruder nature of this state. The results obtained for the 2.467-MeV state are intriguing because negative parity states are expected at higher excitation energy. A comparison of the present ^{30}Mg results with two new shell model calculations, using state-of-the-art interactions, shows two plausible and quite different scenarios. The EEdf1 interaction suggests that many-particle many-hole configurations dominate the low-lying structure of ^{30}Mg and therefore would place ^{30}Mg into the N=20 island of inversion. On the contrary, the SDPF-U-MIX interaction supports a picture of shape coexistence with a dominant spherical configuration in the ground state of ^{30}Mg. The present data suggest that the transion into the island of inversion in the Mg chain is more complex than expected.

Acknowledgements

The authors acknowledge the excellent support provided by the technical staff of LPC and GANIL. B. F. D. and M. C. F. acknowledge financial support from the Ramón y Cajal RYC-2010-06484 and RYC-2012-11585 and from the Spanish MINECO grant No. FPA2015-71690-P. W.N. Catford acknowledges financial support from the STFC grant number ST/L005743/1. This work is partly supported by MINECO (Spain) grant FPA2014-57196 and Programme Centros de Excelencia Severo Ochoa SEV- 2012-0249.

References

[1] Caurier E, Nowacki F and Poves A 2014 *Phys. Rev.* C **90** 014302.
[2] Nowacki F, Poves A, Caurier E and Bounthong B 2016 *Phys. Rev. Lett.* **117** 272501.
[3] Tsunoda N, Otsuka T, Shimizu N, Hjorth-Jensen M, Takayanagi K and Suzuki T 2017 *Phys. Rev* C **95** 021304(R).
[4] Anne R 1997 *Nuc. Inst. and Meth.* B **126** 279
[5] Bianchi L et al. 1989 *Nucl. Instr. Meths.* A **276** 509
[6] Simpson J et al. 2000 *Act. Phys. Hung.* **11** 159
[7] Bauman P et al. 1989 *Phys. Rev.* C **39** 626
[8] Deacon A N et al. 2010 *Phys. Rev.* C **82** 034305
[9] Klotz G et al. 1993 *Phys. Rev.* C **47** 2502
[10] Shimoda T et al. 2014 *Hyper. Inter.* **225**
[11] Mach H et al. 2005 *Eur. Phys. J.* A **25** s01 105-109
[12] Schwerdtfeger W et al. 2009 *Phys. Rev. Lett.* **103** 012501
[13] Tostevin J A et al. 1999 *J. Phys.* G **25** 735
[14] Tostevin J A et al. 2001 *Nucl. Phys.* A **682** 320c
[15] Takeuchi S et al. 2009 *Phys. Rev.* C **79** 054319
[16] Gade A et al 2008 *Phys. Rev.* C **206** 77044

Measuring nuclear reaction cross sections to extract information on neutrinoless double beta decay

M. Cavallaro[1*], F. Cappuzzello[1,2], C. Agodi[1], L. Acosta[3], N. Auerbach[4], J. Bellone[1,2], R. Bijker[3], D. Bonanno[5], D. Bongiovanni[1], T. Borello-Lewin[6], I. Boztosun[7], V. Branchina[2,5], M. P. Bussa[8,9], S. Calabrese[1,2], L. Calabretta[1], A. Calanna[1], D. Calvo[8], D. Carbone[1], E.R. Chávez Lomelí[3], A. Coban[7], M. Colonna[1], G. D'Agostino[1,2], G. De Geronimo[10], F. Delaunay[8,11], N. Deshmukh[1], P.N. de Faria[12], C. Ferraresi[8], J.L. Ferreira[1,12], P. Finocchiaro[1], M. Fisichella[8], A. Foti[2,5], G. Gallo[1,2], U. Garcia[13], G. Giraudo[8], V. Greco[1,2], A. Hacisalihoglu[1], J. Kotila[14], F. Iazzi[8,15], R. Introzzi[8,15], G. Lanzalone[1,16], A. Lavagno[8,9], F. La Via[1,17], J.A. Lay[18], H. Lenske[19], R. Linares[12], G. Litrico[1], F. Longhitano[5], D. Lo Presti[2,5], J. Lubian[12], N. Medina[6], D. R. Mendes[12], A. Muoio[1], J.R.B. Oliveira[6], A. Pakou[20], L. Pandola[1], H. Petrascu[21], F. Pinna[8,15], S. Reito[5], D. Rifuggiato[1], M.R.D. Rodrigues[6], A. D. Russo[1], G. Russo[2,5], G. Santagati[1], E. Santopinto[13], O. Sgouros[20], S.O. Solakcı[7], G. Souliotis[22], V. Soukeras[20], A. Spatafora[1,2], D. Torresi[1], S. Tudisco[1], R.I.M. Vsevolodovna[13], R.J. Wheadon[8], A. Yildirin[7], V.A.B. Zagatto[1,12]

[1]Istituto Nazionale di Fisica Nucleare, Laboratori Nazionali del Sud, Catania, Italy
[2]Dipartimento di Fisica e Astronomia, Università di Catania, Italy
[3]Universidad Nacional Autónoma de México
[4]School of Physics and Astronomy Tel Aviv University, Israel
[5]Istituto Nazionale di Fisica Nucleare, Sezione di Catania,
[6]Universidade de Sao Paulo, Brazil
[7]Akdeniz University, Antalya, Turkey
[8]Istituto Nazionale di Fisica Nucleare, Sezione di Torino
[9]Università di Torino, Torino, Italy
[10]Stony Brook University, USA
[11]Université de Caen and Laboratoire de Physique Corpusculaire de Caen, France
[12]Universidade Federal Fluminense, Niteroi, Brazil
[13]Istituto Nazionale di Fisica Nucleare, Sezione di Genova
[14]University of Jyväskylä, Jyväskylä, Finland
[15]Politecnico di Torino, Italy
[16]Università degli Studi di Enna "Kore", Enna, Italy
[17]CNR-IMM, Sezione di Catania, Italy
[18]University of Seville, Spain
[19]University of Giessen, Germany
[20]University of Ioannina, Ioannina, Greece
[21]IFIN-HH, Romania
[22]University of Athens and HINP, Athens, Greece

*manuela.cavallaro@lns.infn.it

Abstract. Neutrinoless double beta decay (0νββ) is considered the best potential resource to access the absolute neutrino mass scale. Moreover, if observed, it will signal that neutrinos are their own anti-particles (Majorana particles). Presently, this physics case is one of the most important research "beyond Standard Model" and might guide the way towards a Grand Unified Theory of fundamental interactions.
Since the 0νββ decay process involves nuclei, its analysis necessarily implies nuclear structure issues. In the NURE project, supported by a Starting Grant of the European Research Council (ERC), nuclear reactions of double charge-exchange (DCE) are used as a tool to extract information on the 0νββ Nuclear Matrix Elements. In DCE reactions and ββ decay indeed the initial and final nuclear states are the same and the transition operators have similar structure. Thus the measurement of the DCE absolute cross-sections can give crucial information on ββ matrix elements. In a wider view, the NUMEN international collaboration plans a major upgrade of the INFN-LNS facilities in the next years in order to increase the experimental production of nuclei of at least two orders of magnitude, thus making feasible a systematic study of all the cases of interest as candidates for 0νββ.

1. The physics case

The potential observation of 0νββ decay, besides establishing the Majorana nature of neutrinos, has the potential to shed light on the absolute neutrino mass scale. To this purpose, it is critical that the associated Nuclear Matrix Elements (NME) are known with sufficient accuracy. The 0νββ decay rate can be expressed as a product of independent factors: the phase-space factors, the NME and a function of the masses of the neutrino species. Thus the knowledge of the NME can give information on the neutrino mass scale, if the 0νββ decay rate is measured. From a comparison of the results of different NME calculations, obtained within various nuclear structure frameworks [1–4], significant differences are found, even of a factor 2 or 3, which makes the present situation still not satisfactory. In addition, some assumption common to different competing calculations, like the unavoidable truncation of the many body wave-function, could cause overall systematic uncertainties.

The use of heavy-ion double charge exchange reactions (HI-DCE) as a tool in order to access quantitative information relevant for 0νββ decay NME has been recently proposed [5, 6]. These reactions are characterized by the transfer of two charge units, leaving the mass number unchanged, and can proceed by a sequential nucleon transfer mechanism or by meson exchange. Despite the fact that 0νββ decay and HI-DCE reactions are mediated by different interactions, they present a number of similarities. Among that, the key aspects are that initial and final nuclear states are the same and the transition operators in both cases present a superposition of isospin, spin-isospin and rank-two tensor components with a relevant momentum (100 MeV/c or so) available.

2. The technique

In a pioneering experiment, performed at the INFN-LNS laboratory, we studied the DCE reaction $^{40}Ca(^{18}O,^{18}Ne)^{40}Ar$ at 270 MeV, with the aim to measure the cross section at zero degrees [5]. In the experiment an advantageous condition was set, using a beam of ^{18}O and a double magic target as ^{40}Ca and choosing the bombarding energy to mismatch the competing transfer reactions leading to the same outgoing channel. The key tools in the experiment were the high resolution Superconducting Cyclotron beams and the MAGNEX spectrometer [7-9], a modern high resolution and large acceptance magnetic system [10-13] characterized by high resolution in energy, mass and angle. The high-order solution of the equation of motion is the key feature of MAGNEX, which guarantees the above mentioned performances and its relevance in the research of nuclear structure and reaction mechanisms [14-19]. In the "pilot experiment" we have shown that high resolution and statistically significant experimental data can be measured for DCE processes and that precious information towards NME determination could be at our reach.

The NURE (NUclear REactions for neutrinoless double beta decay) project has been selected for receiving funding in the call Starting Grant 2016 of European Research Council (ERC). The project has a duration of 5 years and plans to carry out a campaign of experiments by using heavy-ion accelerated beams impinging on two targets of interest as candidate nuclei for the ββ decay. The DCE channel will be explored by the (^{18}O,^{18}Ne) reaction for the β+β+ direction in the target and the (^{20}Ne,^{20}O) for the β -β-. Moreover, the complete net involving the multi-step transfer processes characterized by the same initial and final nuclei will be also studied in the same experimental setup. In particular, the (^{18}O,^{16}O) and (^{20}Ne,^{22}Ne) two-neutron transfer, (^{18}O,^{20}Ne) and (^{20}Ne,^{18}O) two-proton transfer, (^{18}O,^{18}F) and (^{20}Ne,^{20}F) single charge-exchange will be measured.

3. The optimization of the experimental apparatus

During the first months of the project, tests on the facility have been done in order to find the best compromise between energy resolution and count rate. In particular, a work on the optimization of the gas section of the MAGNEX focal plane detector [20] have been performed. A new design for the multiplication region, based on six position sensitive proportional wires has been developed and installed.

A special care has been devoted on the target production technologies, in order to optimize uniformity and thickness of the target films. Energy loss and spreading calculations have been done to find the optimal compromise between target thickness and requested resolution in the final energy spectra. This is a crucial point since high energy resolution is necessary to separate the 0^+ ground state (which is the transition of interest) from the excited states of the final populated nucleus. On the other hand, a relevant target thickness is important to increase the count rate at the detector, since the expected cross-sections are very low (of the order of hundreds of nb/sr).

4. The NUMEN program

The possibility to extract data-driven information on 0νββ nuclear matrix elements is an ambitious line of research which has been conceived and is developing in a broader context, characterized by a longer time scale and a larger collaboration: the NUMEN program [6, 21].

In the present experimental conditions, due to the limitation arising from the tiny cross-sections of the processes of interest, only very few systems can be measured within the 5-years ERC project. In order to systematically explore all the nuclei candidates for 0νββ, a beam intensity at least two orders of magnitude higher than the present must be achieved [22]. The challenge is to measure a rare nuclear transition under a very high rate of heavy ions produced by the beam-target interaction. As a consequence, major upgrades of the detector technologies (3D ion tracker, particle-identification wall, gamma-ray array, ...) must be developed [23, 24]. Also the target technology must be upgraded, to avoid the damage of the thin films due to the high temperature involved [25, 26]. The front-end and readout electronics must take into account the high number of channels and the expected rate of detected events. Moreover, a deep and complete investigation of the theoretical aspects connecting nuclear reaction mechanisms and nuclear matrix elements must be carried out.

All of these issues and other aspects related to the high intensity beam physics are the crucial branches of the NUMEN program.

Acknowledgments

This project has received funding from the European Research Council (ERC) under the European Union's Horizon 2020 research and innovation programme (grant agreement No 714625).

References

[1] Vergados J D *et al.* 2012 *Rep. Progr. Phys.* **75** 106301.

[2] Vogel P 2012 *Journal of Physics G: Nuclear and Particle Physics* **39** 124002
[3] Engel J and Menendez J 2017 *Rep. Prog. Phys.* **80** 046301
[4] Barea J *et al.* 2015 *Phys. Rev. C* **91** 034304
[5] Cappuzzello F *et al.* 2015 *Eur. Phys. J. A* **51**:145
[6] Cappuzzello F *et al.* 2015 *Journal of Physics Conference Series* **630** 012018
[7] Cappuzzello F *et al.* 2016 *Eur. Phys. J. A* **52**: 167
[8] Cunsolo A *et al.* 2007 *Eur. Phys. J. A ST* **150** 343
[9] Cappuzzello F *et al.* 2014 *Nucl. Instr. And Meth. A* **763** 314
[10] Lazzaro A *et al.* 2008 *Nucl. Instr. and Methods A* **591** 394
[11] Lazzaro A *et al.* 2008 *Nucl. Instr. and Methods A* **585** 136
[12] Lazzaro A *et al.* 2007 *Nucl. Instr. and Methods A* **570** 192
[13] Lazzaro A *et al.* 2009 *Nucl. Instr. and Methods A* **602** 494
[14] Cappuzzello F *et al.* 2015 *Nature Communications* **6** 6743
[15] Pereira D *et al.* 2012 *Phys. Lett. B* **710** 426
[16] Carbone D *et al.* 2014 *Phys. Rev. C* **90** 064621
[17] Carbone D *et al.* 2017 *Phys. Rev. C* **95** 034603
[18] Cavallaro M *et al.* 2013 *Phys. Rev. C* **88** 054601
[19] Soukeras V *et al.* 2015 *Phys. Rev. C* **91** 05701
[20] Cavallaro M *et al.* 2012 *Eur. Phys. J. A* **48**:59
[21] https://web.infn.it/NUMEN
[22] Calabretta L *et al.* 2017 *Modern Physics Letters A* **32** 17
[23] Carbone D *et al.* 2016 *Results in Physics* **6** 863
[24] Muoio A *et al.* 2016 *EPJ Web of Conferences* **117** 10006
[25] Pinna F *et al.* *Applied Surface Science* in press
[26] Iazzi F *et al.* 2017 *WIT Transactions on Engineering Sciences* **116** 61

IOP Conf. Series: Journal of Physics: Conf. Series **966** (2018) 012022 doi:10.1088/1742-6596/966/1/012022

Ab initio calculation of one-nucleon halo states

D M Rodkin[1,2] and Yu M Tchuvil'sky[1,2,3]

[1]Dukhov Research Institute for Automatics, 127055, Moscow, Russia
[2]Pacific National University, 680035, Khabarovsk, Russia
[3]Skobeltsyn Institute of Nuclear Physics, Lomonosov Moscow State University, 119991, Moscow, Russia

E-mail: tchuvl@nucl-th.sinp.msu.ru

Abstract. We develop an approach to microscopic and ab initio description of clustered systems, states with halo nucleon and one-nucleon resonances. For these purposes a basis combining ordinary shell-model components and cluster-channel terms is built up. The transformation of clustered wave functions to the uniform Slater-determinant type is performed using the concept of cluster coefficients. The resulting basis of orthonormalized wave functions is used for calculating the eigenvalues and the eigenvectors of Hamiltonians built in the framework of ab initio approaches. Calculations of resonance and halo states of ^{5}He, ^{9}Be and ^{9}B nuclei demonstrate that the approach is workable and labor-saving.

1. Introduction

The development of ab initio approaches is one of the basic lines of the advancement of nuclear science. Various versions of ab initio No-Core Shell Model (NCSM) using realistic microscopic NN- and 3N-forces [1 – 5] occupy a prominent place in nuclear structure calculations. A typical basis of this model contains all possible nucleon configurations on equal terms up to a certain truncation border. At the same time, well-known properties of nuclei demonstrate "inequality" of different nucleon configurations. For this reason, methods discarding non-relevant components and, due to that, allowing one to involve additional critically important states are also popular [6 – 8]).

One of the major properties of light nuclei is clustering displaying itself in a certain degree of separation of a nucleus to two or more fragments (core, clusters, and loosely bound nucleons). Moreover an accurate description of nuclei fragmentation is crucial for calculation of nuclear reactions. The description of systems manifesting significant fragment separation by use of ordinary shell-model methods leads to calculation problems: a too large shell-model basis is required.

The goals of this work may be formulated as follows. The first is to build up a basis of translationally-invariant A-nucleon wave functions of an arbitrary channel manifesting two-fragment separation $A = A_1 + A_2$. The second is to assemble the partial bases corresponding to various channels, which are distinguished by the internal states of the fragments A_1 and A_2, together in a unified orthogonal basis. The third one is to add to this clustered basis a (relatively small) number of eigenvectors obtained in ordinary shell-model calculations (called "polarization terms" in monograph

[9]) and in such a way to build a final basis suitable to describe the states manifesting fragmentation (cluster) properties and, in particular, the entrance and exit channels of nuclear reactions.

It should be noted that an approach aimed at an accurate description of reactions induced by light nuclei collisions has been developed in Ref. [10]. This approach also explores matrix mathematics of microscopic Hamiltonians corresponding to the No-Core Shell Model plus Resonating Group Method (i. e. accounting for the clustering) basis. It was called NCSM/RGM. The differences of this scheme and ours are the following. First, we use the concept of cluster coefficients (CCs) to extract the zero vibrations of the centre of mass (CM) of the system. The cluster coefficients formalism is well developed in refs. [11 – 15]. It provides the means to work with a broad variety of excited and rather heavy fragments. Second, applying this formalism we obtain a convenient purely algebraic approach.

The polarization terms were introduced into NCSM/RGM in refs. [16 – 20]. The new model received the name No-Core Shell Model with Continuum (NCSMC). Our approach remains algebraic after involving the polarization terms. Thus, in some sense, the interrelation between the NCSMC and our approach is analogous to the interrelation between the conventional RGM [9, 21, 22] and its algebraic version [23]. In certain examples, in particular in the cases where multi-cluster or multi-channel problems are considered, our approach looks promising.

In this work we present a mathematical formalism suitable for the description of a general two-fragment system, whose applicability is limited by computer power only.

As an example, we discuss the simplest type of fragmentation – the formation of one-nucleon halo states and one-nucleon resonances. Peculiar features of these states make them a very interesting object for the studies of various nuclear processes.

2. Construction of the basis

Let us consider a two-fragment system A_1+A_2. The cluster-channel terms of the basis are built in the form:

$$\Psi_A = \frac{1}{W}\hat{A}\{\Psi_{A_1}\Psi_{A_2}\varphi_{nlm}(\vec{\rho})\}_{JM_J}, \tag{1}$$

where $A = A_1 + A_2$, $\hat{A}$ is the antisymmetrizer, Ψ_{A_i} is a translationally-invariant wave function (WF) of the fragment; $\varphi_{nlm}(\vec{\rho})$ is the oscillator WF of the relative motion.

These terms are labeled by the quantum numbers of the internal cluster states and indices *n, l, m*. The main problem is to present function (1) as a linear combination of the Slater determinants (SDs) containing one-nucleon WFs of the oscillator basis

$$\phi_{\nu\lambda\mu\frac{1}{2}m_s}(r,\sigma) = R_{\nu\lambda}(r)Y_\lambda^\mu(\theta,\varphi)\chi_{1/2}^{m_s}(\sigma). \tag{2}$$

For these purposes function (1) is multiplied by the function of the zero vibrations of the CM $\Phi_{000}(\vec{R})$. This operation is commutative with the antisymmetrization. Then the Talmi-Moshinsky transformation for particles of different masses [24] is performed:

$$\Phi_{000}(\vec{R})\varphi_{nlm}(\vec{\rho}) = \sum_{N_1,L_1,M_1,N_2,L_2,M_2}\left\langle\begin{matrix}000\\ nlm\end{matrix}\middle|\begin{matrix}N_1,L_1,M_1\\ N_2,L_2,M_2\end{matrix}\right\rangle\Phi^{A_1}_{N_1,L_1,M_1}(\vec{R}_1)\Phi^{A_2}_{N_2,L_2,M_2}(\vec{R}_2) \tag{3}$$

Thus, WF (1) is expressed in the form

$$\Phi_{000}(\vec{R})\Psi_A = \frac{1}{W}\sum_{N_1,L_1,M_1,N_2,L_2,M_2}\left\langle\begin{matrix}000\\ nlm\end{matrix}\middle|\begin{matrix}N_1,L_1,M_1\\ N_2,L_2,M_2\end{matrix}\right\rangle\hat{A}\{\Phi^{A_1}_{N_1,L_1,M_1}(\vec{R}_1)\Psi_{A_1}\Phi^{A_2}_{N_2,L_2,M_2}(\vec{R}_2)\Psi_{A_2}\}_{JM_J}. \tag{4}$$

The critical procedure is to transform the products of the CM and internal WFs corresponding to each fragment into a superposition of SDs

$$\Phi^{A_i}_{N_i,L_i,M_i}(\vec{R}_i)\Psi_{A_i} = \sum_k X^{A_i(k)}_{N_i,L_i,M_i}\Psi^{SD}_{A_i(k)}. \tag{5}$$

Overlap $X^{A_i(k)}_{N_i,L_i,M_i} = <\Psi^{SD}_{A_i(k)} | \phi_{nl}(\vec{R}_{A_i})\Psi_{A_i}>$ is called a cluster coefficient. Mathematics of these objects is presented in detail in [15]. There is a large number of methods elaborated for the calculations of CCs. The most general scheme is based on the method of the second quantization of the oscillator quanta. In this scheme the WF of the CM motion is presented as

$$\Phi^{A_i}_{N_i,L_i,M_i}(\vec{R}_i) = N_{N_i,L_i}(\hat{\vec{\mu}}^\dagger)^{N_i-L_i} Y_{N_i,L_i}(\hat{\vec{\mu}}^\dagger)\Phi^{A_i}_{000}(\vec{R}_i), \tag{6}$$

where $\hat{\vec{\mu}}^\dagger$ is the creation operator of the oscillator quantum, and its norm is expressed as

$$N_{N_i,L_i} = (-1)^{N_i-L_i}\sqrt{\frac{4\pi}{(N_i-L_i)!!(N_i+L_i+1)!!}}. \tag{7}$$

Thus the CC turns out to be reduced to a matrix element of the tensor operator presented in (6):

$$<\Psi^{SD}_{A_i(k)} | \phi_{N_i,L_i}(\vec{R}_{A_i})\Psi_{A_i}> = N_{N_i,L_i}\left\langle \Psi^{SD}_{A_i(k)} \middle| (\hat{\vec{\mu}}^\dagger)^{N_i-L_i} Y_{N_i,L_i}(\hat{\vec{\mu}}^\dagger) \middle| \Phi^{A_i}_{000}(\vec{R}_i)\Psi_{A_i}\right\rangle. \tag{8}$$

Contrary to pioneering work [25] in which translationally-invariant WFs were written in Jacobi coordinates, the formula

$$\Psi_{A_i} = \Psi^{shell}_{A_i} / \Phi^{A_i}_{000}(\vec{R}_i) \tag{9}$$

is considered here as the definition of these functions.

Algebraic methods provide possibilities to obtain compact expressions of the CCs in a lot of particular cases. As an example, a general expression for the cluster coefficients of light d-, t-, h-, and α-clusters (X = 2, 3, 4) in their lowest shell-model configurations takes the form:

$$<\prod_{i=1}^{X} n_i(n0):000 |\phi_{(n0)}(R_C)\Psi_C> = X^{-n/2}\left(n!/\prod_{i=1}^{X} n_i!\right)^{1/2}\left(X!\prod_{j=1}^{k}\alpha_i!\right)^{1/2}. \tag{10}$$

Symbol α denotes here the multiplicity of each value of n_i in the configuration $\{n_i\}$. The SU(3)-coupling of the one-nucleon WFs is implied here. For the rearrangement of this configuration to the ordinary SD expression the Clebsh-Gordan coefficients of the SU(3) group having a very simple structure are required.

The next step is to build a basis of orthonormalized WFs which includes both the polarization terms and the terms of several cluster channels. The related norm kernel is obtained by the diagonalization of matrix

$$\left\| \begin{matrix} \left[\left\langle \Psi^{(j)}_{pol} \middle| \Psi^{(j)}_{pol}\right\rangle\right] & \left[\left\langle \Psi^{(j)}_{pol} \middle| \hat{A} \middle| \prod_{i=1,2}\Phi^{A_i}_{N_i,L_i,M_i}(\vec{R}_i)\Psi_{A_i}\right\rangle\right] \\ \left[\left\langle \Psi^{(j)}_{pol} \middle| \hat{A} \middle| \prod_{i=1,2}\Phi^{A_i}_{N_i,L_i,M_i}(\vec{R}_i)\Psi_{A_i}\right\rangle\right] & \left[\left\langle \prod_{i=1,2}\Phi^{A_i}_{N'_i,L'_i,M'_i}(\vec{R}_i)\Psi_{A_i} \middle| \hat{A}^2 \middle| \prod_{i=1,2}\Phi^{A_i}_{N_i,L_i,M_i}(\vec{R}_i)\Psi_{A_i}\right\rangle\right] \end{matrix} \right\| \tag{11}$$

in which the square brackets denote the submatrices and the cluster-channel terms symbolically presented in (11) as the products are, in fact, expressed in the form of superpositions of SDs with the help of formula (5). The eigenvectors of the matrix normalized by its eigenvalues give the desirable basis. Each term of the basis takes the form of a SDs linear combination. Thus the form of such a term is identical to the form of the terms of the ordinary shell-model basis.

Therefore the next step – the computation of the matrix elements of both the kinetic and the potential energy in the discussed basis – is identical in design to the ordinary shell-model computation. Thus, an arbitrary microscopic (ab initio or effective, including two-, three-, etc. nucleon forces) Hamiltonian may be explored. The calculations of the matrix elements of the operators, the estimates of error bars etc. are also analogous to those in the shell-model. The only difference is the list of the basis vectors. This list is significantly shorter compared to the shell-model one. The limitations on the use of the approach are imposed by the dimensionality of the basis vectors. At the same time these restrictions seem to be not so strong compared to the limitations of the canonical NCSMC in the case, where long inter-cluster distances are considered.

This approach as a whole is very flexible due to the possibilities to vary: A-, A_1-, A_2- nucleon shell-model spaces determined by the corresponding truncation border parameters ($N^{(i)}_{max}$), determining the maximal value of the total number of the oscillator quanta in each system, as well as the maximal value of n. For example, even the insertion of "unphysical" channels containing the WFs of clusters in their lowest shell-model configurations into the basis turns out to be a step forward in the spectral computations. The basis built before procedure (10) may be orthogonal, non-orthogonal and, if necessary, overloaded. This gives a way to take into account various halo, cluster and other properties of a system. The multi-dimensional ($N^{(i)}_{max}$,n,ħω)-dependence of the computation results provides a possibility for their extrapolation to the infinite basis space along each discrete variable separately and thus for more reliable estimates of their error bars compared to the shell-model calculation results.

3. One-nucleon halo and resonance systems, results and discussion

To demonstrate the capabilities of our approach, in this section we present the results of total and nucleon binding (resonance) energies calculations. The ground states of five- and nine-nucleon systems are considered. Ab initio Hamiltonian Daejeon16 [26] was chosen for these calculations. Widely-used code Antoine is explored for NCSM computations of the internal WFs of the clusters and the polarization terms. For all WFs of nuclei under study the value of the oscillator parameter ħω=15 MeV was chosen. The computational procedure of each example involves about 10^6 SDs.

3.1 Five-nucleon system. Lowest resonance state of ^{5}He nucleus

The two channels of ^{4}He+n fragmentation corresponding to 0^+_1 and 0^+_2 states of the core nucleus respectively are considered in these calculations. The truncation border parameter for the polarization terms of nucleus ^{5}He is chosen to be $N^{(5He)}_{max}$ =7(9). For the clustered terms, the truncation border of the core nucleus ^{4}He WF is $N^{(4He)}_{max}$ =2.

The results of the calculations are presented in Tab. 1.

Table 1 Total and neutron binding energies (MeV) of ^{5}He nucleus.

		NCSM	n_{max} =7	n_{max} =9	n_{max} =11	Exp.
$N^{(5)}_{max}$ =7	E_{5He}	26.37	26.40	26.51	26.57	
	E_n	–1.85	–1.82	–1.71	–1.65	
$N^{(5)}_{max}$ =9	E_{5He}	26.83	–	26.87	26.92	27.41
	E_n	–1.50	–	–1.46	–1.41	–0.89

IOP Conf. Series: Journal of Physics: Conf. Series **966** (2018) 012022 doi:10.1088/1742-6596/966/1/012022

The contents of the table demonstrate that the contribution of the additional cluster terms to the total binding energy of ^{5}He turns out to be relatively small compared to this energy. Nevertheless, the shift of the resonance energy value by 0.20 MeV or 0.09 MeV respectively caused by the inclusion of the additional cluster terms is not negligible. Having been achieved by using rather modest computational means, this result looks reasonable and therefore the prospects to merge the presented method of one-nucleon resonance states study with the approach developed in [27] are expected to be good.

3.2 Nine-nucleon system

A six-channel model is used for the description of the ^{8}Be+n and ^{8}Be+p fragmentations of ^{9}Be and ^{9}B nuclei. These channels contain $0^+{}_1$, $0^+{}_2$; $1^+{}_1$, $1^+{}_2$; $2^+{}_1$, $2^+{}_2$ states of ^{8}Be respectively. The truncation border parameter $N^{(9)}{}_{max}$=9 is chosen for the performed NCSM calculations, the value of the truncation border parameter of ^{8}Be in all cluster channels is chosen to be 4, i. e. the lowest configurations of the core nucleus in the channel WFs are considered.

The results of the calculations are presented in Tab. 2.

Table 2 Total and neutron (proton) binding energies (MeV) of ^{9}Be, ^{9}B nuclei and Coulomb energy difference.

	NCSM ($N^{(9)}{}_{max}$=9)	$n_{\max}$ =5	$n_{\max}$ =7	$n_{\max}$ =9	Exp.
$E_{9\mathrm{Be}}$	53.51	37.66	37.74	37.76	58.16
E_{n}	1.34	1.46	1.54	1.56	1.66
$E_{9\mathrm{B}}$	51.58	35.74	35.84	35.91	56.31
E_{p}	–0.59	–0.47	–0.37	–0.29	–0.19
ΔE	1.93	1.93	1.90	1.85	1.85

These results allow one to make the following conclusions. First, the calculations using a basis of cluster-channel WFs containing relatively lower configurations of core nucleus result in a significant underestimation of the total binding energies. Thus, taking into account the polarization terms is necessary for an accurate reproduction of these values. At the same time the "differential" quantities: the nucleon binding and resonance energies, the Coulomb energy differences etc. turn out to be well-reproduced in the case, where one and the same truncation border parameter of the core nucleus WFs is used for the construction of the cluster-channel basis and for the computation of the binding energy of the core nucleus. Moreover, it is the differential characteristics of the reaction channels that determine the cross-sections of the respective nuclear reactions. Thus, the method developed here provides a convenient and labor-saving means of ab initio studies of resonance and direct nuclear reactions.

4. Summary

We demonstrate that the formalism of cluster coefficients is a convenient tool to construct a basis that takes into consideration the cluster properties of nuclear systems. This basis looks promising for further ab initio calculations of their characteristics because the adoption of this basis allows one to get rid of most of the components, inherent for the shell model, which do not contribute to the wave functions of the clustered states.

The developed approach is versatile. The basis of this approach includes the shell-model WFs together with various cluster-channel ones, i. e. different versions of the basis conform to the properties of systems under study.

Thus, the approach is suitable for the ab initio computation of clustered systems, halo systems and narrow resonances. The calculations of resonance and halo states of ^{5}He, ^{9}Be and ^{9}B nuclei demonstrate the prospects of this approach for nuclear reaction studies.

Our approach may be considered, in parallel with the approaches presented in [10, 16 – 20], as a step towards realizing an ab initio version of a general theory developed by K. Wildermuth and Y.C. Tang that received the name "A Unified Theory of the Nucleus" [9].

Acknowledgments
The work was supported by Russian Science Foundation (RSF), grant No. 16-12-10048.

References

[1] Navratil P *et al* 2007 *Phys. Rev. Lett.* **99** 042501
[2] Maris P, Vary J P and Shirokov A M 2009 *Phys. Rev.* C **79** 014308
[3] Maris P, Shirokov A M and Vary J P 2010 *Phys. Rev.* C **81** 021301(R)
[4] Maris P and Vary J P 2013 *Int. J. Mod. Phys.* E **22** 1330016
[5] Barret B R, Navratil P and Vary J P 2013 *Progr. Part Nucl. Phys.* **69** 131
[6] Lui L *et al* 2012 *Phys. Rev.* C **86** 014304
[7] Abe T *et al* 2012 *Phys. Rev.* C **86** 054301
[8] Dytrych T *et al* 2013 *Phys. Rev. Lett.* **111** 252501
[9] Wildermuth K and Tang Y C 1977 *A Unified Theory of the Nucleus* (Braunschweig: Veiweg)
[10] Quaglioni S and Navratil P 2009 *Phys. Rev.* C **79** 044606
[11] Smirnov Yu F and Tchuvil'sky Yu M 1977 *Phys. Rev.* C **15** 84
[12] Smirnov Yu F and Tchuvil'sky Yu M 1983 *Czech. J. Phys.* B **33** 1215
[13] Tchuvil'sky Yu M *et al* 1995 *Phys. Rev.* C **51** 784
[14] Volya A and Tchuvil'sky Yu M 2015 *Phys. Rev.* C **91** 044319
[15] Nemetz O F *et al* 1988 *Nucleon Clusters in Atomic Nuclei and Multi-Nucleon Transfer Reactions* (Kiev: Naukova Dumka)
[16] Baroni S, Navratil P and Quaglioni S 2013 *Phys. Rev. Lett.* **110** 022505
[17] Baroni S, Navratil P and Quaglioni S 2013 *Phys. Rev.* C **87** 034326
[18] Romero-Robredo C *et al* 2014 *Phys. Rev. Lett.* **99** 032503
[19] Langhammer J *et al* 2015 *Phys. Rev.* C **91** 021301
[20] Dohet-Eraly J *et al* 2016 *Phys. Lett.* B **757** 430
[21] Wheeler J A 1937 *Phys. Rev.* **52** 1083
[22] Tang Y C, LeMere M and Thompson D R 1978 *Phys. Rep.* **47** 167
[23] Filippov G F and Okhrimenko I P 1980 *Phys. Atom. Nucl.* **32** 480
[24] Smirnov Yu F 1962 *Nucl. Phys.* **39** 346
[25] Kurdyumov I V *et al* 1970 *Nucl. Phys.* A **145** 593
[26] Shirokov A M *et al* 2016 *Phys. Lett.* B **761** 87
[27] Shirokov A M, Mazur A I, Mazur I A and Vary J P 2016 *Phys. Rev.* C **94** 064320

IOP Conf. Series: Journal of Physics: Conf. Series **966** (2018) 012023 doi:10.1088/1742-6596/966/1/012023

Shell Model Far From Stability: Island of Inversion Mergers

F. Nowacki[1,2,3], **A. Poves**[4],

[1] Université de Strasbourg, IPHC, 23 rue du Loess 67037 Strasbourg Cedex, France
[2] CNRS, UMR 7178, 67037 Strasbourg Cedex, France
[3] Dipartimento di Matematica e fisica, Università degli Studi della Campania "Luigi Vanvitelli", Viale Abramo Lincoln 5, I-81100 Caserta, Italy
[4] Departamento de Física Teórica e IFT-UAM/CSIC, Universidad Autónoma de Madrid, E-28049 Madrid, Spain

E-mail: `frederic.nowacki@iphc.cnrs.fr`

Abstract. In this study we propose a common mechanism for the disappearance of shell closures far from stabilty. With the use of Large Scale Shell Model calculations (SM-CI), we predict that the region of deformation which comprises the heaviest Chromium and Iron isotopes at and beyond N=40 will merge with a new one at N=50 in an astonishing parallel to the N=20 and N=28 case in the Neon and Magnesium isotopes. We propose a valence space including the full pf-shell for the protons and the full sdg shell for the neutrons, which represents a come-back of the the harmonic oscillator shells in the very neutron rich regime. Our calculations preserve the doubly magic nature of the ground state of ^{78}Ni, which, however, exhibits a well deformed prolate band at low excitation energy, providing a striking example of shape coexistence far from stability. This new Island of Inversion (IoI) adds to the four well documented ones at N=8, 20, 28 and 40.

1. Introduction

The limits of the nuclear stability and the origin of the chemical elements in the universe are burning issues in nuclear physics. These fundamental questions pose a major challenge to our theoretical understanding of nuclei, in particular towards the neutron drip line. The link between the nuclear structure of neutron-rich nuclei and nucleosynthesis is well established; for instance, the modeling of the composition of neutron-star crusts requires the knowledge of the binding energies and the structure of the nuclides near the N=50 and N=82 shell closures [1]. Intense experimental programs are devoted to structure studies of neutron-rich exotic nuclei and recently the details of the shape coexistence in ^{68}Ni and the existence of a new IoI surrounding ^{64}Cr have been explored [2, 3, 4, 5, 6, 7, 8, 9, 10, 11] and theoretical described [13, 14]. In this study, we describe the occurrence of a new Island of Inversion (IoI) at N=50 and an equivalent phenomenon of IoI's merging occuring for N=40 and N=50 similar to the one observed between N=20 and N=28 [12]. In the meantime, there have been new experimental studies on the chromium and iron isotopes up to N=42 and N=46 respectively, which seem to support it as well [15], and interesting theoretical an experimental explorations of the physics close to ^{78}Ni [16, 17, 18], which have helped us a lot in the present investigation. The trends in the evolution of the spherical mean field which favour the appearance of the IoI's have also been discussed in ref. [19].

2. Model space and interaction

The valence designed for the study of ^{78}Ni region is defined of the pf shell for protons and the sdg shell for neutrons using an inert core of ^{60}Ca. When full space calculations are not feasible, the configurationnal space is extended up to eight particle-hole excitations across the $Z = 28$ and $N = 50$ gaps ensuring convergence of the wave functions. The dimensions of the matrices to be diagonalized are at the edge of present computer capabilities for solving the problem exactly, in several cases reaching $2{\times}10^{10}$ Slater determinants, using Antoine and Nathan shell model codes [20, 21]. The effective interaction is based on realistic Two-Body Matrix Elements (TBME) with phenomenological constrains to incorporate missing three body effects. These contrainsts of monopole type are chosen to monitor

- cross-shell excitations across the proton $Z = 28$ gap through reproduction of the E2 strength in ^{80}Zn
- and for the neutrons to reproduce the zirconium two-neutron separation energies S_{2n}.

In Fig. 1, we show the comparison between calculations and experimental S_{2n} energies including recent ISOLTRAP measurements [22]. The quality of the obtained spectroscopic known data can also be seen in 2 energies for neutron rich zinc and germanium isotopes (Fig. 2) including most recent data[23, 24] from RIKEN. We name the interaction PFSDG-U hereafter and more details can be obtained in [25]. In order to assess more clearly the deformation properties in our study, we will complement our results, prior to full diagonalisation with Constrained Hartree-Fock Calculations (CHF) within the same valence space and effective interaction.

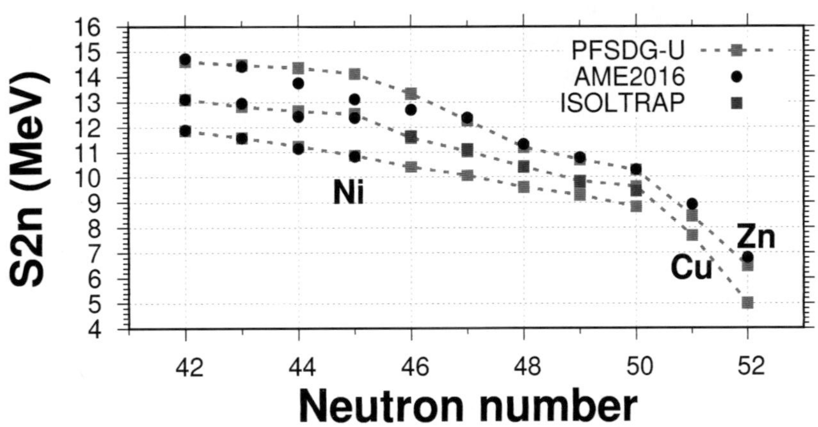

Figure 1. Two-neutron separation energies compared to PFSDG-U calculations (blue points are from [22]).

3. ^{78}Ni: shape coexistence

Fig. 3 shows the Potentiel Energy Surface obtained from CHF calculations and the low-lying spectrum resulting from the diagonalisation. At the mean field level, ^{78}Ni has a double closed structure characterized by a spherical minimum corresponding to the closed shell configuration but several minima appear at relatively low excitation energy. From the diagonalisation, the structure is much more complex, with a deformed structure with a 0^+ band-head around 2.5

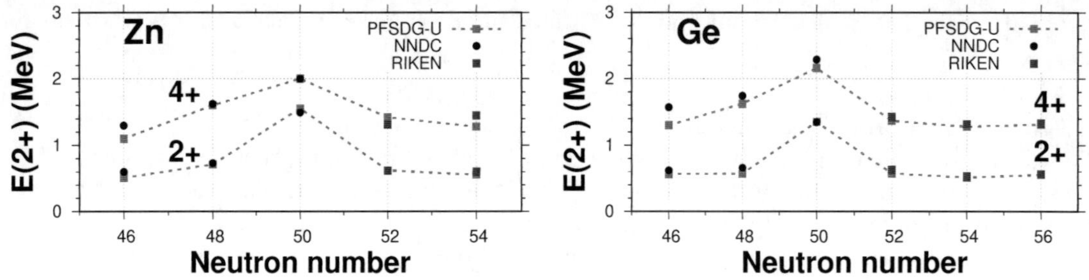

Figure 2. Systematics of 2^+ and 4^+ excitation energies compared to PFSDG-U calculations (blue points are from [23, 24]).

MeV. The Yrast sequence also belongs to this rotationnal band up to spin 6^+ where the crossing appears with "normal" $1p1h$ states. for this band (Tab. 1) shows the necessity to go beyond ab-initio or standard "LNPS" $(pf - g_{9/2}d_{5/2})$ space for the description of low-lying states in ^{78}Ni which do not predict deformed intruder states as Yrast states. On the other hand the 2_2^+ state of "$1p1h$" nature lies around 3.1 MeV, in the energy range of recent ab-initio predictions [26].

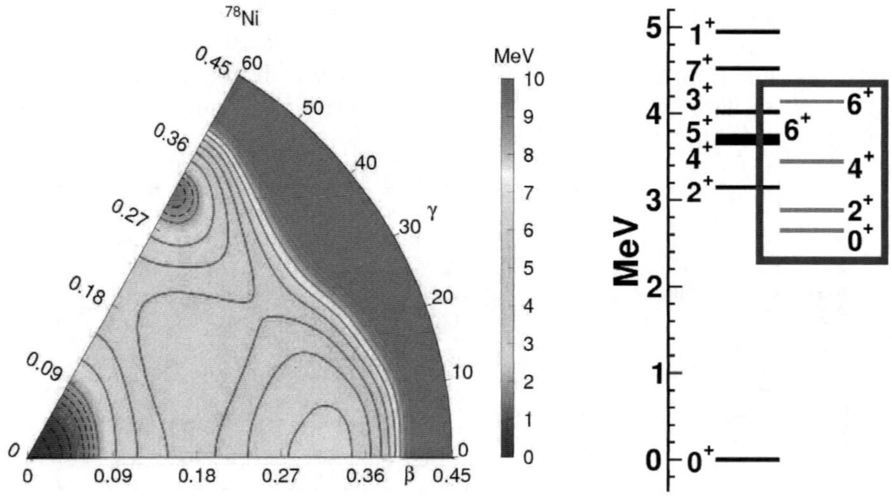

Figure 3. Potential energy Surface and Energy spectrum for ^{78}Ni with the PFSDG-U effective interaction. The low-lying deformed structure is emphasized with a blue rectangle.

4. Chromium and Iron isotopes: deformation

We move now to open proton shell nuclei and investigate the chromium and iron isotopes. The PES here reflects the transition from the nickel spherical cases to clear prolate deformed minima for the chromium chain with an intermediate situation in the iron chain. From the full diagonalisation results, the deformation is present in both isotopic chains. The most deformed case is the one of ^{74}Cr were the β deformation parameter extracted from the instrinsic quadrupole moment derived from the Bohr-Mottelson model [27] amounts to almost 0.4 signing a clear strong deformation regime.

IOP Conf. Series: Journal of Physics: Conf. Series **966** (2018) 012023 doi:10.1088/1742-6596/966/1/012023

Table 1. Occupancies of the low-ling deformed structure in ^{78}Ni

p	$f_{\frac{7}{2}}$	$f_{\frac{5}{2}}$	$p_{\frac{3}{2}}$	$p_{\frac{1}{2}}$	
	5.4	1.0	1.0	0.5	
n	$g_{\frac{9}{2}}$	$d_{\frac{5}{2}}$	$s_{\frac{1}{2}}$	$g_{\frac{7}{2}}$	$d_{\frac{3}{2}}$
	7.0	1.2	0.8	0.5	0.4

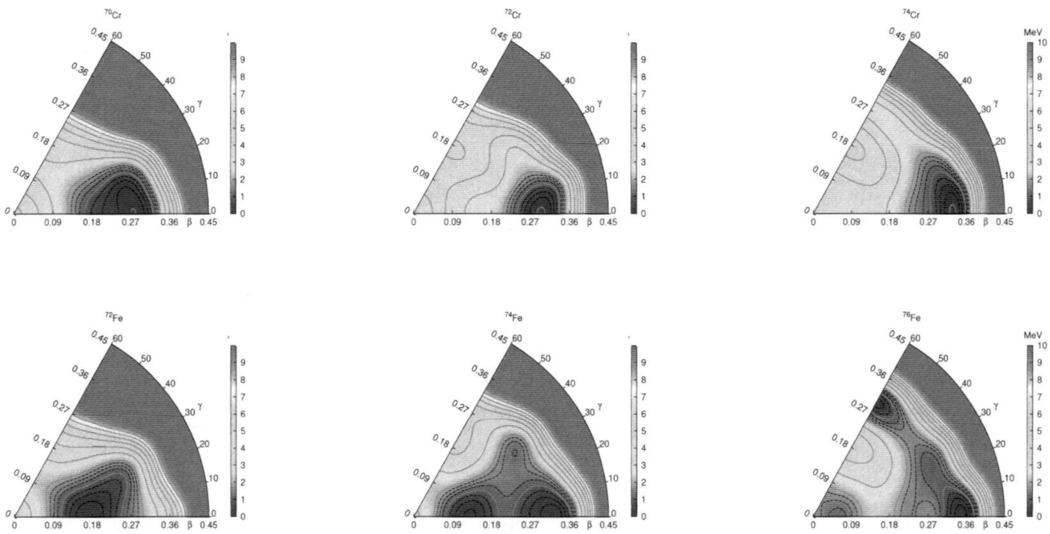

Figure 4. Potential Energy Surfaces for $^{70-74}$Cr (upper panel) and $^{72-76}$Fe (lower panel).

Table 2. E2 properties of the yrast states of the N=50 isotopes. Energies in MeV, B(E2)'s in e^{2}fm^4, Q's in efm^2.

	E*(2_1^+) (MeV)	Q_s (e.fm^2)	BE2↓ (e^2.fm^4)	$\overline{Q_i^m}$ (e.fm^2)	β^m
^{70}Cr	0.30	-41	420	340	0.26
^{72}Cr	0.23	-48	549	407	0.30
^{74}Cr	0.24	-51	630	552	0.39
^{72}Fe	0.44	-36	316	289	0.21
^{74}Fe	0.47	-39	330	308	0.22
^{76}Fe	0.35	-39	346	320	0.25

Finally, we show in Figure 5 (right panel) the evolution of the 2^+ excitation energies for the nickel and chromium chains. The present calculations are complemented towards N=40 with the results obtained using the LNPS interaction and valence space [13]. It is seen that the magic peaks at N=40 and N=50 in the nickels disappear completely in the chromiums; the fingerprint of the onset of deformation and of the entrance in the IoI's. The same is indeed true for the iron chain. In this figure, we also compare this evolution with the one for the calcium, silicon and magnesium chains from our previous study, using the SDPF-MIX hamiltonian [12]. The profiles observed between N=20-28 and N=40-50 indicates strong similarities in the physics developing far from stability. The transitions from magicity with shape coexistence to deformation appear to be a common scenario for the IoI's.

5. Conclusions

In conclusion, the deformation mechanism observed at N=40 is predicted to extend up to N=50. Shape coexistence in doubly magic ^{78}Ni turns out to be the portal to a new IoI at N=50, which merges with the well established one at N=40 for the isotopes with Z≤26. With this new addition, the archipelago of IoI's in the neutron rich shores of the nuclear chart counts now five members: N=8, 20, 28, 40, and 50.

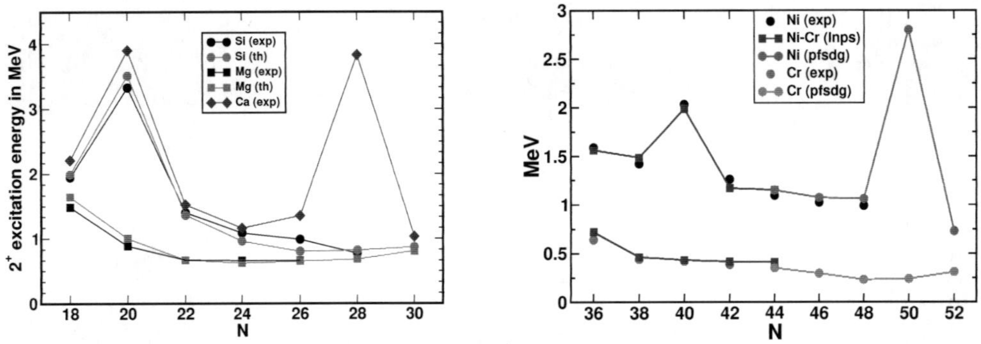

Figure 5. The chart of nuclides below ^{100}Sn showing the Islands of Inversion and their mergers.

Acknowledgments. This work is partly supported by the exchange program of Università degli Studi della Campania "Luigi Vanvitelli" (F. N.) and by an USIAS Fellowship (A. P.) of the Université de Strasbourg.

References

[1] R. N. Wolf, et al. 2013 *Phys. Rev. Lett.* **110** 041101
[2] O. Sorlin, et al. 2002 *Phys. Rev. Lett.* **88**, 092501
[3] M. Hannawald, *et al.* 1999 *Phys. Rev. Lett.* **82** 1391
[4] W. Rother, *et al.* 2011 *Phys. Rev. Lett.* **106** 022502
[5] J. Ljungvall, *et al.* 2010 *Phys. Rev.* C **81** 061301
[6] A. Gade, *et al.* 2010 *Phys. Rev.* C **81** 051304
[7] A. Dijon, *et al.* 2012 *Phys. Rev.* C **85** 031301
[8] F. Recchia, *et al.* 2013 *Phys. Rev.* C **88** 041302(R)
[9] S. Suchyta, *et al.* 2014 *Phys. Rev.* C **89** 021301(R)
[10] T. Baugher, *et al.* 2014 *Phys. Rev.* C **86** 011305(R)
[11] H. L. Crawford, *et al.* 2013 *Phys. Rev. Lett.* **110** 242701
[12] E. Caurier, F. Nowacki, and A. Poves 2014 *Phys. Rev.* C **90** 014302
[13] S. M. Lenzi, F. Nowacki, A. Poves, and K. Sieja 2010 *Phys. Rev.* C **82** 054301
[14] L. Coraggio, A. Covello, A. Gargano, and N. Itaco 2014 *Phys. Rev.* C **89** 024319
[15] C. Santamaria, *et al.* 2015 *Phys. Rev. Lett.* **115** 192501
[16] K. Sieja, and F. Nowacki 2012 *Phys. Rev.* C **85** 051301
[17] A. Gottardo, et al. 2016 *Phys. Rev. Lett.* **116** 182501
[18] X. F. Yang, et al. 2016 *Phys. Rev. Lett.* **116** 182502
[19] O. Sorlin 2014 *EPJ Web of Conferences* **66** 01016

IOP Conf. Series: Journal of Physics: Conf. Series **966** (2018) 012023 doi:10.1088/1742-6596/966/1/012023

[20] Caurier E, Martinez-Pinedo G, Nowacki F, Poves A and Zuker A P 2005 *Rev. Mod. Phys* **77** 427
[21] Caurier E and Nowacki F 1999 *Acta. Phys. Polonica* B **30** 705
[22] A. Welker et al. 2017 *Phys. Rev. Lett.* **115** 192502
[23] C. Shand et al. 2017 *Phys. Lett.* B **773** 492
[24] M. Lettman et al. 2017 *Phys. Rev.* C **96** 011301(R)
[25] F. Nowacki, A. Poves, E. Caurier, B. Bounthong 2016 *Phys. Rev. Lett.* **117** 272501
[26] G. Hagen, G. R. Jansen, and T. Papenbrock 2016 *Phys. Rev. Lett.* **117** 172501
[27] Bohr A and Mottelson B 1953 *Math. Fis. Medd. Dan. Vid. Selsk* **27** n° 16

IOP Conf. Series: Journal of Physics: Conf. Series **966** (2018) 012024 doi:10.1088/1742-6596/966/1/012024

β-decay properties in the Cs decay chain

G. Benzoni

Istituto Nazionale di Fisica Nucleare, Sezione di Milano, I-20133 Milano, Italy

E-mail: giovanna.benzoni@mi.infn.it

R. Lică

ISOLDE-EP, CERN, CH-1211 Geneva 23, Switzerland
"Horia Hulubei" National Institute of Physics and Nuclear Engineering, RO-077125 Bucharest, Romania

M.J.G. Borge

ISOLDE-EP, CERN, CH-1211 Geneva 23, Switzerland
Instituto de Estructura de la Materia, CSIC, E-28006 Madrid, Spain

L.M. Fraile

Grupo de Física Nuclear, Facultad de CC. Físicas, Universidad Complutense, CEI Moncloa, 28040 Madrid, Spain

for the IDS collaboration.

Abstract.
The study of the decay of neutron-rich Cs isotopes has two main objectives: on one side β decay is a perfect tool to access the low-spin structures in the daughter Ba nuclei, where the evolution of octupole deformed shapes can be followed, while, on the other hand, the study of the gross properties of these decays, in terms of decay rates and branching to delayed-neutron emission, are fundamental inputs for the modelling of the r-process in the Rare-Earth Elements peak. Results obtained at CERN-ISOLDE are discussed within this framework and compared to existing data and predictions from state-of-the-art nuclear models.

1. Introduction

Barium isotopes ($Z = 56$) are located in a region of the Segrè chart characterized by a variety of shape-related phenomena, including shape coexistence and possible static octupole deformation. High order deformations can have a strong influence on γ-decay rates and on quasi-particle energies, which are, in turn, inputs for the various theoretical models developed to describe these nuclei [1, 2].

The interest in this mass region is not only based on the occurrence of exotic shapes, but also on the fact that these nuclei contribute to the peak of Rare-Earth Elements (REE). The origin of this peak, characteristic of the r-process, is still under debate: it can originate either via the

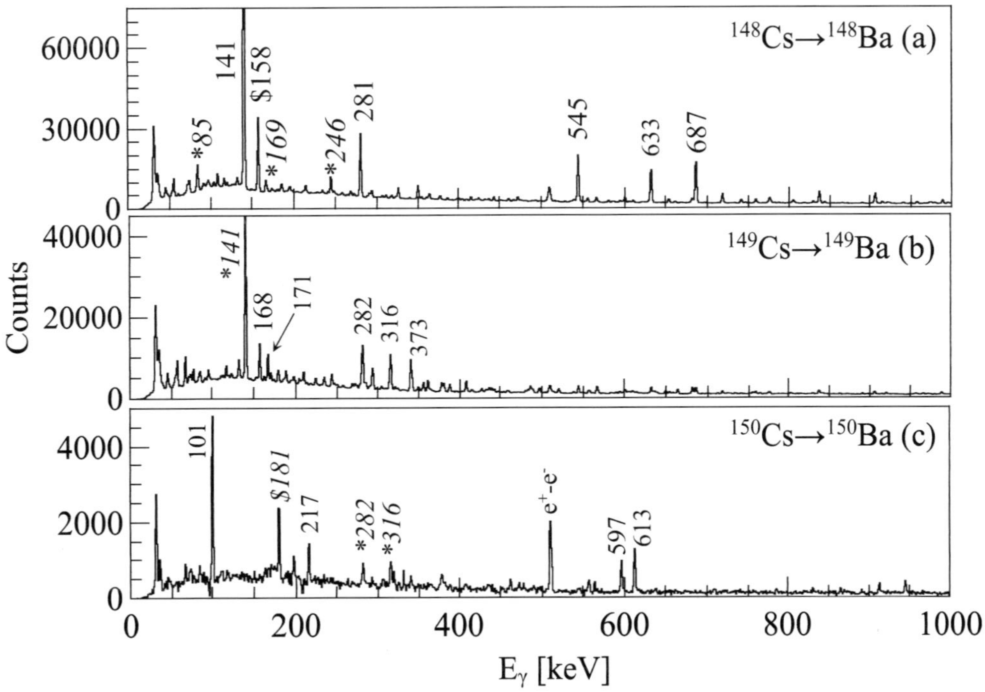

Figure 1. γ spectra following the ^{148}Cs→^{148}Ba decay (top panel), following the decay ^{149}Cs→^{149}Ba (central panel) and following the decay ^{150}Cs→^{150}Ba (bottom panel). Transitions attributed to the main decay branches ($^{148-150}$Cs →$^{148-150}$Ba) are indicated by their energies, while those corresponding to the βn branch by asterisks. The transition indicated by the dollar symbol belongs to the decay ^{149}Ba→^{149}La. See text for more details.

r-process freeze-out stage when we have an equilibration of (n,γ) $\rightleftharpoons$ (γ,n) processes, or it can be fuelled by the fission of very heavy nuclei formed in merging stars [3, 4, 5].

In order to correctly model this process one has to provide a series of experimental quantities (fission, neutron capture, and β-decay rates), which are not easily accessible, given the very exotic nature of the parent nuclei.

2. Experimental details

In this work we study the β decays of $^{148-150}$Cs to $^{148-150}$Ba isotopes. The parent Cs isotopes have been produced at ISOLDE (CERN) [6, 7] by fission of a nano-structured UCx target induced by the 1.4-GeV proton beam delivered by the PS-Booster (PSB). The Cs atoms, thermally diffused out of the target matrix, were surface ionised and separated in the ISOLDE General Purpose Separator (GPS). The proton beam current ranged between $1.5 - 2\,\mu$A. Intensities of the exotic beams reaching the experimental setup were measured to be 5.5×10^2 ions/μC for mass A=148, 100 ions/μC for A=149 and 2 ions/μC for A=150 beam, taking into account a beamline transmission of ~70%. The beam extraction was started 6 ms after the arrival of the proton pulse. Cs ions were implanted on an aluminised mylar tape at the center of the detection setup. The data acquisition recorded the arrival time of each proton pulse, using it as a reference for building decay curves. The tape was moved every 20-25 proton pulses (corresponding to a PSB super-cycle of ~1 minute) in order to remove the unwanted daughter activity.

The experimental set-up consisted of the Isolde Decay Station (IDS), equipped with 3 fast-responding plastic scintillator detectors, to detect the β particles, 4 HPGe Clover detectors for the detection of γ rays following internal decay in the daughter nuclei, and 3 small-volume conic $LaBr_3$(Ce) detectors to perform lifetime measurements of specific nuclear states. The β detection array efficiency was 20% while the total HPGe efficiency, after add-back, amounted to 6% at 0.6 MeV. The digital processing of the energy signals yields resolutions at 1.3 MeV of the order of 2.3 keV for the HPGe detectors and 40 keV for the $LaBr_3$(Ce) ones. To increase the efficiency of the γ array no anti-Compton shield has been used, with HPGe crystals placed at few centimetres from the implantation point. The collected spectra show peaks due to partial depositions of energy by γ rays interacting via Compton or pair-production mechanisms.

Further details on the experimental set-up and analysis techniques can be found in Ref. [8, 9].

3. Results

Total γ spectra have been produced requiring a coincidence (ΔT) between the proton signal and the emission of a β particle, registered in the plastic scintillator. Different conditions, depending on the expected half-life, have been used for the three decays. In panel (a) of Fig. 1 we show the spectrum related to the decay ^{148}Cs→^{148}Ba, collected requiring ΔT<400 ms, in panel (b) the one corresponding to ^{149}Cs→^{149}Ba, collected in the same condition, and in panel (c) the one from the decay ^{150}Cs→^{150}Ba, collected in an interval ΔT<200 ms.

The main transitions attributed to each decay are indicated in the figure by their energy, while the ones marked by an asterisk, and whose energy is indicated in italics font, belong to the βn channel, which sizeably contributes to these decays, as it will be discussed later on. These transitions were used to extract the decay half-lives, which have been published in [8].

The extraction of the probability for emitting delayed neutrons (expressed in terms of P_n values) is usually based on the ratio between the population of known states in the grand-daughter nuclei ^{A}Y and $^{A-1}$Y, taking into account the emission probability from the daughter, if known.

In the case of ^{148}Cs→^{148}Ba, being the βn successor (^{147}Ba) an even-odd nucleus its decay is characterised by a large fragmentation onto many final states. Three transitions are clearly seen in the spectrum at energies 85, 169 and 246 keV. The line at 158 keV belongs to the decay ^{148}La →^{148}Ce, coming as background from previous implantations, since the tape was moving every minute.

The ^{148}Cs neutron emission probability of P_n=38(4)% was determined from the ratio between the total number of ^{147}Ba and ^{148}Ba nuclei produced, estimated using the adopted values I_{abs}(^{147}La; 167.4 keV) = 15.3(16)% and I_{abs}(^{147}La; 196.1 keV) = 6.7(7)% [12].

In general, the decay from ^{149}Cs is fragmented onto many, closely lying, final states, and a strong contribution from delayed emission of neutrons is present. The βn branch seems to be mainly populating the 2^+ state in ^{148}Ba, even if a contribution from the $4^+ \rightarrow 2^+$ transition at 281 keV might be present. This second transition is hidden behind a γ transition at similar energy (282 keV) in the decay of ^{149}Ba.

The neutron emission probability, P_n = 25(4)% for ^{149}Cs, was extracted from the ratio of the total number of ^{148}Ba and ^{149}Ba nuclei using absolute γ-ray intensities in ^{149}Ba and βn daughters ^{148}La and ^{148}Ce. Literature values were used for I_γ (^{148}La; 415.8 keV) = 3.6(1)% [13] and I_γ (^{148}Ce; 158.5 keV) = 56(1)% [14].

In panel (c) two transitions can be identified as originating from the βn channel in the decay ^{150}Cs→^{150}Ba, at 282 and 326 keV. A third transition belongs to the successor's decay ^{149}Ba→^{149}La, indicated by the dollar symbol. In the case of the decay ^{150}Cs→^{150}Ba, the collected statistics is limited, and no previous experimental information on daughter decay are available. A ^{150}Cs neutron emission probability of P_n <44(11)% was determined from the ratio between the total number of ^{149}Ba [8] and ^{150}Ba nuclei produced, estimated using the 316.6-

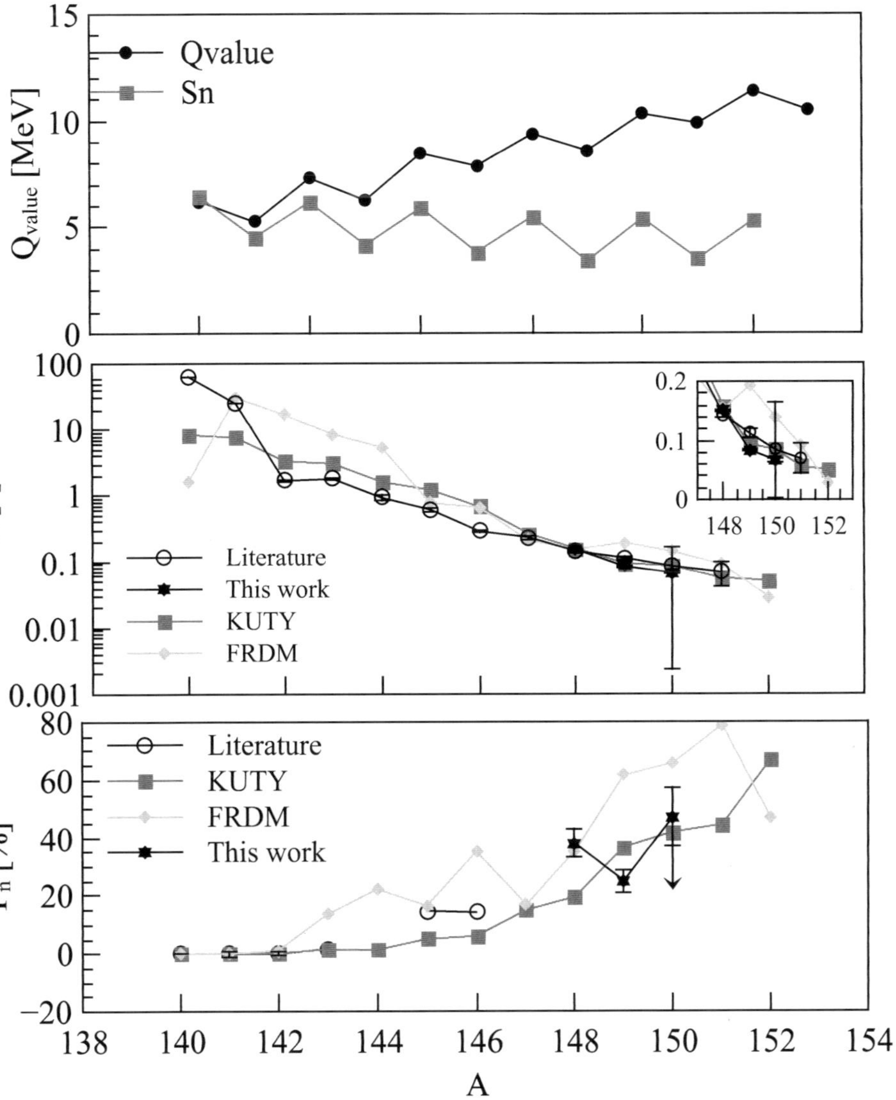

Figure 2. Top panel: Evolution of Q_β window (filled circles) and neutron separation energy (filled squares) as a function of mass in the Cs isotopic chain with A=140-152. The values are taken from the ENSDF database [10]. Central panel: Evolution of β-decay $T_{1/2}$: experimental values, shown by open circles, from masses A<144 are taken from the ENSDF database, while those for A≥144 from the recent compilation given in Ref. [11]. The experimental data obtained in this work are shown with filled stars. The data are compared with predictions obtained with KUTY (squares) and FRDM (diamonds) as described in the text. Bottom panel: evolution of P_n values (same symbols used in the middle panel).

and 101.1-keV transitions, respectively. For the transition at 101.1 keV a conversion coefficient of 1.86(3) [15] has been used, assuming a pure E2 character. We consider this as an upper limit, owing to the fact that we could not follow the decay to the grand-daughter.

4. Discussion

In this section we compare the extracted half-lives and P_n values with literature data and predictions coming from two of the most credited mass models available on the market.

The first is the finite-range droplet model (FRDM) which is based on quasi-particle random-phase approximation (QRPA). This model has been recently updated to include contributions also from first-forbidden transitions, in the framework of the statistical gross theory [16], which are expected to contribute $\approx 50\%$ of the total decay rate, with increased importance towards the most exotic species. The second model is based on the KUTY mass formula with the second generation of β-decay gross theory (GT2)[17]. Both models include the possibility of having deformed ground states, which are established to be present in this mass region [8, 9].

In the central panel of Fig. 2, we show the evolution of experimental and theoretical half-lives: The values extracted in this work, shown in the two panels with filled stars, are compared to experimental data (open circles) taken from literature: those referring to masses A<144 are taken from the ENSDF database [10], while for A$\geq$ 144 are from a recent measurement performed in Riken (J) during the EURICA campaign [11].

In the inset we expand the region where the two dataset can be compared: $T_{1/2}$ data extracted from our work are in general lower than the values extracted in Riken, even if they agree within the error bars. The large error bar for A=150 refers to the value reported in [11].

The data are compared with predictions obtained used the KUTY mass formula, shown in filled red squares and from the FRDM model in green diamonds.

Apart from the sudden jump seen for mass A=142, the $T_{1/2}$ data show a smooth decreasing trend, which is reproduced by both theories, with some slight deviations in the case of FRDM predictions.

At increasing mass we see, in the top panel of Fig. 2, that the Q_{value} window increases steeply, while the neutron separation energy, S_n, decreases smoothly. This opens the competition with the β-delayed neutron-emission channel, leading to the population of (A-1) species. The P_n values obtained in this work are again shown in filled stars, while the ones taken from literature with open circles. They are again compared with predictions from the previously described models.

We see that, up to ^{143}Cs $\rightarrow^{143}$Ba decay, the branching is negligible, while it becomes sizeable at mass around A=145. One should stress that experimental data points do not exist for all the decay chain. The comparison with theoretical predictions shows a better agreement with the description given by KUTY, while FRDM tends to overestimate the value by a large factor. The presence of strong non-axial deformations might affect the predictions for half-lives and P_n values.

Acknowledgments

The authors are indebted to the late Prof. Henryk Mach whose experience helped in designing and performing the experiment. The IDS collaboration acknowledges financial support from: Istituto Nazionale di Fisica Nucleare, the Italian"Programmi di Ricerca Scientifica di Rilevante Interesse Nazionale" (PRIN) contract number 2001024324 01302, the European Union seventh framework through ENSAR, contract no. 262010, the European Unions Horizon 2020 Framework research and innovation programme under grant agreement no. 654002 (ENSAR2), the FATIMA-NuPNET network via the PRI-PIMNUP-2011-1338 project, the Romanian IFA Grant CERN/ISOLDE and Romanian PN-II-RU-TE-2014-4-2003, the Spanish MINECO projects reference numbers FPA2013-41467-P, FPA2015- 64969-P, FPA2015-65929 and FIS2015-63770, Spanish Grants No. FIS-2014-53434-P MINECO and Programa Ramon y Cajal 2012 No. 11420, Spanish MINECO grant IJCI-2014-19172 and the MINECO project FPA2014-52823-C2-1-P the German BMBF under contract 05P15PKCIA , contract 05P15PKFNA, and "Verbundprojekt 05P2015", the FWO-Vlaanderen (Belgium) and the IAP Belgian Science Policy (BriX network

IOP Conf. Series: Journal of Physics: Conf. Series **966** (2018) 012024 doi:10.1088/1742-6596/966/1/012024

P7/12). V.Ch. and Z.P. acknowledge support by the polish grant of Narodowe Centrum Nauki nr 2015/18/M/ST2/00523.

[1] Butler P and Nazarewicz W 1996 *Rev. Mod. Phys.* **68** 349
[2] Robledo L M *et al.* 2010 *Phys. Rev. C* **81** 034315
[3] Mathews G J and Cowan J J 1990 *Nature* **345** 491
[4] R Surman J Engel J R B and Meyer B S 1997 *Phys. Rev. Lett.* **79** 1809
[5] M R Mumpower G C M and Surman R 2012 *Phys. Rev. C* **86** 035803
[6] Kugler E 2000 *Hyperfine Interactions* **129** 23
[7] Catherall J *et al.* 2017 *J. Phys. G: Nucl. Part. Phys.* **44** ..
[8] Lică R *et al.* 2017 *J. Phys. G: Nucl. Phys.* **44** 054002
[9] Lică R *et al.* submitted to *Phys. Rev. C*
[10] URL `http://www.nndc.bnl.gov/ensdf/`
[11] Wu J *et al.* 2017 *Phys. Rev. Lett.* **118** 072701
[12] Shmid M *et al.* 1981 *Proc.Int.Conf.Nuclei Far from Stability, Helsingor, Denmark* **2** 576
[13] Chung C *et al.* 1984 *Phys. Rev. C* **29** 592–602
[14] Gill R L *et al.* 1983 *Phys. Rev. C* **27** 1732–1744
[15] Kibédi T *et al.* 2008 *Nucl. Instr. and Meth. A* **589** 202–229
[16] Möller P, BPfeifer and Kratz K L 2003 *Phys. Rev.C* **67** 055802
[17] Koura H, Tachibana T, Uno M and Yamada M 2005 *Prog. Theor. Phys.* **113** 305

New proton drip-line nuclei relevant to nuclear astrophysics

L.S. Ferreira

Centro de Física e Engenharia de Materiais Avançados CeFEMA, Instituto Superior Técnico, Universidade de Lisboa, Avenida Rovisco Pais, P1049-001 Lisbon, Portugal

Abstract. We discuss recent results on decay of exotic proton rich nuclei at the proton drip line for Z< 50, that are of great importance for nuclear astrophysics models. From the interpretation of the data, we assign their properties, and impose a constraint on the separation energy which has strong implications in the network calculations.

1. Introduction

Nuclear structure far from stability plays a crucial role in the processes that lead to the formation of the elements [1, 2] In the specific case of the proton drip-line, its location constrains the path of the rapid proton capture reactions (rp) that lead to the nucleosynthesis of medium heavy elements in explosive astrophysical scenarios such as in supernovae, and X-ray bursters [3], and to determine the reaction rates needed for the calculation of the abundances [4].

In such scenarios, the density and temperature are so high, that rapid proton capture can occur, and unstable nuclei will be generated up to and beyond the proton drip-line. The path for these reactions depends on the level structure and existence of specific resonances in proton rich nuclei. The existence of isomeric states, for example, might change to higher masses the location of waiting points that held the process. Therefore, the link between reactions and properties of exotic nuclei and astrophysical observables for explosive scenarios is very strong.

However, direct experiments with unstable nuclei are still challenging, creating an obstacle to our understanding of their structure, and implying that a great part of the nuclear structure at the extremes of stability, still relies on theoretical extrapolations. It is then crucial to have a solid theory, with great independence from parameterizations, which would be capable to interpret the experimental data and predict new features.

The observation of proton radioactivity [5, 6], and its theoretical interpretation [7, 8] has made possible to access the nuclear structure properties in the neutron deficient region of the nuclear chart, for nuclei with charges $50 < Z < 81$. It has also provided an indirect way to determine separation energies [9], an important input in the network calculations, and to the understanding of the rp process.

Proton radioactivity from nuclei with $Z < 50$ is also very important to determine the time scale of the (rp) capture path, which in this region runs along the proton-drip line, but up to now, proton emission from the ground state has only been observed in two nuclei, ^{93}Ag and ^{89}Rh [10].

The observation and theoretical interpretation of these exotic decays, specially if they encompass the waiting points isotopes will help to understand their nuclear structure, that

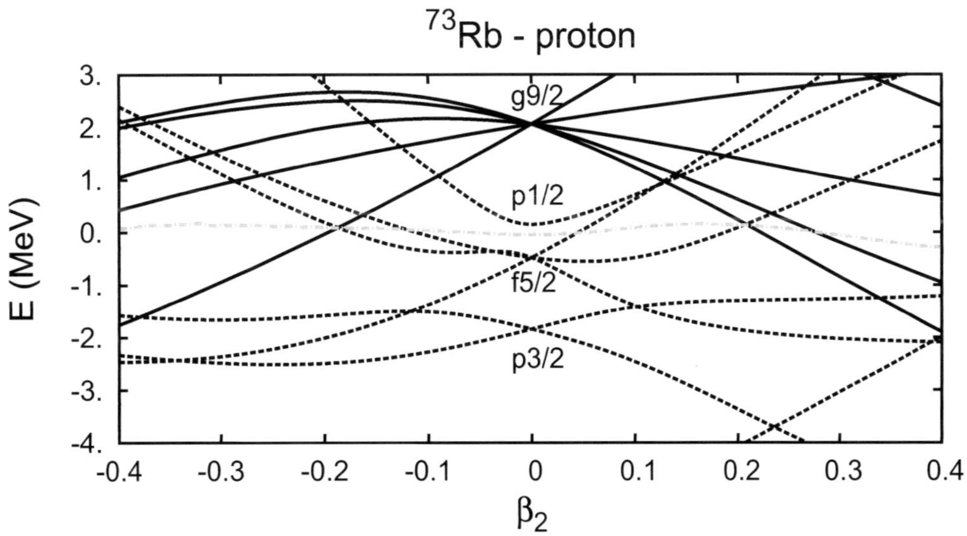

Figure 1. Nilsson proton resonances in 73Rb as a function of deformation.

controls how the capture process evolves. A good example the nucleus ^{72}Kr, a waiting point isotope whose properties have not been determined by direct measurements, but where the knowledge of the proton separation energies, and half-lives of the neighbour Rb isotopes recently produced at Riken [11, 12], would allow to establish the most probable path through ^{72}Kr.

Theoretical models developed to explain decay by proton emission, provide consistent interpretation of the experimental data, helping to determine some nuclear properties. They can explain very well the measured half-lives for spherical as well as deformed nuclei, predicting the nuclear shape parameters and quantum numbers of the decaying states. Microscopic studies of deformed proton emitters allow to test the details of the nuclear wave function to which the half-lives are quite sensitive.

The most consistent non-relativistic theoretical approach for proton emission is the non-adiabatic quasiparticle approach [7], which is very successful in bringing out several interesting features of deformed odd-even proton emitters including the triaxially deformed ones [13]. Its extension to odd-odd nuclei [14], suggested that it is possible to test the residual neutron proton interaction [15] and to determine the neutron Nilsson level even if the neutron does not participate actively in the decay [16]. Electromagnetic transitions can also be described within the model in a consistent way [17, 18].

It is the purpose of this work to discuss recent developments in the study of proton rich nuclei with $Z < 50$, and their relevance to nuclear astrophysics, and to the study of fundamental symmetries. From the interpretation of their decay data by proton emission, we will identify properties of their spectra and shape, and deduce constraints to the astrophysical processes.

2. Decay of ^{73}Rb

Particle decay is the only way to obtain information on separation energies of very short lived nuclei at the limits of stability, but not always the energy of the escaping nucleon is measured, and only information on the half life is obtained. Therefore, theoretical models can play an important role in providing some hint on the range for these energies.

The case of ^{73}Rb is an interesting example in this context. In a multi-fragmentation experiment at RIBF Riken [11, 12], where a 124Xe beam of 345 MeV/u was impinged on a

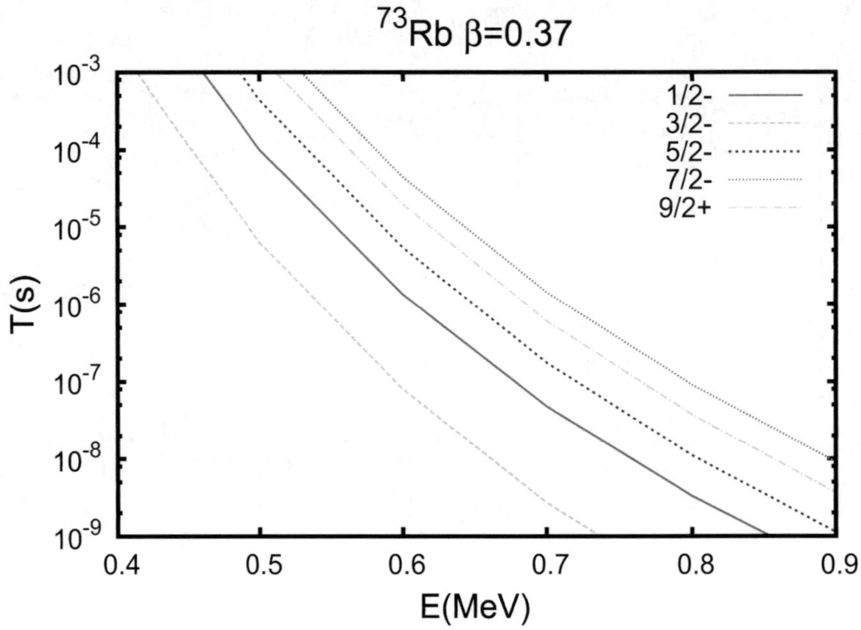

Figure 2. Theoretical half-lives, as a function of the proton energy, for decay of the lowest energy states, at a fixed positive deformation $\beta_2 = 0.37$

beryllium target, different neutron deficient nuclei were produced with charges below Z=50. The nucleus ^{73}Rb was not directly observed, but from the observation of the decay of neighbouring nuclei, it was possible to derive an upper limit for the half-life of the ground-state of this nucleus.

We have performed calculations for decay by proton emission of ^{73}Rb, within the non-adiabatic quasiparticle model. According to mass formula estimates of Möller-Nix [19], ^{73}Rb is expected to have a quite large quadrupole deformation, of the order of $\beta_2 \approx 0.37$. As a first step in our calculation we have determined the proton Nilsson resonances in the daughter nucleus ^{72}Kr, that are shown in Fig 1 as a function of deformation. They were obtained as solutions of the Schrödinger equation with outgoing boundary conditions in the single particle mean field with the parameterization of Ref. [20].

In order to obtain the spectrum of the decaying nucleus, we have to diagonalize the full Hamiltonian, containing the Nilsson contribution, the pairing residual interaction, and the rotational contribution, which includes the spectrum of the core and the coupling of the proton to the experimental spectrum of the even-even core ^{72}Kr, which is known experimentally [21]. The latter guarantees that the rotational excitation of the daughter nucleus is correctly implemented. The residual pairing interaction is taken into account in the calculation in a consistent way, by diagonalizing the Coriolis interaction between the quasi-particle states.

Following this procedure, the calculation of the spectrum of ^{73}Rb gives various possibilities for the angular momentum and parity of the lowest states in energy, which are the possible candidates for the ground-state. We found that for positive parity states the $9/2^+$ is the lowest one, for all positive and negative values of the deformation. In the case of negative parity states, and large deformation $\beta > 0.3$ the calculation predicts as the lowest state the $3/2^-$ or the $5/2^-$.

The region in the nuclear chart of Rb isotopes, presents drastic changes of deformation oscillating from very large prolate to very large oblate shapes. Whereas $^{70-72}$Rb isotopes are oblate, heavier ones above^{73}Rb, and lighter ions below ^{69}Rb, are prolate. The same occurs in the Kr region where ^{73}Kr has $\beta \approx 0.38$, and ^{72}Kr has $\beta \approx$ -0.35. Besides these rapid changes

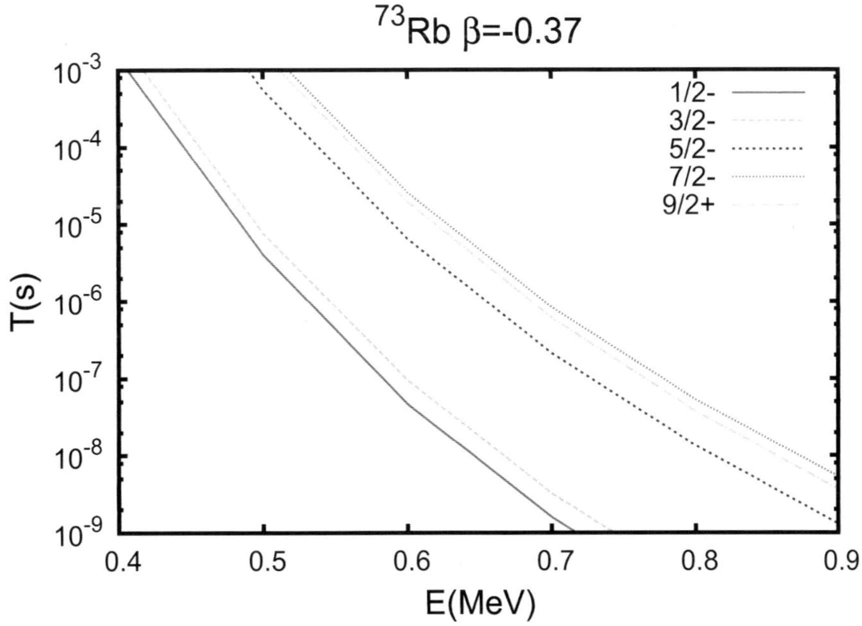

Figure 3. As in Fig 2, for a negative deformation β=-.37.

of shape, shape coexistence is also possible, making these region a quite challenging probe for theoretical models.

The next step is the calculation of the half-lives for decay from the lowest states close to the Fermi surface, previously obtained by the diagonalization of the Hamiltonian. these are the states that should have a large spectroscopic factor. We have followed the procedure described in Ref.[7], and the results are presented in Fig.2, and Fig.3.

Since the energy of the emitted proton is not known, the figures are done as a function of the proton escaping energy, and due to the great ambiguity in the attribution of a definite deformation, the calculation was done for the deformation predicted by Möller-Nix [19], and also for a large oblate deformation β = -.37, which is the deformation of the daughter nucleus. As it can be seen from the figures, for the same deformation and different energies, the half life just scales.

From the RIBF Riken experiment of Ref. [11, 12], an upper limit of 80 ns was found for the half-life. From the analysis of Fig 2., a decaying state consistent with this value, would need that the proton escaping energy should be above 600 KeV. If the prediction of Möller-Nix [19] for the deformation is correct, and ^{73}Rb has a prolate deformation, the $3/2^-$ would be a good candidate as the decaying state. The conclusions are very much the same, as it can be seen from Fig 3., if the nucleus has instead a large oblate deformation of the same magnitude, or even for large deformations different from the Möller-Nix value.

The lower bound found in our calculation for the proton energy, imposes a constraint on the proton separation energy in ^{73}Rb, coming directly from the experiment. According to the recent evaluation of atomic masses, AME2016 [22, 23], the proton separation energy in this isotope is Sp=-570(200). Our results are in agreement with these predictions, and are the first time that a direct experimental constraint of S_p is found. This has important implications for nuclear astrophysics, since the half-life of a waiting point depends exponentially on the S_p of the intermediate nucleus.

In x-ray bursts, the temperature is quite high, therefore, the rapid proton capture processes for the formation of the elements, proceeds through a sequence of fast (p,γ) reactions, but the flow can be slowed down if there are nuclei in the path that have very long β decay half-lives, and very small proton separation energies. Then, β decay will compete with the proton capture and the low Q-value, will suppress the p capture rate. The burst observables [24] will be affected by this behaviour, but to interpret these observables, the half lives of the waiting points have to be well determined.

The nucleus ^{72}Kr, is one of the waiting points in the sequence, so the proton separation energy in ^{73}Rb is very important to obtain the half-life of this isotope. Also, since ^{73}Rb is unbound, it might also be possible a competition between two-proton capture and beta-decay. The constraint on S_p through network calculations, can tell us if ^{72}Kr is a strong waiting point with an rp-process flux not larger than expected. X-ray bursts calculations with this limit on S_p [25], seem to indicate that the two-proton capture thorough ^{72}Kr is inhibited.

3. Conclusions

In conclusion, the theoretical interpretation of the experimental data on proton decay from proton rich nuclei, provides the possibility to predict the properties of the decaying state, and impose directly from the experiment, constraints on the separation energy. This information, is quite relevant for nuclear astrophysics studies, specially when the decay process involves a waiting point, as in the case adiscussed in this work.

References

[1] Woods P J, and Davids C N 1997 *Ann. Rev. Nucl. and Part. Science* **47** 541
[2] Arnould M, Goriely S 2003 *Phys. Rep.* **384** 184
[3] Schatz H, Rehm K E 2006 *Nucl. Phys. A* **777** 601
[4] Schatz H, Aprahamian A, Barnard V, Bildsten L, Cumming A,Ouellette M, Rauscher T, Thielemann F -K, and Wiescher M 2001 *Phys. Rev. Lett.* **86** 3471
[5] Sonzogni A 2002 *Nucl. Data Sheets* **95** 1
[6] Blank B, Ploszajczak M 2008 *Rep. Prog. Phys.* **71** 046301
[7] Fiorin G, Maglione E and Ferreira L S 2003 *Phys. Rev.* C **67** 054302
[8] Ferreira L S, Maglione E, Ring P 2016 *Phys Lett. B* **753** (237
[9] Mazzocchi C *et al.* 2007 *Phys. Rev. Lett.* **98** 212501
[10] Ćeliković I, Lewitowicz M, Gernhäuser R, Krücken R, Nishimura S, Sakurai H, Ahn D, Baba H, Blank B, Blazhev A, *et al.* 2016 *Phys. Rev. Lett.* **116** 162501
[11] Sinclair L *et al.* 2014 *Acc. Prog. Rep.* **47**
[12] Suzuki H, Söderström P -A, Lorusso G, Sinclair L, Davies P, Ferreira L S, Maglione E, *et al.* 2017 *Phys. Rev. Lett.* **119** 192503
[13] Arumugam P, Ferreira L S and Maglione E *Phys. Lett. B* 2009 **680** 443.
[14] Patial M., Arumugam P., Jain A. K., Maglione E. and Ferreira L. S. 2013 *Phys. Rev. C* **88** 054302
[15] A O Macchiavelli, *et al.* 2000 *Phys.Rev. C* **61** 041303
[16] Ferreira L S and Maglione E 2001 *Phys. Rev. Lett.* **86** 1721
[17] Procter M G, *et al.* 2013 *Phys. Lett. B* **725** 79
[18] Taylor M, Cullen D M, Procter M G, Smith A J, *et al.* 2015 *Phys. Rev. C* **91** 044322
[19] Möller P, Nix J R, and Kratz K -L 1997 *At. Data Nucl. Data Tab.* **66** 131
[20] Esbensen H and Davids C N 2000 *Phys. Rev. C* **63** 014315
[21] Abriola D and Sonzogni A A 2010 *Nucl. Data Sheets* **111** 1
[22] Huang W, Audi G, Wang M, Kondev F G, Naimi S, and Xu X 2017 *Chin. Phys. C* **41** 030002
[23] Wang M, Audi G, Kondev F, Huang W J, Naimi S, and Xu X 2017 *Chin. Phys. C* **41** 030003
[24] Parikh A., Jose J., Sala G., Iliadis C. 2013 *Prog. Part. Nucl. Phys.* **69** 225
[25] Sinclair L, Ph.D. thesis, University of York (2016)

IOP Conf. Series: Journal of Physics: Conf. Series **966** (2018) 012026 doi:10.1088/1742-6596/966/1/012026

Shell-model studies of the ^{34g,m}Cl(p,γ)^{35}Ar rp reaction rates

W. A. Richter[1,2,3] and B. A. Brown[4]

[1] *University of Stellenbosch, South Africa*
[2] *Botswana International University of Science and Technology, Palapye, Botswana*
[3] *iThemba LABS, South Africa*
[4] *Department of Physics and Astronomy, and National Superconducting Cyclotron Laboratory, Michigan State University, East Lansing, USA*

Contact e-mail: richter@sun.ac.za

Abstract. The thermonuclear ^{34g,m}Cl(p,γ)^{35}Ar reaction rates are unknown because of a lack of experimental data, and thus one has to rely on theory to estimate the rates. At energies in the resonance region near the proton-emission threshold many negative-parity states appear. We present results of calculations in a full (0+1)$\hbar\omega$ model space which addresses this problem. The rates for transitions from the ground state of ^{34}Cl as well as from the isomeric first excited state of ^{34}Cl are explicitly calculated. Energies, spectroscopic factors and proton-decay widths are calculated for input into the reaction rate.

1. Introduction

A classical nova is a thermonuclear explosion on the surface of a white dwarf star accreting hydrogen-rich gas from a companion star in a binary system. In such explosive stellar environments, which include classical novae and x-ray bursters, thermonuclear radiative capture reactions on unstable nuclei determine the path of nucleosynthesis toward the proton drip line. The thermonuclear ^{34g,m}Cl(p,γ)^{35}Ar reaction rates are unknown at nova temperatures due to a lack of experimental nuclear physics data for the resonances up to about 800 keV above the ^{35}Ar proton separation energy [1]. Uncertainties in these rates translate to uncertainties in ^{34}S production in models of classical novae on oxygen-neon white dwarfs. ^{34}S abundances have the potential to aid in the classification of presolar grains. A fast thermonuclear reaction rate leads to the destruction of ^{34}Cl and bypasses the production of ^{34}S, the beta decay daughter of ^{34}Cl ($T_{1/2}$ = 1.53 s). The isotopic ^{32}S/^{34}S ratio depends strongly on the ^{34}Cl(p,γ)^{35}Ar reaction rate.

Current nova models treat the ^{33}S(p,γ)^{34}Cl and ^{34}Cl(p,γ)^{35}Ar rates as single, total rates, without separately considering the ground state ^{34g}Cl and the isomeric first excited state ^{34m}Cl (Ex = 146.36(3) keV, $T_{1/2}$ = 2.5 min). However, Schatz et al. have found that the ^{32}Cl(p,γ)^{33}Ar reaction rate is dominated entirely by capture on the first excited state in ^{32}Cl over the whole temperature range relevant in x-ray bursts [2]. Therefore it is important to constrain not only the ^{34g}Cl(p,γ)^{35}Ar reaction rate, but also the ^{34m}Cl(p,γ)^{35}Ar rate. Estimates based on shell-model calculations are complicated by high level density and the presence of negative-parity states in the resonance region near the proton-emission threshold. We present results of calculations in a full (0+1)$\hbar\omega$ model space which addresses this problem, using the interaction sdpfmu [3] and NuShellX [4]. The basis consists of a complete (0+1)$\hbar\omega$ basis made from all possible excitations of one nucleon from 1s-0d to 1p-0f. Such calculations were carried out recently for the first time for the ^{30}P(p,γ)^{31}S reaction [5]. We explicitly calculate the rates for transitions from the ground state of ^{34}Cl as well as from the isomeric first excited state of ^{34}Cl.

In a study by Fry et al. [1] seventeen new ^{35}Ar levels have been detected in the energy region E_x = 5.9 - 6.7 MeV and their excitation energies have been determined, but not spins and parities. Because of the paucity of such information we are obliged to rely on shell-model calculations. We have calculated energies, spectroscopic factors and proton-decay widths for input into the reaction rate.

2. Results for the reaction rates

Fig. 1 shows the total *rp* reaction rate versus temperature (solid line) T9 (GigaK), and the contributions for positive and negative parity states for transitions from the ground state of ^{34}Cl. It is evident that the negative parity states dominate the reaction rate by up to three orders of magnitude at the lower temperatures. The contribution of each of the dominant final states (both parities) is shown in the lower panel. The rate is mainly due to two resonances, $3/2^-(3)$ at 6.052 MeV and $1/2^-(2)$ at 6.501 MeV.

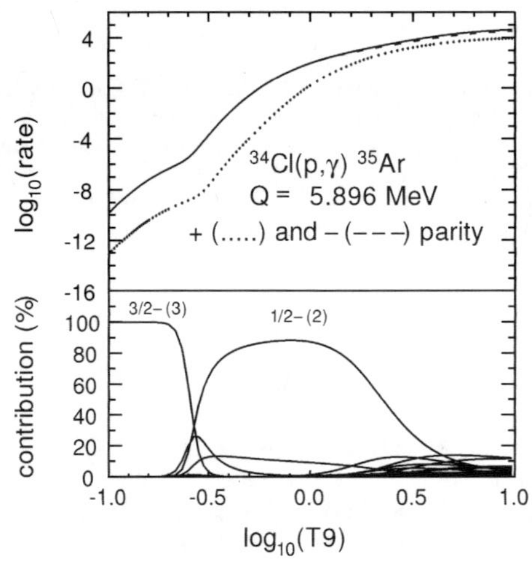

Figure 1. The total *rp* reaction rate (solid line) versus temperature T9 (GigaK), and the contributions for positive and negative parity states for transitions from the ground state of ^{34}Cl(top panel), and the contribution of each of the final states (lower panel).

Fig. 2 shows the same results as for Fig. 1 but for transitions from the first excited state. Again the negative parity states dominate the rate by up to two orders of magnitude, and the rate is mainly due to two resonances, $5/2^-(5)$ at 6.278 MeV and $5/2^-(7)$ at 6.695 MeV.

Fig. 3 shows the total rate including positive and negative parity (solid line) and transitions from both the ground and first excited state of ^{34}Cl. The relative populations of the two states have been taken into account through the stellar enhancement factor. The stellar enhancement factor is defined as the total rate divided by the rate based on the ground state. The rate from the ground state is dominant up to about 1 GK and the rate from the isomeric state dominates for higher temperatures. The rate for a statistical Hauser-Feshbach model [6] is based on transitions from the ground state, and should be compared with our result from the ground state (dashed line).

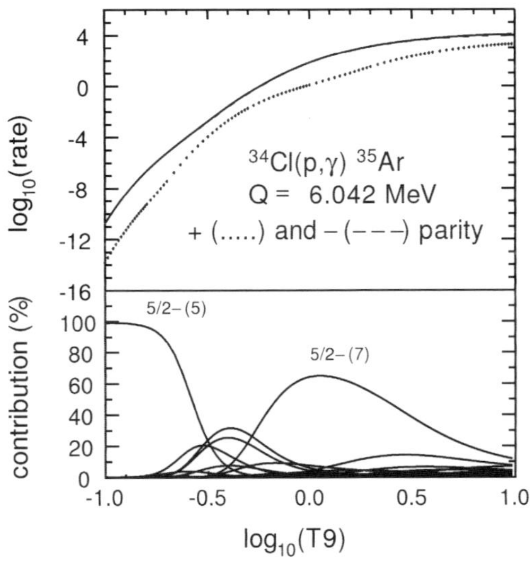

Figure 2. The total *rp* reaction rate (solid line) versus temperature T9 (GigaK), and the contributions for positive and negative parity states for transitions from the first excited state of ^{34}Cl(top panel, and the contribution of each of the dominant final states (lower panel).

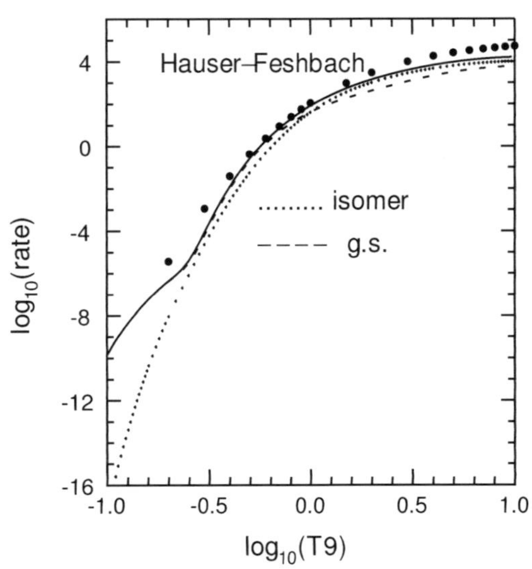

Figure 3. The total *rp* reaction rate (solid line), with the relative populations of the ground and first excited isomeric states of ^{34}Cl taken into account through the stellar enhancement factor, versus temperature T9 (GigaK). The contributions from the ground state and the isomeric state are also shown. The rate from the statistical Hauser-Feshbach model is indicated by large dots.

3. Conclusions

Our theoretical analysis shows that the reaction rates for transitions from both the ground state of ^{34}Cl and the first excited state are dominated by negative parity states by between

IOP Conf. Series: Journal of Physics: Conf. Series **966** (2018) 012026 doi:10.1088/1742-6596/966/1/012026

two and three orders of magnitude. For transitions from the ground state the dominance is for most of the temperature range, but the contribution of the positive parity states become more significant at the higher temperature range. For transitions from the first excited state the difference between the contributions of negative and positive parity states as a function of temperature is more constant. The contributions to the total rate from the isomeric first excited state of ^{34}Cl become important and dominant above about half of the temperature range considered. The statistical Hauser-Feshbach rate differs from our ground-state rate at lower and higher temperatures by up to about a order of magnitude, but is close to our result for a small region in between. We plan to use Starlib [7] to obtain the uncertainties for these rates.

The calculations also identify the most prominent resonances in the reaction rates, and the analysis should serve as a guide for experiments as the spin-parity assignments of the most prominent resonances and their relative strengths are given.

Acknowledgements
This work is partly supported by NSF Grant PHY-1404442, the Joint Institute for Nuclear Astrophysics NSF Grant PHY08-22648, and the National Research Foundation of South Africa Grant No. 105608.

References

[1] Fry C, Wrede C, Bishop S, Brown B A, Chen A A, Faestermann T, Hertenberger R, Parikh A, Pérez-Loureiro D, Wirth H-F, García A and R Ortez 2015 *Phys. Rev. C* **91** 015803
[2] Schatz H, Bertulani C A, Brown B A, Clement R R C, Sakharuk A A and Sherrill B M 2005 *Phys. Rev. C* **72** 065804
[3] Utsuno Y, Otsuka T, Brown B A, Honma M, Mizusaki T and Shimizu N 2012 *Phys. Rev. C* **86** 051301
[4] Brown B A and Rae W D M 2014 *Nuclear Data Sheets* **120** 115
[5] Brown B A, Richter W A and Wrede C 2014 *Phys. Rev. C* **89** 062801(R)
[6] Iliades C, D'Auria J M, Starfield S, Thompson W J and Wiescher M 2001 *Astrophys. J. Suppl.* **134** 151
[7] Longland R, Iliadis C, Champagne A E, Newton J R, Ugalde C, Coc A and Fitzgerald R 2010 *Nucl. Phys. A* **841** 1

IOP Conf. Series: Journal of Physics: Conf. Series **966** (2018) 012027 doi:10.1088/1742-6596/966/1/012027

Quantum self-organization and nuclear collectivities

T Otsuka

Department of Physics and Center for Nuclear Study, University of Tokyo, 7-3-1 Hongo, Bunkyo, Tokyo 113-0033 Tokyo, Japan
RIKEN Nishina Center, 2-1 Hirosawa, Wako, Saitama 351-0198, Japan
Instituut voor Kern- en Stralingsfysica, K. U. Leuven, Celestijnenlaan 200D, B-3001 Leuven, Belgium

Y Tsunoda, T Togashi, N Shimizu

Center for Nuclear Study, University of Tokyo, 7-3-1 Hongo, Bunkyo, Tokyo 113-0033 Tokyo, Japan

T Abe

Department of Physics, University of Tokyo, 7-3-1 Hongo, Bunkyo, Tokyo 113-0033 Tokyo, Japan

E-mail: otsuka@phys.s.u-tokyo.ac.jp

Abstract. The quantum self-organization is introduced as one of the major underlying mechanisms of the quantum many-body systems. In the case of atomic nuclei as an example, two types of the motion of nucleons, single-particle states and collective modes, dominate the structure of the nucleus. The outcome of the collective mode is determined basically by the balance between the effect of the mode-driving force (*e.g.*, quadrupole force for the ellipsoidal deformation) and the resistance power against it. The single-particle energies are one of the sources to produce such resistance power: a coherent collective motion is more hindered by larger gaps between relevant single particle states. Thus, the single-particle state and the collective mode are "enemies" each other. However, the nuclear forces are demonstrated to be rich enough so as to enhance relevant collective mode by reducing the resistance power by changing single-particle energies for each eigenstate through monopole interactions. This will be verified with the concrete example taken from Zr isotopes. Thus, when the quantum self-organization occurs, single-particle energies can be self-organized, being enhanced by (i) two quantum liquids, e.g., protons and neutrons, (ii) two major force components, e.g., quadrupole interaction (to drive collective mode) and monopole interaction (to control resistance). In other words, atomic nuclei are not necessarily like simple rigid vases containing almost free nucleons, in contrast to the naïve Fermi liquid picture. Type II shell evolution is considered to be a simple visible case involving excitations across a (sub)magic gap. The quantum self-organization becomes more important in heavier nuclei where the number of active orbits and the number of active nucleons are larger. The quantum self-organization is a general phenomenon, and is expected to be found in other quantum systems.

1. Introduction

The underlying mechanisms of the quantum many-body structure of atomic nuclei have been studied over decades as one of the most important objectives of nuclear physics. It has then been understood commonly that there are two types of dominant motion of nucleons in the atomic nucleus: single-particle states and collective modes. Regarding the single-particle states, Mayer and Jensen introduced the shell structure and associated magic numbers [1–3]. The nuclear shell model developed on these concepts has been extremely successful in the description of the structure of many nuclei (see for example [4–6]). The collective modes include various cases. Among them, the deformation of the nuclear shape has been studied since Rainwater [7], and Bohr and Mottelson [8, 9]. The nuclear shapes have been one of the major focuses of the nuclear structure physics, including spherical, vibrational and rotational ones [10]. The relation between the single-particle states and the collective modes has naturally become of much interest, as described in [10] as "the problem of reconciling the simultaneous occurrence of single-particle and collective degrees of freedom and exploring the variety of phenomena that arise from their interplay".

The atomic nucleus is a many-body quantum system comprised of protons and neutrons, which is often considered to be a Landau's Fermi liquid. In a somewhat simplified expression of this picture, protons and neutrons of a nucleus are in a mean potential which is like a rigid "vase", and those nucleons are like free particles moving in this vase, interacting weakly among themselves. The single-particle energies (SPE) of such a system exhibit the shell structure, and are split in general. If the splitting is large enough, the many-body structure is dominated by the SPEs: nucleons occupy the lowest single particle states in the ground state, the next lowest configuration gives us the first excited state, and so forth. The correlations due to the interaction between nucleons may contribute, but their effects are suppressed by the SPE splittings more or less. If the energy gain from such correlations overcomes the relevant SPE splittings, a collective mode dominates the structure of the ground and low-lying states. Although the understanding of the relation between the single-particle states and the collective modes has been pursued in many ways, it seems to remain an open problem. For instance, G.E. Brown had kept, throughout his life, the question, " how single particle states can coexist with collective modes " as quoted from "Fermi liquid theory: A brief survey in memory of Gerald E. Brown" in [11]. We shall present a novel mechanism which is closely related to this problem.

2. Nuclear shapes and quantum phase transition

We shall focus on the quadrupole deformation of the nuclear shape in this paper, while the scope is general. Figure 1 exhibits the excitation energy of 2^+_1 state, or the 2^+_1 level, for Sm and Zr isotopes as a function of the neutron number, N. In the Sm chain, the 2^+ level comes down rather gradually, similarly to many other isotopic chains. As shown in Fig. 1, a higher 2^+ level corresponds to a spherical shape and its surface oscillation, while a lower 2^+ level implies an ellipsoidal deformed shape and its rotation. On the other hand, in the Zr chain, the 2^+ level drops down abruptly from N=58 to 60. Due to the abrupt change, this phenomenon can be called a quantum phase transition [13]. Likewise, the ground-state structure of the Zr isotopes is changed drastically between N=58 and 60, also from the spherical to the strongly deformed shapes. The Monte Carlo Shell Model (MCSM) describes both situations including the abrupt change with the same Hamiltonian [13, 14].

3. Quantum Phase Transition in Zr isotopes

We shall look into the structure changes in Zr isotopes. The left panel of Fig. 2 shows the occupation numbers of proton orbits for some states. The $g_{9/2}$ orbit is almost empty in the 0^+_1 state of ^{98}Zr, whereas it is occupied by about 3.5 protons in its 0^+_2 state. Note that this 0^+_1 (0^+_2) state is spherical (deformed). Such changes, including the numbers of proton holes in the pf orbits, result in substantial shifts of the neutron (effective) SPEs as shown schematically in the middle panel of Fig. 2. The proton-neutron monopole interaction (wavy line in the figure) generates those shifts. The right panel depicts the actual neutron effective SPEs. One notices a

IOP Conf. Series: Journal of Physics: Conf. Series **966** (2018) 012027 doi:10.1088/1742-6596/966/1/012027

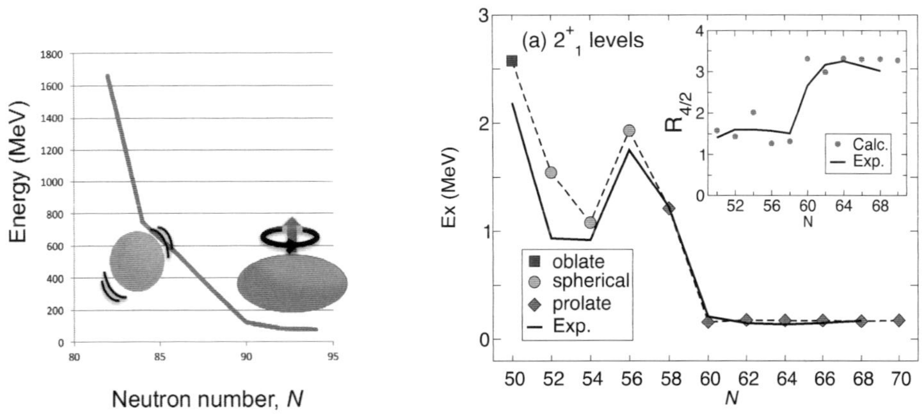

Figure 1. Systemtic changes of the 2_1^+ level in (left) Sm and (right) Zr isotopes, as functions of N. Data taken from [12] for Sm. The right panel is based on [13].

substantial change in SPE for the different states considered. One sees that the spacing between the $d_{5/2}$ and $g_{7/2}$ orbits is nearly 5 MeV for the 0_1^+ state of ^{98}Zr, whereas it is reduced to about 2 MeV in 0_2^+ state. Such a reduced splitting is found also in the 0_1^+ state of ^{100}Zr which is also strongly deformed.

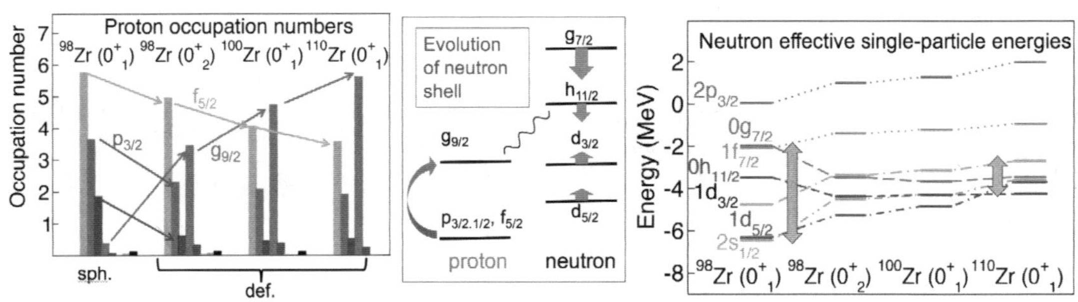

Figure 2. (left) Occupation numbers of proton orbits of Zr isotopes. (middle) Schematic illustration of the changes of neutron (effective) single-particle energies in Zr isotopes. (right) Actual values of neutron (effective) single-particle energies obtained in the calculation of [13]. Left and right panels are based on [13].

We now discuss why the SPEs are so different between spherical and deformed states. First, the nuclear deformation at low excitation energy is a Jahn-Teller effect [15], which means that the collective motion causing the deformation occurs as a consequence of coherent contributions from some relevant orbits near the Fermi energy. For such coherent effects, larger splittings of SPEs weaken the coherence, leading to less collectivity. On the other hand, the monopole interaction can change the effective SPEs depending on the occupancy of the other nucleons. If the monopole interaction were uniform, no configuration dependence would appear, and this change should be absent. The tensor-force component of the nuclear force makes the monopole interaction attractive or repulsive, depending on the combination of the orbits [16, 17]. This is certainly against the uniformity, and its effect can be crucial. The central-force component changes its magnitude also depending on the combination of the orbits mainly due to varying overlaps of radial wave functions of single-particle states. Thus, the monopole interaction is

IOP Conf. Series: Journal of Physics: Conf. Series **966** (2018) 012027 doi:10.1088/1742-6596/966/1/012027

indeed far from being uniform, and the selection of favored configurations can move the SPEs of relevant orbits substantially. If relevant SPEs can be made closer to one another, it helps the deformation. We shall formulate this novel mechanism in the next section.

4. Quantum self-organization and its appearance in Sm and Hg isotopes

The nuclear deformation is determined by the balance between the effect of the collective-mode driving force and the resistance power against this collective mode. This property is expressed schematically as

$$\text{deformation} = \frac{\text{quadrupole force}}{\text{resistance power}}. \tag{1}$$

The collective-mode driving force is the quadrupole (or quadrupole-quadrupole) interaction in the case of the ellipsoidal shape. This interaction is one of the major components of the proton-neutron realistic force. A typical example of the resistance power is the pairing interaction, which tends to make the shape more spherical because all time-reversal pairs are equally favored. Keeping the pairing interaction aside, we shall consider another source of the resistance power. That is the SPEs.

We here propose a novel mechanism, **Quantum Self-Organization**. This mechanism means the following property: Atomic nuclei can "organize" their single-particle energies by taking particular configurations of protons and neutrons, optimized for each eigenstate, thanks to orbit-dependences of monopole components of nuclear forces (*e.g.*, tensor force). This results in an enhancement of Jahn-Teller effect, *i.e.*, an enhancement of the collective mode. The deformation and quantum self-organization can be linked in a non-linear way with a positive feedback: once some nucleons are excited to certain orbits, the SPEs are shifted in favor of a larger deformation. A larger deformation can promote such excitations with more nucleons. This cycle continues until a self-consistency is achieved, whereas intermediate situations are skipped. In many cases, massive excitations are involved, and the particle-hole hierarchy is broken, for instance, a 6p-6h deformed state comes right after a 2p-2h near-spherical state, skipping 4p-4h state [18].

The property shown in eq. (1) is somewhat analogous to the relation,

$$\text{electric current} = \frac{\text{voltage}}{\text{resistance}}, \tag{2}$$

where the electric current, voltage and resistance mean the usual quantities regarding the electricity. The higher voltage produces a higher current, but the current can be increased also by decreasing the resistance. The quantum self-organization implies that the atomic nucleus finds particular configurations which decrease the resistance power.

The most favorable configurations and associated SPEs vary for individual eigenstate even within the same type of the collective mode. For instance, prolate, oblate or triaxial shapes belong to the quadrupole deformation, but can appear with different patterns of the SPEs within the same nucleus. The oblate shape is less affected by the quantum self-organization, because smaller numbers of nucleons on unique-parity orbits are the key element of the oblate shape in most cases. In those cases, the organization of many orbits are rather irrelevant, and the quantum self-organization may not occur to a sizable extent. This feature has been verified with concrete cases. On the other hand, many orbits contribute coherently to the prolate deformation, and the quantum self-organization can produce crucial effects. This has been confirmed by changing the monopole interactions, for instance, closer to the uniform one.

We present a concrete example by taking the case of the prolate band in ^{68}Ni [18]. The monopole interaction between the neutron $1g_{9/2}$ orbit and the proton $1f_{5/2}$ orbit is more attractive than that between the neutron $1g_{9/2}$ orbit and the proton $1f_{7/2}$ orbit mainly due to the robust property of the monopole interaction of the tensor force, and this difference serves as the major origin of the quantum self-organization in this particular case: more neutrons in the $1g_{9/2}$ orbit reduces the $1f_{7/2}$-$1f_{5/2}$ spin-orbit splitting for protons [16, 17, 19]. The effect of this difference on the deformation can be seen quantitatively by replacing the strengths of these monopole interactions with the average of their original values, *i.e.*, the same value. Likewise, we reset the monopole interaction between the neutron $1f_{5/2}$ and the proton $1f_{7/2}$ orbits and that between the neutron $1f_{5/2}$ and the proton $1f_{5/2}$ orbits. These modifications correspond basically to the removal of the tensor-force monopole contributions, and are nothing but the suppression of the present effects of the quantum self-organization. The resulting Potential Energy Surface is shown in Fig. 3 for the axially symmetric deformation compared with that obtained from the original Hamiltonian. Around the spherical minimum the energy curves of the two calculations are similar, however, when going to stronger deformation values, the two approaches differ substantially. In particular, the prolate profound local minimum, seen in the original calculation (red solid line), is pushed up by about 4 MeV, if the quantum self-organization is suppressed as described above (blue dashed line). Thus, the quantum self-organization is a part of the crucial mechanisms producing the nuclear deformation.

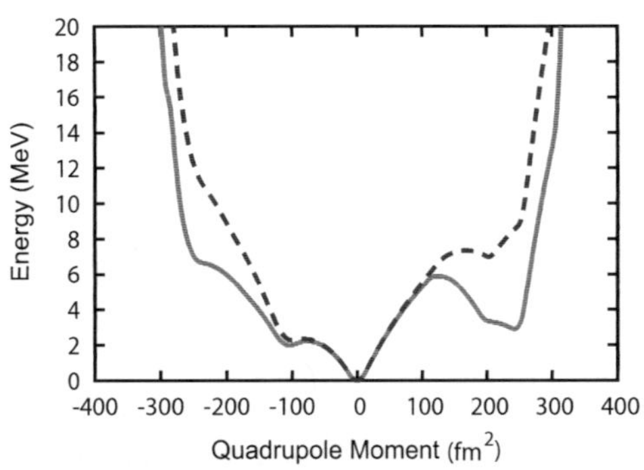

Figure 3. Potential Energy Surface with the axially symmetric deformation for ^{68}Ni. The red solid line denotes the energy of the constrained Hartree-Fock calculation with the original Hamiltonian. The blue dashed line implies the same calculation except that the quantum self-organization is suppressed (see the text). Figure taken from [18].

At this point, we mention that the SPE being discussed corresponds somehow to the spherical terms in the Nilsson model [10] which are comprised of the $\ell\ell$ and ℓs terms as well as the harmonic-oscillator-quanta term. As their strengths are independent of the deformation, the present effect is not included in the Nilsson model.

Type-II shell evolution [18] has been discussed, for instance, in Co/Ni region [19–21], where neutrons are excited from the pf shell to $g_{9/2}$ across the N=40 sub-magic gap. The neutrons in $g_{9/2}$ and neutron holes in $f_{5/2}$ provide similar sizable monopole effects, as discussed above. A smaller $1f_{7/2}$-$1f_{5/2}$ spin-orbit splitting for protons reduces the resistance power against deformation, pulling down the prolate band as seen in Fig. 3. Type II shell evolution was introduced as the particle-hole excitation over a magic or sub-magic gap. Clearly, this kind of mechanism is a very simple and visible case of the quantum self-organization. On the other hand, the quantum self-organization can occur certainly in more complex ways. Such a complex way may be found in the shape transition of Sm isotopes (see Fig. 1), where no magic or sub-magic gap is involved. We can see the spherical-vibrational-rotational shape evolution in MCSM calculations, as will be reported in detail elsewhere.

Likewise, the shape coexistence in Hg/Pb isotopes have been studied. In those cases, the quantum self-organization gives intriguing contributions on the pattern of the shape coexistence, as reported also elsewhere.

5. Summary and Perspectives

We presented a novel mechanism on the relation between single-particle states and collective modes. A summary is given below.

- The atomic nuclei are not necessarily like simple rigid vases containing almost free nucleons, in contrast to the naïve Fermi liquid picture.

- Nuclear forces are rich enough to change single-particle energies for each eigenstate, leading to the quantum self-organization.
- Single-particle energies can be self-organized, being enhanced by
  (i) two quantum liquids (*e.g.*, protons and neutrons)
  (ii) two major force components
  e.g., quadrupole interaction : to drive collective mode
  monopole interaction : to control resistance
- Type II shell evolution is a simple visible case involving excitations across (sub)magic gap.
- Actual cases such as shape coexistence, quantum phase transition, octupole vibration/deformation, super deformation, *etc.* can be studied with this scope.
- The quantum self-organization becomes more important in heavier nuclei where the number of active orbits and the number of active nucleons are larger. With larger numbers of them, the effects of the organization can be more significant. This feature may be linked to fission and superheavy elements. On the other hand, the quantum self-organization may not be so visible in light nuclei except for particular cases like intruder bands or cluster (or multiple particle-hole excited) states.
- Time-dependent version of quantum self-organization may be of another interest for reactions and fission.

Acknowledgements
The presentation and this proceedings are supported by JSPS and FWO under the Japan-Belgium Research Cooperative Program. T.O. thanks Prof. P. Van Duppen for valuable discussions and comments. The MCSM calculations were performed on K computer at RIKEN AICS (hp140210,hp150224,hp160211,hp170230). This work has been supported in part by the HPCI Strategic Program (The origin of matter and the universe) and "Priority Issue on Post-K computer" (Elucidation of the Fundamental Laws and Evolution of the Universe) from MEXT and JICFuS.

References

[1] Mayer M G 1949 *Phys. Rev.* **75** 1969
[2] Haxel O, Jensen J H D and Suess H E 1949 *Phys. Rev.* **75** 1766
[3] Mayer M G and Jensen J H D 1955 *Elementary Theory of Nuclear Shell Structure* (New York: Wiley)
[4] de Shalit A and Talmi I 2004 *Nuclear Shell Theory* (Dover)
[5] Heyde K L G 1990 *The Nuclear Shell Model* (Springer)
[6] Caurier E, Martínez-Pinedo G, Nowacki F, Poves A and Zuker A P 2005 *Rev. Mod. Phys.* **77** 427
[7] Rainwater 1950 *Phys. Rev.* **79** 432
[8] Bohr A 1952 *Mat. Fys. Medd. Dan. Vid. Selsk.* **26**(14)
[9] Bohr A and Mottelson B R 1952 *Mat. Fys. Medd. Dan. Vid. Selsk.* **27**(16)
[10] Bohr A and Mottelson B R 1975 *Nuclear Structure* vol 2 (New York: Benjamin)
[11] Schaefer T 2014 *Nucl. Phys. A* **928** 180
[12] NNDC 2016 Nudat 2 `http://www.nndc.bnl.gov/nudat2/`
[13] Togashi T, Tsunoda Y, Otsuka T and Shimizu N 2016 *Phys. Rev. Lett.* **117** 172502
[14] Kremer C, Aslanidou S, Bassauer S, Hilcker M, Krugmann A, von Neumann-Cosel P, Otsuka T, Pietralla N, Ponomarev V Y, Shimizu N, Singer M, Steinhilber G, Togashi T, Tsunoda Y, Werner V and Zweidinger M 2016 *Phys. Rev. Lett.* **117** 172503
[15] Jahn H A and Teller E 1937 *Proc. R. Soc. A* **161** 220
[16] Otsuka T, Suzuki T, Fujimoto R, Grawe H and Akaishi Y 2005 *Phys. Rev. Lett.* **95** 232502
[17] Otsuka T, Suzuki T, Honma M, Utsuno Y, Tsunoda N, Tsukiyama K and Hjorth-Jensen M 2010 *Phys. Rev. Lett.* **104** 012501
[18] Otsuka T and Tsunoda Y 2016 *J. Phys. G* **43** 024009
[19] Tsunoda Y, Otsuka T, Shimizu N, Honma M and Utsuno Y 2014 *Phys. Rev. C* **89** 031301(R)
[20] Leoni S, Fornal B *et al.* 2017 *Phys. Rev. Lett.* **118** 162502
[21] Morales A, Benzoni G *et al.* 2017 *Phys. Lett. B* **765** 328

IOP Conf. Series: Journal of Physics: Conf. Series **966** (2018) 012028 doi:10.1088/1742-6596/966/1/012028

Spes: An intense source of Neutron-Rich Radioactive Beams at Legnaro

A Andrighetto[1], M Manzolaro[1], S Corradetti[1], D Scarpa[1], A Monetti[1], M Rossignoli[1], M Ballan[1], F Borgna[1], F D'Agostini[1], F Gramegna[1], G Prete[1], G Meneghetti[2], M Ferrari[3] and A Zenoni[3]

[1] INFN - Laboratori Nazionali di Legnaro, Viale dell'Universitá 2, 35020 Legnaro (Pd), Italy
[2] Universitá degli Studi di Padova, Dipartimento di Ingegneria Industriale, Via Venezia 1, 35131 Padova, Italy
[3] Universitá degli Studi di Brescia, Dipartimento di Ingegneria Meccanica e Industriale, Via Branze 38, 25123 Brescia, Italy

E-mail: alberto.andrighetto@lnl.infn.it

Abstract. The Isotope Separation On-Line (ISOL) method for the production of Radioactive Ion Beams (RIB) is attracting significant interest in the worldwide nuclear physics community. Within this context the SPES (Selective Production of Exotic Species) RIB facility is now under construction at INFN LNL (Istituto Nazionale di Fisica Nucleare Laboratori Nazionali di Legnaro). This technique is established as one of the main techniques for high intensity and high quality beams production. The SPES facility will produce n-rich isotopes by means of a 40 MeV proton beam, emitted by a cyclotron, impinging on a uranium carbide multi-foil fission target. The aim of this work is to describe the most important results obtained by the study of the on-line behavior of the SPES production target assembly. This target system will produce RIBs at a rate of about 10^{13} fissions per second, it will be able to dissipate a total power of up to 10 kW, and it is planned to work continuously for 2 week-runs of irradiation. ISOL beams of 24 different elements will be produced, therefore a target and ion source development is ongoing to ensure a great variety of produced isotopes and to improve the beam intensity and purity.

1. Introduction

The SPES facility has the primary goal of producing intense exotic beams, as described in [1], [2]. The radioactive nuclei will be produced by the interaction of a 40 MeV proton beam, at currents up to 200 μA, with a properly designed target. The specific radioactive ion beam (RIB) will be selected making use of the isotope separation on-line (ISOL) method and delivered to several experimental halls in which nuclear physics experiments will be performed. The C70 H- cyclotron of Best Cyclotron Systems, with a maximum current of 0.750 mA rowing two exit ports will be used as a proton driver accelerator with variable energy (35-70 MeV) [2]. Two proton beams can be delivered at the same time, sharing the total current of 0.750 mA. In an ISOL-based facility, the fission fragments are generated inside the grains of the ceramic material constituting the target [3]. They subsequently diffuse towards the surface of the grains; once there, they move towards the ion source by random transport in high vacuum (effusion). In order to facilitate both the diffusion and the effusion mechanisms, the temperature of the target is kept as high as possible, in the order of 2000 °C. In order to manage a target with up to 10 kW

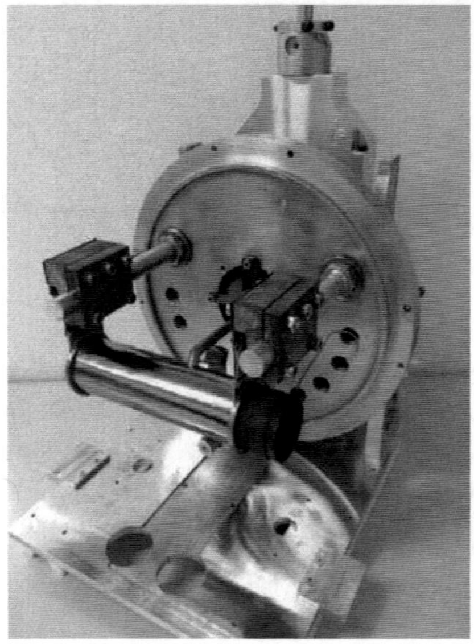

Figure 1. The SPES Target-Ion Source Unit, without the water cooled aluminum cover.

power deposition, the development of new techniques for the target production and handling are necessary [4]. For this reason, the SPES target design has been optimized in order to dissipate the foreseen beam power and maximize the release efficiency and at the same time to make use of devices (typically ion sources) developed in collaboration with other laboratories. The Target-Ion Source assembly (TIS), see figure 1, incorporates many of the engineering features used in other ISOL RIB facilities design. The SPES target-ion source chamber is indeed designed for easily removal or reconnection to the beam transport system, to allow for both the substitution of the target material and ion source maintenance, as required in a high radioactive environment. The unit is coupled to the RIB line and to the proton driver beam by means of two quick connectors and two pumping ports which can be sealed off with valves. The handling system must be designed in order to move the target chamber unit, housed in a lead container, into the area where the target will be irradiated, and to connect it to the beam lines. After the irradiation stage, the target chamber must be removed from the beam lines, stored in the lead container and finally sent out of the hot area for further handling and long term storage.

The ions produced with the ion sources are accelerated towards the ion extraction electrode by a potential up to 40 kV. After passing through a High-Resolution Mass Separator (HRMS), the selected isotopes will be stopped inside a charge breeder and extracted with increased charge (n^+). A final mass selector will be installed before reaching the PIAVE-ALPI re-accelerator [5], to clean the beam from the contaminations introduced by the charge breeder itself. The re-accelerator complex will deliver ion beams at energies of 10 AMeV and higher, for masses in the region of A = 130 amu, with an expected rate on the secondary target of 10^8 pps. The aim of this paper is to describe the main devices of the SPES RIB source: the general design of the target assembly, the target material, the ion sources and the TIS handling system.

2. Target and Ion Source complex

The Target Ion Source (TIS) system constitutes the working core of the facility [4]. It converts the 40 MeV, 200 μA proton beam coming from the cyclotron into a radioactive ion beam. The TIS system is constituted by the production target and the ion source devices, linked together by

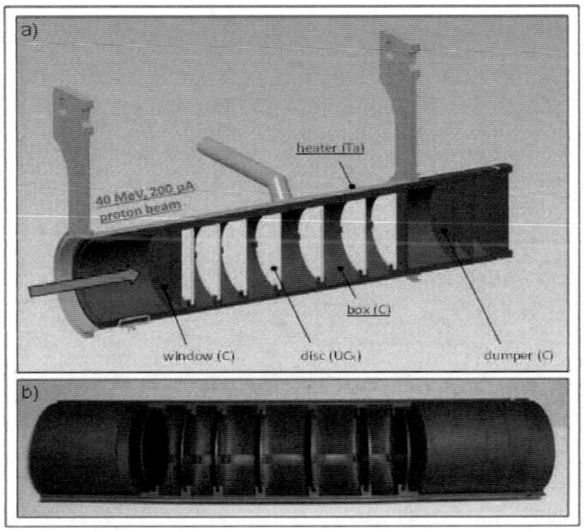

Figure 2. a) The SPES production target. b) Picture of the graphite box containing the discs and both the window and the dumper assemblies.

means of a tubular transfer line made of tantalum [6]. All the aforementioned components work in general at very high temperatures, close to 2000 °C. The SPES production target (see figure 2) is composed of 7 uranium carbide discs characterized by a diameter and a thickness of 40 and 1.3 mm, respectively. They are spaced in the axial direction in order to dissipate by thermal radiation the power of the primary beam passing through them. Two thin graphite windows are positioned at the proton beam entrance, preventing the undesired loss of radioactive isotopes and an excessive temperature drop for the first and the second target discs. Four circular graphite dumpers, with thickness ranging from 0.8 up to 10 mm, stop the proton beam at the rear side of the production target. All the discs, windows and dumpers are closed inside a tubular graphite box, that is located under vacuum inside a water-cooled chamber and has to maintain the average temperature of 2000 °C. All the SPES TIS systems presented in figure 2 are closed inside a water-cooled vacuum chamber, capable to dissipate the amount of power associated to both the target and the ion source, and to guarantee a vacuum level of approximately 10^{-6} mbar. By means of commercial fast couplings, it provides the water for the cooling system and the electrical current needed to heat by Joule effect the production target and the ion source.

3. The SPES Ion Sources

In this section the ionization mechanisms that will be implemented in the SPES ion sources, that are surface ionization, laser ionization and electron impact ionization are described. They correspond to three different ion sources, that define, as a consequence, three different TIS systems, as reported in figure 3. The SPES Surface Ion Source (SSIS) [7] is a tubular hot-cavity ion source capable to produce efficiently 1^+ ions for the elements with ionization potential smaller than 7 eV, mainly for the alkali and the alkaline earth metals (such as Rb, Cs, Sr and Ba). It is composed of a Re tubular ionizing cavity connected to one side to a tantalum support and to the other side to the transfer line. During operation both the SSIS and the transfer line are resistively heated at temperature levels close to 2000 °C by the same power supply. The geometry of the SPES Laser Ion Source (SLIS) [8] is similar to the SSIS, but in this case the material used to manufacture the hot-cavity is tantalum, characterized by a lower work function value respect to rhenium, in order to reduce the undesired contaminants coming from the surface

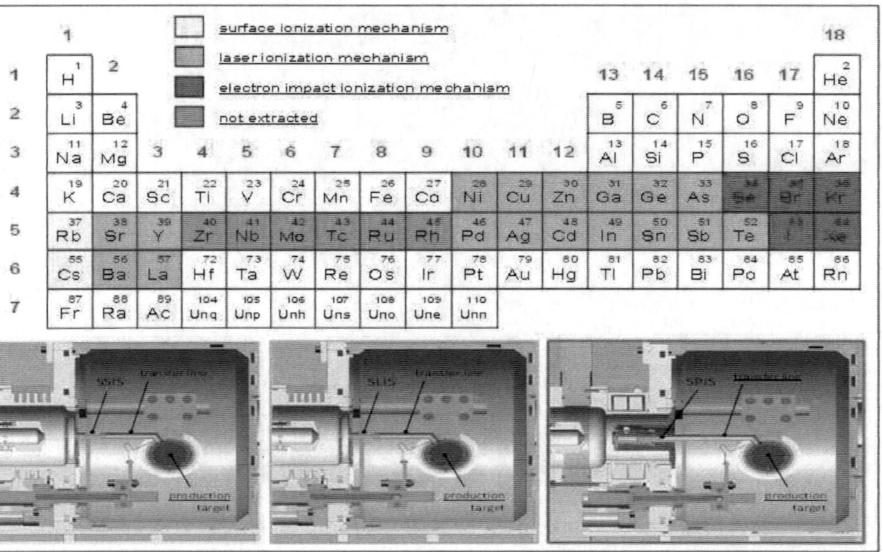

Figure 3. Main fission products in the SPES production target and representation of the three main target ion source systems.

ionization. In the resonant laser ionization process atoms are stepwise-excited leading finally to continuum. It typically consists of two or three steps, everyone characterized by a precise absorption energy level for the atom, i.e. a precise emission wavelength for the lasers. This ionization process is chemically selective and it can produce an extremely pure ion beam. The laser beams enter the beam line thanks to a view port positioned at the first mass separator magnet. The SPES Plasma Ion Source (SPIS) [9], [10] is a Forced Electron Beam Induced Arc Discharge (FEBIAD) ion source. It is a non-selective device, and is particularly used for the ionization of noble gases. The two main components of the SPIS are the cathode and the anode, made of tantalum and molybdenum, respectively. The former is heated at high temperature by Joule effect, allowing the production of electrons by thermionic emission on the surface facing the anode, whereas the latter is kept at approximately 150 V with respect to the rest of the source, confining the plasma from which the beam is extracted.

4. Target discs production

The properties required for the SPES target material [11], as in the case of other ISOL facilities, are directly related to the efficiency of the processes undergoing between the isotopes production and their release, which can be divided into two distinct phases:

- The diffusion of the isotopes, generated inside the ceramic material grains, towards their surface. This mechanism is governed by Ficks laws.
- The effusion in high vacuum towards the ion source, consisting of random walks and bounces on the surrounding surfaces, each one characterized by a specific sticking time which contributes to the overall delay.

The obtainment of a fast release of isotopes depends on a large number of parameters, both relative to the material properties and to the target operative conditions. The characteristics required for a material to work with a high efficiency as an ISOL target can be summarized as [12]:

IOP Conf. Series: Journal of Physics: Conf. Series **966** (2018) 012028 doi:10.1088/1742-6596/966/1/012028

- A sufficiently high cross-section for the reaction between the target constituting atoms and the primary beam.
- The target material must be able to work for a sufficient amount of time at high temperatures in high vacuum.
- The material should exhibit high thermal emissivity and conductivity.
- The target should be highly porous to favor the release of isotopes towards the ion source.

The research on materials for the SPES target represents an important part of the entire facility development, both in terms of material structural properties and target behavior under irradiation in specifically designed on-line tests. The most used ISOL target material is by far uranium carbide, since it fulfills all of the requisites above highlighted. The most common type of uranium carbide produced, tested and used in ISOL facilities is commonly referred to as UC_x, indicating it is composed of different phases: uranium dicarbide (UC_2), graphite (C), and a minor amount of uranium monocarbide (UC). In the last years, the synthesis and characterization of uranium carbide thin discs have been successfully carried out, and the production methodology can be considered ready to be used to produce real UC_x targets when SPES will be operative. The synthesis is based on the reaction between a proper uranium source, typically uranium dioxide, and graphite:

$$UO_2 + 6C \rightarrow UC_2 + 2C + 2CO(1) \quad (1)$$

which is made to occur at high temperature (up to 1800 °C) in high vacuum (10^{-6} mbar). In order to characterize the target behavior during irradiation, two target prototypes, constituted by seven UC_x thin discs, with the same layout of the SPES target but with scaled dimensions were tested at the OLTF (On-Line Test Facility) of Oak Ridge National Laboratory (ORNL) [13], [14], [15]. The on-line irradiation tests performed at OLTF with a 50 nA, 40 MeV primary proton beam demonstrated the capability of the SPES target prototypes to produce and release a set of more than 75 isotopes, with half-lives ranging from seconds up to few hours.

5. The TIS Handling system

The target handling system consists of three devices: in one hand there is the coupling table positioned on the front-end apparatus which is in charge of coupling and uncoupling the target chamber to the beam lines. On the other hand, there are two independent handling devices (called vertical and horizontal devices) that will move the target chamber to and from the irradiation bunker zone. The horizontal one will be used as the primary handling device while the vertical one is intended as a backup solution. Figure 4 shows the location of each device on the facility layout. The coupling table is located inside the bunker chamber where a high level of radiation is present. The system is operated using only electro-pneumatic actuators, mechanical switches and passive linear potentiometers. Four pneumatic motors with screw-type transmission system will move the chamber, the protonic beam line vacuum pipe and two vacuum valves in order to perform the coupling and uncoupling operations.

The horizontal handling device is formed by an Automatic Guide Vehicle (AGV) that will move the target chamber to and from the bunker area and a cartesian handling system, located on the top of the AGV, that will move the target chamber to and from the coupling table once the AGV is in position. The Cartesian device is implemented using a screw type system and electrical motors. A PLC will be installed on the AGV in order to perform the operation, and it will communicate with a user interface using wireless communication. The whole system will be operated on batteries. Therefore, the system will work in a completely wireless way.

The vertical handling device will access the bunker through a hole on the bunker roof. It will

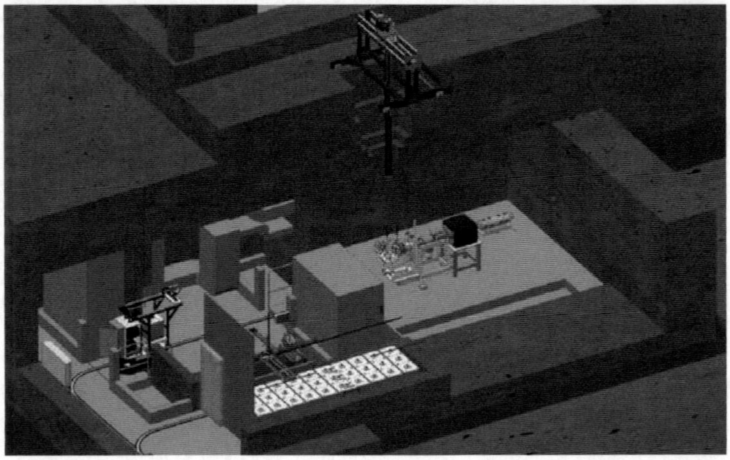

Figure 4. Layout of the SPES target chamber handling system.

grab the target chamber using an interface tool placed on top of the chamber.
The control of the whole system will be completed automatized using PLCs (Programmable Logic Controllers). In order to increase its reliability, the control system is designed using redundancy as much as possible. Furthermore, the system will allow the switch over to manual operation, in case an operator should intervene in case of emergency or failure on the automatic system.

6. Conclusions

The SPES project at INFN-LNL is one of the main nuclear physics developments in Italy for the next years. It is organized as a wide collaboration among the INFN Divisions and Laboratories, Italian Universities. SPES is a second generation ISOL facility and is an up to date project in this field with a very competitive throughput representing a step forward for the implementation of the European project EURISOL. The relevance of the project is not only related to nuclear physics research but also to astrophysics and applied physics. The main field of applied research is the possible production of radiopharmaceuticals for medicine. The ability to simultaneously operate the ALPI Linac and the two-exit-port Cyclotron will result in a large improvement in the research capabilities at LNL. The first exotic beam at SPES is expected within 2019.

References

[1] Andrighetto A, Bisoffi G, Bassato G, Calabretta L *et al* 2012 *AIP Conf. Proc.* **1491** 58-61
[2] Prete G, Andrighetto A, Manzolaro M, Corradetti S *et al* 2014 *Eur. Phys. J. Web of Conf.* **66** 11030
[3] Corradetti S, Biasetto L, Innocentini M D M, Carturan S *et al* 2016 *Ceram. Int.* **42** 17764-72
[4] Monetti A, Andrighetto A, Petrovich C, Manzolaro M *et al* 2015 *Eur. Phys. J.* A **51** 128
[5] Bisoffi G, Prete G, Andrighetto A, Andreev V *et al* 2016 *Nucl. Instrum. Meth.* B **376** 402-7
[6] Manzolaro M, Meneghetti G, Andrighetto A, Vivian G *et al* 2016 *Rev. Sci. Instrum.* **87** 02B502
[7] Manzolaro M, Andrighetto A, Meneghetti G, Rossignoli M *et al* 2013 *Nucl. Instrum. Meth.* B **317** 446-9
[8] Scarpa D, Vasquez J, Tomaselli A, Grassi D *et al* 2012 *Rev. Sci. Instrum.* **83** 02B317
[9] Manzolaro M, Andrighetto A, Meneghetti G, Monetti A *et al* 2014 *Rev. Sci. Instrum.* **85** 02B918
[10] Manzolaro M, Meneghetti G, Andrighetto A and Vivian G 2016 *Rev. Sci. Instrum.* **87** 033303
[11] Corradetti S, Andrighetto A, Manzolaro M, Scarpa D *et al* 2014 *Eur. Phys. J. Web of Conf.* **66** 11009
[12] Corradetti S, Carturan S, Biasetto L, Andrighetto A and Colombo P 2013 *J. Nucl. Mater.* **432** 212-21
[13] Scarpa D, Biasetto L, Corradetti S, Manzolaro M *et al* 2011 *Eur. Phys. J.* A **47** 32
[14] Corradetti S, Biasetto L, Manzolaro M, Scarpa D *et al* 2011 *Eur. Phys. J.* A **47** 119
[15] Corradetti S, Biasetto L, Manzolaro M, Scarpa D *et al* 2013 *Eur. Phys. J.* A **49** 56

IOP Conf. Series: Journal of Physics: Conf. Series **966** (2018) 012029 doi:10.1088/1742-6596/966/1/012029

Enhanced collectivity along the $N = Z$ line: lifetime measurements in ^{44}Ti, ^{48}Cr, and ^{52}Fe

K. Arnswald[1], P. Reiter[1], L. Coraggio[2], B. Birkenbach[1], A. Blazhev[1], T. Braunroth[1], A. Dewald[1], C. Fransen[1], B. Fu[1], A. Gargano[2], H. Hess[1], R. Hirsch[1], N. Itaco[2,3], S. M. Lenzi[4], L. Lewandowski[1], J. Litzinger[1], C. Müller-Gatermann[1], M. Queiser[1], D. Rosiak[1], D. Schneiders[1], M. Seidlitz[1], B. Siebeck[1], T. Steinbach[1], A. Vogt[1], K. Wolf[1], and K. O. Zell[1]

[1] Institut für Kernphysik, Universität zu Köln, Zülpicher Straße 77, D-50937 Köln, Germany
[2] Istituto Nazionale di Fisica Nucleare, Sezione di Napoli, I-80126 Napoli, Italy
[3] Dipartimento di Matematica e Fisica, Università degli Studi della Campania "Luigi Vanvitelli", viale A. Lincoln 5, I-8110 Caserta, Italy
[4] Dipartimento di Fisica dell'Università and INFN, Sezione di Padova, I-35141 Padova, Italy

E-mail: `konrad.arnswald@ikp.uni-koeln.de`

Abstract. Lifetimes of the 2_1^+ states in ^{44}Ti, ^{48,50}Cr, and ^{52}Fe were determined with high accuracy exploiting the recoil distance Doppler-shift method. The reduced E2 transition strengths of ^{44}Ti and ^{52}Fe differ considerably from previously known values. A systematic increase in collectivity is found for the $N = Z$ nuclei compared to neighboring isotopes. The B(E2) values along the Ti, Cr, and Fe isotopic chains are compared to shell-model calculations employing established interactions for the $0f1p$ shell, as well as a novel effective shell-model Hamiltonian starting from a realistic nucleon-nucleon potential. The theoretical approaches underestimate the B(E2) values for the lower-mass Ti isotopes. Strong indication is found for particle-hole cross-shell configurations, recently corroborated by similar results for the neighboring isotone ^{42}Ca. A detailed manuscript has meanwhile been published in *Physics Letters B* [1].

1. Motivation

The $0f1p$ shell above the ^{40}Ca core is a fertile region for shell-model studies. Several interactions like KB3G [2], FPD6 [3], and the GXPF1A [4] are available and well established, yielding good agreement with various experimental observables, such as energy spectra or – with the inclusion of Coulomb contributions – isospin-dependent effects like mirror-energy differences. In fact, a considerable number of nuclei near the $N = Z$ line is accessible via stable-beam fusion-evaporation reactions with reasonable production yields. This opens the door for high-precision measurements of lifetimes and transition strengths to be compared to theory. This work focuses on the $N = Z$ nuclei ^{44}Ti, ^{48}Cr, and ^{52}Fe. Reduced transition strengths are sensitive signatures to describe collective excitations in this region of the Segrè chart. Along the isotopic chains, (sub-)shell closures can be identified by a drop of the $B(\mathrm{E2}; 2_1^+ \rightarrow 0_{\mathrm{g.s.}}^+)$ value which increases towards midshell. In the chain of titanium isotopes such a drop was previously observed at

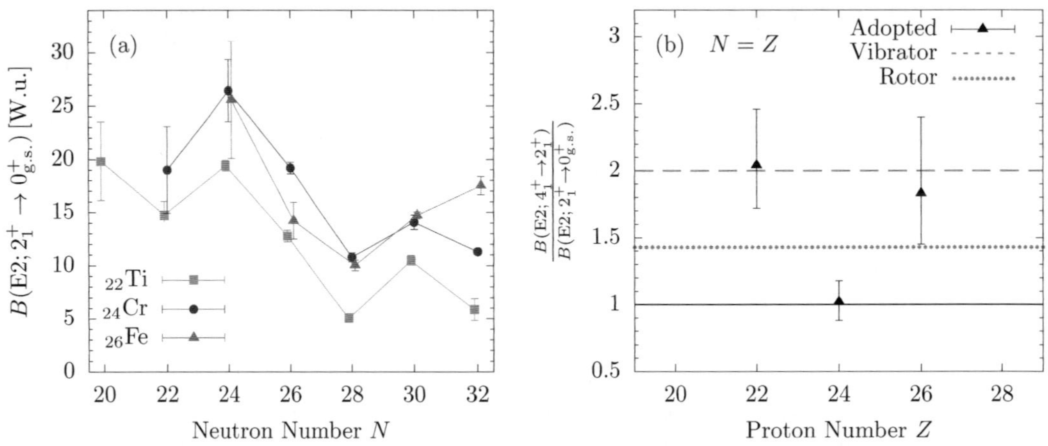

Figure 1. (a) Adopted $B(\mathrm{E2}; 2_1^+ \rightarrow 0_{\mathrm{g.s.}}^+)$ values of Ti (green squares), Cr (blue circles), and Fe (red triangles) isotopes. Data are taken from Refs. [5, 6]. Data points of the same element are connected and the Ti and Fe chains are slightly shifted horizontally with respect to the Cr data to guide the eyes. (b) Adopted $B_{4/2}$ ratios for even-even $N = Z$ nuclei between ^{40}Ca and ^{56}Ni. Limits for an ideal vibrator (long dashed purple line), an ideal rotor (dotted green line), and the limit for a non-collective behavior ($B_{4/2} = 1$) (black line) are indicated. See text for details.

$N = 22$ (cf. Fig. 1 (a)) which is not expected with regard to the shell closure at $N = 20$. For ^{48}Cr and ^{52}Fe the B(E2) values are only known with considerable errors.

Furthermore, reduced transition strengths in these nuclei are of high interest to investigate the systematics of $B_{4/2}$ ratios, as recently discussed in Ref. [7]. The $B_{4/2}$ ratio (sometimes also referred to as RE4) is defined as follows:

$$B_{4/2} = \frac{B(\mathrm{E2};\ 4_1^+ \rightarrow 2_1^+)}{B(\mathrm{E2};\ 2_1^+ \rightarrow 0_{\mathrm{g.s.}}^+)} \tag{1}$$

In rotational and vibrational nuclear models as well as in shell-model calculations the value of $B_{4/2}$ is always larger than one. $B_{4/2}$ ratios deduced from previously adopted B(E2) values evolve dramatically in the $0f_{7/2}$ shell along $N = Z$ from a vibrational behavior for ^{44}Ti and ^{52}Fe to a single-particle value at the $0f_{7/2}$ mid-shell for ^{48}Cr (cf. Fig. 1(b)).

2. Experiment

Lifetime measurements of the 2_1^+ states in ^{44}Ti, ^{48,50}Cr, and ^{52}Fe were performed at the Institute for Nuclear Physics, University of Cologne employing the Cologne plunger device [8]. Ion beams were provided by the FN tandem accelerator at 62, 26, and 86 MeV inducing $^{\mathrm{nat.}}$Mg(^{23}Na, $x$np)^{44}Ti, ^{40}Ca(^{10}B, np)^{48}Cr, ^{27}Al(^{28}Si, αp)^{50}Cr, and ^{27}Al(^{28}Si, 2np)^{52}Fe fusion-evaporation reactions, respectively. Excited recoil nuclei left the target with velocities of 1 to 3% of the speed of light and were finally stopped within a 9.6-mg/cm^2 thick Au foil. Emitted γ rays were detected by a setup of twelve high-purity germanium (HPGe) detectors. The HPGe detectors were placed in rings centered at polar angles of $\theta_0 = 0°$ (1 detector), $\theta_1 = 45°$ (6 detectors), and $\theta_2 = 143°$ (5 detectors) with respect to the beam axis. During the experiments, data were recorded for different target-to-stopper distances. For each target-to-stopper distance coincident γ rays were sorted into $\gamma\gamma$ matrices correlating groups of the different detector rings.

3. Results

In the analysis lifetimes were determined using the recoil distance Doppler-shift (RDDS) technique in combination with the differential decay-curve method (DDCM) [8]. To reduce systematic uncertainties caused by unobserved side feeding, the DDCM was applied for $\gamma\gamma$-coincidence data. Lifetimes of the 2_1^+ states were obtained from the intensity ratios of shifted (SH) and unshifted (US) components (i) of the depopulating transition for gates on the shifted part of feeding transitions and (ii) of the direct feeding transitions for coincidence requirements on the unshifted component of the $2_1^+ \rightarrow 0_{\text{g.s.}}^+$ transition. Exemplary gated γ-ray spectra of the $2_1^+ \rightarrow 0_{\text{g.s.}}^+$ transition in ^{50}Cr are shown in Figs. 2 (a, b). The lifetime τ is deduced from the weighted mean of each lifetime τ_i obtained at a distance i in the sensitive range. The resulting plots for the 2_1^+ lifetime of ^{50}Cr are given in Figs. 2 (c, d). Corresponding spectra and lifetime curves of ^{44}Ti, ^{48}Cr, and ^{52}Fe are published in Ref. [1]. Results of the determined lifetimes and corresponding B(E2) values are summarized in Table 1.

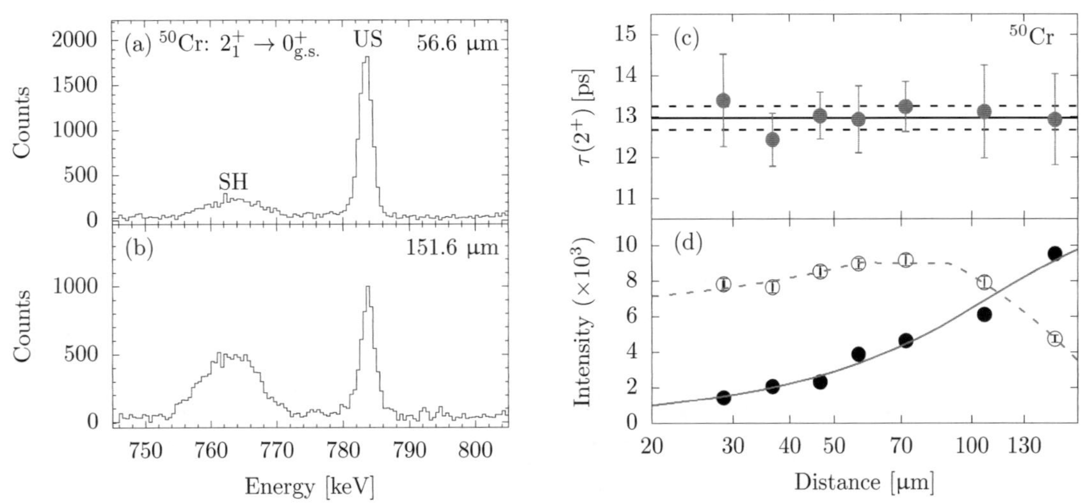

Figure 2. Exemplary γ-ray energy spectra of the $2_1^+ \rightarrow 0_{\text{g.s.}}^+$ decay at 783 keV in ^{50}Cr, obtained by a cut on the shifted component of the $4_1^+ \rightarrow 2_1^+$ transition at $E_\gamma = 1098$ keV. Short (a) and large (b) target-to-stopper distances at backward angles are shown. (SH) and (US) label the Doppler-shifted as well as the unshifted components, respectively. τ curves of the 2_1^+ state (c) for ^{50}Cr are presented. The weighted mean value of the lifetime is represented by a black solid line; dashed lines mark the statistical uncertainty. (d) Shifted and unshifted intensities of the $2_1^+ \rightarrow 0_{\text{g.s.}}^+$ transition are marked with full black circles and open black circles, respectively. The polynomial fit function to the given intensities is presented in solid and dashed red. Lifetimes and γ-ray intensities are presented in dependence of the target-to-stopper distances. Note the logarithmic distance scale.

4. Discussion

The newly determined $B(\text{E2}; 2_1^+ \rightarrow 0_{\text{g.s.}}^+)$ value for ^{44}Ti yields $22.2_{-1.8}^{+2.2}$ W.u. which is significantly larger than the adopted value of $14.7_{-0.6}^{+1.3}$ W.u.. Specifically, it now gives a pronounced minimum for the $B(E2) = 19.8(37)$ W.u. value of its semi-magic neighboring isotope ^{42}Ti. In ^{44}Ti, the hitherto adopted value includes results deduced from lifetime experiments employing the Doppler-shift attenuation method (DSAM) following α-capture and α-transfer reactions on a

Table 1. 2_1^+ lifetimes and corresponding B(E2) values in ^{44}Ti, ^{48,50}Cr, and ^{52}Fe from the present experiment as well as previously adopted values taken from Ref. [5]. The results are compared to four different shell-model calculations (for details see text).

Nucleus	Experiment				Theory			
	Lifetime [ps]		B(E2) [W.u.]		$B(\text{E2}; 2_1^+ \rightarrow 0_{\text{g.s.}}^+)$ [W.u.]			
	Present	Previous	Present	Previous	GXPF1A	KB3G	FPD6	Realistic
^{44}Ti	2.7(2)	$4.03^{+0.18}_{-0.32}$	$22.2^{+2.2}_{-1.8}$	$14.7^{+1.3}_{-0.6}$	11.4	12.9	15.2	13.0
^{48}Cr	12.2(7)	12.4(14)	$26.9^{+1.7}_{-1.4}$	26.4(29)	24.2	24.4	30.3	26.2
^{50}Cr	13.0(4)	$13.15^{+0.41}_{-0.29}$	19.6(6)	19.23(58)	19.9	19.9	25.2	22.2
^{52}Fe	7.0(4)	11.3(14)	$23.0^{+1.3}_{-1.1}$	14.2 (18)	19.2	18.8	25.3	23.7

^{40}Ca target [9, 10]. Challenging aspects of these experiments were the lack of precise knowledge of the stopping power, unknown feeding contributions, and the decreasing sensitivity of the DSAM for lifetimes longer than 1 ps. The present value provides a solution of the puzzling question of the robustness of the $N = 20$ shell closure (see Fig. 3(a)). The $B(\text{E2}; 2_1^+ \rightarrow 0_{\text{g.s.}}^+) = 26.9^{+1.7}_{-1.4}$ W.u. of ^{48}Cr and $B(\text{E2}; 2_1^+ \rightarrow 0_{\text{g.s.}}^+) = 19.6(6)$ W.u. of ^{50}Cr are in good agreement with previously known values. In the case of ^{48}Cr the relative uncertainty was further reduced and amounts to 5.8%. Both values follow the overall trend along the $Z = 24$ isotopes (cf. Fig. 3(b)). Similar to ^{44}Ti, also the $B(\text{E2}; 2_1^+ \rightarrow 0_{\text{g.s.}}^+)$ value of $23.0^{+1.3}_{-1.1}$ W.u. in ^{52}Fe exceeds the previously known value of 14.2(18) W.u. [11] and suggests a stronger collectivity at $N = Z = 26$. Both B(E2) values, the previously known as well as the present one, are consistent with the expectation of an increased collectivity with respect to the $N = 28$ shell closure, however, the present slope is more pronounced (see Fig.3(c)). To sum up, the $B(\text{E2}; 2_1^+ \rightarrow 0_{\text{g.s.}}^+)$ values in ^{44}Ti and ^{52}Fe are considerably larger than the previously measured ones. Thus, an enhanced collectivity close to the doubly-magic nuclei ^{40}Ca and ^{56}Ni at $N = Z$ is found.

Furthermore, shell-model calculations were performed to compare the experimental results with theory. Shell-model calculations were performed using the NuShellX@MSU code [12]. For comparison three established interactions were employed, namely the KB3G [2], FPD6 [3], and GXPF1A [4]. Single-particle excitations are limited to the full $0f1p$ valence space. For the calculation of the matrix elements, effective charges $e_\pi = 1.5\,e$ and $e_\nu = 0.5\,e$ were employed. Furthermore, realistic shell-model calculations, using the shell-model code Antoine [13], have been performed [14]. The calculated $B(\text{E2}; 2_1^+ \rightarrow 0_{\text{g.s.}}^+)$ values are reported in Table 1. An overview on the evolution of the B(E2) values for Ti, Cr, and Fe isotopes is shown in Fig. 3.

For each interaction the theoretical B(E2) values along the chain of titanium isotopes are in poor agreement with the experimental ones close to $N = Z$ (see Fig. 3(a)). A large deviation is observed for $N = 20$ where the theoretical values are considerably smaller than the experimentally determined values. This discrepancy decreases with an increasing number of valence neutrons toward $N = 26$. However, the sub-shell closure at $N = 28$ is also not reproduced by the calculations. In the neighboring even-even isotone ^{42}Ca a recent investigation revealed a similar result [15]. Shell-model results underpredict the enhanced experimental B(E2) values. It was suggested that the reduced theoretical values arise from $2\hbar\omega$ sd-shell excitations, which are not included in the $0f1p$ model space [16]. Evidence for the importance of core excitations is also found in the study of the schematic α-cluster model where excitation energies of all rotational bands in ^{44}Ti are reproduced using an $\alpha +^{40}$ Ca(I^π) system [17]. Moreover, their calculated $B(\text{E2}; 2_1^+ \rightarrow 0_{\text{g.s.}}^+) = 18$ W.u. value [17] is in a better agreement with the present experimental results.

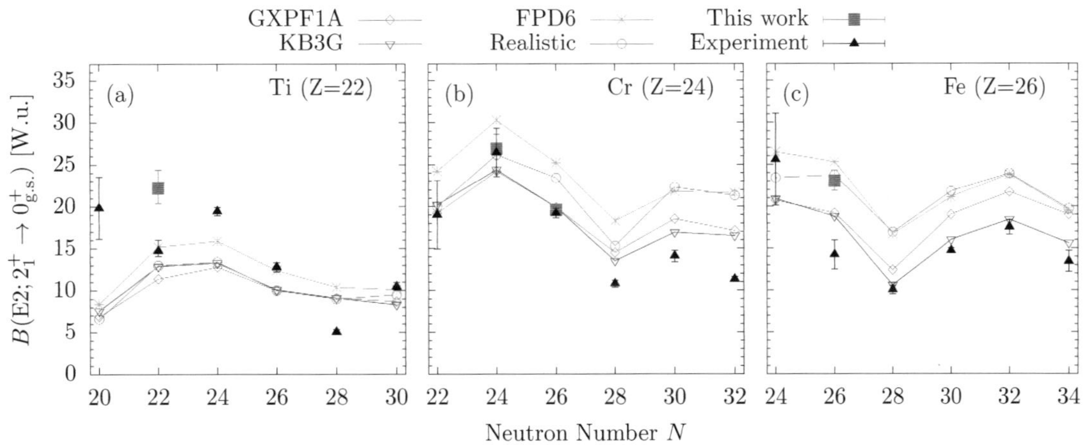

Figure 3. Experimental and theoretical $B(\text{E2}; 2_1^+ \rightarrow 0_{\text{g.s.}}^+)$ values in dependence of the neutron number for (a) titanium, (b) chromium, and (c) iron isotopes. $B(\text{E2})$ values from this work are marked with red squares. Black triangles mark the previously adopted $B(\text{E2})$ values [5, 6]. Shell-model calculations based on the GXPF1A (orange open diamonds), KB3G (blue open triangles), and FPD6 (green stars) interactions, as well as a shell-model calculation using an effective Hamiltonian starting from a realistic potential (purple open circles) are shown.

In the chain of chromium isotopes the different shell-model calculations reproduce the trend of the measured $B(\text{E2})$ values for $N < 28$, whereas the measured values are overestimated for $N \geq 28$ (cf. Fig. 3(b)).

Only partial agreement between experiment and theory is found for the iron isotopes (see Fig. 3(c)). For light isotopes at $N = 24, 26$ results from the realistic calculation and the FPD6 interaction are consistent with the present experimental data. However, nice agreement is achieved between the KB3G interaction and experiment for $N \geq 28$. It is noteworthy, that no calculation describes the huge experimentally observed drop in $B(\text{E2})$ values between $N = Z = 26$ and the shell closure at $N = 28$; it is underestimated by each interaction.

Considering the new $B(\text{E2}; 2_1^+ \rightarrow 0_{\text{g.s.}}^+)$ values, the $B_{4/2}$ values can be reevaluated using the new $B(\text{E2}; 2_1^+ \rightarrow 0_{\text{g.s.}}^+)$ values from Table 1 and the $B(\text{E2}; 4_1^+ \rightarrow 2_1^+)$ values from Refs. [18, 19, 20] (labeled Present) as well as the adopted $B(\text{E2}; 2_1^+ \rightarrow 0_{\text{g.s.}}^+)$ values from Ref. [5] and those of the $4_1^+ \rightarrow 2_1^+$ transition from Refs. [18, 19, 20] (labeled Previous). The $B_{4/2}$ values are shown in Fig. 4. The present evaluation of the $B_{4/2}$ ratios shows considerable differences for ^{44}Ti and ^{52}Fe with respect to the previous results. The observed strong variations of the $B_{4/2}$ ratios along the chain of isodiapheres, which were discussed in Ref. [7], are not confirmed. Moreover, the evaluated vibrator-like behavior for ^{44}Ti and ^{52}Fe is now suggested to be more rotor-like. The puzzle of the minimum value for mid-shell ^{48}Cr close to unity could not be resolved and remains subject of future studies.

5. Summary

These proceedings report on lifetime measurements in $N = Z$ nuclei ^{44}Ti, ^{48}Cr, and ^{52}Fe. RDDS experiments yield an increased $B(\text{E2}; 2_1^+ \rightarrow 0_{\text{g.s.}}^+)$ value in ^{44}Ti and ^{52}Fe. Enhanced collectivity is indicated for all investigated $N = Z$ nuclei. Results from state-of-the-art shell-model calculations are ambiguous; none of the four shell-model interactions reproduces the $B(\text{E2}; 2_1^+ \rightarrow 0_{\text{g.s.}}^+)$ values for the investigated Ti, Cr, and Fe isotopes in a consistent way. The observed discrepancy between experimental and theoretical $B(\text{E2})$ values for ^{44}Ti can

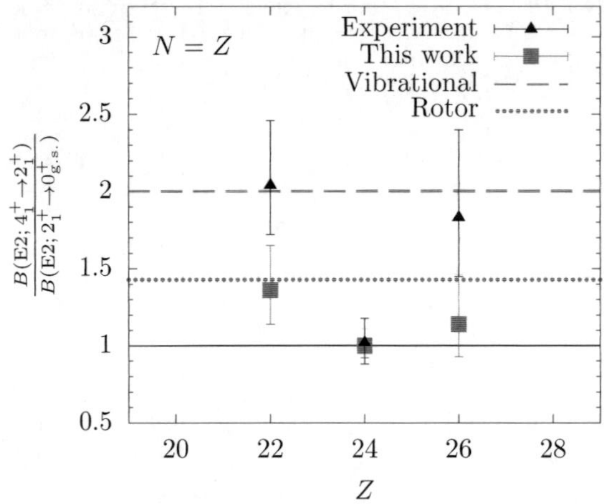

Figure 4. $B_{4/2}$ ratios for even-even $N = Z$ nuclei between ^{40}Ca and ^{56}Ni. Results from this work are given in red squares and the adopted values are shown in black triangles. Same color code as in Fig. 1 (b) is used. See text for details.

be traced back to the importance of multi-particle multi-hole cross-shell configurations similar to recent observations in the neighboring isotone ^{42}Ca. The different calculations reproduce the experimental B(E2) values for ^{48}Cr well, whereas for ^{52}Fe the experimental value is only favored by the realistic and the FPD6 interaction. An extended description of these results was submitted for publication after the conference and can be found in Ref. [1].

References

[1] Arnswald K *et al.* 2017 *Phys. Lett. B* **772** 599 – 606
[2] Poves A *et al.* 2001 *Nucl. Phys. A* **694** 157 – 198
[3] Richter W A *et al.* 1991 *Nucl. Phys. A* **523** 325 – 353
[4] Honma *et al.* 2005 *Eur. Phys. J. A* **25** 499–502
[5] Pritychenko B *et al.* 2016 *At. Data Nucl. Data* **107** 1 – 139
[6] Seidlitz M *et al.* 2011 *Phys. Rev. C* **84**(3) 034318
[7] Hertz-Kintish D *et al.* 2014 *Phys. Rev. C* **90**(3) 034307
[8] Dewald A *et al.* 2012 *Prog. Part. Nucl. Phys.* **67** 786 – 839
[9] Dixon W R *et al.* 1973 *Nucl. Phys. A* **202** 579 – 592
[10] Schielke S *et al.* 2003 *Phys. Lett. B* **567** 153 – 158
[11] Yurkewicz K L *et al.* 2004 *Phys. Rev. C* **70**(3) 034301
[12] Brown B A *et al.* 2014 *Nucl. Data Sheets* **120** 115 – 118
[13] Caurier E *et al.* 2005 *Rev. Mod. Phys.* **77**(2) 427–488
[14] Coraggio L *et al.* 2012 *Ann. Phys.* **327** 2125 – 2151
[15] Hadyńska-Klęk K *et al.* 2016 *Phys. Rev. Lett.* **117**(6) 062501
[16] Brown B A 2001 *Prog. Part. Nucl. Phys.* **47** 517 – 599
[17] Ohkubo S *et al.* 1998 *Phys. Rev. C* **57**(5) 2760–2762
[18] Chen J *et al.* 2011 *Nucl. Data Sheets* **112** 2357 – 2495
[19] Burrows T W 2006 *Nucl. Data Sheets* **107** 1747 – 1922
[20] Dong Y and Junde H 2015 *Nucl. Data Sheets* **128** 185 – 314

IOP Conf. Series: Journal of Physics: Conf. Series **966** (2018) 012030 doi:10.1088/1742-6596/966/1/012030

Structure of two-neutron halo in light drip-line nuclei

Toshio Suzuki[1,2] **and Takaharu Otsuka**[3]

[1] Department of Physics, College of Humanities and Sciences, Nihon University, Sakurajosui 3-25-40, Setagaya-ku, Tokyo 156-8550, Japan

[2] National Astronomical Observatory of Japan, Osawa 2-21-1, Mitaka, Tokyo 181-8588, Japan

[3] RIKEN Nishina Center for Accelerated-Based Science, Hirosawa 2-1, Wako, Saitama 351-0198, Japan

E-mail: suzuki@phys.chs.nihon-u.ac.jp

Abstract. The formation of two-neutron halo is described using the neutron-neutron (n-n) interaction constrained by the low-energy n-n scattering limit. This method is tested for loosely-bound two neutrons in ^{24}O, where a good agreement with experimental data is found. It is applied to halo neutrons in ^{22}C, where the ^{20}C-core is treated as a correlated-core with mixing of $2s_{1/2}$ orbit. This n-n interaction is shown to be strong enough to produce a two-neutron halo, locating ^{22}C on the drip line, while ^{21}C remains unbound. A unique relation between the two neutron separation energy, S_{2n}, and the radius of neutron halo is presented. The estimated halo radius is found to be consistent with a recent experimental data. The appearance of Efimov states is discussed. The present method is extended to apply to the case of mixed configurations for the two-neutron halos such as ^{17}B and ^{11}Li.

1. Introduction

Two-neutron halos are found in light drip-line nuclei such as ^{10}He, ^{11}Li, ^{14}Be, ^{17}B, ^{19}B and ^{22}C since it was first found in ^{11}Li [1]. When one neutron is removed from these nuclei, they become unbound. The two-neutron halos in ^{11}Li, ^{14}Be and ^{17}B have rather complex structure, that is, with mixtures of two or more orbits, but those in ^{10}He and ^{22}C have simple structure with almost pure $s^2_{1/2}$ components. Here, we first study how the two-neutron halo in ^{22}C is formed from the low energy limit of neutron-neutron (n-n) interaction.

We use a three-body model which assumes a ^{20}C-core and two valence neutrons interacting each other. This is a Borromean situation [2,3], which have been considered in previous theoretical works [4-7] but with different methods and interactions. The ^{20}C-core is treated as a correlated-core with mixing of $2s_{1/2}$ orbit. ^{20}C is known to have $2s^2_{1/2}$ components whose spectroscopic factor is about 1 [8]. Relations between the two-neutron separation energy, S_{2n}, and the halo radius are derived. The halo radius is estimated by taking into account a condition for ^{21}C to be unbound, and compared with experimental data. A comment on Efimov states is made. In sect. 2, a three-body model is explained and applied to ^{24}O. S_{2n} and halo radius in ^{22}C are studied using the correlated ^{20}C-core. Spectra of ^{22}C are also predicted. In sect. 3, the three-body model is extended to the case of mixed configurations. The method is applied to ^{18}O and ^{17}B as well as ^{11}Li. A summary is given in sect. 4.

2. Three-body model with low energy n-n interaction

We use a three-body model consisting of a core and two neutrons. Neutron-core interaction is taken to be a one-body Woods-Saxon (WS) potential. The energy of a neutron in the WS potential , ϵ, is positive (negative) if it is unbound (bound). For the n-n interaction, v_{nn}, one in the low energy limit which reproduces both the scattering length, a_{nn}, and the effective range, r_{nn}, is used [9]. A gaussian radial form with the range of 1.795 fm is adopted, whose strength is adjusted to give a_{nn} = -18.9 fm and r_{nn} =2.75 fm. The two-neutron energy, E_{nn}, is given by $E_{nn} = 2\epsilon + < v_{nn} >$, where $< v_{nn} >$ is the matrix element of v_{nn} for the two-neutron wave function and is negative. The three-body system is bound if E_{nn} is negative. Two-neutron separation energy is defined by S_{2n} = $-E_{nn}$, and as one-neutron separation energy is given by $S_{1n}=-(\epsilon + < v_{nn} >)$, $S_{2n} = 2S_{1n} + < v_{nn} >$. As $\epsilon = -(S_{2n} + < v_{nn} >)/2$, the condition for core+one-neutron state to be unbound is $\epsilon > 0$, that is, $S_{2n} < - < v_{nn} > = |< v_{nn} >|$.

Here, for a neutron the attraction from the other neutron is treated as an additional one-body potential, $w(r)$, obtained by convolution of v_{nn} by the neutron wave functions. S_{1n} and bound neutron wave function, $\phi(r)$, are obtained by solving the following Schrodinger equation;

$$\{h + w(r)\}\phi(r) = -S_{1n}\phi(r) \tag{1}$$

where h is the one-body Hamiltonian for core-neutron system with the WS potential. The neutron wave function obtained is used for the convolution of v_{nn}. Thus, after this iterative procedure, S_{1n}, $< v_{nn} >$ and S_{2n} are obtained self-consistently when they are converged.

In the three-body model, ^{24}O is treated as ^{22}O-core and two neutrons. ^{22}O is bound with ϵ =-2.73 MeV. The WS potential is fixed to reproduce this value for ϵ. Then, S_{2n} and S_{1n} are obtained by the above procedure; S_{2n} =6.94 MeV and S_{1n} =4.21 MeV which are very close to the experimental values of S_{2n} =6.92 MeV and S_{1n} =4.19 MeV, respectively [10].

The occupation number of neutron $2s_{1/2}$ orbit in the correlated ^{20}C-core is about 1 [8]. It is also true for a shell-model calculation with the YSOX interaction [11]. There is a large mixing of components of $\nu 1d^4_{5/2}2s^2_{1/2}$ configurations, which lowers the ground state energy of ^{20}C. On the other hand, the halo state in ^{22}C is dominantly $2s^2_{1/2}$ configuration [8]. How to reconcile this situation? We here adopt the following model. The halo s-orbit is occupied by two neutrons as predicted by the experiments [8]. The orthogonality condition between this halo s-orbit and the s-orbit of the ^{20}C-core state should be satisfied, that is, the core s-orbit is made orthogonal to this halo s-orbit by Gram-Schmidt method. This gives rise to blocking effects on the core states. Energy of the ^{20}C-core of the ground state of ^{22}C is pushed up with respect to the energy of the ^{20}C ground state. This energy shift is denoted as Δ. The two-neutron separation energy is thus reduced; $S_{2n} = -E_{nn} - \Delta$. As the core s-orbit gets halo components, two-body matrix elements of the YSOX interaction with s-components are modified. The single-particle energy of $2s_{1/2}$ orbit outside the ^{4}He-core is also modified. Shell-model calculations are performed by taking into account these modifications, and the energy shift of the ground state of ^{20}C is evaluated to be $\Delta \sim$1-1.5 MeV.

The relation between S_{2n} and the halo radius for the correlated ^{20}C-core case is shown in Fig. 1. The condition for ^{21}C to be unbound gives $S_{2n} < 0.8$ MeV. Fig.1 suggests that the r.m.s. radius of the neutron halo is 6-7 fm, which is quite small compared with the value obtained in Ref. [13]. Recently, the matter radius of ^{22}C was re-measured, which yields 3.44$\pm$0.08 fm [14] much smaller compared with the previous one, 5.4$\pm$0.9 fm [13]. A smaller halo radius of 6.74+0.71/-0.48, which is consistent with the present three-body model with the correlated ^{20}C-core, is estimated from the recent measurement [14] and the radius of ^{20}C [15]. Dependence of the halo radius on S_{2n} is small.

The hypothesis of Efimov [16,17] states implies the appearance of similar states at different scales near threshold. The upper bound on the radius of the halo in the correlated-core model

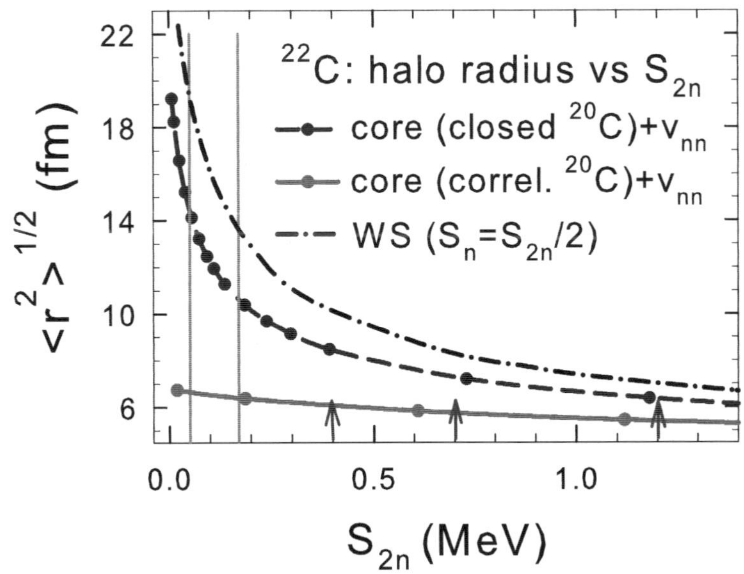

Figure 1. R.m.s raius of the halo neutron as a function of two-neutron separation energy S_{2n}. The solid line indicate the result obtained with the correlated ^{20}C core, while the dashed line the result of the closed-shell ^{20}C. The result obtained from WS potential ($S_{1n} = S_{2n}/2$) without v_{nn} is shown by the dash-dotted line. The range of S_{2n} from Ref. [10] is shown by vertical lines. Green arrows denote values discussed in Ref. [8]. (Taken from Fig. 5 of Suzuki et al. [12])

contradicts this hypothesis. The ground state of ^{22}C is already close to this upper limit, and there are no excited bound states. The state of the two-neutron halo in ^{22}C can be called a single Efimov state for the correlated-core case.

Energy spectra of ^{22}C are obtained for correlated ^{20}C-cores. As the correlation increases by blocking effects and the occupation number of the $2s_{1/2}$ orbit in the core decreases, the excitation energy of the 2_1^+ state is pushed up [18] as shown in Fig. 2. The energy position of the 2_1^+ state could give us information on the degree of the correlation and S_{2n}.

3. Three-body model with mixed configurations

Here, the three-body model is extended to the case of mixed configurations for the halo. For example, the two-neutron state in ^{17}B is described as the linear combination of $2s_{1/2}^2$ and $1d_{5/2}^2$ configurations;

$$\Phi = \alpha \mid 2s_{1/2}^2(0^+) > +\beta \mid 1d_{5/2}^2(0^+) > . \tag{2}$$

Total energy, E_{nn} and halo wave functions ϕ_s and ϕ_d are obtained by solving the following coupled equations,

$$\begin{aligned} \alpha(2\epsilon^s + V_{ss}) + \beta V_{sd} = \alpha E_{nn} \\ \beta(2\epsilon^d + V_{dd}) + \alpha V_{ds} = \beta E_{nn} \end{aligned} \tag{3}$$

where $\epsilon^i = < \phi_i \mid h \mid \phi_i >$ and $V_{ij} = < ij \mid v_{nn} \mid ij >$. E_{nn}, α, β and radial wave functions $\phi_s(r)$ and $\phi_d(r)$ are obtained self-consistently by solving the equations in Eq. (3).

We first test this method in ^{18}O, where one neutron in ^{17}O is bound. When only $1d_{5/2}^2$ configuration is considered, S_{2n} in ^{18}O is obtained to be S_{2n} =10.84 MeV by using observed

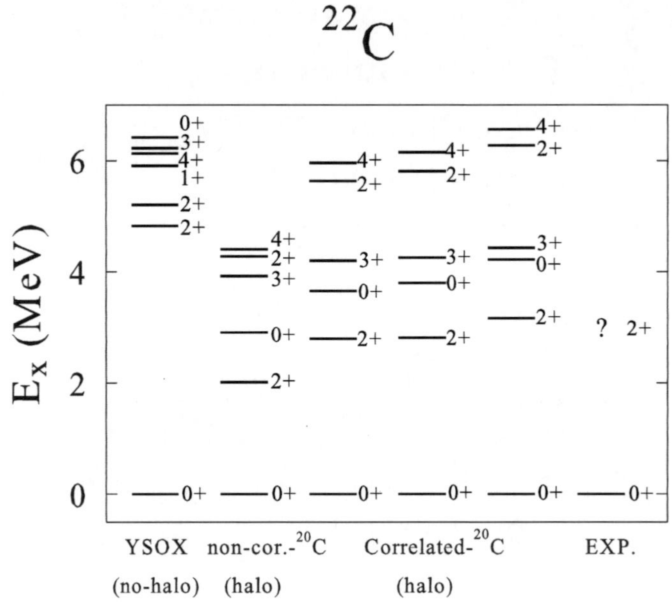

Figure 2. Energy spectra of ^{22}C obtained by shell-model calculations with modified YSOX interaction. The first column is the case for the original YSOX Hamiltonian without the halo effects. Second column shows the result for the case without the correlation in ^{20}C-core. The third, fourth and fifth ones are results with correlated-core.

value of S_{1n} =4.143 MeV. When both $1d^2_{5/2}$ and $2s^2_{1/2}$ configurations are taken into account, S_{2n} is obtained to be S_{2n} =11.29 MeV and comes closer to the experimental value, $S^{exp}_{2n} = 12.188$ MeV, with $|\ \beta\ |^2$ =0.94. Calculated values of S_{2n} in the present model are consistent with those of shell-model calculation with USD [19] as shown in Fig. 3. The experimental value is reproduced when the full $(sd)^2$ configurations are included. We thus find that the present three-body model works well for the two-component configurations.

Now we apply the method to ^{17}B whose two-neutron halo is assumed to consist of $2s^2_{1/2}$ and $1d^2_{5/2}$ components. Calculated results for the relation between S_{2n} and halo radius are shown in Fig. 4. Comparing with experimental information on S_{2n} [10,20] and halo radius [15,21], the probability for the $2s^2_{1/2}$ configuration is estimated to be $|\ \alpha\ |^2 \approx 35\%$, that is, more dominance of 1d-orbit in the halo. This probability is consisitent with the value of $|\ \alpha\ |^2 = 0.39 \pm 0.19$ in Ref. [21], but small compared with the value of $|\ \alpha\ |^2 = 0.69 \pm 0.20$ in Ref. [22].

We next study the structure of the drip-line nucleus ^{19}B. The two-neutron state in ^{19}B is dominantly $1d^2_{5/2}$ character; $P(2s^2_{1/2}) < 20\%$ [23]. Assuming a pure $1d^2_{5/2}$ configuration for the two-neutron state, the three-body model is applied. The relation between the valence neutron radius and S_{2n} is shown in Fg. 5. The valence neutron radius is estimated to be R_n = 4.35-4.45 fm from the experimental restriction on S_{2n}.

Two-neutron halo in ^{11}Li is investigated by the three-body model with two components for the halo, $2s^2_{1/2}$ and $1p^2_{1/2}$. The relation between the neutron halo radius and S_{2n} is shown in Fig. 6. The probability for $2s^2_{1/2}$ configuration is found to be about $P(2s^2_{1/2}) = 50\%$, which is consisitent with the experimental data, $P(2s^2_{1/2}) > 44\%$ [25].

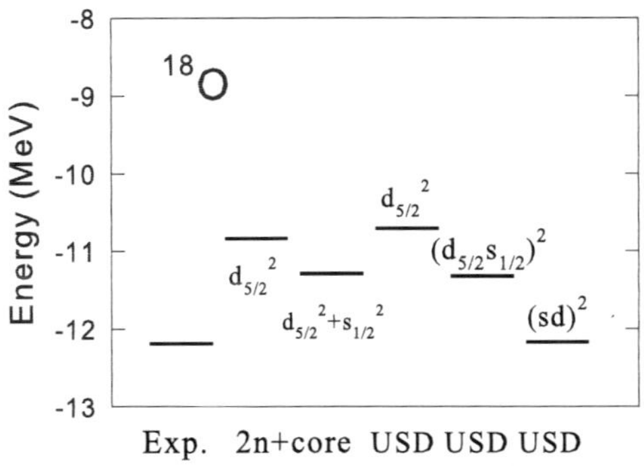

Figure 3. Energy of the ground state of ^{18}O obtained by the present three-body model and shell-model calculations with USD as well as the experimental value. In the three-body model, pure $1d^2_{5/2}$ and $1d^2_{5/2}+2s^2_{1/2}$ configurations are considered. In the shell-model calculations, the case with full $(sd)^2$ space with inclusion of $1d_{3/2}$ orbit is also considered.

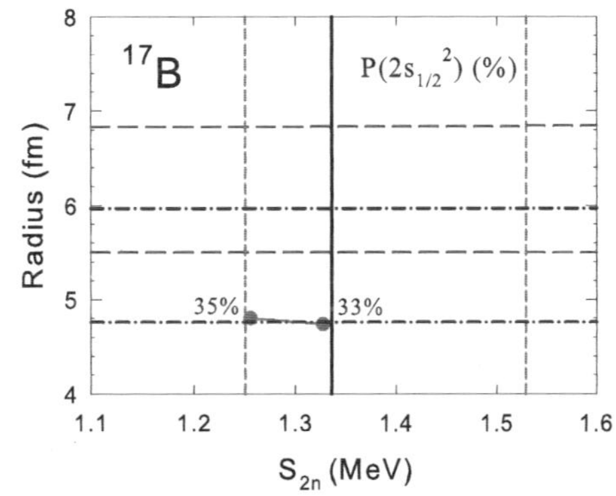

Figure 4. R.m.s radius and two-neutron separation energy S_{2n} for ^{17}B obtained by the present three-body model. The probability for the $2s^2_{1/2}$ component is also denoted. The solid and dashed vertical lines indicate experimental S_{2n} of Ref. [10] and [20], respectively. The dashed and dash-dotted horizintal lines show experimental halo radius of Ref. [21] and [15], respectively.

4. Summary

We studied the formation of two-neutron halo in light drip-line nuclei using n-n interaction fixed by the low-energy limit of n-n scattering. S_{2n} in ^{24}O is found to be well reproduced by the present three-body model. The structure of the ground state of ^{22}C is studied with a correlated ^{20}C-core. A unique relation between S_{2n} and halo radius is presented. S_{2n} is constrained to be less than 0.8 MeV from the condition that ^{21}C is unbound. The two-neutron halo state is successfully obtained in ^{22}C while ^{21}C remains unbound. The halo radius is obtained to be 6-7 fm for the correlated core, which is consistent with the recent measurement [14]. Because of the existence of the upper limit of the halo radius, there are no excited bound states. The spectra for ^{22}C are also studied. The energy of the first 2^+ state is pushed up as the s-state in the core loses more $2s_{1/2}$ component. The position of the 2^+_1 state can be a good measure of the correlation of the ^{20}C-core and the value of S_{2n}.

The three-body model is extended to treat two valence neutron cases with mixed configurations. The model is found to be successful to obtain S_{2n} in ^{18}O and applied to two-neutron halos in ^{17}B and ^{11}Li. The probability for $2s^2_{1/2}$ component is evaluated by constraints from experimental information on S_{2n} and halo radius. $P(2s^2_{1/2})$ for ^{17}B and ^{11}Li are estimated to be about 35% amd 50%, respectively.

Acknowledgments

This work was supported by Grants-in-Aid for Scientific Research (20244022, 22540290, 15K05090). It was supported also in part by Priority Isuue on post-K computer (hp160211) and CNS-RIKEN joint project for large-scale nuclear structure calculations.

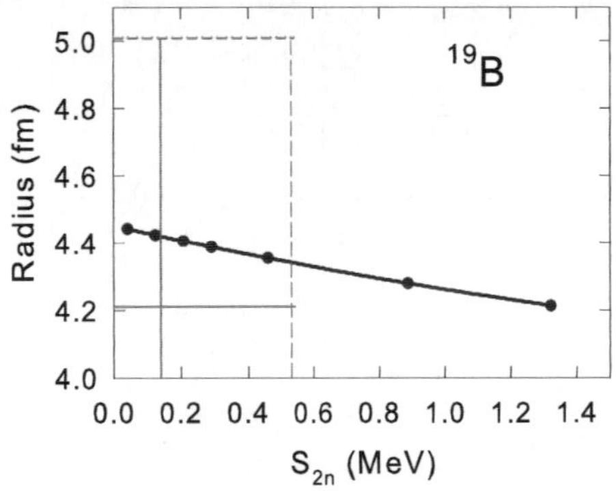

Figure 5. Valence neutron radius as a function of S_{2n} in ^{19}B obtained by the three-body model assuming a pure $1d^2_{5/2}$ configuration. The vertical lines indicate experimental value of S_{2n}; S_{2n} = 0.14±0.39 MeV [21]. The horizontal lines show experimental neutron radius; R_n = 4.21+0.80/-085 fm obtained from matter radii of ^{19}B and ^{17}B [21].

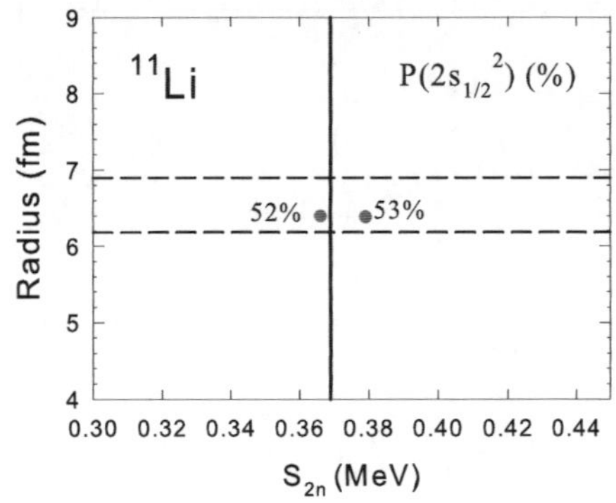

Figure 6. Neutron halo radius as a function of S_{2n} for ^{11}Li. Two components of $2s^2_{1/2}$ and $1p^2_{1/2}$ configurations are assumed for the two-neutron halo state. The experimental value of S_{2n} =369keV and neutron halo radius, R_n = 6.54±0.38 fm, are taken from Ref. [10] and Ref. [24], respectively.

References

[1] Tanihata I et al. 1985 *Phys. Rev. Lett.* **55** 2676
[2] Zhukov M V, Danilin B V, Fedorov D V, M. Bang J M, Thompson I J, and Vaagen J S 1993 *Phys. Rep.* **231** 151
[3] Hansen P G, Jensen A S and Jonson B 1995 Ann. *Rev. Nucl. Part. Sci.* **45** 591
[4] Horiuchi W and Suzuki Y 2006 *Phys. Rev. C* **74** 034311
[5] Yamashita M T et al. 2011 *Phys. Lett. B* **697** 90
[6] Kucuk Y and Tostevin J A 2014 *Phys. Rev. C* **89** 034607
[7] Acharya B, Ji C and Phillips D R 2013 *Phys. Lett. B* **723** 196
[8] Kobayashi N et al. 2012 *Phy. Rev. C* **86** 054604
[9] Machleidt R 2001 *Phys. Rev. C* **63** 024001; Miller G A, Nefkens M K and Alaus I 1990 *Phys. Rep.* **194** 1; Howell C R et al. 1998 *Phys. Lett. B* **444** 252; Gonzalez Trotter D E et al. 1999 *Phys. Rev. Lett.* **83** 3788
[10] NNDC http://www.nndc.bnl.gov/nudat2/chartNuc.jsp
[11] Yuan C, Suzuki T, Otsuka T, Xu F and Tsunoda N 2012 *Phys. Rev. C* **85** 064324
[12] Suzuki T, Otsuka T, Yuan C and Alahari N 2016 *Phys. Lett. B* **753** 199
[13] Tanaka K et al. 2010 *Phys. Rev. Lett.* **104** 062701
[14] Togano Y et al. 2016 *Phys. Lett. B* **761** 412
[15] Ozawa A et al. 2001 *Nucl. Phys. A* **693** 32
[16] Efimov V 1970 *Phys. Lett. B* **33** 563
[17] Braaten A and Hammer H W 2006 *Phys. Rep.* **428** 259
[18] Suzuki T, Otsuka T, Yuan C and Alahari N 2016 INPC2016
[19] Brown B A and Wildenthal B H 1988 *Ann. Rev. Nucl. Part. Sci.* **38** 29
[20] Audi et al. 1997 *Nucl. Phys. A* **624** 1
[21] Suzuki T et al. 1999 *Nucl. Phys. A* **658** 313
[22] Suzuki T et al. 2002 *Phys. Rev. Lett.* **89** 012501
[23] Gaudefroy L et al. 2012 *Phys. Rev. Lett.* **109** 202503
[24] Egelhof P 2001 *Prog. Part. Nucl. Phys* **46** 307
[25] Sanetullaev A et al. 2016 *Phys. Lett. B* **755** 481

IOP Conf. Series: Journal of Physics: Conf. Series **966** (2018) 012031 doi:10.1088/1742-6596/966/1/012031

LOW-ENERGY MAGNETIC RADIATION ENHANCEMENT WITHIN THE NUCLEAR SHELL MODEL

S. Karampagia[1,2], B. A. Brown[1,3] and V. Zelevinsky[1,3]

[1] National Superconducting Cyclotron Laboratory, Michigan State University, East Lansing, MI 48824, USA

[2] Department of Physics, Grand Valley State University, Allendale, MI 49401, USA

[3] Department of Physics and Astronomy, Michigan State University, East Lansing, MI 48824, USA

E-mail: karampso@gvsu.edu

Abstract.
The γ-ray strength function, the average reduced probability of absorbing or emitting a γ-ray of a given energy, is an indispensable quantity for calculations of astrophysical interest. Experimental studies of the γSF have revealed an enhancement of this quantity in the low E_γ energy region, which cannot be described by none of the known resonances or by semiclassical models. To understand the origin of the low-energy enhancement we have calculated the M1 transition probabilities, both in the emission and absorption regions, for the ^{49,50}Cr and ^{48}V nuclei in the $f_{7/2}$ shell-model basis. We find that the M1 strength distribution peaks at zero transition energy and falls off exponentially, independentely of the excitation energy or spin range selected. The form of this exponential is the same across all three different nuclei studied within this model space. We also show that the slope of the exponential is proportional to the strength of the $T = 1$ pairing matrix elements.

1. Introduction

An important quantity that gives information on nuclear structure and is critical for calculations of nuclear reaction rates is the γ-ray strength function (γSF) [1], namely the average reduced γ-decay probability of a particular multipole type.

The γSF is dominated by the giant electric dipole (E1) resonance, which arises due to collective oscillations of protons against neutrons in the nucleus and is located at $E_\gamma \approx 78 \cdot A^{-1/3}$ MeV. For the description of this resonance, a standard Lorentzian shape [2, 3] can be used, however, the generalized Lorentzian [5] is usually prefered, by taking also into account the temperature dependence of the γSF (Kadmensky-Markushev-Furman model) [4]. In the low E_γ region, a damped electric dipole resonance is expected to be observed for neutron-rich nuclei, namely the pigmy dipole resonance [6, 7] (PDR), interpreted as the oscillation of the neutron skin against the proton plus neutron core.

The γSF also has contributions of M1 character, which can be attributed to either spin-flip excitations of the nucleus [8, 9], typically at energy $E_\gamma \approx 8$ MeV, or to counter rotation of protons against neutrons in deformed nuclei (scissors mode) which generates a resonance around

3 MeV [9]. The M1 contributions to the γSF are also described by a Lorentzian line [5, 10] with parameters based on the recommended systematics [11].

Measurements of the γSF dating back to 2004 [12] have revealed that it increases as the γ-ray energy decreases. This low-energy enhancement has been found in many nuclei extending from the *pf* shell, ^{56,57}Fe [13, 14], ^{50,51}V [15], ^{43,44,45}Sc [16, 17], ^{60}Ni [18], ^{44,45,46}Ti [19, 20, 21], and to heavier nuclei, $^{93-98}$Mo [22, 23], ^{105,106}Cd [24], ^{105}Pd [25], ^{197,198}Au [26], ^{151,153}Sm [27], ^{89}Y [28], ^{73,74}Ge [29]. Calculations of the radiative neutron capture integrating the low energy enhancement of the γSF show that these cross sections can potentially increase considerably when approaching the neutron drip line [30].

The newly discovered behavior of the low-energy part of the γSF needs a new physical interpretation. There are two questions that arise concerning the low-energy upbend: (i) the character of this resonance and its multipolarity and (ii) the mechanism which produces it. It has been found experimentally [14, 28] that E2 transitions are of minor importance, whereas dipole transitions dominate in the low-E_γ enhancement region. Unfortunately, there are still no available experimental data distinguishing between the electric or magnetic character of the transition or providing an explanation of its origin.

The first theoretical attempt to explain the low energy enhancement suggested that this is of E1 character and is attributed to transitions from thermally unblocked states of the single-particle spectrum to the continuum [31]. At the same time, various groups performed shell-model calculations [28, 29, 32, 33] and found that the M1 contribution of the γSF has an enhancement for low E_γ energies and a maximum for E_γ=0 MeV, closely following the experimentally observed enhancement. The shell model studies suggest two interpretations of the low-energy upbend. According to the first interpretation [32], protons and neutrons, occupying high spin orbitals couple their spins forming a band, within which enhanced M1 transitions take place. The similarity of this mechanism to the "shears" mechamism previously found in nuclei [34], led to the suggestion that the the low-energy enhancement should take place mainly near closed shells, where the shears mechanism is strong. The second interpretation [33] proposes that strong M1 transitions originate from high spin diagonal single particle orbitals and that these M1 transitions will contribute to the low-energy M1 γSF of all nuclei.

In this text we give arguments in favor of the second explanation. To test the suggestion that the low-energy enhancement comes from transitions between high spin orbitals, we calculate B(M1) values for ^{49}Cr, ^{50}Cr, and ^{48}V in the small model space of $f_{7/2}$ using the OXBASH shell-model code [35] that allows us to calculate not only the emission, but also the absorption spectrum. We take into account only transitions between states with $T = T_z$ and we find a characteristic peak of the γSF at zero energy, falling off exponentially below and above that point. The characteristic upbend of the low-energy M1 γSF is present independently of the nucleus, the range of excitation energies used, or the spin values of the initial states. We find that the low-energy enhancement is well approximated by an exponential function, similarly to previous studies [32], with the same parameters for all studied nuclei and that the slope of the exponential fall off is determined mainly by the $T = 1$ (pairing) part of the Hamiltonian.

The slope of the exponent could be affected by the masses of the nuclei studied or by the presence of deformation. Details of the shell-model calculations presented in Section 2 could also play a noticeable role, for example mixing of different orbitals can quench the calculated low-energy enhancement. In Section 3 we suggest the interpretation of the results. We finish with the conclusions in Section 4.

2. Shell model calculations and results

As stated in the Introduction we calculate the transition rates B(M1) restricting ourselves to the $f_{7/2}$ orbital space. The nuclear states are obtained with the F742 Hamiltonian from [36] that reproduces the known low-lying energies in the nuclei of this region. Experimentally, the quantity of interest is the γ-ray strength function γSF defined by [1] $f^{i}_{\rm ML}(E_\gamma) = \rho_i \frac{\langle \Gamma_{\gamma i}(E_\gamma)\rangle}{E_\gamma^{2L+1}}$, where L is the multipolarity of the transition, ρ_i the level density of the initial states and Γ_γ the partial radiative M1 width given by $\Gamma_{\gamma i,M1}(E_\gamma) = \frac{16\pi}{9}\left(\frac{E_\gamma}{\hbar c}\right)^3 B({\rm M1})(E_\gamma)_i$, where the index i specifies initial spin values and the initial energy region E_i. Combining the two expressions, a new form of the γSF is derived, $f_{\rm M1}(E_\gamma) = a\,\langle B({\rm M1})(E_\gamma)\rangle_i \rho_i(E_i)$, where $a = \frac{16\pi}{9(\hbar c)^3} = 11.5473 \cdot 10^{-9}\mu_N^{-2} \cdot {\rm MeV}^{-2}$. We are going to show our results in terms of the average $\langle B({\rm M1})(E_\gamma)\rangle$. Actually the γSF and $\langle B({\rm M1})(E_\gamma)\rangle$ turn out to have very similar shapes. To determine $\langle B({\rm M1})(E_\gamma)\rangle$, we first calculate the B(M1) values and then sort them according to increasing transition energy, E_γ. The results are grouped in energy bins of 0.2 MeV width. For each bin the average B(M1) value is found. This procedure guarantees that the average B(M1) value at given energy, E_γ, does not dependent on the bin size.

For all ranges of initial energies and spins, strong low-energy M1 enhancement is observed as shown in Figs. 1 and 2 for ^{50}Cr, in both the emission and the absorption spectrum. This result is very different from the Brink-Axel hypothesis in which the strength function for excited states is related to the absorption strength function in the ground state. In contrast, the low-energy distribution is a generic feature for excited states that cannot be obtained from information on the ground state since it peaks at zero energy. The red straight lines represent the exponential expression $B_0\, e^{-|E_\gamma|/T_B}$, where we follow the notation of reference [19], with parameters $B_0 = 0.75\, \mu_N^2$ and $T_B = 1.33$ MeV. The exponential expression with these parameters can describe well not only ^{50}Cr, but also the M1 distribution produced from different excitation energies of ^{48}V and ^{49}Cr as well. In Fig. 3 B(M1) diagrams for all three nuclei at E_i = 6-8 MeV can be seen and they all follow the same exponential form.

We find also that different parts of the Hamiltonian don't contribute equally to the M1 distribution. Our two-body configuration interaction Hamiltonian has only eight non-zero matrix elements, four corresponding to isospin $T = 0$ and four to isospin $T = 1$. Conserving the isospin symmetry we follow the procedure of [37], where we vary the strength of one of these two groups using the numerical coefficients, k_0 and k_1,

$$H = h + k_0 V(T=0) + k_1 V(T=1), \tag{1}$$

where the part h contains the single-particle energies. The most important matrix elements in the $V(T = 1)$ part are the pairing matrix elements, $J^\pi T = 0^+1$ followed by the matrix elements with $J^\pi T = 2^+1$. We find that the shape of the M1 distribution depends very little on the $T = 0$ interaction, as shown in the upper panels of Fig. 4, but there is a strong dependence on the strength of the $T = 1$ interaction. The reduction by half of the value of the latter makes the M1 distribution considerably steeper.

3. Discussion

The approximation of the low-energy M1 strength by an exponential function, as was used here, has already been proposed in [19]. For the nuclei used in the context of this study, it is found that the slope, T_B, and the height, B_0, of the exponential functions, fitted on the $\langle B({\rm M1})(E_\gamma)\rangle$, are almost constant for all three nuclei. Nuclei away from this mass region will have different parameters. For example, in [19] the $\langle B({\rm M1})(E_\gamma)\rangle$ of ^{94,95,96}Mo and ^{90}Zr were calculated for a model space which allows for both positive and negative parity states. The slopes of the exponents were different for positive and negative parity and more steep than the ones found in the present study.

As shown previously, the pairing interaction strongly affects the M1 distribution in a way that weaker pairing gives a steeper slope to the $\langle B({\rm M1})(E_\gamma)\rangle$. Since pairing in average depends on the mass number A as $\alpha_p/A^{1/2}$ [38], in the A =90-96 region, pairing is 25% weaker than in

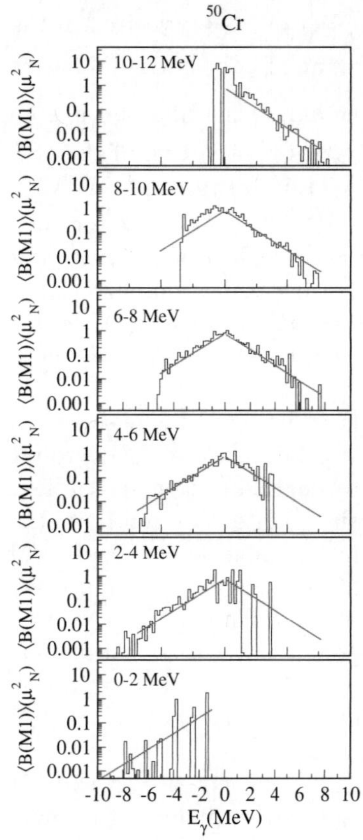

Figure 1. Average $B(\mathrm{M1})$ values as a function of γ-ray energy E_γ for ^{50}Cr and various 2 MeV ranges of initial energies E_i. The lowest panel is for 0-2 MeV, the highest for 10-12 MeV. Each M1 distribution is compared to the same exponent, red line, with parameters $B_0 = 0.75\ \mu_N^2$ and $T_B = 1.33$ MeV.

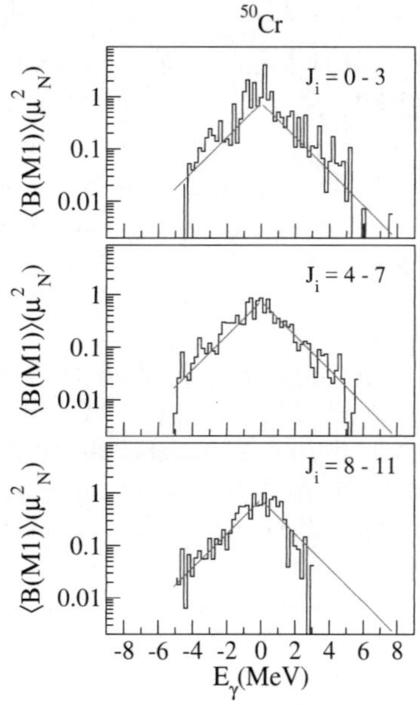

Figure 2. Average $B(\mathrm{M1})$ values for ^{50}Cr as a function of γ-ray energy E_γ for different initial spin ranges. Each M1 distribution is compared to the same exponent, red line, with the same parameters as in Figure 1.

A =48-50, thus the slope of the M1 distribution is expected to be steeper. Another factor that affects the calculated slope is the mixing of orbitals. In [39] the γSF for ^{48}V was calculated using the GX1A interaction [40, 41] in the *pf* model space. Nucleons were allowed to occupy either only the $f_{7/2}$ orbital or more orbitals of the *pf* model space and compared with that obtained within the $f_{7/2}$ space. It turns out that the mixing of different orbitals with $f_{7/2}$ quenches the low-energy strengths.

In a recent study [42], where the average reduced M1 transition probability, $\langle B(\mathrm{M1})(E_\gamma)\rangle$, was calculated using the shell model in a series of iron isotopes (^{60,64,68}Fe), it was found that the low-energy enhancement and the scissors mode are correlated. The sum of the strength of these two modes is constant, with the strength moving from the low-energy spike to the 3 MeV-resonance as deformation increases.

To address the role of the E1 transitions in the low-energy enhancement, a broad model space was employed, consisting of the *sd-pf-gds* orbitals [43]. No low-energy enhancement was found for the $E1$ strengths when studying nuclei in the A =43-45 region. On the contrary, the low-energy dipole strength function had a strong M1 contribution.

The exponential form, like that describing the low-energy enhancement, seems to be generic

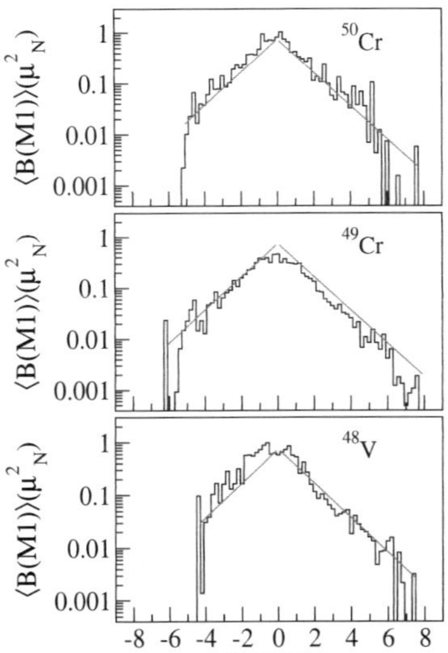

Figure 3. Average $B(\mathrm{M1})$ values as a function of γ-ray energy E_γ for ^{49}Cr, ^{50}Cr, and ^{48}V for initial energies, E_i, in the interval 6-8 MeV. Each M1 distribution is compared to the same exponent, red line, with parameters used in Figures 1 and 2.

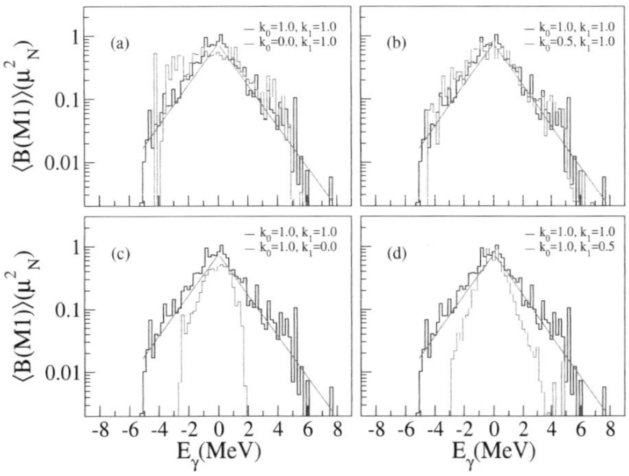

Figure 4. Average $B(\mathrm{M1})$ values as a function of γ-ray energy E_γ (black line) for ^{50}Cr for initial energy, E_i, in the interval 6-8 MeV, compared with the average $B(\mathrm{M1})$ values derived using (a) $k_0 = 0$, $k_1 = 1.0$, (b) $k_0 = 0.5$, $k_1 = 1.0$, (c) $k_0 = 1.0$, $k_1 = 0$, (d) $k_0 = 1.0$, $k_1 = 0.5$, red line. The green line is the exponential fit with $B_0 = 0.75\ \mu_N^2$, $T_B = 1.33$ MeV.

for the problems which have a bilinear combination of random operators. An analog can be found in the statistical distribution of the off-diagonal matrix elements of a realistic many-body Hamiltonian used in the full shell-model calculations in a finite orbital space which show onset of quantum chaos. This was studied in detail for the *sd* shell model space in [44] and examples can be found there in Figs. 8 and 9 and the Appendix. Contrary to the standard embedded ensembles of random matrices with Gaussian-like distribution of matrix elements [45], in these applications we typically have a distribution close to the exponential (sometimes prefactors are present, mostly important for the smallest matrix elements). This situation is supposed to emerge when the random quantities are matrix elements of multipole operators, while the main terms of the many-body Hamiltonian are their bilinear combinations, like multipole-multipole forces. For the components changing the seniority, the mean transition energy is of the order of the pairing gap Δ, about 1.5 MeV, for this group of nuclei. This estimate agrees with the effective temperature T_B found above.

This physics cannot support the Brink-Axel hypothesis which can be approximately valid for the excitations of general macroscopic nature. In the case of the giant dipole resonance, the main part is played by the local dipole polarization of the nuclear medium, which is essentially a universal property of nuclear matter. Such an excitation can be erected on top of any shell-model state, in the Brink-Axel spirit. In the case considered above, low-energy properties, such as isovector pairing and spin-orbit splitting of specific single-particle orbitals, are crucial.

4. Conclusion

We have performed shell-model calculations in the $f_{7/2}$ model space producing the full absorption and decay schemes of ^{48}V, ^{49}Cr, and ^{50}Cr. The results indicate a strong low-E_γ B(M1) component, in accordance with experimental and theoretical findings. The low-energy enhancement is essentially a one-partition phenomenon, which can be attributed to transitions stemming from diagonal high-spin orbitals. The low-energy upbend is independent of the initial spin and energy and it can be well fitted by an exponentential with the same parameters for all energy and spin ranges. The $T = 1$ matrix elements, with the pairing being the most important part, are responsible for the exponential shape of the B(M1) distribution. We also discussed the possible factors which affect the effective temperature of the low-energy enhancement. The slope of the low-energy upbend could change depending on the mass of the nucleus involved, through pairing interaction, but also due to the mixing of different orbitals or truncations of the orbital space. A reference to the interesting correlation of the low-energy upbend with the scissors-like resonance was made which also affects the low-energy B(M1) slope. We included more general arguments on relation of the phenomenon under consideration to the general problem of onset of chaotic dynamics.

Acknowledgements

We acknowledge support from the NSF grant PHY-1404442.

References

[1] Bartholomew G A, Earle E D, Ferguson A J, Knowles J W and Lone M A 1973 *Adv. Nucl. Phys.* **7** 229
[2] Axel P 1962 *Phys. Rev.* **126** 671
[3] Berman B L and Fultz S C 1975 *Rev. Mod. Phys.* **47** 713
[4] Kadmenskiĭ S G, Markushev V P and Furman V I 1983 *Sov. J. Nucl. Phys.* **37** 165
[5] Kopecky J and Uhl M 1990 *Phys. Rev. C* **41** 1941
[6] Igashira M, Kitazawa H, Shimizu M, Komano H and Yamamuro N 1986 *Nucl. Phys.* **A457** 301
[7] Krtička M, Bečvář F, Honzátko J, Tomandl I, Heil M, Käppeler F K, Reifarth R, Voss F and Wisshak K 2004 *Phys. Rev. Lett.* **92** 172501
[8] Bohr A and Mottelson B R 1998 *Nuclear Structure* Vol 2 (Singapore: World Scientific)
[9] Heyde K, von Neumann-Cosel P and Richter A 2010 *Rev. Mod. Phys.* **82** 2365
[10] Kopecky J, Uhl M and Chrien R E 1993 *Phys. Rev. C* **47** 312
[11] Capote R *et al.* 2009 *Nucl. Data Sheets* **110** 3107
[12] Voinov A, *et al.* 2004 *Phys. Rev. Lett.* **93** 142504
[13] Algin E, *et al.* 2008 *Phys. Rev. C* **78** 054321
[14] Larsen A C *et al.* 2013 *Phys. Rev. Lett.* **111** 242504
[15] Larsen A C *et al.* 2006 *Phys. Rev. C* **73** 064301
[16] Larsen A C *et al.* 2007 *Phys. Rev. C* **76** 044303
[17] Bürger A *et al.* 2012 *Phys. Rev. C* **85** 064328
[18] Voinov A, *et al.* 2010 *Phys. Rev. C* **81** 024319
[19] Sued N U H *et al.* 2009 *Phys. Rev. C* **80** 044309
[20] Guttormsen M *et al.* 2011 *Phys. Rev. C* **83** 014312
[21] Larsen A C *et al.* 2012 *Phys. Rev. C* **85** 014320
[22] Guttormsen M *et al.* 2005 *Phys. Rev. C* **71** 044307
[23] Wiedeking M *et al.* 2012 *Phys. Rev. Lett.* **108** 162503
[24] Larsen A C *et al.* 2013 *Phys. Rev. C* **87** 014319
[25] Eriksen T K *et al.* 2014 *Phys. Rev. C* **90** 044311
[26] Giacoppo F *et al.* 2015 *Phys. Rev. C* **91** 054327
[27] Simon A *et al.* 2016 *Phys. Rev. C* **93** 034303
[28] Larsen A C *et al.* 2016 *Phys. Rev. C* **93** 045810
[29] Renstrøm T*et al.* 2016 *Phys. Rev. C* **93** 064302
[30] Larsen A C and Goriely S 2010 *Phys. Rev. C* **82** 014318
[31] Litvinova E and Belov N 2013 *Phys. Rev. C* **88** 031302(R)
[32] Schwengner R, Frauendorf S and Larsen A C 2013 *Phys. Rev. Lett.* **111** 232504

[33] Brown B A and Larsen A C 2014 *Phys. Rev. Lett.* **113** 252502
[34] Frauendorf S 2011 *Rev. Mod. Phys.* **73** 463
[35] Rae W D M, Etchegoyen A, Godwin N S and Brown B A 1984 *OXBASH (The Oxford-Buenos-Aires-MSU Shell Model Code)* Michigan State University Cyclotron Laboratory Report Number 524
[36] Kutchera W, Brown B A and Ogawa K 1978 *Nuov. Cim.* **1** No. 12
[37] Sen'kov R and Zelevinsky V 2016 *Phys. Rev. C* **93** 064304
[38] Bohr A and Mottelson B R 1969 *Nuclear Structure* Vol 1 (Singapore: World Scientific).
[39] Karampagia S, Brown B A and Zelevinsky V 2017 *Phys. Rev. C* **95** 024322
[40] Honma M, Otsuka T, Brown B A and Mizusaki T 2004 *Phys. Rev. C* **69** 034335
[41] Honma M, Otsuka T, Brown B A and Mizusaki T 2005 *Eur. Phys. J. A* **25** Suppl. 1 499
[42] R. Schwengner R, S. Frauendorf S and B. A. Brown B A 2017 *Phys. Rev. Lett.* **118** 092502
[43] K. Sieja K 2017 *Phys. Rev. Lett.* **119** 052502
[44] Zelevinsky V, Brown B A, Frazier N and Horoi M 1996 *Phys. Rep.* **276** 85
[45] Kota V K B 2014 *Embedded Random Matrix Ensembles in Quantum Physics* Lecture Notes in Physics vol 884 (Heidelberg: Springer)

NUCLEAR LEVEL DENSITY, QUANTUM CHAOS AND RELATED PHYSICS

V. Zelevinsky[1,2] **and S. Karampagia**[2,3]

[1] Department of Physics and Astronomy, Michigan State University, East Lansing, MI 48824, USA

[2] National Superconducting Cyclotron Laboratory, Michigan State University, East Lansing, MI 48824, USA

[3] Grand Valley State University, Allendale, MI 49401, USA

E-mail: zelevins@nscl.msu.edu

Abstract.
The knowledge of the level density is necessary for understanding nuclear reactions and many applications of nuclear science.We use the moments method based on the statistical properties of mesoscopic systems (interactions between the constituents and resulting quantum chaos) and the many-body Hamiltonian in a truncated orbital space avoiding the diagonalization of huge matrices. We briefly discuss the method, general results in comparison with other approaches and available data, the underlying physics, collective enhancement phenomenon, and constant temperature description.

1. Introduction

It is not necessary to explain the importance of reliable knowledge of the level density for any mesoscopic system of interacting constituents. In nuclei, the level density is a regular part of treating various nuclear reactions, including technological and medical applications and nucleosynthesis in cosmos. In fact, the concept of the level density appeared historically as the first application of statistical ideas in nuclear theory [1, 2, 3].

The nucleus is a self-supporting conglomerate of strongly interacting nucleons. The starting point is the mean field created by averaging the action of individual particles. In the rough approximation, the particles independently occupy various orbitals in this field when the only limitation comes from Fermi-statistics. The excited states (particles and holes) are generated by combining individual promotions of nucleons to higher orbitals. The higher is the excitation the larger the number of possible ways to reach approximately the same total energy, and the level density grows exponentially. This combinatorics was first exploited by Bethe [4] who presented the estimate of the Fermi-gas level density. For a Fermi gas in a spherical box, the level density is a smooth function of excitation energy; its growth can be described by increasing temperature uniquely related to the energy. For a more realistic mean field with the shell structure of single-particle orbitals, the level density is modulated by the density of those orbitals. In a spherical field, the orbitals carry angular momentum quantum numbers and it is important to know the level density for specific classes of many-body states (total spin, isospin, parity). The idea of Bethe was to assume the random coupling of individual angular momenta which results in the

Gaussian distribution of the total spin projection with the width determined by the effective moment of inertia (the so-called spin cut-off parameter).

The next stage of the development is well documented in the review article [5]. The Fermi-gas model now is enriched by the real shell-model structure and pairing effects which significantly change the level density at least at not very high energy. The changes (like "back-shifted Fermi-gas formula" [6]) are phenomenological and contain empirical parameters for each nucleus. Later the pairing correlations were included on a microscopic level through the analog of the BCS or Hartree-Fock-Bogoliubov theory so that the mean field becomes self-consistent and the system superfluid [7, 8]. The level density predicted in this way typically has a smooth background and significant bumps related to the subshell structure.

If we would have a reliable description of interactions inside a nucleus, we would solve the quantum many-body problem and find the theoretical level density. The shell model with effective interactions is just an appropriate instrument. Unfortunately, the orbital states have to be truncated, otherwise the computational problem becomes not tractable. The shell model Monte Carlo approach [9] does not require huge diagonalization and was repeatedly used for finding the level density. It handles however only a limited set of possible interparticle interactions. The modern versions of the shell model (configuration interaction) cover a broad interval of excitation energy and the complete set of possible interaction matrix elements. The real problem is to be able to extract the level density avoiding full diagonalization.

2. Quantum chaos and moments method

With excitation energy growing and diminishing level spacings, the stationary wave functions become extremely complicated superpositions of the mean field states (particles and holes). Assuming the interaction being gradually turned on, the wave functions with the fixed global quantum numbers go through the process of multiple avoided crossings. Very fast (even with interaction still weaker than its realistic strength) the states reach the high degree of complexity. The energies inside the given class of states form an "aperiodic crystal" with the weakly fluctuating distances. This is the stage close to the mathematical limit of quantum chaos [10, 11]. There are broad studies of realistic examples which show [13, 14, 15] that indeed the system is locally close to the mathematical limit of the Gaussian Orthogonal Ensemble (GOE).

For our applications we do not need the exact fulfillment of the GOE properties. It is sufficient to satisfy the informal physical description given long ago [10]: in a vicinity of given energy the neighboring states have very similar physical properties ("look the same") and matrix elements of "simple" physical operators of the same order of magnitude. Then we expect that it is possible to apply the methods of statistical spectroscopy [16, 17].

As possible to check in exactly solvable shell-model versions, the level density in each class is a smooth bell-shaped curve with the maximum in the middle. Because we work in a truncated space, the significant part of this curve characterizes just the applied model rather than realistic physics as at some energy the missing states outside our orbital space should enter. Luckily enough, in practice the initial part of the curve provides a good description of reality up to some excitation energy. In this "realistic region" the complexity of physics allows to find the level density by calculating just the first statistical moments of the distribution.

Starting with the spherical mean field we divide all basis states in partitions p where the given particle number is distributed in same way over available orbitals. Let $D_{\alpha p}$ be the dimension of the class of states with exact quantum numbers α inside the partition p. For each partition we calculate the lowest moments (traces) of the Hamiltonian. The first moment is just a centroid,

$$M^{(1)}_{\alpha p} = \frac{1}{D_{\alpha p}} \mathrm{Tr}_{\alpha p} H \equiv \overline{H}_{\alpha p}. \tag{1}$$

The second moment is the mean square fluctuation,

$$M_{\alpha p}^{(2)} = \frac{1}{D_{\alpha p}} \,\mathrm{Tr}_{\alpha p} H^2 - [\overline{H}_{\alpha p}]^2 \equiv \sigma_{\alpha p}^2. \tag{2}$$

The first moment accounts for the mean field energy of the partition and the diagonal matrix elements of the interaction Hamiltonian. The second moment is the mean sum of the off-diagonal matrix elements including the interactions between different partitions. Both moments can be actually read just from the Hamiltonian matrix without any diagonalization procedure.

It was shown long ago [13] that the level density of a partition very fast converges to the Gaussian distribution. Therefore the resulting level density for a given set α of quantum numbers is given by a sum over partitions of the Gaussians $G_\eta(x - x_0; \sigma)$, defined by the eqs. (1) and (2),

$$\rho_\alpha(E) = \sum_p G_\eta(E - (\overline{H}_{\alpha p} - E_{\mathrm{g.s.}}); \sigma_{\alpha p}). \tag{3}$$

Although the procedures are developed for including the higher moments of the Hamiltonian, in practice this is not necessary. The third moment would cut the unphysical tails of the Gaussians. Instead we introduce [18] the parameter η that effectively does the same so that $G(x - x_0, \sigma)$ is set to zero at $|x - x_0| < \eta\sigma$. This is the only parameter that is not determined directly by the Hamiltonian; the shell-model study of the strength functions suggests $\eta \approx 2.4 - 2.8$. The ground state energy $E_{\mathrm{g.s.}}$ should be determined separately either by the calculation (for example, with the Lanczos algorithm) or by application of the exponential convergence method [19]. Figure 1 demonstrates the good agreement of the level density calculated by the two moments procedure with the exact *sd*-shell solution for ^{28}Si. The fourth moment would improve the result around the center of the curve which is anyway outside the physical region of the *sd*-model.

The moments method was recently used for the nuclei of *sd* and *pf* shells, where the reliable two-body interaction Hamiltonians do exist. The discussion of many features of the method and relevant references can be found in [20]; the details of the computational algorithm are given in [21]. For all nuclei studied, the level density in all classes of states is a smooth function with the subshell bumps significantly levelled out. As shown in [20], this is the result of the presence of many incoherent collision-like interactions which are absent in the mean-field, BCS or shell-model Monte Carlo approaches and which in fact form the realistic level density.

3. New development

Having at our disposal a realistic model of the many-body level density we can not only count on the better description of nuclear reactions. It becomes possible to turn to some questions of deep meaning. The first desire is to find an analytic approximation that would agree with the exact level density and reduce the theoretical task to predicting the parameters. The analysis of the back-shifted Fermi-gas approach has shown [20] the long expected Rosenzweig effect [5] of the systematic change of the Fermi-gas temperature along the shell filling. Another observation was that the angular momentum coupling described by the random process does not work in the cases similar to ^{44}Ca. Here the shell model operates with only four protons, and therefore $T = 2$ states and only the $T = 1$ component of the two-body forces. This is not sufficient for the randomness of angular momentum coupling.

Currently the best description seems to be given [22] by a constant temperature model: the level density at not very high excitation energy E is well described as

$$\rho(E) = \frac{1}{T} e^{(E - E_0)/T}. \tag{4}$$

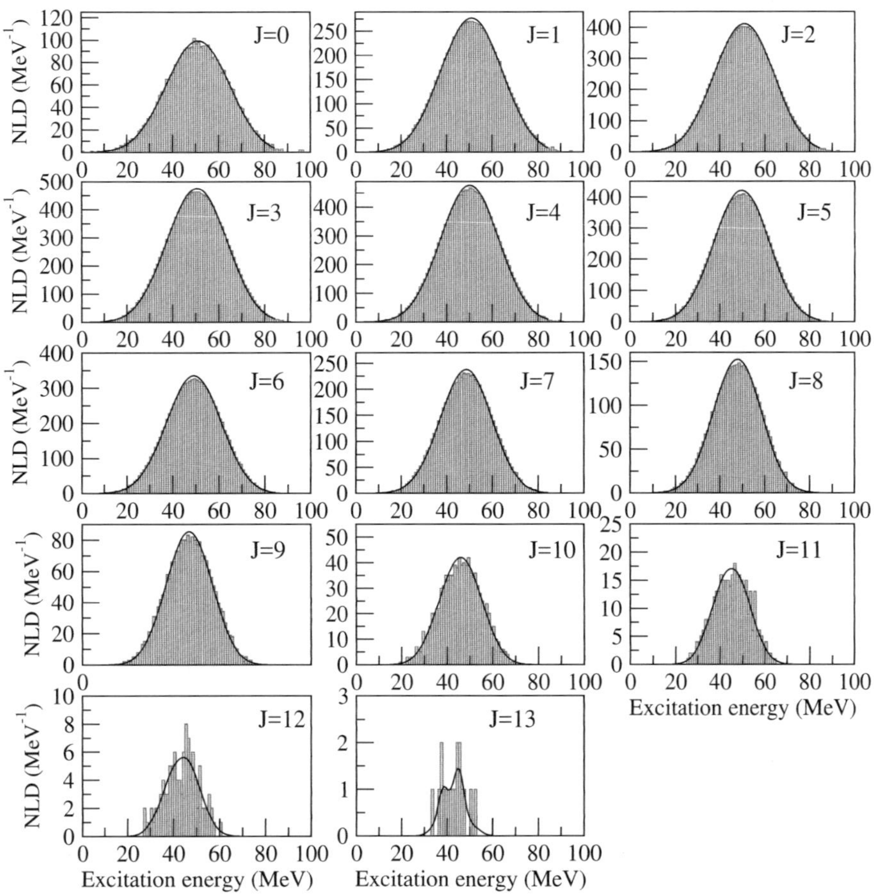

Figure 1. The level density of ^{28}Si in the *sd* orbital space calculated with the moments method (smooth dashed curve) and by exact diagonalization (solid line with small wiggles).

Here T is a fixed parameter that plays the role of the effective temperature of the ground and low-lying states. Fig. 2 demonstrates the high quality of the fit for the level density by this model for various nuclei. The idea of the ground state temperature is not new. It was long ago suggested by Moretto [23] as a sign of the phase transition from paired to normal Fermi liquid. However, the shell-model studies [24] invariably show a slow fall of the pairing correlators as a function of energy in various classes of states instead of sharp phase transitions. The ground state temperature was also discussed [25] as the fluctuational part of the mean field obtained by averaging over chaotic interactions. This interpretation is in qualitative agreement with the systematics of the parameter T, Fig. 3. All isotopes reveal the odd-even staggering and clear minima for $N = Z$ and nearby isotopes. Here pairing and possibly quarteting correlations lead to more "deeply frozen" low-energy configurations. It is also useful to study the changes of the level density varying the interaction. In this way one can see [26] the special role of the interactions mixing single-particle orbitals and inducing the shape deformation along with the related collective enhancement of low-energy level density.

All these properties, as well as dependence on various components of the interaction and on nuclear spin, should be studied in more detail. The consideration has to be generalized to different orbital spaces. For the mixture of shells of different parity one needs to get rid of he spurious states related to the center-of-mass excitation; there is a well established procedure that excludes such admixtures [27]. In the conclusion we nave to mention that the values of the parameter T for all nuclei in the sd shell and for all spin classes are tabulated and can be used by experimentalists [28].

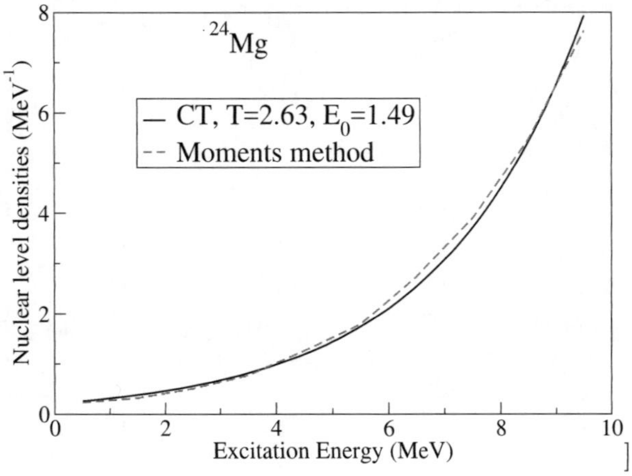

Figure 2. The level density calculated in the moments method and its description in the constant temperature model, eq. (4).

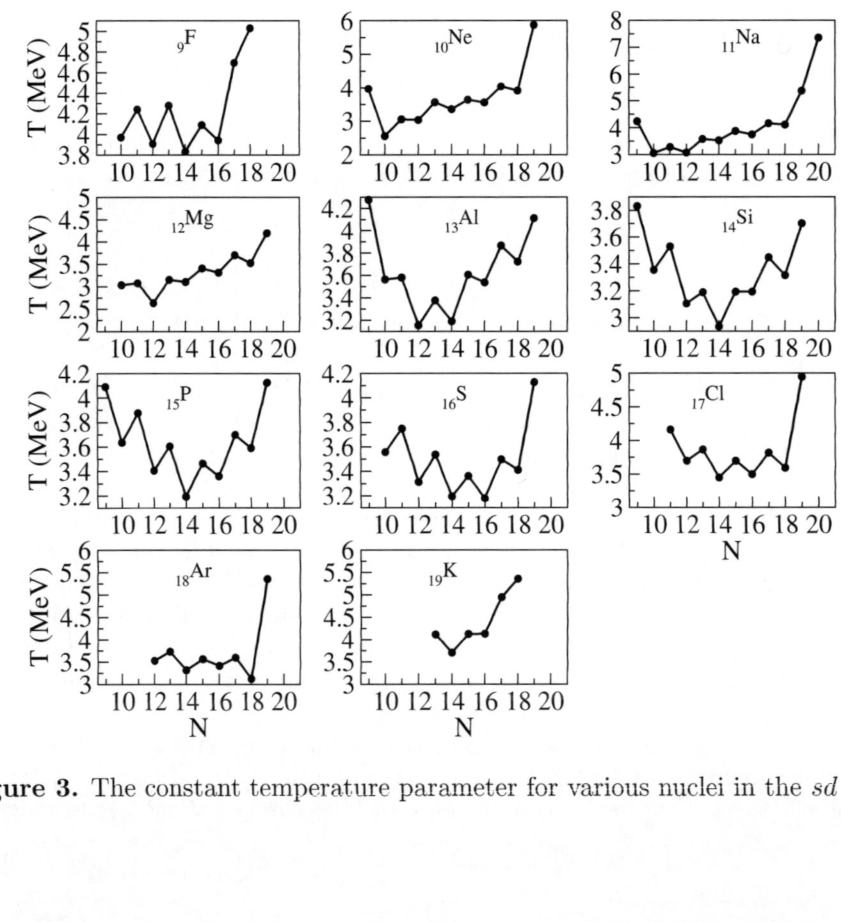

Figure 3. The constant temperature parameter for various nuclei in the sd shell.

The authors are grateful to N. Auerbach, B.A. Brown, and R.A. Sen'kov for discussions. V.Z. acknowledges the support from the BSF US-Israel grant and the NSF grant PHY-1404442.

References

[1] Bohr N 1936 *Nature* **137** 344
[2] Bohr N and Kalckar F 1937 *Kgl. Dansk. Vid. Selsk. Mat.-Fys. Medd.* **14** No. 10
[3] Blatt J M and Weisskopf V F 1952 *Theoretical Nuclear Physics* Wiley & Sons, N.Y/; 1991 Dover Publications, Mineola, N.Y.
[4] Bethe H A 1936 *Phys. Rev.* **50** 332; 1937 *Rev. Mod. Phys.* **9** 69
[5] Ericson T 1960 *Adv. Phys.* **9** 425
[6] Gilbert A and Cameron A G W 1965 *Can. J. Phys.* **43** 1446
[7] Goriely S, Hilaire S and Koning A J 2008 *Phys. Rev. C* **78** 064307
[8] Hilaire S, Girod M, Goriely S and Koning AJ 2012 *Phys. Rev. C* **86** 064317
[9] Alhassid Y, Bonett-Matiz M, Liu S and Nakada H 2015 *Phys. Rev. C* **92** 024307
[10] Percival I C 1973 *J. Phys. B* **6** L229
[11] *Statistical Theories of Spectra: Fluctuations* ed. Porter C E 1965, Academic Press, N.Y.
[12] Hurt N E *Quantum Chaos and Mesoscopic Systems* Springer 1997
[13] Brody T A, Flores J, French J B, Mello P A, Pandey A and Wong S S M 1981 *Rev. Mod. Phys.* **53** 385
[14] Flambaum V V, Gribakina A A, Gribakin G F and Kozlov M G 1994 *Phys. Rev. A* **50** 267
[15] Zelevinsky V, Brown B A, Horoi M and Frazier N 1996 *Phys. Rep.* **276** 85
[16] *Spectral Distributions in Nuclei and Statistical Spectroscopy*, eds. Kota V K B and Haq R U, World Scientific, Singapore, 2010
[17] Wong S S M *Nuclear Statistical Spectroscopy* 1986 Oxford University Press
[18] Horoi M, Ghita M and Zelevinsky V 2004 *Phys. Rev. C* **69** 041307(R)
[19] Horoi M, Volya A and Zelevinsky V 1999 *Phys. Rev. Lett.* **82** 2064
[20] Sen'kov R A and Zelevinsky V 2016 *Phys. Rev. C* **93** 064304
[21] Sen'kov R A, Horoi M and Zelevinsky V G 2013 *Computer Physics Communications* **184** 215
[22] Karampagia S and Zelevinsky V 2016 *Phys. Rev. C* **94** 014321
[23] Moretto L G 1975 *Nucl. Phys. A* **243** 77;
Moretto L G, Larsen A C, Giacoppo F, Guttormsen M S and Siem S 2015 *J. Phys. Conf. Ser.* **580** 012048
[24] Horoi M and Zelevinsky V 2007 *Phys. Rev. C* **75** 054303
[25] Zelevinsky V 1993 *Nucl. Phys. A* **555** 109
[26] Karampagia S, Renzaglia A and Zelevinsky V 2017 *Nucl. Phys. A* **962** 46
[27] Horoi M and Zelevinsky V 2007 *Phys. Rev. Lett.* **98** 262503
[28] Karampagia S, Sen'kov R and Zelevinsky V 2017 *Atomic Data and Nuclear Data Tables* in print

IOP Conf. Series: Journal of Physics: Conf. Series **966** (2018) 012033 doi:10.1088/1742-6596/966/1/012033

The AGATA Campaign at GANIL

Silvia M. Lenzi[1] and Emmanuel Clement[2]

[1] Dipartimento di Fisica e Astronomia dell'Università and INFN, Sezione di Padova, Padova, Italy

[2] GANIL, CEA/DRF - CNRS/IN2P3, Bd Henri Becquerel, BP 55027, F-14076 Caen Cedex 5, France

Abstract. The Advanced Gamma Tracking Array, AGATA, is presently in its construction phase in which the European γ-spectroscopy research community is involved since several years. This powerful HPGe array offers unique possibilities for the study of rare phenomena in nuclei by detailed gamma-ray spectroscopy. The physics campaign in GANIL foresees different setups, with AGATA coupled to different spectrometers, to study nuclear structure properties of nuclei all across the nuclear chart, from light nuclei to very heavy species, using stable and radioactive beams. After a brief description of the AGATA concept, some recent results are presented together with the very interesting opportunities for nuclear structure research in the forthcoming years with AGATA at GANIL.

1. Introduction

The Advanced Gamma Tracking Array, AGATA [1], is a new-generation high-resolution γ-ray spectrometer, based on the novel technique of γ-ray tracking. Since several years, the European γ-spectroscopy research community is involved in the construction of this high-efficiency HPGe-detector array which has started its physics campaign in 2010 in LNL (Italy), and continued in GSI (Germany) in 2012, in configurations that included a small fraction of the 180 detectors planned for the complete array. In 2014 AGATA was installed in GANIL (France), for a new physics campaign with an increased number of detectors.

The GANIL facility delivers high-intensity beams of stable nuclei up to Uranium as well as radioactive-ion beams from SPIRAL1. This combination allows to study a large variety of nuclear properties all along the nuclear chart with γ-ray spectroscopy techniques, coupling AGATA to different complementary instrumentation.

The physics campaign in GANIL started in 2015 and has been organized in a bottom-up approach, following calls of letters of intent that have been discussed yearly in dedicated AGATA@GANIL workshops. Among the proposed physics cases to be studied in the GANIL campaign we can mention: the evolution of the shell structure in neutron- and proton-rich nuclei and the effect of three-body forces, exotic nuclear shapes and shape-coexistence phenomena, proton-neutron pairing and isospin symmetry along the $N = Z$ line, clusterization phenomena, the effect of the coupling to the continuum in weakly-bound systems, the structure of super-heavy nuclei, properties of nuclear astrophysics interest, etc.

The AGATA array installed at GANIL [2] consists of triple (TC) and double (DC) clusters, organized in a compact geometry that will reach up to 45 detectors. Four different setups have been identified since the first collaboration workshop held in GANIL for the AGATA campaign in 2013 to study different nuclear properties. By coupling AGATA to the large-acceptance magnetic spectrometer VAMOS++ [3], heavy-ion beams of high intensity can be used in inverse kinematics in multi-nucleon, deep inelastic and induced-fission reactions. Proton-rich nuclei can be selected in fusion-evaporation reactions by coupling AGATA to the new neutron array NEDA [4] plus the Neutron Wall [5]. To study very heavy nuclei the use of VAMOS in gas-filled mode [6, 7] is foreseen. Light and medium-mass exotic nuclei can be studied in transfer or Coulomb-excitation reactions induced by radioactive beams from SPIRAL1 and detected in the new charge-particle array MUGAST.

2. The first sub-campaign with AGATA coupled to VAMOS++

The setup with AGATA coupled to the large-acceptance magnetic spectrometer VAMOS++ [3] took data from 2015 to 2017 (Fig. 1). The clusters were grouped at the most backward angles, symmetrically around the optical axis of the VAMOS++ spectrometer [2]. This arrangement maximizes the peak-to-total ratio, the Doppler effect correction and the efficiency for lifetime measurements. Indeed, thanks to the performances of both the pulse-shape analysis and the the γ-ray tracking, consisting in a major improvement with respect to the EXOGAM array, the whole solid angle covered by the AGATA clusters can be used for lifetime measurements based on the analysis of Doppler shifts. Some experiments aimed at measuring lifetimes by means of the Recoil Doppler Distance Shift method for which plunger devices were placed at the target position [8, 9]. Arrays of scintillator detectors such as FATIMA [10] and PARIS [11] were also used to measure lifetimes and high γ rays, respectively. From 2015 to 2017, the beamtime devoted to AGATA experiments has been of about 150 days.

A wide physics program has been covered by the performed experiments that aimed at investigating nuclear structure properties such as the evolution of the shell structure far from stability and the development of deformation near magic numbers, exotic shapes and shape-

Figure 1. The setup shows AGATA coupled to the magnetic spectrometer VAMOS++.

IOP Conf. Series: Journal of Physics: Conf. Series **966** (2018) 012033 doi:10.1088/1742-6596/966/1/012033

coexistence phenomena and lifetimes measurements of astrophysics interest.

One of the current problems in nuclear structure physics is the characterization of the islands of inversion, formed near the magic numbers. These are new regions of deformation where the rotational bands are constructed on well deformed intrinsic states whose main configuration involves intruder orbitals from the above main shell. While a signature of deformation is given by the energy of the first excited states, the measurement of their lifetimes allows a better understanding of their properties. By using the OUPS plunger device [8], M. Klintefjord and collaborators [12] have measured the lifetimes of several nuclei in the Island of Inversion around N=40 via the multi-nucleon transfer ^{238}U + ^{64}Ni reaction in inverse kinematics. Lifetimes of the 4^+ states in ^{62,64}Fe and the $11/2^-$ states in ^{61,63}Co and ^{59}Mn were measured. The extracted experimental $B(E2)$ values have been compared with large-scale shell model calculations using the LNPS interaction [13] and with beyond mean-field calculations using the Gogny D1S interaction [14]. An increase of quadrupole collectivity approaching $N = 40$ is observed in all isotopes. In the same framework, lifetimes in neutron-rich Ti isotopes, that are predicted to lie at the border of this Island of Inversion, have also been measured recently in an experiment by Ch. Fransen and collaborators with the Cologne plunger [9]. Above $Z = 28$, lifetimes in ^{73,75}Ga have been measured with the plunger technique for the first time with the ^{208}Pb + ^{76}Ge reaction by I. Celikovic and coworkers. The analysis of these two experiments is in progress.

Proton-rich nuclei towards ^{100}Sn have been populated for the first time by multi-nucleon transfer reactions in two experiments with AGATA and VAMOS++. The first aimed at measuring the lifetimes of the first excited states in ^{106,108}Sn with the plunger technique. These nuclei were populated using a ^{106}Cd beam on a ^{58}Ni target. Previous measurements of the $B(E2; 2^+ \rightarrow 0^+)$ in these isotopes present large error bars and depart from the parabolic behaviour expected from shell model calculations in the truncated *gds* space. This unconventional reaction mechanism for studying proton rich nuclei directly populates low spin states with large probability avoiding the isomeric 6_1^+ state, thus allowing the measurement of the lifetime of the 2^+ states with the differential plunger method. Preliminary results suggest a reduction of the $Z = 50$ shell gap [15].

The second experiment using the same method measured the lifetimes of the even-even $N = 50$ isotones ^{92}Mo and ^{94}Ru. These nuclei present similarities with Ni isotopes, in the sense that neutrons in Ni isotopes are filling the same orbitals than protons in $N = 50$ isotones. Indeed, the energy of the lowest 2^+ and 4^+ states in $^{70-76}$Ni isotopes are quite similar to those in ^{92}Mo,^{94}Ru,^{96}Pd and ^{98}Cd. For the transition probabilities $B(E2; 2^+ \rightarrow 0^+)$ shell model calculations by Lisetskiy *et al.* [16] in the fpg space predict a similar trend for the $N = 50$ isotones and $Z = 28$ isotopes, while a relative inversion of the trend is predicted for the $B(E2; 4^+ \rightarrow 2^+)$ values, probably owing to the strong pairing predicted for the $N = 50$ isotones near ^{100}Sn. Very scarce information was available in this mass region before this experiment was performed. Preliminary results seem to confirm the theoretical predictions and will be published soon [17].

On looking at the evolution of the shell structure far from stability, a very interesting problem is that related to the size and stability of the gap at the magic number $N = 50$ in very neutron-rich systems, together with the energy of the single-particle levels around it. To investigate this mass region the ideal reaction mechanism that exploits the optimum characteristics of the AGATA plus VAMOS++ setup is the fusion-fission reaction in inverse kinematics. Exotic species can be selected by properly choosing the angle of VAMOS++ with respect to the beam line.

Two experiments have been performed with this technique. The first one aimed at the measurement of excited states involving cross-shell configurations in $N = 50$ isotones above ^{78}Ni. These states are predicted to follow the same trend as that of the gap as a function of the proton number [18, 19]. A local minimum in the effective $N = 50$ gap, defined as $\Delta = S_{2n}(N = 52) - S_{2n}(N = 50)$, has been found at $Z = 32$ and an increase is predicted going from $Z = 32$ to $Z = 30$. This would allow to probe the robustness of the $Z = 28$ gap

Figure 2. The setup shows AGATA coupled to the FATIMA array for the 2017 run.

near ^{78}Ni. The analysis of these data is in progress and a publication of new results in ^{81}Ga is in preparation [20]. Interestingly, other nuclei near $N = 60$ have been populated in the same reaction allowing to identify new excited states in ^{96}Kr that show a sudden transition from strongly deformed to almost spherical shapes in the $N = 60$ isotones [21]. To better characterize the interplay between single-particle and collective degrees of freedom at $N = 50$, a similar reaction, using the plunger device, has been performed to measure the lifetimes in these nuclei. The analysis is in progress and preliminary results in ^{84}Ge suggest an increase of collectivity with respect to the neighboring isotones and isotopes [22].

Close to ^{132}Sn, γ-ray spectroscopy after fusion-fission was performed in 2016 by N. Alahari and coworkers to explore this mass region and look at the evolution of structure both in angular momentum and isospin. Some of these nuclei are characterized by the presence of isomeric states and therefore, to be able to perform prompt-delayed γ-ray coincidences, EXOGAM detectors were placed at the focal plane of VAMOS++ [23].

The structure of nuclei in the region around ^{208}Pb has also been studied by measuring lifetimes with the plunger techniques by G. Georgiev, D. Ralet and coworkers. There are preliminary results for ^{207}Pb and the analysis of other nuclei is in progress.

Taking profit of the position sensitivity of AGATA, lifetimes in the order of magnitude of few femtoseconds were measured with the Doppler shift attenuation method by C. Michelagnoli and collaborators. The state of interest is a resonant state in ^{23}Mg at 7786 keV excitation energy, just 205 keV above threshold, that decays also by proton emission. This measurement is of astrophysical interest as the lifetime and proton branching ratio, measured in the same experiment, will provide a precise determination of the rate of the ^{22}Na$(p,\gamma)^{23}$Mg reaction.

The run of 2017 included in the setup the coupling of AGATA and VAMOS++ to large arrays of $LaBr_3$ scintillator detectors (Fig. 2) for lifetime and high energy γ-ray measurements from the PARIS [11] and FATIMA [10] collaborations for the study of subjects related to the effect of three-body forces in light nuclei, shape transitions in rear-earth nuclei, shape coexistence in fission fragments and alpha clusterization in heavy nuclei.

3. Future Campaigns

The next foreseen experimental configuration will see AGATA coupled to the NEDA/Neutron-Wall array [4, 5] and the charged-particle detector DIAMANT [24] for a very ambitious scientific program to study neutron-deficient nuclei and the structure of $N \sim Z$ nuclei.

The approved experimental program at GANIL aims at studying proton-neutron pairing effects, isospin symmetry breaking in mirror nuclei, shape-coexistence phenomena at $N = Z$ near $A = 70$ via lifetime measurements, shell structure near ^{100}Sn together with octupole correlations and clusterization phenomena in light nuclei. The final mechanical design of NEDA + Neutron-WALL foresees the use of 54 self-produced NEDA detectors at forward angles and 14 Neutron-WALL detectors at around 90 degrees.

Two letters of intent have been submitted to the GANIL PAC proposing two different detectors to be coupled to AGATA: MUGAST, that will use radioactive beams from SPIRAL1, and VAMOS in gas-filled mode [6, 7]. Both setups are in construction and will be ready to take data from 2019. Several proposals have been presented and discussed at the AGATA Workshop organized in GANIL in 2016. For VAMOS in gas-filled mode the main interest is focused in studying the spectroscopy of very heavy nuclei populated mainly in fusion-evaporation reactions induced by stable beams, while SPIRAL1 beams will be used in transfer and Coulomb excitation reactions to study single-particle and collectivity, neutron-proton pairing, shell evolution and subjects of nuclear astrophysics interest with the charged-particle array MUGAST.

4. Conclusions and Perspectives

The AGATA physics campaign in GANIL, started in 2015 has attracted the interest of a large community for the study of nuclear structure and reactions dynamics of nuclei far from the valley of stability by means of gamma spectroscopy. The high intensity stable beams delivered by the GANIL cyclotrons up to U, together with the radioactive beams of SPIRAL1 allow to use different reaction mechanisms to populate the nuclei of interest. The sensitivity of AGATA largely increases by the coupling to particle detectors. In particular, the first campaign of experiments has seen AGATA coupled to the magnetic spectrometer VAMOS++ and, in some cases the plunger device and scintillator detectors from the FATIMA and PARIS collaborations. The following campaigns will use AGATA coupled to NEDA/Neutron WALL and the light charged-particle detector array DIAMANT, VAMOS in gas-filled mode and the charged-particle array MUGAST which will be used with radioactive ion beams.

References

[1] Akkoyun S *et al.* 2012 *Nucl. Instr. and Meth.* **A 668** 26
[2] Clement E *et al.* 2017 *Nucl. Instr. and Meth.* **A 855** 1
[3] Rejmund M *et al.* 2011 *Nucl. Instr. and Meth.* **A 646** 184
[4] Hüyük T *et al.* 2016 *Eur. Phys. J. A* **52** 55; Valiente-Dobón JJ 2017 *et al.*, in preparation (2017).
[5] Skeppstedt Ö 1999 *et al.*, *Nucl. Instr. and Meth.* **A 421** 531
[6] Schmitt C *et al.* 2010 *Nucl. Instr. and Meth.* **A 621** 558
[7] Theisen Ch *et al.* 2014 *Nucl. Instr. and Meth.* **A 747** 69
[8] Ljungvall J *et al.* 2012 *Nucl. Instr. and Meth.* **A 679** 61
[9] Dewald A *et al.* 2012 *Nucl. Instr. and Meth.* **A 679** 61.
[10] Regis J-M *et al.* 2010 *Nucl. Instr. and Meth.* **A 622** 83
[11] Maj A 2009 *Acta Physica Polonica B* **40** 565.
[12] Klintefjord M *et al.* 2017 *Phys. Rev. C* **95** 024312.
[13] Lenzi S M, Nowacki F, Poves A and Sieja K 2010 *Phys. Rev. C* **82** 054301.
[14] Gaudefroy L, Obertelli A, Pru S, Pillet N, Hilaire S, Delaroche J P, Girod M and Libert J 2009 *Phys. Rev. C* **80** 064313

IOP Conf. Series: Journal of Physics: Conf. Series **966** (2018) 012033 doi:10.1088/1742-6596/966/1/012033

[15] Siciliano M *et al.*, in preparation
[16] Lisetskiy A F *et al.* 2005 *Eur. Jour. Phys. A* **25** S01
[17] Pérez Vidal R *et al.*, in preparation
[18] Bender M *et al.* 2008 *Phys. Rev. C* **78** 054312
[19] Sieja K and Nowacki F 2012 *Phys. Rev. C* **85** 051301(R)
[20] Macquart G *et al.*, in preparation
[21] Dudouet J *et al.* 2017 *Phys. Rev. Lett.* **118** 162501
[22] Delafosse C *et al.*, in preparation
[23] Kim Y H *et al.* 2017 *Eur. Phys. J. A* **53** 162
[24] Scheurer J *et al.* 1997 *Nucl. Instr. and Meth.* **A 385** 501

IOP Conf. Series: Journal of Physics: Conf. Series **966** (2018) 012034 doi:10.1088/1742-6596/966/1/012034

Theoretical interpretation of the nuclear structure of ^{88}Se within the ACM and the QPM models.

I.N. Gratchev[1,2], G. Thiamova[1], P. Alexa[3], G.S. Simpson[1] and M. Ramdhane[1]

[1]LPSC, Université Grenoble Alpes, CNRS/IN2P3, Institut National Polytechnique de Grenoble, F-38026 Grenoble Cedex, France
[2]Institut Laue-Langevin, 71 avenue des Martyrs, 38042 Grenoble Cedex 9, France
[3]Department of Physics, Technical University of Ostrava, 17. listopadu 15, CZ-70833 Ostrava, Czech Republic

E-mail: grachev@lpsc.in2p3.fr

Abstract. The four-parameter algebraic collective model (ACM) Hamiltonian is used to describe the nuclear structure of ^{88}Se. It is shown that the ACM is capable of providing a reasonable description of the excitation energies and relative positions of the ground-state band and γ band. The most probable interpretation of the nuclear structure of ^{88}Se is that of a transitional nucleus. The Quasiparticle-plus-Phonon Model (QPM) was also applied to describe the nuclear motion in ^{88}Se. Preliminarily calculations show that the collectivity of second excited state 2_2^+ is weak and that this state contains a strong two-quasiparticle component.

1. Introduction

Recently, the nuclear structure of the very neutron-rich nucleus ^{88}Se was studied in Ref [1]. The γ decay scheme in ^{88}Se has been constructed using data from an neutron-induced fission of ^{235}U. The measurement was performed with the EXILL array of Ge detectors at the PF1B cold-neutron beam facility of the Institut Laue-Langevin (ILL) [2], Grenoble. The level scheme of ^{88}Se was established using the γ-γ-γ coincidence technique. A low 2_1^+ energy hints at the onset of quadrupole deformation and the identification of possible members of a 2_2^+ band provides evidence for γ vibrations.

2. ACM model

The algebraic collective model (ACM), introduced as a computationally tractable version of the Bohr model (BM) [3], which describes collective quadrupole vibrational modes in deformed nuclei. Detailed information of the ACM can be found in Ref. [4]. This model is characterized by a well defined algebraic subgroup chain and it allows collective-model calculations to be made using a simple routine procedure.

A great advantage of the ACM is that fully converged calculations can be performed easily for a range of Hamiltonians [5]. This enables one to determine the extent to which a particular BM Hamiltonian can realistically describe experimental data. Such calculations for various BM Hamiltonians prepare the way for more general, but still solvable, algebraic collective

IOP Conf. Series: Journal of Physics: Conf. Series **966** (2018) 012034 doi:10.1088/1742-6596/966/1/012034

models that include intrinsic degrees of freedom, similarly to the unified model of Bohr and Mottelson [6, 7, 8, 9, 10].

A general purpose ACM Hamiltonian is given, for example, in the form

$$\begin{aligned} \widehat{H}(M,\alpha,\kappa,\chi) &= -\frac{\nabla^2}{2M} + \frac{1}{2}M[(1-2\alpha)\beta^2 + \alpha\beta^4] \\ &- \chi\beta\cos 3\gamma + \kappa\cos^2 3\gamma \end{aligned} \tag{1}$$

where

$$\nabla^2 = \frac{1}{\beta^4}\frac{\partial}{\partial\beta}\beta^4\frac{\partial}{\partial\beta} + \frac{1}{\beta^2}\widehat{\Lambda}^2 \tag{2}$$

is the Laplacian on the 5-dimensional collective model space [4].

The above Hamiltonian, expressed in terms of the quadrupole deformation parameters β and γ serves as a useful starting point for a description of a wide range of nuclear collective spectra. Moreover, it has been shown, that this kind of Hamiltonian can be exploited in the study of a phase transition between axial and triaxial deformations [11] or in the study of the second-order phase transition of a model nucleus, from a spherical to a deformed phase, with α as a control parameter [8]. For $\alpha = 0$ the potential is that of a spherical harmonic oscillator, $\frac{1}{2}M\beta^2$, while for $\alpha > 0.5$ it has a minimum for a non-zero value of β, which increases with α increases. Moreover, as the effective mass parameter M of the Hamiltonian (1) increases, the kinetic energy decreases and the result is a decrease of the vibrational β fluctuations of the model about its equilibrium deformation. Thus, the value of the parameter α controls the β deformation of the model and the parameter M controls its rigidity. By adjusting the parameters α and M a model with any equilibrium value of the β deformation and any degree of rigidity may be constructed. In Ref. [5] it has been shown that parameter values in the range $0 < \alpha < 2.0$ and $10 < M < 100$ are sufficient to describe the β deformations and rigidities of the observed nuclear collective states.

The last fourth term of the ACM Hamiltonian (1), proportional to $\cos^2 3\gamma$, allows a triaxial minimum to appear in the potential energy surface (PES). A delicate competition between all the first three terms in the potential and the second and the third one in particular will determine whether the potential energy minimum will remain axially symmetric (the term proportional to $\cos 3\gamma$ dominates) or will be driven to a triaxial minimum by the last term.

3. ACM calculations for transitional nuclei

The detailed spectroscopy of the nuclei in the isotopic chain of Se, Ge, Sr and Kr clearly reveals the existence of γ-soft collective structure. Within the ACM one thus may be tempted to test the validity of the Wilets-Jean (WJ) limit that is characterised by a completely β-rigid and γ-independent potential which leads to characteristic degeneracies of some angular momentum states (e.g. 3_1^+, 4_2^+). After having applied the WJ limit to ^{88}Se we have found the ground state energies were correctly described but the γ-bandhead was about 200 keV higher than the experimental one and its excited doublet of states $(3_1^+, 4_2^+)$ was almost 1 MeV higher than the state $(3^+, 4^+)$ observed at 1798.6 keV. We have also tested the triaxial rigid rotor ACM which also provides a very good description of the ground-state band energies and the two known γ-band energies are very well described, see Fig. 1b. In this interpretation, the state $(3^+, 4^+)$ observed at 1798.6 keV would be described as a 3_1^+ state and the calculated 4_2^+ state would be located at a very high energy. This is, however, not expected for γ-soft nuclei. Even though the rigid triaxial rotor scenario cannot be excluded based on the available experimental data it is most probably not very realistic for a nucleus with only 4 neutrons beyond $N = 50$.

As a weak quadrupole deformation of $\beta = 0.22(2)$ was derived for ^{88}Se [1], we have tested a scenario where this nucleus is assumed to be close to the transitional point between the spherical and deformed phase described by $\alpha = 0.5$. Firstly, a parameter value α should be determined

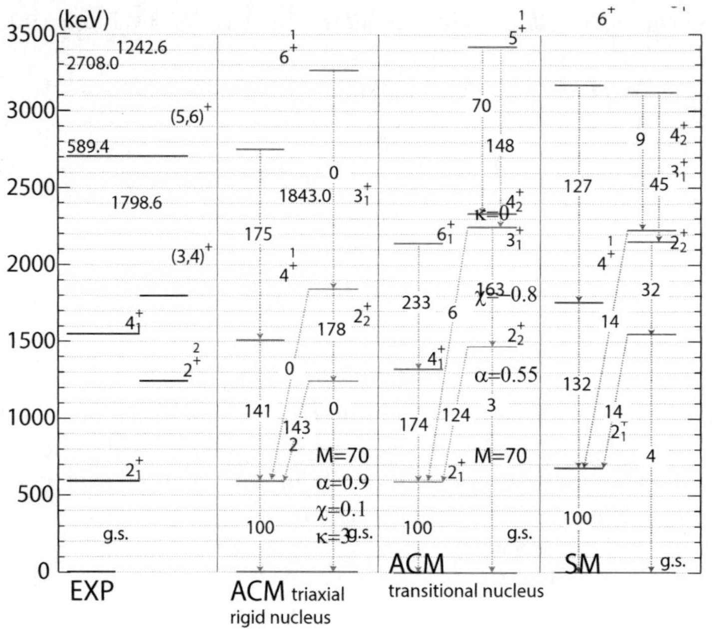

Figure 1. Comparison of the experimental level scheme obtained in this study for ^{88}Se with the result of the calculation using the ACM and the shell model (SM). In these level schemes the arrows represent the transition probability $B(E2)$ normalized to the transition to the ground state $2_1^+ \to g.s.$

in order to provide an optimum description of the observed spectrum. Fig. 2 shows that using the parameters $M = 70, \chi = \kappa = 0$ and in the interval $1 < \alpha < 4$ the calculated energies remain close to those of the WJ limit. When the parameter value α tends to 0.5, the energies of the calculated states 2_2^+, 4_1^+ and 3_1^+, 4_2^+ tend to the experimental values 1242.5 and 1798.6 keV. Thus, in order to model a transitional nucleus in the framework of the ACM model, we have chosen the following nuclear potential parameters $M = 70$, $\alpha = 0.55$, $\chi = -0.8$ and $\kappa = 0$. The non-zero value of the parameter χ is needed to introduce a small axial deformation in this studied nucleus. The minimum of the corresponding potential is not localized around a fixed value of β. The potential is very β-soft as seen in Fig. 3.

The ACM spectrum obtained with the above parameters is shown in Fig. 1c where it is compared to its experimental counterpart and with shell model (SM) calculations [1]. It is seen in both models that the calculated splitting of the levels 3_1^+ and 4_2^+ is almost the same. The position of the 2_2^+ state is reasonably well described.

It should be noted that for very β-soft potentials large centrifugal stretching effects cannot be avoided. These manifest themselves by ground-state band energies that fall systematically below their experimental counterparts (see Fig. 1c).

The transitional interpretation doesn't favour very strong transitions that would be described within the WJ limit as $\Delta v = 2$, where v is the seniority quantum number. This conclusion has been confirmed by SM calculations. In particular, the reduced transition probability $B(E2)$ of transitions $4_2^+(v = 3) \to 2_1^+(v = 1)$ or $2_2^+(v = 2) \to g.s(v = 0)$ with $\Delta v = 2$ is not strong with respect to the transitions with $\Delta v = 1$.

Some of the transition rates, e.g. $B(E2; 2_2^+ \to 2_1^+)$ are much larger than the SM ones but it should be stressed that the values of these transitions change in the vicinity of the transitional

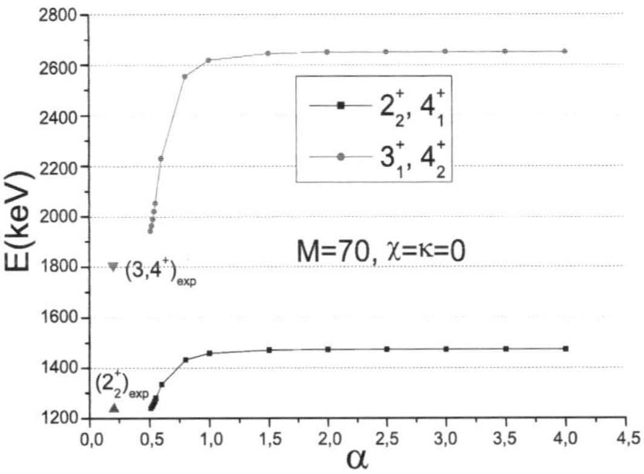

Figure 2. Energy doublets $2_2^+, 4_1^+$ and $3_1^+, 4_2^+$ as a function of parameter α. The triangles represent the experimental energy values of excited states 2_2^+ and $(3_1, 4_2)^+$ of ^{88}Se.

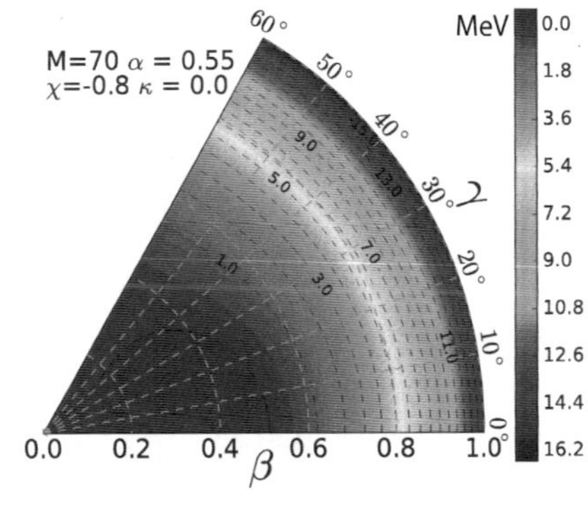

Figure 3. Potential energy surface (PES) using in the ACM calculations for the transitional nucleus ^{88}Se. The contour lines are separated by 0.5 MeV.

point $\alpha = 0.5$. This is discussed in detail in Ref. [12].

In the current interpretation, the ACM and SM indicate that a spin of 5_1^+ or 6_1^+ can be attributed to the state at 2708.0 keV, experimentally identified in this work. Angular correlations with sufficient statistics could confirm with certainty this hypothesis on this spin assignment.

4. QPM model

A detailed description of the QPM can be found, for example, in Refs [13, 14]. In this work, the even-even nucleus ^{88}Se has been studied, this is why only the vibrating even-even core has been taken into account in the model.

The intrinsic Hamiltonian can be written as

$$\hat{H}_{int} = \hat{H}_{av} + \hat{H}_{pair} + \hat{H}_{l,res} \tag{3}$$

where $\hat{H}_{av}$ is an axially symmetric quadrupole and hexadecapole-deformed average mean-field, which can be approximated by the Nilsson single-particle Hamiltonian, $\hat{H}_{pair}$ is a monopole pairing interaction, and $\hat{H}_{l,res}$ is a long-range residual multipole-multipole interaction.

$$\hat{H}_{pair} = -G_n \sum_{ij \in n} a_i^\dagger a_{\bar{i}}^\dagger a_{\bar{j}} a_j - G_p \sum_{ij \in p} a_i^\dagger a_{\bar{i}}^\dagger a_{\bar{j}} a_j \tag{4}$$

and

$$\hat{H}_{l,res} = -1/2 \sum_{\tau=p,n} \sum_{\lambda\mu} \kappa_{\lambda\mu}^{\tau} M_{\lambda\mu}^{\tau\dagger}(\hat{r}) M_{\lambda\mu}^{\tau}(\hat{r}) \tag{5}$$

where $a_i^\dagger$ and a_i are particle creation and annihilation operators of the single-particle state i; G_n and G_p are the neutron and proton pairing strength constants; $M_{\lambda\mu}^{\tau}$ is the multipole-multipole operator of the multipolarity λ, and its projection μ, and $\kappa_{\lambda\mu}$ stands for the corresponding long-range residual interaction strength constant.

In practical calculations with deformed nuclei the strength constants should explicitly depend on μ (the projection of λ on the symmetry axis) and should be fixed by the requirement to obtain good agreement between the energies of the lowest RPA solutions and the corresponding experimental phonon energies in even-even nuclei [15, 16].

The Nilsson-potential parameters used in our calculations are those recommended in Soloviev's monograph [15]. Initial values of the quadrupole- and hexadecapole-deformation parameters and of the ground-state pairing gaps are taken from the finite-range liquid-drop model plus shell-correction method calculations [17]. Experimental energies or the systematics of the quadrupole and octupole one-phonon states in the even-even cores are used to deduce the strengths of the multipole-multipole interactions $\kappa_{\lambda\mu}$, where no distinction is made between interaction strengths for protons and neutrons.

5. The microscopic structure of a low-lying state 2_2^+ in ^{88}Se obtained with the QPM

In this section, we will study the nature of the state 2_2^+ at 1242.5 keV in ^{88}Se. This nucleus consists of 34 protons and 54 neutrons or 4 neutrons above the magic shell $N = 50$. The aim of this theoretical study is to show which Nilsson orbits contribute to the collective motion in ^{88}Se. In order to perform QPM calculations, the shape parameters $\varepsilon_2 = 0.166$ and $\varepsilon_4 = -0.02$

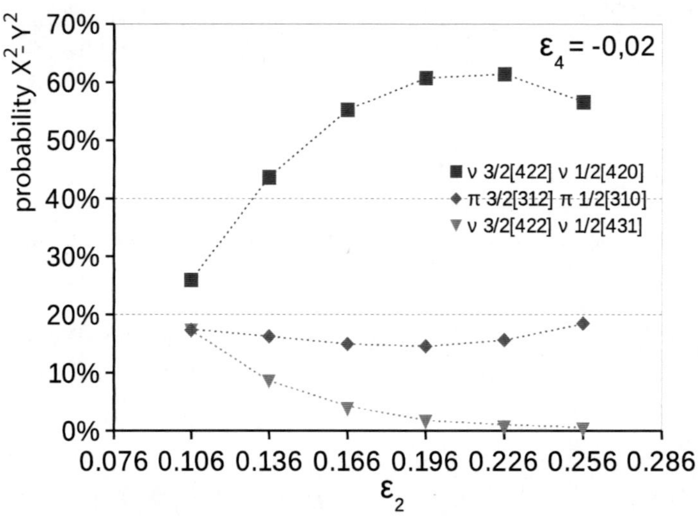

Figure 4. Evolution of the probability of the two quasiparticles presence in the state 2_2^+ as a function of the quadrupole deformation ε_2. The scale of the horizontal axis has been chosen in such a way as to show the probability evolution around the value $\varepsilon_2 = 0.166$ given in Ref. [17]. The calculations were performed for the parameters $\varepsilon_4 = -0.02$, $\Delta_n = 0.957$ MeV and $\Delta_p = 1.185$ MeV given in Ref. [17].

of the ground state were taken from Ref. [17]. To see the stability of the results with respect to changes of the quadrupole deformation parameter ε_2 we have performed several calculations with ε_2 ranging from 0.106 to 0.256, see Fig. 4. In this figure, the probability to find a given configuration in the excited state 2_2^+ is represented by a value $X^2 - Y^2$, where X and Y represent forward and backward amplitudes defining the quasiparticule-RPA phonon operator $\hat{Q}^\dagger_{\lambda\mu i}$

$$\hat{Q}^\dagger_{\lambda\mu i} = \sum_{\nu,\nu'} X^i_{\nu,\nu'} \hat{\alpha}^\dagger_\nu \hat{\alpha}^\dagger_{\nu'} - \sum_{\nu,\nu'} Y^i_{\nu,\nu'} \hat{\alpha}_{\nu'} \hat{\alpha}_\nu . \qquad (6)$$

We observed that the two two-quasiparticle components $\pi 3/2[312] \otimes \pi 1/2[310]$ and $\nu 3/2[422] \otimes \nu 1/2[431]$ are essentially independent of the deformation parameter ε_2 within the chosen range. On the other hand, the two-quasiparticle configuration $\nu 3/2[422] \otimes \nu 1/2[420]$ is sensitive to variations of the quadrupole deformation ε_2, as shown in Fig. 4. The configuration $\nu 3/2[422] \otimes \nu 1/2[420]$ is always dominant except for the case of a low deformation $\varepsilon_2 \lesssim 0.136$.

For $\varepsilon_2 = 0.22(2)$, empirically derived in Ref. [1], the QPM predicts two dominant configurations $\nu 3/2[422] \otimes \nu 1/2[420]$ and $\pi 3/2[312] \otimes \pi 1/2[310]$ with amplitudes of 60% and 15% respectively. Thus, the excited state of the nucleus ^{88}Se is mainly composed of two-quasiparticle excitations $\nu 3/2[422] \otimes \nu 1/2[420]$ and $\pi 3/2[312] \otimes \pi 1/2[310]$. The Nilsson orbitals $\nu 3/2[422]$ and $\nu 1/2[420]$ originate from the spherical orbits $\nu d_{5/2}, g_{7/2}$ respectively.

6. Conclusions

The ACM favours a shape transitional interpretation of the nuclear structure of ^{88}Se. Within this interpretation ^{88}Se is described as a nucleus close to the transitional point between a spherical and an axially deformed shape.

The QPM has been used to study the microscopic structure of the excited state 2_2^+. We have found that the degree of collectivity of this state is low because it has a dominant two quasi-particle structure with main configurations $\nu 3/2[422] \otimes \nu 1/2[420]$ and $\pi 3/2[312] \otimes \pi 1/2[310]$. It should be noted that the QPM calculations have been done with the standard Nilsson parameters recommended in the Soloviev's monograph [15] may not be perfectly adapted for the $A = 80-90$ mass region. A migration of the neutron $g_{7/2}$ orbit to higher energies might be expected when the proton $g_{9/2}$ orbit is empty. This is in agreement with the shell model calculations [1] where the occupation of the neutron orbit $g_{7/2}$ is low. On the other hand there are indications [18] that even with the migration of the neutron $g_{7/2}$ orbit toward higher energies the 1/2[420] Nilsson orbit will play a role for the deformations in question. Thus, a comparison of the SM and QPM calculations raises important questions that will be investigated in more detail in future work.

References

[1] Gratchev I N, Simpson G S *et al.* 2017 *Phys. Rev. C* **95**, 051302.
[2] Urban W, Jentschel M *et al.* 2013 *Journal of Instrumentation* **8**, P03014.
[3] Bohr A, Mat. Fys. Medd. K. Dan. Vidensk. Selsk. 1952 **26**, No.14.
[4] Rowe D J and Wood J L, *Fundamentals of Nuclear Models; Foundational Models*, 2010 World Scientific, Singapore, Vol. 2.
[5] Rowe D J, Welsh T A *et al.* 2009 *Phys. Rev. C* **79**, 054304.
[6] Bohr A and Mottelson R B, *Nuclear Structure*, Vol. II, (Benjamin Reading, Massachusetts 1975).
[7] Rowe D J, *Nuclear Collective Motion: Models and Theory* (Methuen, London, 1970; reprinted by World Scientific, Singapore, 2010).
[8] Turner P S and Rowe D J, 2005 *Nucl. Phys.* **756**, 333.
[9] Caprio M A, 2011 *Phys. Rev. C* **83**, 064309.
[10] Rosensteel G and Rowe D J, 1977 *Phys. Rev. Lett.* **38**, 10.
[11] Iachello F, 2003 *Phys. Rev. Lett.* **91**, 132502.
[12] Gratchev I N, PhD, Thesis, Université Grenoble Alpes, (2017).
[13] Liang C F, Paris P, *et al.* 1991 *Phys. Rev. C* **44**, 676.
[14] Aas A J *et al.* 1996 *Nucl. Phys. A* **611**, 281.
[15] Soloviev V G, *Theory of Complex Nuclei*, (Pergamon, Oxford 1976).
[16] Iudice N Lo, Sushkov A V *et al.* 2004 *Phys. Rev. C* **70**, 064316.
[17] Moeller P, Nix J R *et al.* 1995 Data Nucl. Data Tables **59**, 185.
[18] Woehr A, PhD, Thesis, Johannes-Gutenberg Universitaet Mainz, (1992).

IOP Conf. Series: Journal of Physics: Conf. Series **966** (2018) 012035 doi:10.1088/1742-6596/966/1/012035

Investigation of the γ-decay behavior of ^{52}Cr with the γ^3 setup at HIγS

J. Wilhelmy[1], P. Erbacher[2], U. Gayer[3], J. Isaak[4], B. Löher[5], M. Müscher[1], S.G. Pickstone[1], N. Pietralla[3], P. Ries[3], C. Romig[3], D. Savran[5], M. Spieker[1], W. Tornow[6], V. Werner[3], A. Zilges[1] and M. Zweidinger[3]

[1] Institute for Nuclear Physics, University of Cologne, Germany
[2] Institute for Applied Physics, University of Frankfurt, Germany
[3] Institute for Nuclear Physics, TU Darmstadt, Germany
[4] Research Center for Nuclear Physics, Osaka, Japan
[5] Research Division, GSI, Darmstadt, Germany
[6] Department of Physics, Duke University, USA

E-mail: wilhelmy@ikp.uni-koeln.de

Abstract. The γ-ray strength function is an important input parameter for the calculation of nucleosynthesis processes. To study the dipole response in more detail, the γ-decay behavior of the *fp* shell nucleus ^{52}Cr was investigated with the high-efficiency γ^3 setup at the High Intensity γ-ray Source facility at TUNL in Durham, USA. The highly intense quasi mono-energetic γ-ray beam allows for excitations selective in multipolarity (J=1 and J=2) and energy. The γ^3 setup is a multi-detector array consisting of HPGe and $LaBr_3$ detectors with high efficiency and enables the measurement of γ-γ coincidences. Experimental results of ^{52}Cr will be presented and discussed in this contribution.

1. Introduction

The γ-ray strength function (SF) is an important input parameter for theoretical calculations of nucleosynthesis processes under extreme conditions, e.g., supernovae explosions or binary star mergers. One relevant physics setting is the photon bath in which the seed nuclei undergo radiative capture reactions and photoabsorption reactions. An increase of the SF at lower energies (upbend) has been claimed for several nuclei in different mass regions and it is still under debate if this is a general phenomenon [1, 2, 3, 4, 5, 6, 7, 8, 9, 10]. The SF is dominated by E1, M1, and E2 strength. With parameters constrained by the isovector giant dipole resonance (IVGDR), the E1 part of the SF is in first order well described by the Generalized Lorentzian Model (GLO) [11]. However, at around 5 to 8 MeV, additional E1 strength on top of the low-energy tail of the IVGDR is observed in nuclei with excess neutrons [12]. This resonance-like structure, also denoted as Pygmy Dipole Resonance (PDR), is studied with various probes [12]. The Nuclear Resonance Fluorescence method (NRF) is very powerful to selectively excite dipole states in an environment of states with different spins. Using linearly polarized γ-ray beams, parity quantum numbers can be determined [13] with the γ^3 coincidence setup at HIγS. For several medium to heavy-mass nuclei reduced transition strengths were investigated in (γ,γ')-

experiments. Low-lying dipole strength was also studied recently in the A $\approx$ 50 nuclear mass region in some Ca isotopes [14, 15, 16], Ni isotopes [17, 18], and in the nuclide ^{52}Cr [19].
At higher energies the M1 Gamow-Teller giant resonance is observed, e.g., at around 8 to 9 MeV for nuclei in this mass region [20]. These 1p-1h excitations across closed major shells correspond predominately to the $1f_{7/2} \rightarrow 1f_{5/2}$ spin-flip excitation in this mass region. The B(M1) strength was investigated for different neighboring isotopes of ^{52}Cr, e.g., by (γ,γ')-experiments on ^{50}Cr, ^{56}Fe and ^{60}Ni, as well by (e,e$'$)-experiments for the N = 28 isotones ^{50}Ti, ^{52}Cr and ^{54}Fe [21, 22, 23, 24]. For the N = 28 nuclei it was shown that the magnetic dipole strength significantly influences, e.g., neutrino-nucleus cross sections [25].

2. Experimental setup

2.1. The High Intensity Gamma-ray Source

At the High Intensity Gamma-ray Source (HIγS), almost fully polarized γ-ray beams are generated by Laser Compton backscattering (LCB) inside the Free-electron Laser (FEL) optical cavity [26]. In this optical cavity, the eV-photons are generated by magnetic undulators and reflected by optical mirrors. The next incoming electron bunch collides with the low-energy photons and the relativistic electrons boost them to MeV-energies. Because Compton scattering conserves the polarization of the primary laser photons, the resulting γ-ray beam is almost fully polarized [27]. The resulting γ-beam energy profile is quasi-monoenergetic. After the scattering process the γ-ray beam is collimated and provided for experiments at the γ^3 setup. For ^{52}Cr twelve energy settings were measured for about 5 hours each.

2.2. The γ^3 setup

The high efficiency γ-γ coincidence setup (γ^3 setup) is a multi-detector array. Highly efficient $LaBr_3$ detectors and HPGe detectors with high energy resolution can be combined depending on the demands of an individual experiment [28]. In the present experiment, four $LaBr_3$ detectors were combined with four HPGe detectors to take advantage of both, high efficiency and high energy resolution. Two HPGe detectors and two $LaBr_3$ detectors were placed at $\Theta = 90^\circ$ with respect to the beam axis. This allows a direct measurement of parity quantum numbers, because of the angular distribution of the E1 and M1 transitions. Additionally, two HPGe and two LaBr detectors were placed under backward angles.

3. Analyis

3.1. Parity quantum numbers

For the determination of parity quantum numbers an experimental asymmetry ϵ is defined [13, 29]:

$$\epsilon = \frac{N_\parallel - N_\perp}{N_\parallel + N_\perp} \tag{1}$$

$N_\parallel$ and $N_\perp$ denote the γ-ray intensities parallel and perpendicular to the plane of polarization, respectively, corrected by the detector efficiencies. The detector efficiencies were determined with standard calibration sources (^{56}Co and ^{226}Ra) up to 3 MeV. Additionally, for higher energies, Monte-Carlo simulations using GEANT4 were performed. Because of the opening angle of the detectors, the experimental asymmetry is not -1 (negative parity quantum numbers) or +1 (positive parity quantum numbers), but around $\pm$ 0.9. This still allows a clear assignment of parity quantum numbers. For ^{52}Cr the parity quantum numbers of 41 levels were determined.

3.2. γ-decay behavior

The γ-decay behavior was investigated via HPGe singles spectra and projected HPGe spectra from LaBr-HPGe coincidence data. Because of the quasi-monoenergetic γ-ray beam elastic

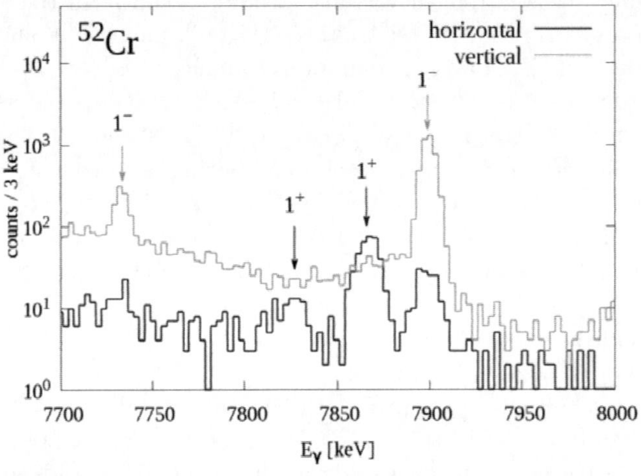

Figure 1. Two HPGe detector spectra in plane (black) and out of plane (grey) are shown. The γ-ray beam energy was 7.9 MeV

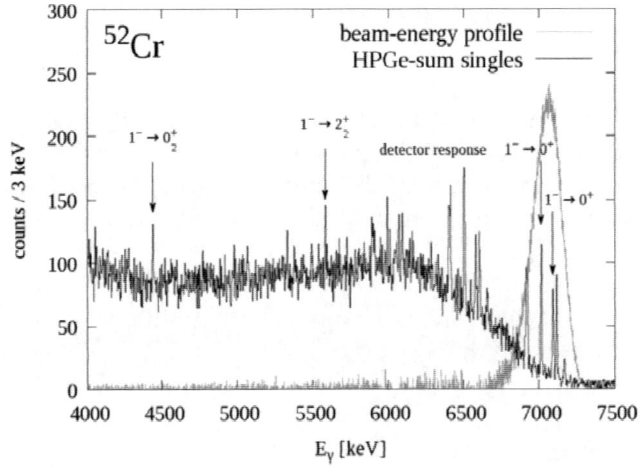

Figure 2. The summed spectrum of all HPGe detectors for a γ-ray beam energy of 7.1 MeV. Transitions to the groundstate and first excited states are clearly visible. The γ-ray beam energy profile (grey) has a FWHM of 260 keV.

transitions (depopulations directly to the groundstate) and inelastic transitions (depopulation to excited states) can be clearly distinguished as demonstrated in Fig. 2. The energy resolution of the highly efficient LaBr detectors is sufficient to apply gates on the depopulating γ-rays of the first excited states, e.g., the $2_1^+ \rightarrow 0_1^+$ transition. Hence, several inelastic transitions to the first excited states could be observed in the gated HPGe spectra. In a previous bremsstrahlung measurement at the Darmstadt-High Intesity Photon setup (DHIPS), reduced transition strength were determined for ^{52}Cr, but mainly for transitions to the groundstate only [19]. Because only dipole states within a small energy range are excited due to the quasi-monoenergetic γ-ray beam, in our experiment no background from higher lying excitations decreases the sensitivity. Therefore, weaker excitations could also be investigated within this experiment. Due to this and the investigation of transitions to lower lying excited states, the total measured B(E1)↑ strength and B(M1)↑ strength up to 9.5 MeV has increased by 20% and 25%, respectively.

4. Discussion

4.1. E1-strength distribution

For ^{52}Cr, seven weaker E1 transitions and 14 γ-decay branching ratios to the first excited states could be observed in this experiment for the first time. Compared to the previous bremsstrahlung measurement [19], the total measured B(E1)↑ strength increased from $51.2(16) \times 10^{-3}\, e^2\text{fm}^2$ to $61(3) \times 10^{-3}\, e^2\text{fm}^2$.

In ^{52}Cr the major part of the observed B(E1)↑ strength is carried by four 1^- states around 8 MeV (7.732 MeV, 7.896 MeV, 8.087 MeV and 8.175 MeV). With $\sum$ B(E1)↑ $= 35.5(14) \times 10^{-3} e^2\text{fm}^2$, they contribute more than half of the total measured B(E1)↑ strength below 9.5 MeV. Whereas the two higher-lying 1^--states show a significant γ-decay branching, the excitations at 7.732 MeV and 7.896 MeV show very small γ-decay branchings.

4.2. M1-strength distribution

The total $M1$ strength from 0 to 9.5 MeV observed in the present work amounts to $4.0(2)\,\mu_N^2$. Compared to previous NRF experiments, the measured B(M1)↑ transition strength increased by 25% from $3.21(13)\,\mu_N^2$ [19]. The additionally measured $M1$ response results from previously unobserved states and branchings.

Shell-model calculations of ^{52}Cr from Ref. [25] show the splitting into orbital and spin contributions of the $M1$ strength for ^{52}Cr. In the present experiment, for the low-lying 1^+ state at 6.75 MeV, γ-decay branchings to the 2_1^+ and 0_2^+ states were observed. This 1^+ state may couple to these states, which would support the character of a multiphonon 1^+ state that generates orbital $M1$ strength at lower energies.

At energies around 8 MeV and 9 MeV, the $M1$ strength is predicted to be dominated by single-particle excited states of spin-flip type. The main contribution stems from the excitation of neutron and proton $1f_{7/2} \rightarrow 1f_{5/2}$ two-quasiparticle components. The shell-model calculations show a strong accumulation of strength at around 8 MeV for isoscalar and at around 9 MeV for isovector spin-flip $M1$ resonances, see Ref. [25]. Comparing these theoretical results with the experimental data of the present work, the M1-strength distributions are in good agreement. The γ-decay branching ratios of 1^+ states seem to decrease with increasing energy. Two pronounced γ-decay branching ratios for the 1^+ states at 8.016 MeV and 8.583 MeV could be detected, whereas for the strong magnetic dipole excitations around 9 MeV, merely one weak branching for the 1^+ state at 9.212 MeV was measured.

Moreover, the $N = 28$ isotones have different numbers of protons in the $1f_{7/2}$ shell, which may influence strength and fragmentation of the $M1$ strength. The strongest $M1$ transition increases in energy with increasing number of protons along the $N = 28$ shell closure. The excitation energies are 8.563 MeV ($0.632(32)\,\mu_N^2$), 9.141 MeV ($0.919(54)\,\mu_N^2$) and 10.53 MeV ($1.262(71)\,\mu_N^2$) for ^{50}Ti, ^{52}Cr, and ^{54}Fe, respectively [23]. The doubly-magic nucleus ^{48}Ca disagrees with this systematics, because it concentrates the $M1$ strength with $B(M1)\!\uparrow = 4.0(3)\,\mu_N^2$ in one single state at 10.23 MeV.

Acknowledgments

This work was supported by the BMBF (05P2015PKEN9 and 05P15RDEN9), the DFG (ZI 510/7-1 and SFB 1245), the Alliance Program of the Helmholtz Association (HA216/EMMI), as well as the Department of Energy, Office of Nuclear Physics under Grant No. DE-FG02-97ER41033. Julius Wilhelmy is supported by the Bonn-Cologne Graduate school.

References

[1] Larsen A *et al.* 2013 *Phys. Rev. Lett.* **111** 242504
[2] Voinov A and Grimes S 2015 *Phys. Rev.* C **92** 064308
[3] Guttormsen M *et al.* 2005 *Phys. Rev.* C **71** 044307
[4] Larsen A *et al.* 2006 *Phys. Rev.* C **73** 064301
[5] Larsen A *et al.* 2007 *Phys. Rev.* C **76** 044303
[6] Voinov A *et al.* 2010 *Phys. Rev.* C **81** 024319
[7] Larsen A *et al.* 2012 *Phys. Rev.* C **85** 014320
[8] Buerger A *et al.* 2012 *Phys. Rev.* C **85** 064328
[9] Larsen A *et al.* 2013 *Phys. Rev.* C **87** 014319
[10] Wiedeking M *et al.* 2012 *Phys. Rev. Lett.* **108** 162503
[11] Harakeh M and von der Woude A 2001 *Giant Resonances* (Oxford University Press)
[12] Savran D, Aumann T and Zilges A 2013 *Prog. Part. Nucl. Phys.* **70** 210
[13] Pietralla N *et al.* 2002 *Phys. Rev. Lett.* **88** 012502
[14] Hartmann T *et al.* 2004 *Phys. Rev. Lett.* **93** 192501
[15] Isaak J *et al.* 2011 *Phys. Rev.* C **83** 034304
[16] Derya V *et al.* 2014 *Phys. Lett.* B **730** 288
[17] Scheck M *et al.* 2013 *Phys. Rev.* C **87** 051304(R)
[18] Wieland O *et al.* 2009 *Phys. Rev. Lett.* **102** 092502
[19] Pai H *et al.* 2013 *Phys. Rev.* C **88** 054316
[20] Heyde K, von Neumann-Cosel P and Richter A 2010 *Rev. Mod. Phys.* **82** 2365
[21] Pai H 2016 *Phys. Rev.* C **93** 014318
[22] Shizuma T *et al.* 2013 *Phys. Rev.* C **87** 024301
[23] Sober D 1985 *Phys. Rev.* C **31** 2054
[24] Scheck M *et al.* 2013 *Phys. Rev.* C **88** 044304
[25] Langanke K *et al.* 2004 *Phys. Rev. Lett.* **93** 202501
[26] Weller *et al.* 2009 *Prog. Part. Nucl. Phys.* **62** 257
[27] Litvinenko V N *et al.* 1998 *Nucl. Inst. and Meth.* A **407** 8
[28] Löher B *et al.* 2013 *Nucl. Inst. and Meth.* A **723** 136
[29] Kneissel U, Pitz H H, Zilges A 1996 *Prog. Part. Nucl. Phys.* **37** 349

IOP Conf. Series: Journal of Physics: Conf. Series **966** (2018) 012036 doi:10.1088/1742-6596/966/1/012036

Cluster structure of light nuclei

Francesco Iachello

Center for Theoretical Physics, Sloane Laboratory, Yale University,
New Haven, Connecticut 06520-8120, USA

E-mail: francesco.iachello@yale.edu

Abstract. Matter and charge densities of $k\alpha$ structures with k=2 (^{8}Be), k=3 (^{12}C) and k=4 (^{16}O) calculated within the framework of the algebraic cluster model (ACM) are briefly reviewed and explicitly displayed. Their parameters are determined from a comparison with electron scattering data.

1. Introduction

The cluster structure of light nuclei has been the subject of many investigations since the seminal work of Wheeler [1]. In 1965, Brink [2],[3] suggested specific geometric configurations for nuclei composed of k α-particles, here referred as $k\alpha$ nuclei. In particular, the suggested configurations of the ground state were for k=2 (^{8}Be) a dumbbell configuration with Z_2 symmetry, for k=3 (^{12}C) an equilateral triangle with D_{3h} symmetry, and for k=4 (^{16}O) a tetrahedron with T_d symmetry.

2. Algebraic cluster model

In order to understand what are the signatures of specific geometric configurations, and to provide explicit analytic formulas which can be easily compared with experiment, we have recently developed an algebraic approach to clustering based on the algebraic theory of molecules introduced in 1981 [4]. For $k\alpha$ structures, the approach amounts to a bosonic quantization of the Jacobi variables in terms of the algebra of U(3k-2), where k is the number of constituents. An explicit construction of the algebra has been completed for k=2, 3, 4 and is summarized in table 1.

Table 1. Algebraic description of $k\alpha$ structures.

k	Nucleus	U(3k-2)	Symmetry	Jacobi variables
2	^{8}Be	U(4) [4-5]	Z_2	$\boldsymbol{\varrho}$
3	^{12}C	U(7) [6-7]	D_3	$\boldsymbol{\varrho}, \boldsymbol{\lambda}$
4	^{16}O	U(10) [8-11]	T_d	$\boldsymbol{\varrho}, \boldsymbol{\lambda}, \boldsymbol{\eta}$

2.1. Matter and charge densities. Analytic expressions for energies and electromagnetic transition rates have been reported previously [12], [13]. Here we report charge and matter densities of k α-clusters with k=2, 3, 4 and from these derive analytic expressions for form factors in electron scattering. Densities are obtained by assuming that each cluster has a matter and charge distribution of Gaussian form

$$\rho_\alpha(r) = \left(\frac{\alpha}{\pi}\right)^{3/2} e^{-\alpha r^2}. \tag{1}$$

The matter and charge density of k α-clusters located at a distance β from the center of mass with spherical coordinates (β, θ_i, ϕ_i) is then given by [14]

$$\begin{aligned}\rho(r) &= \left(\frac{\alpha}{\pi}\right)^{3/2} \sum_{i=1}^{k} \exp\left[-\alpha\left(r - r_i\right)^2\right] \\ &= \left(\frac{\alpha}{\pi}\right)^{3/2} e^{-\alpha(r^2+\beta^2)} 4\pi \sum_{\lambda\mu} i_\lambda(2\alpha\beta r) Y_{\lambda\mu}(\theta,\varphi) \sum_{i=1}^{k} Y^*{}_{\lambda}(\theta_i,\varphi_i)\end{aligned}, \tag{2}$$

where $i_\lambda = j_\lambda(ix)/i^\lambda$ is the modified spherical Bessel function. The charge density is obtained by multiplying equation (2) by Ze/k, where Z is the total charge, and the matter density by A/k, where A is the total mass. The densities of equation (2) are shown in figures 1-3, respectively.

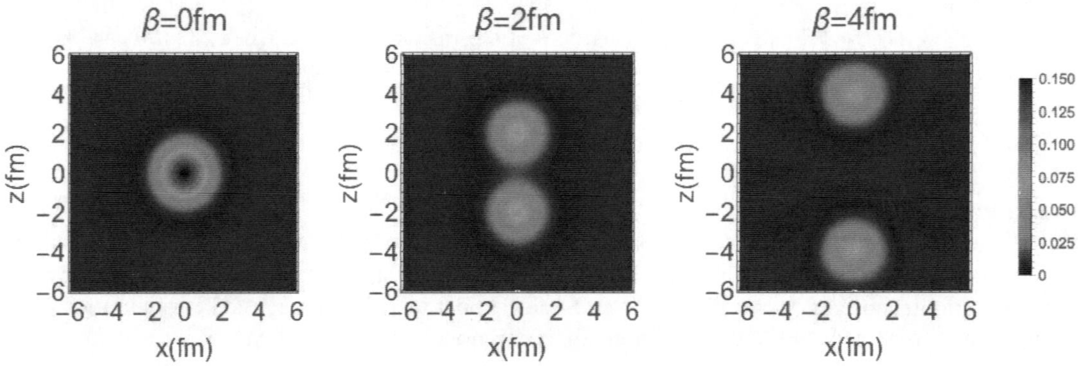

Figure 1. Densities of a k=2 α-cluster as given in equation (2). The value of α=0.56 fm^{-2}. The color scale is in fm^{-3}. Reproduced with permission from [14].

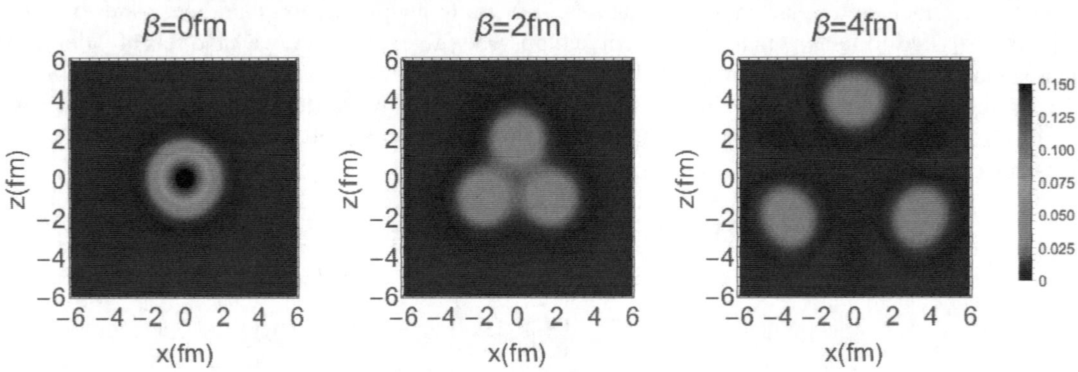

Figure 2. Densities of a k=3 α-cluster as given in Eq.(2). The value of α=0.56 fm^{-2}. The color scale is in fm^{-3}. Reproduced with permission from [14].

One may note that the densities given in figures 1-3 describe clusters with the appropriate symmetry, k=2 (Z_2), k=3 (D_{3h}), and k=4 (T_d), not only in the intermediate region, β~2 fm, but also in the asymptotic region, β→∞.

IOP Conf. Series: Journal of Physics: Conf. Series **966** (2018) 012036 doi:10.1088/1742-6596/966/1/012036

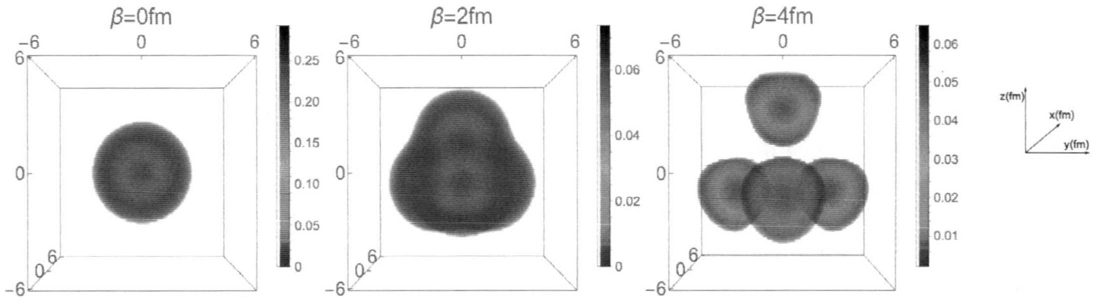

Figure 3. Densities of a k=4 α-cluster as given in equation (2). The value of α=0.56 fm^{-2}. The color scales are in fm^{-3}. The opacity of the colors is tuned to show the inside of the surfaces of equal density. Reproduced with permission from [14].

2.2. Form factors in electron scattering. From the charge densities one can evaluate the form factors in electron scattering. For the α-particle, the elastic form factor is given by the Fourier transform of equation (1),

$$F_\alpha(q^2) = e^{-\frac{q^2}{4\alpha}}, \tag{3}$$

conventionally normalized to 1 for q^2=0. By fitting the elastic electron scattering data [15] one obtains α=0.56(2) fm^{-2}. For $k\alpha$ clusters, one can derive analytic expressions for both elastic and inelastic scattering to members of the ground state rotational band. They are given by

$$F_L(0 \rightarrow L) = e^{-\frac{q^2}{4\alpha}} c_L j_L(q\beta), \tag{4}$$

which are the convolution of the form factor of the α-particle with that of the cluster, again normalized to 1 for L=0 and q^2=0. Here j_L is the spherical Bessel function and the coefficients c_L depend on the symmetry of the cluster. Explicit analytic expressions for the coefficients c_L are:

$$Z_2 \qquad c_L^2 = \frac{2L+1}{4}\left[2+2P_L(\ 1)\right] \tag{5a}$$

$$D_{3h} \qquad c_L^2 = \frac{2L+1}{9}\left[3+6P_L\left(\ \frac{1}{2}\right)\right] \tag{5b}$$

$$T_d \qquad c_L^2 = \frac{2L+1}{16}\left[4+12P_L\left(\ \frac{1}{3}\right)\right] \tag{5c}$$

where P_L is a Legendre polynomial. The value of β in the density can then be obtained by fitting the elastic form factor data at small q^2 (charge radii), except for ^{8}Be where electron scattering cannot be done since this nucleus is unstable. In this case the value of β is obtained from the observed moment of inertia [16], which is given by

$$\Im = Am\left(\beta^2 + \frac{1}{\alpha}\right). \tag{6}$$

The values so obtained are summarized in table 2.

Table 2. Density parameters of $k\alpha$ structures, k=2, 3, 4.

k	Nucleus	β [fm]	α [fm^{-2}]
2	^{8}Be	1.82(4)	0.56(2)
3	^{12}C	1.74(4)	0.52(4)
4	^{16}O	2.07(4)	0.56(4)

The value of α in ^{12}C appears to be slightly different from the free value due to polarization effects. It should be noted that the parameters in table 2 correspond to close packing of the α-particles.
With these values of β and α one can calculate all properties of $k\alpha$ nuclei. As an example, in figure 4, the elastic $0^+_1 \rightarrow 0^+_1$ and inelastic $0^+_1 \rightarrow 3^-_1$, $0^+_1 \rightarrow 4^+_1$ form factors of ^{16}O are shown as a function of momentum transfer q. The agreement with experimental data is excellent, especially for $q \lesssim 3$ fm^{-1}. For larger values of q other effects come into play originating from both meson exchange corrections and from the fact that the α form factor of equation (3) should be corrected as discussed in [15]. It should be noted that while the elastic form factor contains a parameter β which is adjusted as discussed above, the inelastic form factors are predictions of the cluster ACM model and do not contain additional adjustable parameters.

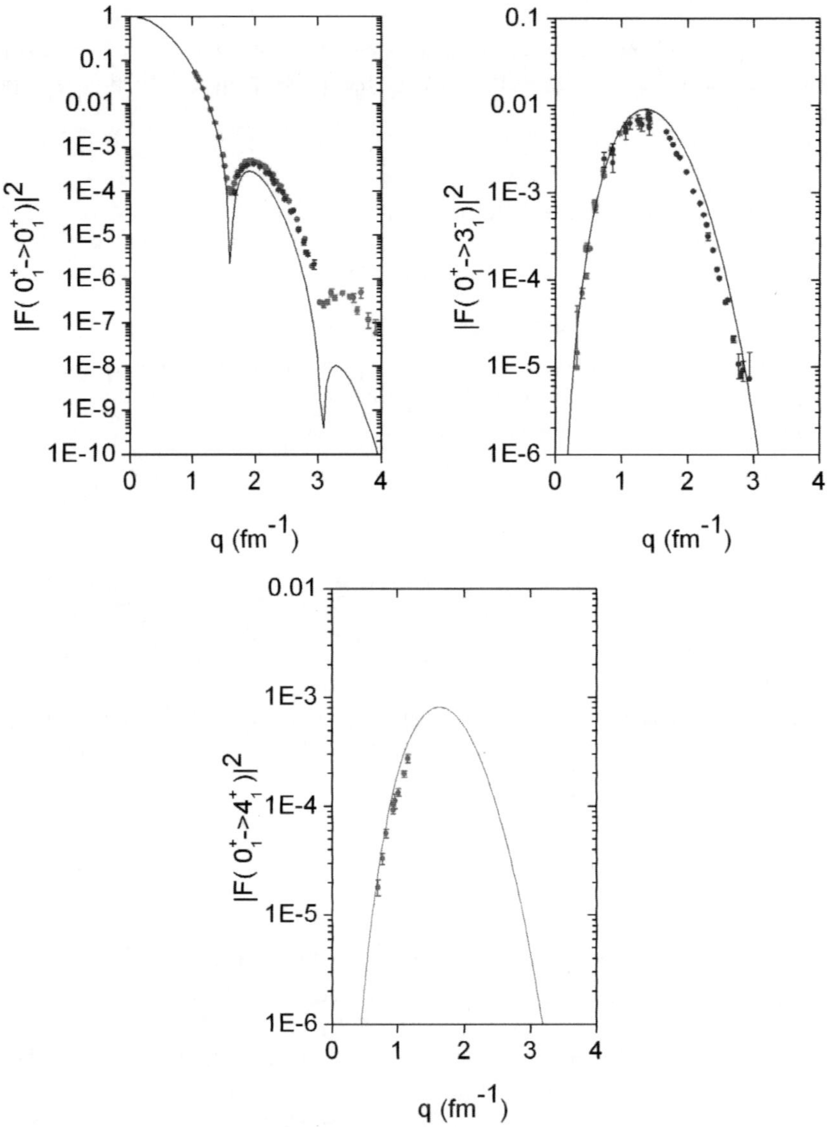

Figure 4. Comparison between the experimental form factors $|F_L(0 \rightarrow L)|^2$ of ^{16}O for the final states 0^+_1, 3^-_1 and 4^+_1 and those obtained with the ACM. Figure adapted from [11].

From the values of α and β one can also calculate the charge radii, given, for all cases k=2, 3 and 4, by the expression

$$\left\langle r^2 \right\rangle^{1/2} = \sqrt{\frac{3}{2\alpha} + \beta^2} \tag{7}$$

and shown in table 3.

Table 3. Charge radii of $k\alpha$ nuclei

k	Nucleus	$\langle r^2\rangle^{1/2}$ (fm) Th.	$\langle r^2\rangle^{1/2}$ (fm) Exp.
2	^{8}Be	2.45	---
3	^{12}C	2.47	2.468(12)
4	^{16}O	2.71	2.710(15)

B(EL) values can also be evaluated in analytic form by making use of

$$B(EL;0\rightarrow L) = (Ze)^2 \frac{\left[(2L+1)!!\right]^2}{4\pi} \lim_{q\rightarrow 0} \frac{F_L(0\rightarrow L)^2}{q^{2L}}, \tag{8}$$

where the normalization of the charge density has been explicitly introduced. From equation (4), one obtains the expressions

Z_2 $$B(EL;0\rightarrow L) = \left(\frac{Ze\beta^L}{2}\right)^2 \frac{2L+1}{4\pi}\left[2+2P_L(-1)\right] \tag{9a}$$

D_{3h} $$B(EL;0\rightarrow L) = \left(\frac{Ze\beta^L}{3}\right)^2 \frac{2L+1}{4\pi}\left[3+6P_L\left(-\frac{1}{2}\right)\right] \tag{9b}$$

T_d $$B(EL;0\rightarrow L) = \left(\frac{Ze\beta^L}{4}\right)^2 \frac{2L+1}{4\pi}\left[4+12P_L\left(-\frac{1}{3}\right)\right] \tag{9c}$$

The comparison with experiments is shown in tables 2-4 of [12] and will not be repeated here.

The density of equation (2), and the expressions equations (4)-(9) which can be derived from it, form the basis for a treatment of structures composed of k α-particles plus x-neutrons or protons. By convoluting the density equation (2) with the nucleon-alpha interaction, one can construct the potential generated by the clusters. By considering the form factors equation (4) and the B(EL) values equation (9), one can evaluate the cluster contribution to $k\alpha$+ x neutrons (or protons) structures. The corresponding results are summarized in the accompanying contribution [17].

Acknowledgements
This work was supported in part by U.S.D.O.E. Grant DE-FG02-91ER40608.

References

[1] Wheeler JA 1937 *Phys. Rev.* **52** 1083
[2] Brink DM 1965 *Proc. Int. School of Physics "Enrico Fermi"*, Course XXXVI, 247
[3] Brink DM, Friedrich H, Weiguny A and Wong CW 1970 *Phys. Lett. B* **33** 143
[4] Iachello F 1981 *Chem. Phys. Lett.* **78** 581
[5] Iachello F 1981 *Phys. Rev. C* **23** 2778 (R)
[6] Bijker R and Iachello F 2000 *Phys. Rev. C* **61** 067305
[7] Bijker R and Iachello F 2002 *Ann. Phys. (N.Y.)* **298** 334
[8] Bijker R 2010 *AIP Conf. Proc.* **1323** 28
[9] Bijker R 2012 *J. Phys. Conf. Series* **380** 012003

[10] Bijker R and Iachello F 2014 *Phys. Rev. Lett.* **112** 152501
[11] Bijker R and Iachello F 2017 *Nucl. Phys. A* **957** 154
[12] Iachello F 2017 *IOP Conf. Series* **863** 012003
[13] Bijker R 2017 *IOP Conf. Series* **863** 012009
[14] Della Rocca V, Bijker R and Iachello F 2017 *Nucl. Phys. A* **966** 158
[15] Sick I, Mc Carthy JS and Whitney RR 1976 *Phys. Lett. B* **64** 33
[16] Tilley DR *et al.* 2004 *Nucl. Phys. A* **745** 155
[17] Della Rocca V and Iachello F 2017 These Proceedings

IOP Conf. Series: Journal of Physics: Conf. Series **966** (2018) 012037 doi:10.1088/1742-6596/966/1/012037

Experimental Tests of the Algebraic Cluster Model

Moshe Gai

LNS at Avery Point, University of Connecticut, 1084 Shennecossett Rd., Groton, CT 06340.

E-mail: moshe.gai@uconn.edu

Abstract. The Algebraic Cluster Model (ACM) of Bijker and Iachello that was proposed already in 2000 has been recently applied to ^{12}C and ^{16}O with much success. We review the current status in ^{12}C with the outstanding observation of the ground state rotational band composed of the spin-parity states of: 0^+, 2^+, 3^-, $4^\pm$ and 5^-. The observation of the $4^\pm$ parity doublet is a characteristic of (tri-atomic) molecular configuration where the three alpha-particles are arranged in an equilateral triangular configuration of a symmetric spinning top. We discuss future measurement with electron scattering, ^{12}C(e,e') to test the predicted B(Eλ) of the ACM.

1. Introduction

The Algebraic Cluster Model (ACM) of Bijker and Iachello [1, 2] received recently a great deal of attention due to its success in predicting the structure of ^{12}C [3] and ^{16}O [4, 5]. Most striking is the observed ground state rotational band in ^{12}C that includes both positive parity (K=0) and negative parity (K=3) state and the $4^\pm$ parity doublet, as shown in Figure 1 taken from [3].

2. The Structure of ^{12}C

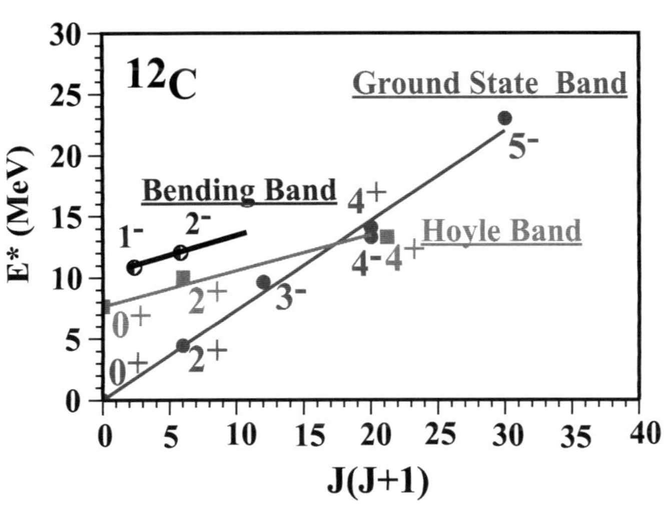

Figure 1. Observed rotational bands of low lying states of ^{12}C including the ground state rotational band with $J^\pi = 0^+$, 2^+, 3^-, $4^\pm$ and 5^- [3]. Both positive and negative parity state and the $4^\pm$ parity doublet are members of the ground state rotational band.

A question of recent interest is the structure of the Hoyle state at 7.6542 MeV in ^{12}C, namely whether the three alpha-particles are arranged in an obtuse triangle geometry [6] or an equilateral triangle arrangement as suggested in [3]. Both these geometries can be calculated in the ACM, but as shown in Fig. 2 taken from [3] the equilateral triangle geometry leads to a Hoyle Band structure similar to the one observed for the ground state band with the spin-parity states of: 0^+, 2^+, 3^-, $4^\pm$, albeit with a factor of 2 larger moment of inertia. In this context a search for the predicted 3^- state (and the second $4^\pm$ parity doublet) in ^{12}C above 12 MeV is essential. Indeed a recent study by the Aarhus group [7] of new states in ^{12}C observed in the decay of the (J^π, T) = (2^+, 1) state at 16.11 MeV in ^{12}C, suggests (one or perhaps two) previously unobserved natural parity spin state(s) between 11.5 and 13 MeV in ^{12}C.

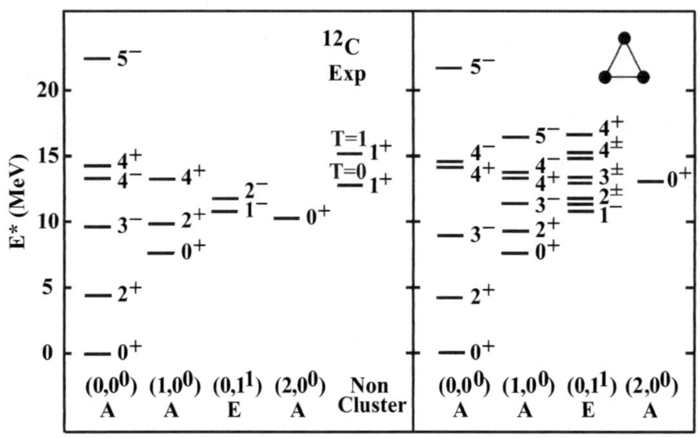

Figure 2. Comparison between the low-lying experimental spectrum of ^{12}C and the energies of the oblate symmetric top. The last column on the left-hand side, shows the lowest observed non-cluster (1^+) levels.

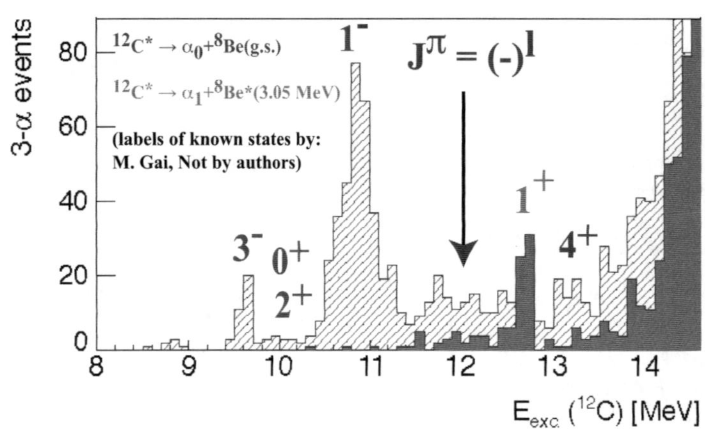

Figure 3. Possible new natural parity states observed by the Aarhus group [7] between 11.5 and 13 MeV in ^{12}C via the decay of (J^π, T) = (2^+, 1) state at 16.11 MeV populated in the ^{11}B(p,γ) reaction at E_p = 175 keV.

IOP Conf. Series: Journal of Physics: Conf. Series **966** (2018) 012037 doi:10.1088/1742-6596/966/1/012037

Another interesting point of interest raised by the ACM is the coalescing of the K=0 positive parity and K=3 negative parity bands into one rotational band as observed for molecules, including the α + ^{14}C nuclear molecule in ^{18}O [8]. In such a rotational band the predicted B(Eλ) are related and can be calculated with one free parameter (β) as shown in Table 1. This parameter (β) itself can be extracted in the ACM from the r.m.s. radius of the ground state of ^{12}C, hence the predictions shown in Table 1 are essentially parameter free. A ^{12}C(e,e') experiment to measure the B(Eλ) shown in Table 1 is planned at the S-DALINAC at Darmstdat during the Summer of 2017.

Table 1. ACM predictions of B(Eλ) in the g.s. band of ^{12}C.

Initial State	Final State	B(Eλ).
Ground State	2^+	$(Ze)^2 \frac{1}{4\pi} \frac{5}{4} \beta^4$
	3^-	$(Ze)^2 \frac{1}{4\pi} \frac{35}{8} \beta^6$
	4^+	$(Ze)^2 \frac{1}{4\pi} \frac{81}{64} \beta^8$

3. Conclusions

The Algebraic Cluster Model represents one of the most exciting developments in the nuclear structure of light nuclei and the cluster model. The remarkable success of the ACM in predicting the observed structures of ^{12}C and ^{16}O is very encouraging and new planed measurements at Darmstadt and Aarhus in ^{12}C are very important tests of this model. Recent application of the ACM to odd nuclei [9] will lead to additional important tests of the ACM.

Acknowledgment

The material presented in this paper is based upon work supported by the U.S. Department of Energy, Office of Science, Office of Nuclear Physics, under Award Number DE-FG02-94ER40870.

References

[1] Bijker R and Iachello F 2000 *Phys. Rev.* **C 61** 067305
[2] Bijker R and Iachello F 2002 *Ann. Phys.* **298** 334
[3] Marin-Lambarri D J, Bijker R, Freer M, Gai M, Kokalova Tz, Parker D J and Wheldon C, 2014 *Phys. Rev. Lett.* **113** 012502
[4] Bijker R and Iachello F 2014 *Phys. Rev. Lett.* **112** 152501
[5] Bijker R and Iachello F 2017 *Nucl. Phys.* **A957** 154
[6] Epelbaum E, Krebs H, Lahde T A, Lee D, and Meissner U -G 2012 *Phys. Rev. Lett.* **109** 252501
[7] Laursen K L, Fynbo H O U, Kirsebom O S, Madsbll K S and Riisager K 2016 *Eur. Phys. J.* **A52** 370
[8] Gai M, Ruscev M, Hayes A C, Ennis J F, Keddy R, Schloemer E C, Sterbenz S M and Bromley D A 1983 *Phys. Rev. Lett.* **50** 239
[9] Della Rocca V and Iachello F contribution to this 12th International Spring Seminar on Nuclear Physics, Sant'Angelo d'Ischia, May 15-19, 2017

Rotating and vibrating tetrahedrons in heavy nuclei

Andreas Heusler
Gustav-Kirchhoff-Strasse 7/1, 6120 Heidelberg, Germany
E-mail: A.Heusler@mpi-hd.mpg.de

Abstract.
In the heavy nucleus ^{208}Pb the 3^- yrast and 4^+ yrast states ($E_x = 2615$ and 4324 keV) are described by a rotating tetrahedron and eight more states at $4 < E_x < 8$ MeV by rotating and vibrating tetrahedrons. The stiffness in three vibration modes is hundred times higher than for corresponding states in ^{16}O by comparing the excitation energies from the prediction by the algebraic cluster model [*Nucl. Phys. A* 957 (2017) 154] and the splitting of a $2^\pm$ doublet predicted to be degenerate.

1. Introduction

The lowest excited state in the heavy nucleus ^{208}Pb has spin 3^- [1, 2]. It is described by the shell model with difficulty and was assumed to be a collective octupole vibration [3]. The 2^+ yrast and 4^+ yrast states were similarly assumed to be collective excitations. All three states are strongly excited by electromagnetic probes [1].

For nearly fifty years a search for double octupole vibrations was pursued. Indeed a 0^+ state in ^{208}Pb was discovered [1] at exactly twice the excitation energy of the 3^- yrast state. Yet among the lowest 150 states identified with high reliability [4] the other members of the double octupole multiplet with spins 4^+ and 6^+ were not found at excitation energies $E_x < 6.2$ MeV.

A perfect tetrahedron built from fermions may rotate and in addition vibrate in the dilation, flattening, and torsion mode as deduced by Wheeler [5]. The Pauli exclusion principle together with the symmetry relations derived from group theory [6] predicts the leading members of the

Table 1. Parameters of rotating and vibrating tetrahedronal states in ^{16}O and ^{208}Pb.

Nucleus	I^π	$(v_1v_2v_3)$	E_x	[Eq. (1)]		[W.u.]	B(E4)/B(E3)
^{16}O	3^-	000	6130	$\kappa_r = 511$	$\kappa_r/\omega_3 = 0.21$	$B(E3) = 13.5(7)$	
^{16}O	0^+	100	6049	$\omega_1 = 4033$	$\omega_1/\omega_3 = 1.65$		
^{16}O	2^+	010	6917	$\omega_2 = 1926$	$\omega_2/\omega_3 = 0.79$	$B(E2) = 3.1(1)$	0.23
^{16}O	1^-	001	7117	$\omega_3 = 2438$	$\omega_3/\omega_3 = 1$	$B(E1) = 0.00035(2)$	$3 \cdot 10^{-5}$
^{208}Pb	3^-	000	2615	$\kappa_r = 218$	$\kappa_r/\omega_3 = 0.12$	$B(E3) = 33.8(6)$	
^{208}Pb	0^+	100	5241	$\omega_1 = 3494$	$\omega_1/\omega_3 = 1.98$		
^{208}Pb	2^+	010	4086	$\omega_2 = 1389$	$\omega_2/\omega_3 = 0.79$	$B(E2) = 8.4(5)$	0.25
^{208}Pb	1^-	001	4842	$\omega_3 = 1762$	$\omega_3/\omega_3 = 1$	$B(E1) = 0.021(3)$	$6 \cdot 10^{-4}$

IOP Conf. Series: Journal of Physics: Conf. Series **966** (2018) 012038 doi:10.1088/1742-6596/966/1/012038

rotation band to be the 0^+ ground state, a 3^- state, a 4^+ state, and a $6^\pm$ doublet. The spin sequence in the dilation mode is the same as for the pure rotation. The leading members of the flattening band are a $2^\pm$ doublet and a $4^\pm$ doublet. The leading members of the torsion band are a 1^- state and a 2^+ state followed by a $3^\pm$ doublet.

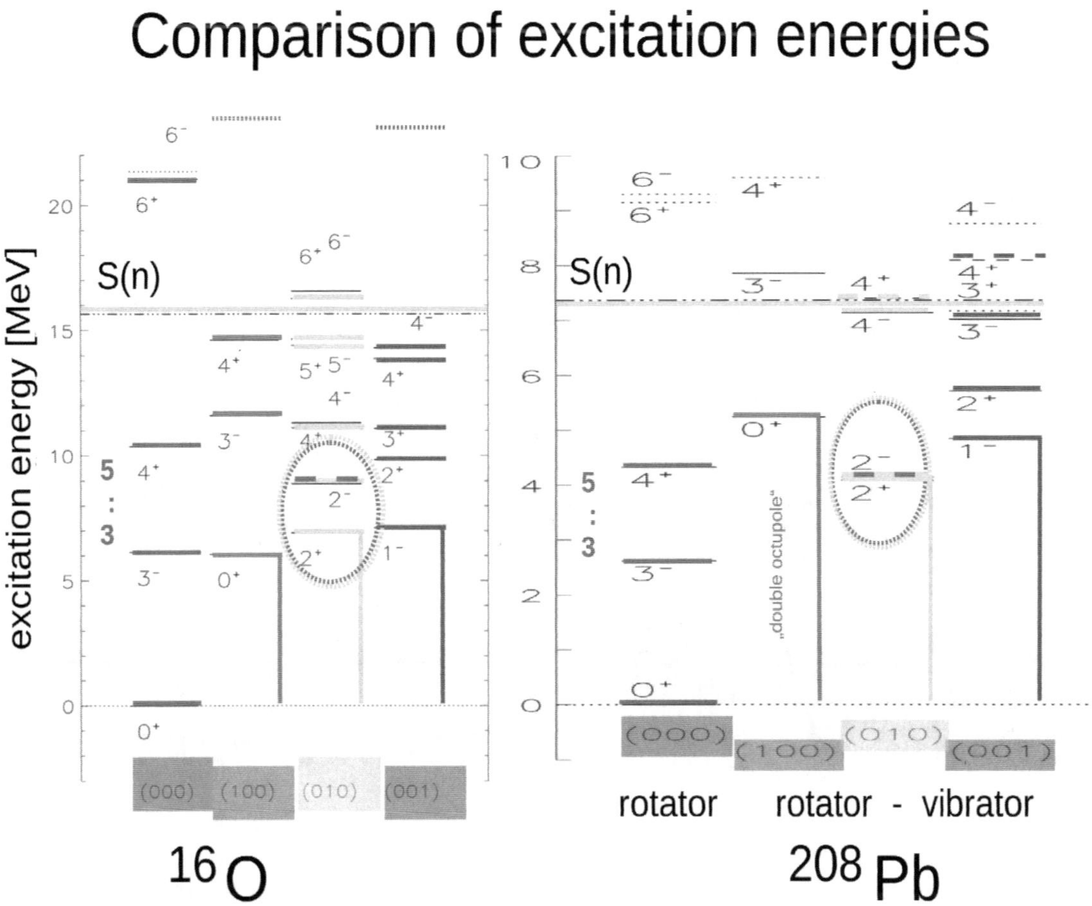

Figure 1. Level scheme of states with rotating and vibrating tetrahedrons (left) for ^{16}O [7], (right) for ^{208}Pb. The values v_1, v_2, v_3 from Eq. (1) are given at bottom for each of the four rotating and vibrating bands. The neutron threshold $S(n)$ is used to adapt the two scales for the excitation energy. The splitting of the $2^\pm$ (010) doublet is marked by a dotted circle.

The excitation energies are calculated [7] with four parameters $\kappa_r, \omega_1, \omega_2, \omega_3$,

$$\begin{aligned} E_x^{calc}(J, v_1, v_2, v_3) &= E_x^{rot}(J) + E_x^{vib}(v_1, v_2, v_3), \\ E_x^{rot}(J) &= \kappa_r J(J+1), \\ E_x^{vib}(v_1, v_2, v_3) &= \omega_1 \left(v_1 + \frac{1-\delta_{0v_1}}{2}\right) + \omega_2 \left(v_2 + \frac{2-2\delta_{0v_2}}{2}\right) + \omega_3 \left(v_3 + \frac{3-3\delta_{0v_3}}{2}\right). \quad (1) \end{aligned}$$

Here δ is the Kronecker symbol and J the angular momentum. v_1, v_2, v_3 count the number of phonons for the fundamental vibrations; they are lateron shortly written as $(v_1 v_2 v_3)$.

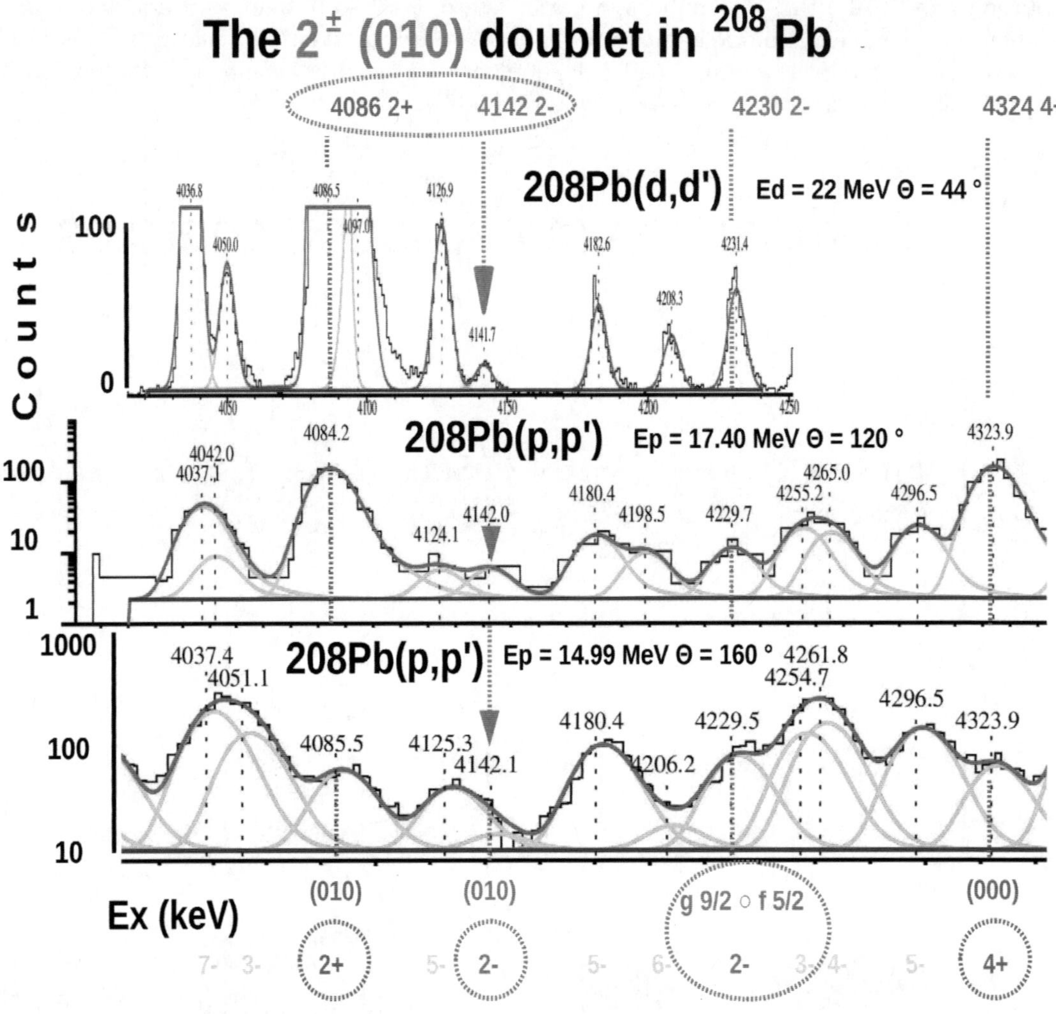

Figure 2. Spectra of (top) ^{208}Pb(d,d'), (middle) ^{208}Pb(p,p') – on $g_{7/2}$ resonance in ^{209}Bi (E^{res} = 17.43 MeV), (bottom) ^{208}Pb(p,p') – on $g_{9/2}$ resonance (E^{res} = 14.92 MeV). The 4142 2^- state was first observed by Wagner *et al* [9]. The 4230 2^- yrare state is described by the one-particle one-hole configuration $g_{9/2}f_{5/2}$ [8].

2. Identification

The ratio of the excitation energies for the 3^- and 4^+ yrast states in ^{208}Pb with $E_x(4^-)/E_x(3^+)$ = 1.6538 deviates from the predicted ratio 5/3 [Eq. (1)] by less than 1%. Therefore the 3^- and 4^+ yrast states are interpreted as the first excited states in the rotation band. The large electromagnetic moments confirm the interpretation. The ratio B(E3, ^{16}O) / B(E3, ^{208}Pb)= 0.40 ± 0.02 agrees within the uncertainties with the ratio of the moments of inertia as given by the ratio of the parameters κ_r of the excitation energies of the 3^- yrast state, $\kappa_r(^{16}\mathrm{O})/\ \kappa_r(^{208}\mathrm{Pb})$ = 0.426 (Fig. 1). The ratios B(E4)/B(E3) for ^{16}O and ^{208}Pb are similar, too (Table 1).

The 4086 2^+ and 4842 1^- states are interpreted as the heads of the flattening and torsion bands. The failure of all shell model calculations to describe these states is thus solved.

IOP Conf. Series: Journal of Physics: Conf. Series **966** (2018) 012038 doi:10.1088/1742-6596/966/1/012038

The search for double octupole vibrations was based on the misleading interpretation of the 2615 3^- state as a vibration; it is now described as a rotating tetrahedron. Therefore the 5241 0^+ state previously described as a "double octupole" state is identified with the head of the dilation mode.

The comparison of the 1^-, 3^-, 2^+, 4^+ yrast states and the 5241 0^+ state in ^{208}Pb to the corresponding states in ^{16}O corroborates the interpretation by rotating and vibrating tetrahedrons (Fig. 1, Table 1). The relative values of the parameters for the three vibration modes are similar for ^{16}O and ^{208}Pb, the ratio ω_2/ω_3 is even exactly the same. The difference of the ratio κ_r/ω_3 by a factor two should be compared to the ratio 13 of the masses. It shows that the nucleus ^{208}Pb is more compact than ^{16}O.

The knowledge of the four parameters allows to predict five more tetrahedronal states in ^{208}Pb by Eq. (1), see Fig. 1 and Table 2. The 7137 4^- and 7020 3^- states are known from $^{207}\mathrm{Pb}(d,p)$ and $^{207}\mathrm{Pb}(d,p\,\gamma)$ [1]. (The spin 1^- for the 7020 state is excluded by the re-interpretation of the few data points for $^{207}\mathrm{Pb}(d,p)$ [10].) The 5715 2^+ state was previously counted as a member of the double octupole multiplet which is shown to be a miscomprehension. The 7838 3^- state is known as a resonance in the neutron capture on ^{207}Pb [11]. It has one of the biggest cross sections among all resonances up to 0.5 MeV above the neutron threshold $S(n) = 7368$ keV.

The 4142 2^- state was observed by $^{208}\mathrm{Pb}(p,p')$ with $E_p = 35$ MeV [9]. By the criteria set up for a firm identification of a nuclear state [4] this level was assumed to be spurious because no other data supported the observation. Recent $^{208}\mathrm{Pb}(d,d')$ spectra confirm the existence of the 4842 2^- state and the ongoing reanalysis of the $^{208}\mathrm{Pb}(p,p')$ data taken in 1968-1969 [8] adds confidence (Fig. 2).

The ratio of the excitation energies for the 3^- and 4^+ yrast states in ^{208}Pb,

$$R_{4/3} = \frac{E_x(4^+)}{E_x(3^-)} = \frac{5}{3}\;0.9923, \tag{2}$$

deviates from the prediction for a rotating tetrahedron by less than 1%.

Table 2. Comparison of states in ^{16}O and ^{208}Pb.

$(v_1v_2v_3)$	I^π	Nucleus	E_x keV	E_x^{calc} keV	δE_x keV	$\delta E_x/E_x$ 10^{-4}	Nucleus	E_x keV	E_x^{calc} [7] keV	$\delta E_x/E_x$ 10^{-4}
(000)	0^+	^{208}Pb	0	0			^{16}O	0	0	
(000)	3^-	^{208}Pb	2615	a			^{16}O	6130		
(000)	4^+	^{208}Pb	4324	4358	−34	−80	^{16}O	10356	10220	−130
(100)	0^+	^{208}Pb	5241	a			^{16}O	6049	6050	0
(100)	3^-	^{208}Pb	7838	7856	−18	−23	^{16}O	11600	10970	−570
(010)	2^+	^{208}Pb	4086	a			^{16}O	6917	7740	−1060
(010)	2^-	^{208}Pb	4142	4086	+56	+135	^{16}O	8872	7740	+1460
(010)	4^+	^{208}Pb					^{16}O	11097	11690	−430
(010)	4^-	^{208}Pb	7137	7136	+1	+2	^{16}O		11690	
(001)	1^-	^{208}Pb	4842	a			^{16}O	7117	6850	+400
(001)	2^+	^{208}Pb	5715	5713	+2	+3	^{16}O	9844	8460	+1650
(001)	3^+	^{208}Pb	7021	7020	−1	−1	^{16}O	1080	10870	−170

(a) The excitation energy defines the parameters $\kappa_r, \omega_1, \omega_2, \omega_3$ by Eq. (1).

IOP Conf. Series: Journal of Physics: Conf. Series **966** (2018) 012038 doi:10.1088/1742-6596/966/1/012038

3. Conclusion

Rotating and vibrating tetrahedrons are identified in the doubly magic nucleus ^{208}Pb. By comparison to ^{16}O the tetrahedron in ^{208}Pb is stiffer by a factor hundred if the relative deviation of the excitation energies from the predictions by the model calculation [Eq. (1)] is considered (Table 2).

The dilation mode head is predicted to be a degenerate $2^{\pm}$ doublet. Indeed close to the 4086 2^+ yrast state in ^{208}Pb, the 4142 state first observed by Wagner *et al* [9] is verified (Fig. 2). The low splitting of 0.2% of the $2^{\pm}$ doublet in ^{208}Pb is explained by the stiffness of the tetrahedron. In contrast, the separation of the 2^- state from the 2^+ state in ^{16}O by 30% indicates a rather floppy tetrahedron.

Data for other heavy nuclei but ^{16}O and ^{208}Pb do not allow to clearly identify rotating and vibrating tetrahedrons. The knowledge of excitation energies, spin, and parity is often rather limited [12]. However a low-lying 3^- state with a large B(E3) value is often found in heavy nuclei [13].

Acknowledgments

Thanks are given to P. von Brentano for encouragement during more than 50 years.

References

[1] Martin M J 2007 *Nucl. Data Sheets* **108** 1583
[2] Elliott L G, Graham R L, Walker J, Wolfson J L 1954 *Phys. Rev.* **93** 356
[3] Hamamoto I 1970 *Nucl. Phys.* A **155** 362
[4] Heusler A, Jolos R V, Faestermann T, Hertenberger R, Wirth H-F, von Brentano P 2016 *Phys. Rev.* C **93** 054321
[5] Wheeler J A 1937 *Phys. Rev.* **52** 1083
[6] Tisza L 1933 *Z. Phys.* **82** 48
[7] Bijker R and Iachello F 2017 *Nucl. Phys.* A **957** 154
[8] Heusler A, Glöckner H-J, Grosse E, Moore C F, Solf J, Brentano P von 2014 *Eur. Phys. J.* A **50** 92
[9] Wagner W T, Crawley G M, Hammerstein G R, McManus H 1975 *Phys. Rev.* C **12** 757
[10] Valnion B D 1998 *PhD Thesis Universität München* (Herbert Utz Verlag, München)
[11] Horen D J, Auchampaugh G F, Harvey J A 1978 *Phys. Lett.* B **79** 39
[12] National Nuclear Data Center, Brookhaven 2017 *Evaluated Nuclear Structure Data File* http://ie.lbl.gov/ensdf/.
[13] Spear R H 1989 *At. Data Nucl. Data Tables* **42** 55

IOP Conf. Series: Journal of Physics: Conf. Series **966** (2018) 012039 doi:10.1088/1742-6596/966/1/012039

Cluster-shell model: I. Structure of ^{9}Be, ^{9}B

Valeria Della Rocca[1,2] **and Francesco Iachello**[2]

[1]Gran Sasso Science Institute, viale Francesco Crispi 7, 67100 L'Aquila, Italy
[2]Center for Theoretical Physics, Sloane Laboratory, Yale University, New Haven, CT 06520-8120, USA

E-mail: `valeria.dellarocca@gssi.it`, `valeria.dellarocca@yale.edu`, `francesco.iachello@yale.edu`

Abstract. A cluster-shell model suitable for the description of light nuclei seen as k α-particles + x nucleons is described. Its application to ^{9}Be and ^{9}B is also shown, for which we explicitly provide a comparison with the available experimental data.

1. Introduction

The cluster description of light nuclei was originally proposed in 1937 by Wheeler [1] and independently by Hafstad and Teller [2] in 1938, as a consequence of the observation of the high binding energy of the ^{4}He nucleus. Specific cluster configurations were suggested by Brink [3, 4]. Following recent experimental developments [5], this model has regained interest. Indeed, it was found that the spectra of ^{8}Be, ^{12}C and ^{16}O are well reproduced by 2, 3 and 4 cluster configurations with Z_2, D_{3h} and T_d symmetry respectively. An extension of this idea consists in the inclusion of an extra particle moving in the potential generated by the clusters. In this contribution, we will first describe the recently introduced cluster-shell model (CSM) [6], which we propose as a possible model to describe light even-odd nuclei with a cluster core, and then show its explicit application in the case of ^{9}Be and ^{9}B.

2. Cluster shell model

In previous works, the charge and matter density $\rho(\vec{r})$ of 2, 3, and 4 α-particles have been found [6]. The convolution:

$$V(\vec{r}) = \int \rho(\vec{r}\,')v(\vec{r} - \vec{r}\,')d^3\vec{r}\,'. \tag{1}$$

of $\rho(\vec{r})$ and the interaction between the α and the nucleon, $v(\vec{r} - \vec{r}\,')$, which we assume to be of Volkov-type [7], allows the determination of the potential $V(\vec{r})$ in which the odd nucleon is moving. Explicit calculation leads to [6]

$$V(\vec{r}) = V_0 e^{-\alpha(r^2+\beta^2)} 4\pi \sum_{\lambda\mu} i_\lambda(2\alpha\beta r) Y_\mu^\lambda(\theta,\phi) \sum_i^k Y_\mu^{*\lambda}(\theta_i,\phi_i) \tag{2}$$

where i_λ is the modified spherical Bessel function. The α-particles are assumed to be placed at positions $r_i = (\beta, \theta_i, \phi_i)$ which are chosen according to the desired symmetry we wish to reproduce.

The single particle energies ϵ_Ω of a particle in the cluster potential in equation (2) are obtained by solving the Schrödinger equation:

$$H = \frac{\vec{p}^2}{2m} + V(\vec{r}) + V_{so}(\vec{r}) \tag{3}$$

where $V_{so}(\vec{r})$ is the spin-orbit interaction, which we assume of the form:

$$V_{so}(\vec{r}) = V_{0,so} \frac{1}{r} \frac{\partial V}{\partial r} \left(\vec{s} \cdot \vec{l} \right) \tag{4}$$

Hence, the problem is essentially that of a particle in potentials that change in a continuous manner from having spherical symmetry to having point group symmetries, Z_2 ($k = 2$), D_{3h} ($k = 3$), T_d ($k = 4$), as the distance of the clusters β increases. Since the dimension of these discrete groups is smaller than the dimension of the rotation group, we expect each level to split according to the splitting of the representation of $O(3)$ in terms of the irreducible representations of the point groups of interest.

We have diagonalized the Hamiltonian H in equation (3) numerically in the harmonic oscillator basis. The correlation diagrams corresponding to $k = 2$, $k = 3$, and $k = 4$ are depicted in figures 1, 2 and 3, respectively [6]. The states are labelled with the notation of the spherical shell model in $\beta = 0$. For $\beta > 0$, in the case of $k = 2$ we have labelled the states with the projection of the angular momentum along the symmetry axis, K, and the parity, P, as well as with the molecular physics notation, while for $k = 3, 4$ we have used the labels of the irreducible representations of D_{3h} and T_d, respectively. In figures 1 to 3 we also show that bigger gaps occur at different energies for different values of β.

3. Application to ^{9}Be

The treatment of ^{9}Be is a straightforward application of the model described so far, where the core is assumed to have the particular configuration depicted in figure 4, i.e. when $k = 2$ and $\beta = 1.82$ fm. The value of β has been derived from the experimental moment of inertia of ^{8}Be. This particular configuration can be obtained choosing in equation (2) $\vec{r}_1 = (r_1, \theta_1, \phi_1) = (\beta, 0, -)$ and $\vec{r}_2 = (r_2, \theta_2, \phi_2) = (\beta, \pi, -)$ as the locations of the two α-particles.

The single particle states and wave functions are obtained as explained in Sect. 2. Using the standard expression for rotational bands for axially symmetric nuclei [8]

$$E(J) = \epsilon_K + \frac{(\hbar c)^2}{2\mathscr{I}} \left[J(J+1) - K^2 + \delta_{(K,1/2)} a (-)^{J+1/2} \left(J + \frac{1}{2} \right) \right] \tag{5}$$

we obtain the energy levels in figure (5), where the moment of inertia $\mathscr{I}$ is the sum of the core and single particle contributions.

The electric transition rates, $B(E\lambda)$, and multipole electric moments, $Q^{(\lambda)}(K, I)$, with $\lambda = 2, 4, \cdots =$ even are dominated by the core contribution, while those with $\lambda = 1, 3, \cdots =$ odd between states with different parity (i.e. from the $K^P = 1/2^+$ band excited states to $K^P = 3/2^-$ ground state) are the sum of collective and single particle contributions. In the latter case, expanding the wave function in the h.o. basis:

$$|\psi\rangle = \sum_{nljm} a_{nljm} |n, \frac{1}{2}, l, j, m\rangle \tag{6}$$

we can evaluate the matrix elements of the operators defining the single particle contribution to moments and electric transition rates. The result has to be summed to the core contribution,

IOP Conf. Series: Journal of Physics: Conf. Series **966** (2018) 012039 doi:10.1088/1742-6596/966/1/012039

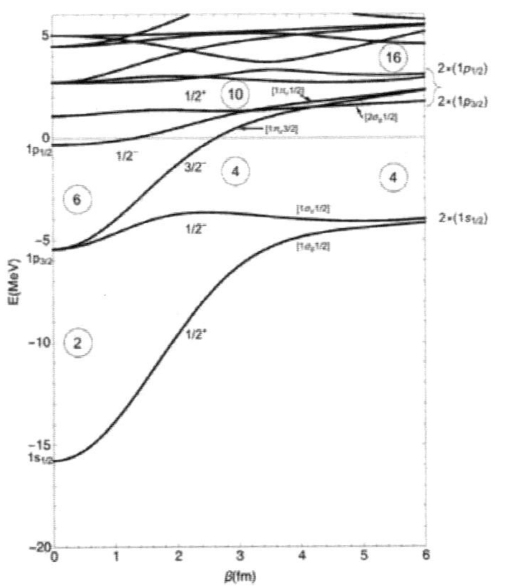

Figure 1. Correlation diagrams of the energy levels of a spin 1/2 particle for the potential in equation (2) as function of the distance β between the clusters and the center of mass for: $k = 2$. The parameters used are: $V_0 = 20$ MeV, $V_{0,so} = 22$ MeVfm^2, $\alpha = 0.1115$ fm^{-2}. Reproduced from Ref. [6] with permission.

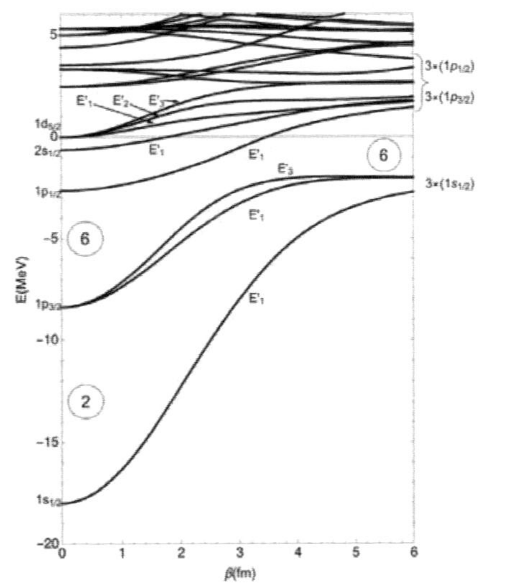

Figure 2. Correlation diagrams of the energy levels of a spin 1/2 particle for the potential in equation (2) as function of the distance β between the clusters and the center of mass for $k = 3$. The parameters used are: $V_0 = 13.3$ MeV, $V_{0,so} = 16.9$ MeVfm^2, $\alpha = 0.0872$ fm^{-2}. Reproduced from Ref. [6] with permission.

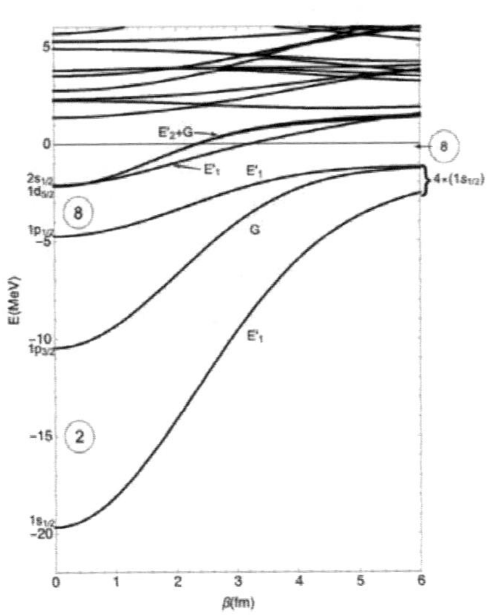

Figure 3. Correlation diagrams of the energy levels of a spin 1/2 particle for the potential in equation (2) as function of the distance β between the clusters and the center of mass for $k = 4$. The parameters used are: $V_0 = 10$ MeV, $V_{0,so} = 13.4$ MeVfm^2, $\alpha = 0.0729$ fm^{-2}. Reproduced from Ref. [6] with permission.

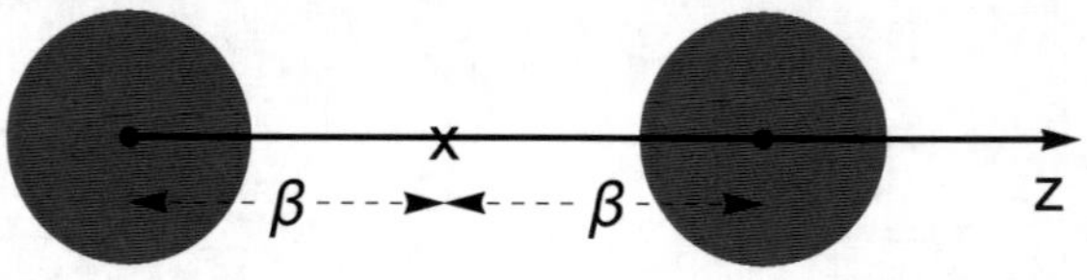

Figure 4. ^{8}Be seen as two α-particles in a dumbbell configuration.

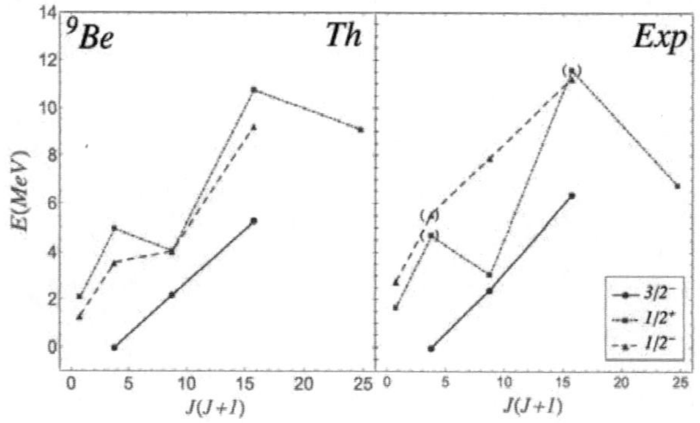

Figure 5. Comparison between (a) theory and (b) experiment of the energy levels of ^{9}Be. Adapted from Ref. [11].

Table 1. Comparison between (e, e') [10], NDT [9], shell model (SM) and cluster shell model (CSM) calculations. Units of Q are in efm^2 and of $B(E2)$ are in e^{2}fm^4.

	(e,e')	NDT	SM	CSM
$Q^{(2)}(3/2^-)$	5.86(6)	5.288(38)	4.35	5.30
$B(E2; 3/2^- \to 5/2^-)$	46.0(5)	40.5(30)	32.2	35.9
$B(E2; 3/2^- \to 7/2^-)$	33(1)	18(8)	12.7	20.0

which is obtained by coupling single particle and core transition operators. Similar considerations hold for magnetic transition rates, $B(M\lambda)$, and magnetic moments, $\mu^{(\lambda)}(K, I)$. The explicit derivation of these quantities is rather long and therefore not reported here, but it can be found in [11, 12].

In table 1 we show the comparison between theory and experiment [9, 10] for the quadrupole moment of the ground state and for $B(E2)$ transitions. The large value of $Q^{(2)}(3/2^-)$ that we have obtained in the CSM is in good agreement with the existing data, a feature that is difficult to explain in a spherical model. Similarly, the large $B(E2)$ values are a proof of collective structure of the system, being these transitions mostly due to the contribution of the core. This is a quite remarkable result, since we did not introduce any effective charge in the calculation.

Although smaller than the data from electron scattering, Ref. [10], the $B(E1)$ transition rates, table 2, are in agreement with the Nuclear Data Tables (NDT) [9].

The magnetic moment of the ground state, $\mu(3/2^-)$, is also in fairly good agreement with the experiments, while $B(M1; 3/2^- \to 5/2^-)$ is smaller, although of the same order of magnitude (see table 3 for comparison).

In figures 6 to 8 we show the longitudinal electric form factors relative to members of the

Table 2. Comparison between (e, e') [10], NDT [9], shell model (SM) and cluster shell model (CSM) calculations. Units are in e^2fm^2.

	(e,e')	NDT	SM	CSM
$B(E1; 3/2^- \rightarrow 1/2^+)$	0.034(3)	0.029(3)	0.0045	0.024
$B(E1; 3/2^- \rightarrow 5/2^+)$	0.029(5)	0.015(12)	0.0039	0.0048

Table 3. Comparison between (e, e') [10], NDT [9], shell model (SM) and cluster shell model (CSM) calculations. Units of magnetic moment are in μ_N and of $B(M1)$ those of Ref. [10], e^2fm^2.

	Exp(e,e')	NDT	SM	CSM
$\mu(3/2^-)$	-1.16(2)	-1.1778(9)	-1.27	-1.13
$B(M1; 3/2^- \rightarrow 5/2^-)$	0.0090(3)	0.0089(9)	0.0068	0.0038

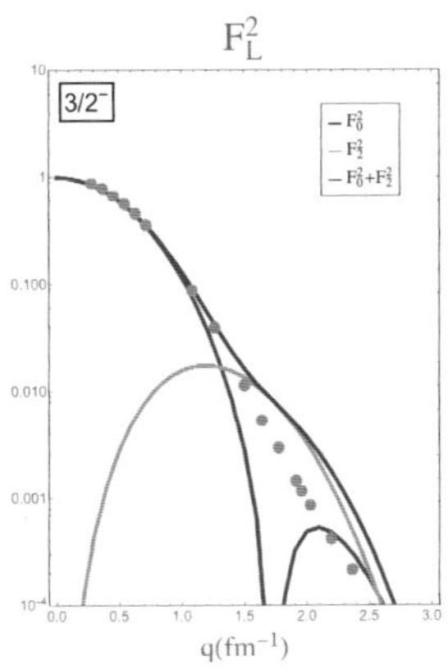

Figure 6. Longitudinal electric form factor of the state $3/2^-$. Adapted from Ref. [12].

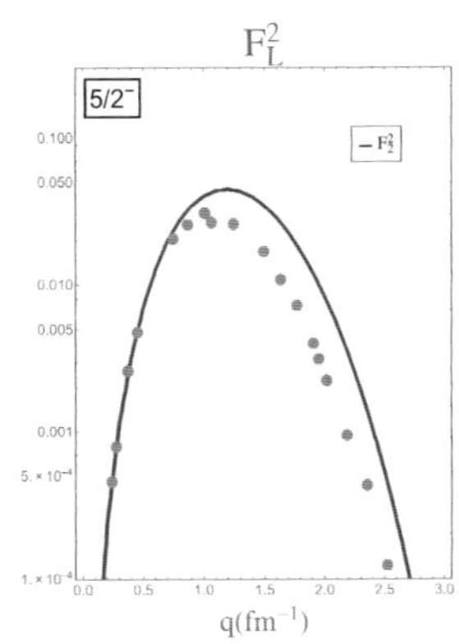

Figure 7. Longitudinal electric form factor of the state $5/2^-$. Adapted from Ref. [12].

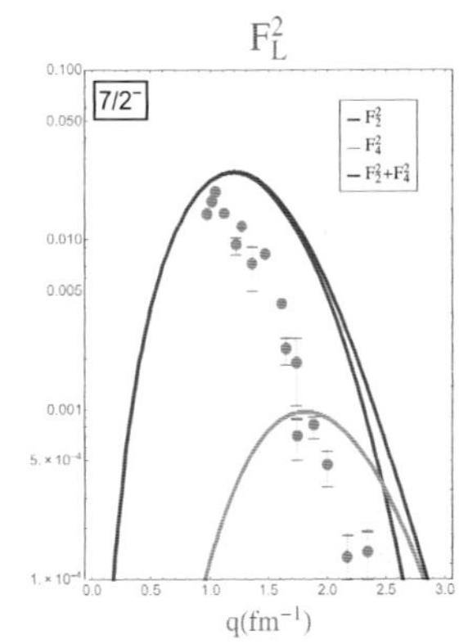

Figure 8. Longitudinal electric form factor of the state $7/2^-$. Adapted from Ref. [12].

ground state band. For small values of the momentum transfer q the CSM reproduces the experimental results well, while as q increases it does not. This is expected, since for high momentum transfer one probes the inner structure of the α-particle.

Our conclusion is that the cluster structure of ^{8}Be is not destroyed when an extra fermion is added, i.e. ^{9}Be can be seen as made of a core of 2α-particles plus an additional neutron.

4. Application to ^{9}B

The treatment of ^{9}B in CSM is identical to that of ^{9}Be, except that equation (3) is replaced by:

$$H = \frac{\vec{p}^2}{2m} + V(\vec{r}) + V_{so}(\vec{r}) + V_C(\vec{r}) \tag{7}$$

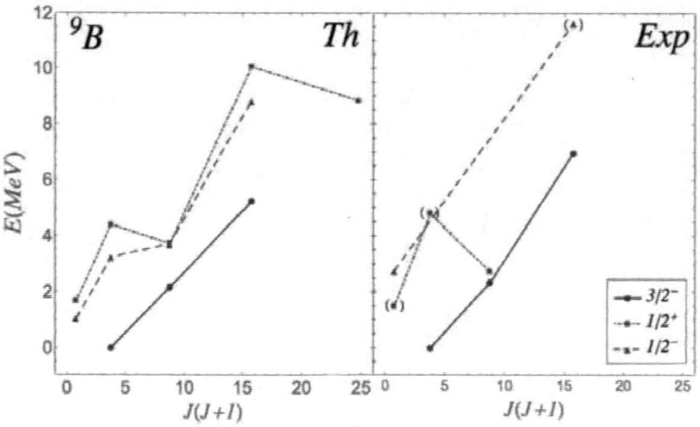

Figure 9. Rotational spectra of ^{9}B. Adapted from Ref. [11].

where $V_C(\vec{r})$ is the Coulomb potential generated by the density distribution of [6]. Figure 9 shows a comparison between the energy levels calculated using the CSM and the experimental data [9]. However, being ^{9}B unstable, a comparison between CSM and experiment for $B(E\lambda)$ and $B(M\lambda)$ rates is not possible at present, due to the lack of experimental data. We have nonetheless evaluated such quantities for future reference and reported them in [12].

Acknowledgments
This work was supported in part by US Department of Energy Grant No. DE-FG-02-91ER-40608, and, in part, by the Gran Sasso Science Institute (GSSI).

References

[1] Wheeler J A 1937 *Phys. Rev.* **52** 1083
[2] Hafstad L R and Teller E 1938 *Phys. Rev.* **54** 681
[3] Brink D M 1965 *Proc. Int. School of Physics "Enrico Fermi"* Course XXXVI p 247
[4] Brink D M, Friedrich H, Weiguny A, and Wong C W 1970 *Phys. Lett. B* **33** 143
[5] Marín−Lámbarri D J, Bijker R, Freer M, Gai M, Kokalova Tz, Parker D J, and Wheldon C 2014 *Phys. Rev. Lett.* **113** 012502
[6] Della Rocca V, Bijker R and Iachello F 2017 *Nucl. Phys. A* **966** 158
[7] Volkov A B 1965 *Nucl. Phys.* **74** 33
[8] Rowe D J 2010 *Nuclear Collective Motion* (Singapore: World Scientific)
[9] Tilley D R et al. 2004 *Nucl. Phys. A* **745** 155
[10] Glickman J P et al. 1991 *Phys Rev. C* **43** 1740
[11] Della Rocca V and Iachello F submitted to *Nucl. Phys. A*
[12] Della Rocca V 2017 *Ph.D. thesis* Gran Sasso Science Institute (L'Aquila, Italy: GSSI)

IOP Conf. Series: Journal of Physics: Conf. Series **966** (2018) 012040 doi:10.1088/1742-6596/966/1/012040

Experimental investigation of exotic clustering in ^{13}B and ^{14}C using the resonance scattering method

A. Di Pietro[a], J. P. Fernández-García[a,b], F. Ferrera [a], P. Figuera[a], M. Fisichella[a], M. Lattuada[a,b], S. Marletta[a], C.Marchetta[a], D. Torresi[a,b], M. Alcorta[c], M. J. G. Borge[d], T. Davinson[e], S. Heinitz[f], A. M. Laird[g], A. C. Shotter[c,e], D. Schumann[f], N. Soic[i], O. Tengblad[h], M. Zadro[i]

[a] INFN-Laboratori Nazionali del Sud, Catania, Italy
[b] Dipartimento di Fisica ed Astronomia, Universitá di Catania, Catania, Italy
[c] TRIUMF, Vancouver, Canada
[d] ISOLDE-CERN, Geneva, Switzerland
[e] School of Physics and Astronomy, University of Edinburgh, Edinburgh, UK
[f] Paul Scherrer Institut, Villigen, Switzerland
[g] Department of Physics, University of York, York, UK
[h] Instituto de Estructura de la Materia, CSIC-Madrid, Spain
[i] Ruder Boskovic Institute, Zagreb,Croatia

E-mail: `dipietro@lns.infn.it`

Abstract.
In order to investigate the existence of molecular and/or exotic cluster configurations in Boron and Carbon n-rich isotopes we undertook two experiments: the first experimental study of exotic ^{9}Li+α cluster states in ^{13}B using the resonance scattering method at TRIUMF (Canada), and, with the same technique, the measurement of ^{10}Be+α scattering at LNS in Catania, where a ^{10}Be radioactive beam was produced for the first time. In order to measure the excitation function in a wide energy range, the beams were stopped in a Helium-flooded chamber. In the case of ^{13}B, the elastic excitation function shows the presence of various peaks in an excitation energy region never explored before. In the case of ^{14}C, our exclusive measurement of elastic scattering data with a high intensity beam, sheds some light on the contradictory previously published results [1, 2].

1. Introduction

Some of the properties of nuclei can be described by assuming a nuclear structure made of a few weakly interacting clusters. Generally the clusters are strongly bound particles and hence αs are the typical cluster particles. In light unstable nuclei different types of clusters may exist: clusters which are not ordinary stable particles (like the αs), but unstable, deformed and easy to break-up particles. The presence of these types of clusters is expected to become more and more favored when nuclei approach the drip-line [3]. Structures made of such exotic clusters are predicted to exist in some of the excited states of Beryllium and Boron n-rich isotopes.
Calculations, performed in the framework of the Antisymmetrized Molecular Dynamics (AMD) [4], describe some of the ^{13}B excited states in terms of ^{4}He-^{9}Li clusters, in the excitation energy region above the decay threshold of the nucleus into the two components under consideration. In

order to investigate the existence of these structures in ^{13}B, we studied the excitation function for the ^{4}He(^{9}Li,α) elastic scattering process. In this paper will be shown the results of the experimental investigation of the ^{9}Li+^{4}He at E_{lab}=32 MeV performed to explore upon the existence of exotic clustering in ^{13}B in the excitation region 14 MeV $\leq E_x \leq$ 20 MeV were these structure are predicted to exist.

An additional phenomenon that may occur in light unstable nuclei, in particular in the n-rich ones, is a type of clustering where the $\alpha - \alpha$ cluster structure as a core persists; however, it is the exchange of neutrons between the α-particle cores that bounds the system, like electrons in the covalent bonding.

The existence of linear chain configurations of α-particles in excited states of 4N-nuclei, such as ^{12}C or ^{16}O, has been searched for a long time; however no evidence has been found so far. In the case of n-rich nuclei, theoretical predictions of a linear chain configuration in ^{14}C of α-particles bound together by neutrons, was made by Suhara and Enýo [5] with AMD. Associated to this configuration, a band is predicted to exist (J_π=0$^+$, 2$^+$, 4$^+$) associated with a linear shape, at a few MeV above the ^{10}Be+α threshold. In order to look for this structure in ^{14}C, three experiments have been performed so far [1, 2, 6], where the ^{10}Be+^{4}He excitation function has been measured. Controversial results on the possible presence of the inelastic scattering contribution in the detected α-particle spectrum have been reported in [1, 2]. This could affect the observation reported in [1] of the rotational band associated to the linear chain configuration in ^{14}C. The present paper will also report on the results of the ^{10}Be+^{4}He resonance scattering experiment performed to shed some light on the controversial results reported in the literature [1, 2].

2. The ^{9}Li+^{4}He reaction

2.1. Experiment

The experiment ^{9}Li+^{4}He was done at TRIUMF (Canada) using a ^{9}Li beam at 32 MeV delivered by the ISACII facility. The target consisted of the TUDA chamber (1.5 m long) filled with an isotopically pure ^{4}He gas at pressures of 650 and 680 Torr. The chamber was separated from the high vacuum beam line by a Kapton window 12 μm thick. Elastically scattered α-particles were detected and discriminated from other reaction processes using both ΔE-E and Time of Flight (ToF) techniques. The detection system consisted of three telescopes each made of one four-quadrants, 50$\times$50 mm^2, 50 μm thick Si as ΔE detector and a 50$\times$50 mm^2, 1000 μm thick Si detector as residual energy detector. One of the telescopes (T1) was placed around 0^0 ($\theta_{c.m.}$=180^0) and a second one (T2) next to it downstream the TUDA chamber. The gas pressure was chosen to be sufficient to stop the beam before it reached the 0^0 detector, but not the recoiling α-particles. A third telescope (T3) was placed closer to the entrance Kapton window, at about half the distance from the window of the other two telescopes, as sketched in figure 1. In this way some information on the angular distribution for the highest energy events (the ones occurring near the entrance window) could be gathered.

A microchannel plate (MCP) detector was placed under vacuum, just upstream the entrance window, in order to give a signal whenever a ^{9}Li beam particle entered into the chamber. The MCP gives a way to count the beam particles, necessary for the cross-section normalisation and, at the same time, provides a time signal for the ToF measurement. The beam intensity during the runs with the MCP detector was kept at around 5$\times$10^5 pps. In some of the runs of the experiment, the MCP detector was switched-off and the beam intensity was increased to 10^7 pps. These runs were normalised to the low intensity ones where the MCP detector was on. The detectors placed around 0^0 (T1 and T2) were detecting also the βs and the β-delayed αs coming from the radioactive decay of the ^{9}Li beam. These were producing a large background. Due to this background, only α-particles with energy $\geq$4 MeV were selected. The background above 4 MeV was less then 1% and in all cases subtracted.

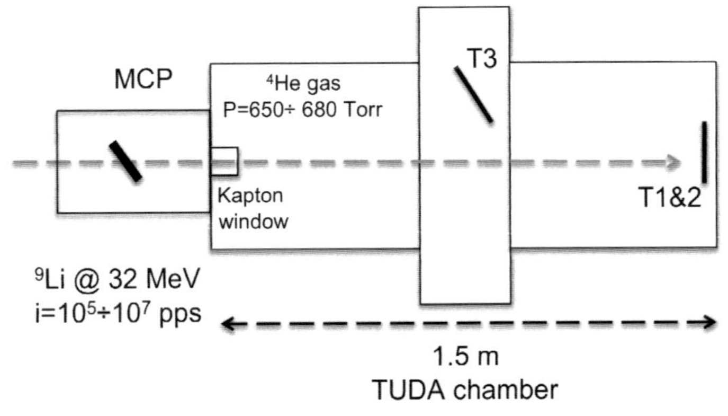

Figure 1. (Colour on line) Sketch of the set-up used for the ^{9}Li+^{4}He experiment

2.2. Results

From the detected-energy spectra of α-particles at the various angles, the ^{13}B excitation energy spectra were obtained. In the thick target experiments, the elastic scattering process can occur at any point of the 1.5 m long gas target, along the beam direction, at energies that go from 32 MeV down to practically 0 MeV. In the case of the elastic scattering process, energy and angles of the recoiling particles are uniquely related to the position in the target at which the process occurs. Therefore, via kinematics and energy loss calculations of the beam and recoiling particles, the excitation function for the elastic scattering process can be obtained from the recoiling α-particle spectra. The data analysis was performed on an event-by-event base, reconstructing, for the events punching through the ΔE detectors, the total energy ($\Delta E+E_{res}$). The total energy was corrected for the energy loss in the dead-layers of the detectors and in the gas in between the two detectors of the telescope. In figure 2 it is shown the ^{13}B excitation energy spectra corresponding to the different measured angles. Due to the extension of the target, the angle of the recoiling α-particles depends on $E_{c.m.}$ so it changes as the beam is slowed down in the ^{4}He gas; $E_{c.m.}$ and $\theta_{c.m.}$ are reconstructed event-by-event from the energy and angle of the detected α-particles.

In figure 2 it is possible to observe at least two large peaks at excitation energies of 16.3 and 19.5 MeV. The peak at 19.5 MeV is very asymmetric, an indication that it could originate from the superposition of several states. This is confirmed by looking at the excitation function measured by the telescope T3 placed at the largest angles ($\theta_{c.m.}\simeq 160^0$). The theoretical analysis is on going to determine spin and parity of these states. According to the AMD predictions high spin states of ^{9}Li+^{4}He type are to be populated. The rapid variation of the cross-section with the angle, seem to confirm these predictions.

3. The ^{10}Be+^{4}He reaction

3.1. Experiment

The experiment was performed at LNS using a radioactive ^{10}Be beam produced in batch-mode. The ^{10}Be radioactive material (0.1 mg of ^{10}BeO prepared at PSI Villigen, Switzerland) was inserted in the cathode of the TANDEM sputtering source. The beam was then accelerated at 47 MeV by the TANDEM accelerator. For this experiment the Inverse Kinematic Thick Target method was used. The target consisted in the CT2000 scattering chamber (2 m diameter) filled with an isotopically pure ^{4}He gas at a pressure of 650 Torr and separated from the beam line

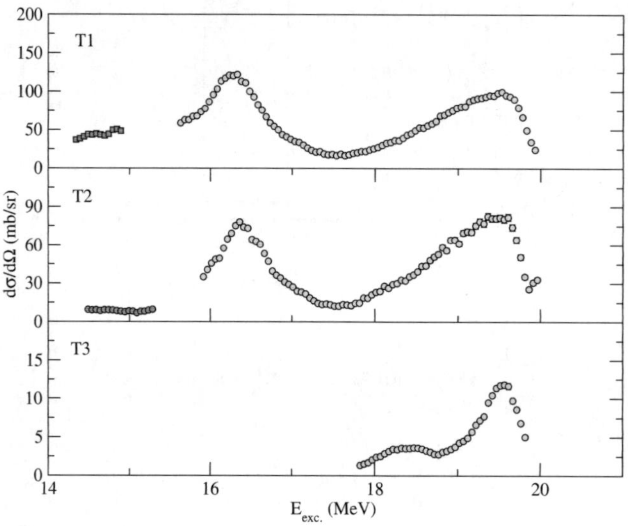

Figure 2. (Colour on line) ^{13}B excitation energy spectra corresponding to the different telescopes i.e. different angles.

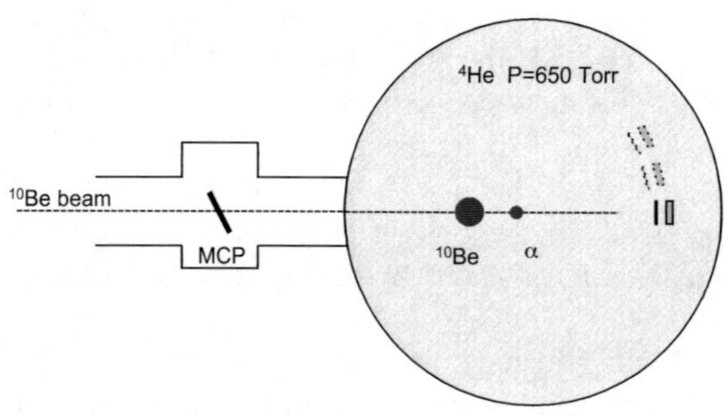

Figure 3. (Colour on line) Sketch of the set-up used for the ^{10}Be+^{4}He experiment

by a ~12 μm thick Kapton window. As for the ^{9}Li+^{4}He both ΔE-E and ToF techniques were used to discriminate elastically scattered αs. The detection system consisted of one telescope made of a 18 μm thick Si detector, as ΔE and a 500 μm thick Si as residual energy detector. In front of the telescope a collimator of ϕ= 6 mm was placed. The telescope was mounted on a rotating arm so that the detection angle could be changed during the measurement. The arm was rotating with respect to the center of the circular chamber, which is the target position in standard experiments. Three angular settings were used: $\theta_{\rm lab}$=0^0 ($\theta_{\rm c.m.}$=180^0), 5^0 and 10^0 measured with respect to the center of the chamber.

As for the ^{9}Li+^{4}He experiment, a microchannel plate detector was placed just upstream the entrance window, in order to have a fast signal whenever a ^{10}Be beam particle entered into the chamber, producing a reference signal for the ToF measurement and allowing to count the

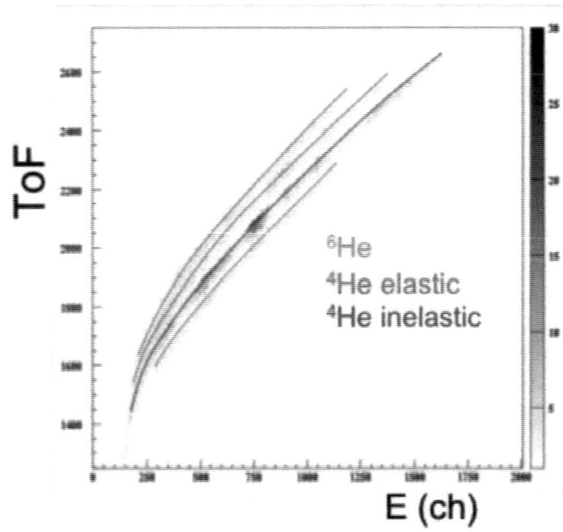

Figure 4. (Colour on line) ToF *vs* E_{res} spectrum. See text for details.

incoming beam particles. The beam intensity during the runs with the MCP detector was kept $<1\times10^7$ pps.
$E_{c.m.}$ and $\theta_{c.m.}$ are reconstructed on an event-by-event basis from the energy and angle of the detected α-particles as done for ^{9}Li+^{4}He analysis.
The gas pressure was chosen to be sufficient to stop the beam before it reached the detector when placed at $\theta_{lab}= 0^0$, but not the recoiling α-particles. In figure 3 a sketch of the experimental set-up is shown.
In figure 4 the ToF-E_{res} 2D-spectrum is shown. In the spectrum it is possible to identify ^{6}He (orange line), ^{4}He coming from elastic scattering (red line), ^{4}He coming from inelastic scattering of the $^{10}Be_{2+}$ state at E_x=3.37 MeV and of the group of states at $E_x\sim$ 6 MeV (blue lines). From the figure, one can confirm what was reported in [3] i.e. that the inelastic scattering contribution is not negligible as reported in [1] and it might affect the conclusions drawn in [1] about the existence of the linear chain configuration in ^{14}C.

3.2. Results
Elastic scattering α-particle spectra were obtained by putting a gate on the corresponding locus on the ToF-E_{res} plot. Similar plots were obtained for the inelastic scattering and ^{6}He data. The residual energy spectra of α-particles of elastic (top) and inelastic scattering (bottom) events selected at $\theta_{lab}=0^0$ are shown in figure 5. As one can see from the figure, the inelastic contribution is as large as the elastic one and shows some structures.
Although in [1] a similar technique as used in the experiment reported here was adopted, the authors claim that the contribution of inelastic scattering events is negligible. They also claim that the locus observed in [2] was not corresponding to true inelastic scattering events therefore, they suggest the data of [2] are not "a credible source for the discussion"[1]. Our results clearly contradict this conclusion, therefore the analysis performed in [1] has to be revised. It could be that, since the data of [1] are measured at lower beam energy, the inelastic scattering is reduced. However, a comparison of the inclusive excitation function measured in the present experiment with the one of [1] shows a perfect overlap, thus confirming the inclusion of inelastic scattering in the case of the experiment reported in [1].

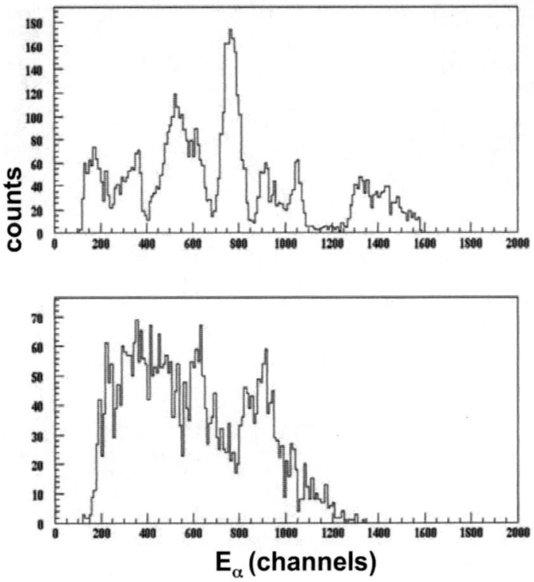

Figure 5. Detected α-particle energy spectra for elastic (top) and inelastic (bottom) scattering events selected at $\theta_{\text{lab}}=0^0$.

4. Conclusions

In this paper details of the Resonant Elastic Scattering experiments ^{4}He(^{9}Li, α) performed at TRIUMF and ^{4}He(^{10}Be, α) performed at LNS, at $\theta_{\text{c.m.}}\sim180^0$, have been discussed.
The first reaction allowed to investigate the excitation function of ^{13}B in the range E_x=14-20 MeV. The excitation function shows two large structures, which are likely to be due to more than two states in ^{13}B. The extraction of the excitation function at all angles, and, the following theoretical analysis, will allow to understand whether the observed structures can be associated to predicted exotic cluster states of ^{13}B of ^{9}Li-^{4}He type.
With the ^{4}He(^{10}Be, α) reaction it was investigated the excitation energy spectrum of ^{14}C in the range E_x=16-24 MeV where it shows the presence of many states. In the present experiment, contrary to findings reported in [1] a large contribution of inelastic scattering events is observed. These processes contribute to a large fraction of the measured alpha production cross-sections. In the light of this result, the R-matrix analysis that led to the determination of the linear-chain rotational band reported in [1], has to be revised.

5. References

[1] H. Yamaguchi *et al.* 2017 *Phys. Lett.* B **766** 11
[2] A. Fritsch *et al.* 2016 *Phys. Rev.* C **93** 014321
[3] H. Horiuchi 2002 *Eur. Phys. J.* **13** 39
[4] Y. Kanada-Enýo and H. Horiuchi 1995 *Phys. Rev.* C **52** 647
[5] T. Suhara and Y. Kanada-Enýo 2010 *Phys. Rev.* C **82** 044301
[6] M. Freer *et al.* 2014 *Phys. Rev.* C **90** 054324

IOP Conf. Series: Journal of Physics: Conf. Series **966** (2018) 012041 doi:10.1088/1742-6596/966/1/012041

On the way to a microscopic derivation of covariant density functionals in nuclei

Peter Ring[1,2]

[1] Physics Department, Technical University of Munich, 85748 Garching, Germany
[2] School of Physics, Peking University, Beijing 100871, People's Republic of China

E-mail: ring@ph.tum.de

Abstract. Several methods are discussed to derive covariant density functionals from the microscopic input of bare nuclear forces. In a first step there are semi-microscopic functionals, which are fitted to ab-initio calculations of nuclear matter and depend in addition on very few phenomenological parameters. They are able to describe nuclear properties with the same precision as fully phenomenological functionals. In a second step we present first relativistic Brueckner-Hartree-Fock calculations in finite nuclei in order to study properties of such functionals, which cannot be obtained from nuclear matter calculations.

1. Introduction

At present, in spite of the important progress made recently in the ab-initio studies of nuclear properties, the majority of heavy nuclei are only accessible through nuclear density functional theory (DFT). Non-relativistic [1] and relativistic [2] versions enable an effective description of the nuclear many-body problem not only for bulk properties, such as binding energies and radii, but also for collective excitations such as rotations and giant resonances, and, by going beyond mean field, for complicated configurations [3] and for sophisticated low-lying spectra in transitional nuclei [4, 5].

At present most of these functionals are purely phenomenological. Of course, one of the main goals in nuclear physics is to build a universal density functional theory based on microscopic calculations [6, 7]. At present, however, attempts to derive such a density functional provide only qualitative results for two reasons: first, the three-body term of the bare interaction is not known well enough and, second, the methods to derive such a functional are not precise enough to achieve the required accuracy. Note that a 1 per mille error in the binding of symmetric nuclear matter leads to an error of several MeV in the binding energy of heavy nuclei, an error which is an order of magnitude larger than required by astrophysical applications. From such considerations it is evident, that successful ab-inito functionals in nuclear physics, at least in foreseeable future will always depend on some phenomenological parameters for fine tuning.

Symmetries play an essential role in the construction of such functionals. One of the underlying symmetries of QCD is Lorentz invariance and therefore covariant density functionals are of particular interest in nuclear physics. This symmetry not only allows to describe the spin-orbit coupling, which has an essential influence on the underlying shell structure, in a consistent way, but it also puts stringent restrictions on the number of parameters in the corresponding functionals without reducing the quality of the agreement with experimental data [2].

Of course, we do not need relativistic kinematics and therefore non-relativistic functionals are very successful too. However, this is only possible at the cost of additional phenomenological parameters connected with the relativistic corrections, as for instance the spin-orbit term. Because of the extremely large relativistic fields and because of the reduced effective mass in these systems such expansions in terms of the velocity do not reach the necessary precision required for such calculations.

2. Phenomenological covariant DFT's

In contrast to non-relativistic functionals of Skyrme [8] or Gogny [9] type, which start from the beginning with roughly a dozen parameters and a simple power-law of $\sim \rho^\alpha$ for the density dependence, relativistic functionals are based on the Walecka model [10] and start with only four parameters, the coupling constants for the three mesons g_σ, g_ω, g_ρ, and the mass m_σ of the σ-meson. These four parameters are clearly connected with basic properties of the effective nuclear force: medium range attraction (g_σ), short range repulsion (g_ω), isospin dependence (g_ω) and the range of the force (m_σ), which determines the surface properties, such as radii in finite nuclei. The details are, however more complicated. As Boguta and Bodmer [11] have pointed out already very early, this linear model fails to describe basic nuclear properties as for instance nuclear compressibility or nuclear deformations [12]. This was the birthday of covariant density functional theory and from that moment all the efforts have been devoted to a better description of the density dependence by additional phenomenological parameters. In a first step non-linear meson couplings have been proposed [11]. Examples are the parameter sets NL3[13] or FSUGold [14]. Modern versions, as for instance DD-ME2 [15] replace the non-linear meson couplings by density dependent coupling constants $g_\sigma(\rho)$, $g_\omega(\rho)$, and $g_\rho(\rho)$. Calculations with finite meson masses are still relatively complicated and therefore modern point-coupling functionals have been developed [16], which include zero-range forces and derivative terms. Again one has essentially only four parameters and their density dependence. All in all one has roughly 10 phenomenological parameters. This seems a similar number as in the non-relativistic case, but there is a difference: in the relativistic models most of the parameters (roughly 6) are used for a sophisticated description of the density dependence, where Skyrme or Gogny use only a power law ρ^α, and even that α is most cases chosen as fixed $\alpha = \frac{1}{3}$ or $\alpha = 1$.

3. Semi-microscopic covariant DFTs

In Coulombic systems an essential input for the derivation of microscopic functionals [17] is the exact numerical solution of the homogeneous electron gas at various densities. Starting from this energy functional $E[\rho]$ additional gradient terms and many other corrections have been added with great success. Therefore it seems to be reasonable to apply a similar concept in nuclear physics [18, 19]. Of course, at present there are no exact solutions of the nuclear matter problem available. One has to rely on approximate solutions, such as sophisticated variational calculations [20] or modern Brueckner-Hartree-Fock methods [21].

The basic idea of microscopic covariant density functionals is therefore to start from RBHF-calculations in nuclear matter [22] of various densities and to map the results to a Walecka model with density dependent parameters [23, 24]. However, this mapping is not unique and therefore there are many open questions until today. In addition, none of these fully microscopic functionals is able reach the accuracy necessary for practical applications.

Much more successful are semi-microscopic functionals, which use microscopic calculations in nuclear matter in order to reduce the number of phenomenological parameters considerably. An example is the parameter set DD-MEδ [25]. Here the density dependence of the corresponding meson-nucleon couplings is adjusted to microscopic calculations in nuclear matter and only four phenomenological parameters $g_\sigma(\rho_{\rm sat})$, $g_\sigma(\rho_{\rm sat})$, $g_\rho(\rho_{\rm sat})$ ($\rho_{\rm sat}$ is the saturation density), and m_σ are fine tuned by a fit to masses and radii of finite nuclei, .

Extended applications of this functional on a global scale [26] have shown that the parameter set DD-MEδ is comparable with all the modern relativistic high precision functionals containing considerably more phenomenological parameters. The deviations between theoretical and experimental binding energies are well below 0.5 %, with some exceptions for light nuclei. It should be emphasized that there are also successful non-relativistic semi-microscopic functionals based on the same concept [19].

We can conclude that, at present, we are able to derive very successful semi-microscopic covariant density functionals [25] from ab-initio results for nuclear matter and neutron matter that contain only few remaining parameters for a fine tuning.

4. The problem of tensor forces

We have to emphasize that all the microscopic input comes at this stage from ab-initio calculations in nuclear matter. On the other hand, there are problems in finite nuclei, where simple density functionals of this type consistently fail. An example are single-particle energies and their distribution. This has been found in an indirect way by global investigations in the area of transitional nuclei, which depend in some cases in a sensitive way on the single particle structure, but also in experimental observations of systematic shifts of single particle energies in isotopic chains. A famous case is the splitting of $1h_{11/2}$ and $1g_{7/2}$ proton configurations in the Sn-region [27], which show a very characteristic behavior as a function of the neutron number. None of the conventional density functionals can reproduce this behaviour on the Hartree level. Only relativistic Hartree-Fock calculations which contain tensor forces in the exchange terms can reproduce this behavior. However, the strength of the tensor has to be fitted in a phenomenological way [28]. Similar results have also been found in density dependent Relativistic Hartree-Fock (RHF) theory [29].

On the other hand, it is an open question whether these shifts in observed single particle energies are an indication of the necessity of effective tensor forces in relativistic density functionals. It is also known that present nuclear density functionals do not describe the exact solutions of the nuclear many-body problem. There are cases, which cannot be taken into account on the mean field level, and where one has to go beyond mean field. An example is particle vibrational coupling which has a considerable influence on the single particle energies. It can also be treated in covariant density functionals [30]. In fact, recently it has been shown by A. Afanasjev and E. Litvinova [31] that a large part of the shifts of $1h_{11/2}$ and $1g_{7/2}$ proton configurations in the Sn-region [27] can be described with the simple functional NL3* including quasiparticle-vibrational coupling.

5. Microscopic calculations in finite nuclei

The concept to derive density functionals from properties of infinite nuclear matter cannot teach us much when it comes to properties of the functionals, which do not show up in nuclear matter calculations as for instance the tensor term in spin non-saturated systems. Here one needs ab-initio calculations in finite nuclei. An obvious way to carry out such investigations is Brueckner theory, the mother of modern density functional theory in nuclei. The advantage of relativistic Brueckner theory is that one does not need three-body forces to get saturation in nuclear matter close to the experimental area. Therefore Relativistic Brueckner-Hartree-Fock (RBHF) theory has been used to describe also finite nuclei [23, 32, 33, 34]. In most of these applications the local density approximation (LDA) has been used, i.e. in a first step the self-energies are calculated in nuclear matter of various densities. In a second step the relativistic Hartree-Fock equations for Walecka-type functionals are solved in finite nuclei with density dependent coupling constants adjusted to the results of nuclear matter calculations at the corresponding density. This provides a mapping of the microscopically obtained nuclear matter results onto RHF-models of Walecka type.

Although such calculations are successful, they cannot reach the accuracy of present day phenomenological covariant density functionals. In particular, the mapping is not unique and therefore the results of different groups deviate form each other considerably. Therefore we started a new project to solve the DBHF equations in large finite basis. First we started in an oscillator basis [35], but then we found we found a large Dirac-Woods-Saxon (DWS) basis more appropriate [36]. There are basis states with positive and negative energies and cut-off-parameters are introduced for the basis, such that convergence is achieved. For details see Ref. [37]

Within relativistic Brueckner-Hartree-Fock (RBHF) theory [38, 39, 40] the relativistic Hartree-Fock (RHF) equations are solved with an effective interaction in the nuclear medium, the so-called G-matrix. Its matrix elements $\langle ab|G(\omega)|cd\rangle$ are determined by the solution of the Bethe-Goldstone (BG) equation in the DWS basis

$$G(\omega) = \bar{V}^N + \bar{V}^N Q_F \frac{1}{\omega - H_0} Q_F G(\omega). \quad (1)$$

Here $\bar{V}^N$ are the anti-symmetrized matrix elements of the relativistic bare force, H_0 are the self-consistent two-particle energies $\epsilon_m + \epsilon_{m'}$ of the RHF-operator at each step of the iteration, and Q_F is the Pauli operator

$$Q_F = \sum_{m<m'} |mm'\rangle\langle mm'|. \quad (2)$$

summing over intermediate states m, m' above the Fermi surface. Since the single particle energies ϵ_m and $\epsilon_{m'}$ refer to the eigenvalues of the RHF-operator, Eq.(1) has to be transformed to the self-consistent relativistic Hartree-Fock basis during the iteration.

Table 1. Ground state properties of ^{16}O obtained with RBHF theory are compared various other approximations and with experiment (from Ref. [41]).

	E (MeV)	r_c (fm)	r_m (fm)	$\Delta E^{ls}_{\pi 1p}$ (MeV)
EXP	−127.6	2.70	2.54(2)	6.3
RBHF, Bonn A	−113.5	2.56	2.42	5.4
RBHF (DWS) [37]	−120.7	2.52	2.38	6.0
DDRHF, PKO1 [29]	−128.3	2.68	2.54	6.4
DDRHF, PKA1 [42]	−127.0	2.80	2.67	6.0
BHF [43], Bonn A	−105.0		2.29	7.5
BHF [44], AV18	−134.2		1.95	13.0
CC [45], N^{3}LO	−120.9		2.30	
NCSM [46], N^{3}LO	−119.7(6)			
NLEFT [47], N^{2}LO	−121.4(5)			

As an application we consider the nucleus ^{16}O. We use the realistic NN interaction Bonn A which has been adjusted to the NN scattering data in Ref. [48]. The ground-state properties are listed in Table 1. The results of our full RBHF calculation are compared with the corresponding experimental data and with several other calculations: BHF are non-relativistic Brueckner calculations. We also show results obtained in RHF-calculations with the phenomenological effective interaction PKO1, which has been fitted to binding energies and charge radii of a set of spherical nuclei. It is seen that the ground-state properties in RBHF theory are improved considerably as compared with the non-relativistic results. Of course the results of the calculations with PKO1 which has been fitted to these data shows only a very small deviation from the experimental values. At present our results show considerable underbinding. The origin for that is not clear and requires further investigations. It could have to do with the potential Bonn A, where the non-relativistic calculations show strong also underbinding, but it could be also caused by the fact, that Brueckner theory is only an approximation where 3-hole lines and higher configurations are neglected. Finally it could be an indication that we need three-body forces, even in the relativistic case.

In Fig. 1 we compare our results in a DWS-basis with non-relativistic BHF calculations [43] and with two versions of LDA [49]. The single-particle energy levels of ^{16}O for protons and neutrons obtained from the full RBHF calculation are plotted in Fig. 2. They are compared with experimental data and (for the protons) with the results of non-relativistic BHF calculations.

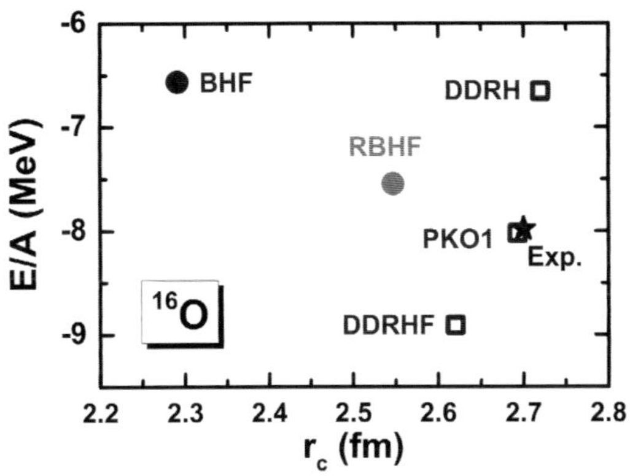

Figure 1. Energy per particle and charge radius of ^{16}O by (relativistic) BHF theories compared with experimental data and other calculations. See text for details (from Ref. [37]).

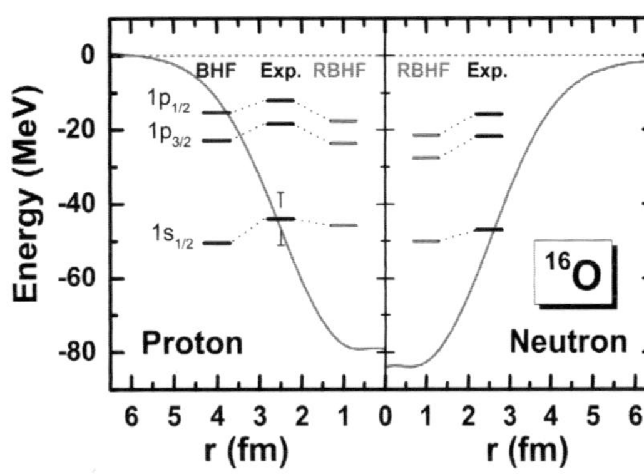

Figure 2. Single-particle spectra for protons and neutrons obtained from the solution of the RHF-equation are compared with experiment. (from Ref. [37]).

6. Conclusions

In summary, the full relativistic Brueckner-Hartree-Fock (RBHF) equations have been solved. The relativistic structure of the two-body matrix elements as well as of the Pauli operator is fully taken into account. The only input is the bare NN-interaction Bonn A adjusted to the scattering phase shifts in Ref. [48]. No other parameter is used. Since nuclear matter calculations within the same framework produce results far away from the Coester line and close to the experimental values of saturation, we neglect at this stage three-body forces.

Despite the good agreement of these results, there is room for improvements. The RBHF-theory presented here is no exact solution of the nuclear many-body problem. So far, rearrangement terms are not taken into account and higher order diagrams in the hole-line expansion are not included.

On the other hand, our method has the potential to investigate heavier nuclei, where exact solutions are impossible, in particular systems without spin saturation and with large neutron excess. In this case we hope to be able to gain a parameter-free, microscopic understanding of open questions in modern phenomenological density functional theories, such as their isospin dependence or the importance of the tensor terms.

Acknowledgments

I would like to express my deep gratitude to all my collaborators, who contributed to the results presented in this contribution, in particular to Anatoli Afanasjev, George Lalazissis, Elena Litvinova, Jie Meng, Takaharu Otsuka, Xavier Roca-Maza, Shihang Shen, and Milena

Serra†. This work was partly supported by the DFG cluster of excellence "Origin and Structure of the Universe" (www.universe-cluster.de).

References

[1] Bender M, Heenen H-P and Reinhard P-G 2003 *Rev. Mod. Phys.* **75** 121
[2] Meng J (Ed.) 2014 *Relativistic Density Functional for Nuclear Structure, Int. Rev. Nucl. Phys.* **10**
[3] Ring P and Litvinova E 2009 *Phys. At. Nucl.* **72** 1285
[4] Nikšić T, Vretenar D, and Ring P 2011 *Prog. Part. Nucl. Phys.* **66** 519
[5] Yao J M, Hagino K, Li Z P, Meng J, and Ring P 2014 *Phys. Rev. C* **89** 054306
[6] Lalazissis G A, Ring P and Vretenar D (Eds.) 2004 *Lecture Notes in Physics* **641**
[7] Drut J, Furnstahl J and Platter L 2010 *Prog. Part. Nucl. Phys.* **64** 120
[8] Vautherin D and Brink D M 1972 *Phys. Rev. C* **5** 626
[9] Dechargé J and Gogny D 1980 *Phys. Rev. C* **21** 1568
[10] Serot B D and Walecka J D 1986 *Adv. Nucl. Phys.* **16** 1
[11] Boguta J and Bodmer A R 1977 *Nucl. Phys. A* **292** 413
[12] Gambhir Y K, Ring P and Thimet A 1990 *Ann. Phys. (N.Y.)* **198** 132
[13] Lalazissis G A, König J and Ring P 1997 *Phys. Rev. C* **55** 540
[14] Todd-Rutel B G and Piekarewicz J 2005 *Phys. Rev. Lett.* **95** 122501
[15] Lalazissis G A, Nikšić T, Vretenar D and Ring P 2005 *Phys. Rev. C* **71** 024312
[16] Nikšić T, Vretenar D and Ring P 2008 *Phys. Rev. C* **78** 034318
[17] Perdew J P and Kurth S 2003 *A Primer in Density Functional Theory*, edited by Fiolhais C, Nogueira F and Marques M A L *(Springer Berlin Heidelberg)* **620** 1
[18] Fayans S 1998 *JETP Lett.* **68** 169
[19] Baldo M, Schuck P and Viñas X 2008 *Phys. Lett. B* **663** 390
[20] Akmal A, Pandharipande V R and Ravenhall D G 1998 *Phys. Rev. C* **58** 1804
[21] Baldo M, Maieron C, Schuck P and Viñas X 2004 *Nucl. Phys. A* **736** 241
[22] Brockmann R and Machleidt R 1990 *Phys. Rev. C* **42** 1965
[23] Brockmann B and Toki H 1992 *Phys. Rev. Lett.* **68** 3408
[24] Hofmann F, Keil C M and Lenske H 2001 *Phys. Rev. C* **64** 034314
[25] Roca-Maza X, Viñas X, Centelles M, Ring P and Schuck P 2011 *Phys. Rev. C* **84** 054309
[26] Agbemava S E, Afanasjev A V, Ray D and Ring P 2014 *Phys. Rev. C* **89** 054320
[27] Schiffer J P et al. 2004 *Phys. Rev. Lett.* **92** 162501
[28] Lalazissis G A, Karatzikos S, Serra M, Otsuka T and Ring P 2009 *Phys. Rev. C* **80** 041301
[29] Long W-H, Van Giai N and Meng J 2006 *Phys. Lett. B* **640** 150
[30] Litvinova E and Ring P 2016 *Int. Rev. Nucl. Phys.* **10** 469
[31] Afanasjev A V and Litvinova E 2015 *Phys. Rev. C* **92** 044317
[32] Fritz R, Müther H and Machleidt R 1993 *Phys. Rev. Lett.* **71** 46
[33] Fritz R and Müther H 1994 *Phys. Rev. C* **49** 633
[34] Shi H L, Chen B Q and Ma Z Y 1995 *Phys. Rev. C* **52** 144
[35] Ring P 2015 *J. Phys. Conf. Ser.* **580** 012005
[36] Zhou S-G, Meng J and Ring P 2003 *Phys. Rev. C* **68** 034323
[37] Shen S H, Hu J N, Liang H Z, Meng J, Ring P and Zhang S Q 2016 *Chin. Phys. Lett.* **33** 102103
[38] Anastasio M R, Celenza L S, Pong W S and Shakin C M 1983 *Phys. Rep.* **100** 327
[39] Brockmann R and Machleidt R 1984 *Phys. Lett. B* **149** 283
[40] ter Haar B and Malfliet R 1987 *Phys. Rep.* **149** 207
[41] Shen S H, Liang H Z, Meng J, Ring P and Zhang S Q 2017 *Phys. Rev. C* **96**, 014316
[42] Long W H, Sagawa H,Van Giai N and Meng J 2007 *Phys. Rev. C* **76** 034314
[43] Müther H, Machleidt R and Brockmann R 1990 *Phys. Rev. C* **42** 1981
[44] Hu B S, Xu F R, Ma Y Z, Wu Q and Sun Z H 2017 *Phys. Rev. C* **95** 034321
[45] Hagen G, Papenbrock T, Dean D J, Hjorth-Jensen M and Asokan B V 2009 *Phys. Rev. C* **80** 021306
[46] Roth R, Langhammer J, Calci A, Binder S and Navrátil P 2011 *Phys. Rev. Lett.* **107**, 072501
[47] Laehde T A et al. 2014 *Phys. Lett. B* **732** 110
[48] Machleidt R 1989 *Adv. Nucl. Phys.* **19** 189
[49] van Dalen E N E and Müther H 2011 *Phys. Rev. C* **84** 024320

IOP Conf. Series: Journal of Physics: Conf. Series **966** (2018) 012042 doi:10.1088/1742-6596/966/1/012042

Realistic Gamow shell model for resonance and continuum in atomic nuclei

F.R. Xu, Z.H. Sun, Q. Wu, B.S. Hu, and S.J. Dai

School of Physics, and State Key Laboratory of Nuclear Physics and Technology, Peking University, Beijing 100871, China

E-mail: frxu@pku.edu.cn

Abstract. The Gamow shell model can describe resonance and continuum for atomic nuclei. The model is established in the complex-moment (complex-k) plane of the Berggren coordinates in which bound, resonant and continuum states are treated on equal footing self-consistently. In the present work, the realistic nuclear force, CD Bonn, has been used. We have developed the full $\hat{Q}$-box folded-diagram method to derive the realistic effective interaction in the model space which is nondegenerate and contains resonance and continuum channels. The CD-Bonn potential is renormalized using the $V_{\text{low-}k}$ method. With choosing ^{16}O as the inert core, we have applied the Gamow shell model to oxygen isotopes.

1. Introduction

Advances in theory and experiment make it possible to study the properties of exotic nuclei far from the stability. However, the description of weakly-bound nuclei is still a challenge to theoretical nuclear physics. Exotic nuclei have low particle thresholds, and cannot be described properly within the standard shell model because of strong coupling to continuum. The weakly-bound property makes dripline nuclei a complex open quantum system (OQS), and therefore the interplay between scattering and bound states should be taken into account explicitly.

The traditional shell model which utilizes the harmonic oscillator (HO) basis does not always provide a good description of loosely-bound nuclei as the HO basis wave functions are well bound and localized. Different from the situation of the HO basis, OQS has small separation energy and large spatial spread, therefore continuum states play an important role. In many cases, the HO basis gives poor convergence for weakly-bound nuclei [1]. Even though super computers have been available for large-scale HO-shell-model calculations, most nuclei around driplines are still unreachable due to huge model dimensions. Methods which unify bound and continuum states are eager.

The importance of continuum has been recognized for a long time. Even in stable nuclei, the continuum can have effects. Tremendous efforts have been devoted to the treatment of the continuum in the framework of multi-configuration interactions, e.g., the continuum shell model (CSM) [2-4], the continuum coupled cluster [5] and the continuum-coupled shell model [6]. The configuration-interaction many-body Gamow shell model (GSM) has also been developed recently [7-10], which can be considered as a generalization of the HO-basis shell model by utilizing the Berggren basis in a complex-energy plane. The Berggren coordinates [11] describes

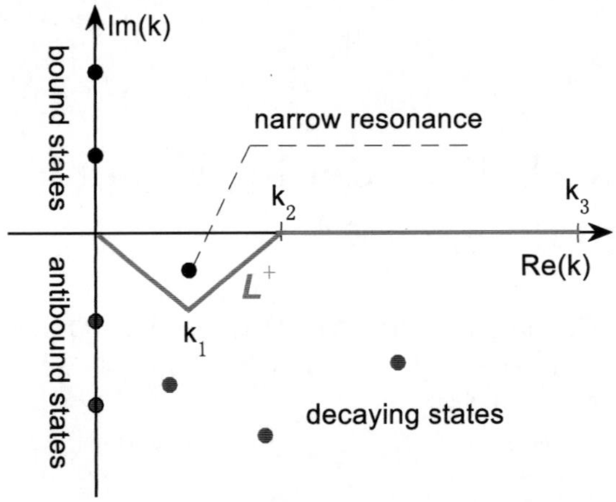

Figure 1. Schematic Berggren complex-k plane.

bound and unbound states in the same framework. With phenomenological interactions, the GSM has been successfully applied to nuclear structure calculations [7-10].

Recently, we are pursuing a GSM calculation based on realistic nuclear forces [12]. The *ab-initio* no-core Gamow shell model (NCGSM) has been proposed [13]. However, the NCGSM is limited to light nuclei due to too large model dimension, e.g., only to light helium isotopes in Ref. [13]. The shell model by taking an inert core can dramatically increase the capability of numerical computations. The core GSM (CGSM) based on realistic forces has been successful [14, 15]. It is a challenge to derive the realistic effective interaction of the model space from a realistic force in the complex-k Berggren basis. Usually, the many-body perturbation theory is adopted to build the interaction, e.g., the $\hat{Q}$-box-plus-folded-diagram method. However, due to the existence of the continuum, the usual degenerate $\hat{Q}$-box approach is inappropriate for GSM. One needs to extend the $\hat{Q}$-box approach to the nondegenerate case. In the recent work [12], we have developed the full $\hat{Q}$-box folded-diagram method which is appropriate for the nondegenerate complex-k space.

2. The core Gamow shell model

In the present CGSM, we choose a doubly magic nucleus as the inert core. For the *sd*-shell nuclei, the ^{16}O is chosen as the core. We take the Woods-Saxon (WS) potential with inclusion of the spin-orbit coupling [16] to provide single-particle basis states in the Berggren complex-k plane. The radial wave functions of the Berggren single-particle states are determined by the time-independent Schrödinger equation,

$$\frac{d^2u(k,r)}{dr^2} = \left[\frac{l(l+1)}{r^2} + \frac{2m}{\hbar}U(r) - k^2\right]u(k,r), \tag{1}$$

where l is the orbital angular momentum of the nucleon motion. The momentum k and wave function $u(k,r)$ can be complex numbers (or function). $U(r)$ is the spherical WS potential generated by the core. We are investigating oxygen isotopes, therefore there is no Coulomb interaction for active valence neutrons. The piecewise perturbation method [17] has been used to solve the Berggren single-particle eigenvalue equation (1). For a resonant state, the energy eigenvalue is a complex number, $\tilde{e}_n = e_n - i\gamma_n/2$, where γ_n stands for the resonance width.

Fig. 1 shows the Berggren complex-k plane. The Berggren single-particle states form a complete set of basis states with discrete bound and resonant states and continuum scattering states [11]. The wave functions of resonant states are not square integrable. The exponential increase and an infinitely oscillating behavior make that the resonant wave function cannot be normalized with conventional techniques. A regularization technique was introduced by Zel'dovich [18] by multiplying a convergence factor $e^{-\epsilon r^2}$ in normalization and taking the limit of $\epsilon \to 0$. In the present work, we use the exterior complex scaling method [19] for the normalizations of resonant states. The scattering states are normalized to Dirac delta [10].

We use the universal WS parameters but with a strength $|V_0|$ reduced by 2.3 MeV to obtain a reasonable $0d_{3/2}$ resonance width compared with data. The WS potential gives the bound $0d_{5/2}$ and $1s_{1/2}$ orbits at the energies of -5.31 and -3.22 MeV, respectively, and a resonant $0d_{3/2}$ at $\tilde{e} = 1.06 - 0.09i$ MeV. The higher orbits $f_{7/2,5/2}$, $ip_{3/2,1/2}$ $(i \geq 1)$, $g_{9/2,7/2}$, $id_{5/2,3/2}$ $(i \geq 1)$ and $is_{1/2}$ $(i \geq 2)$ are continuum states. We limit the $l \leq 4$ orbits for the sd shell-model calculations. $0d_{5/2}$, $1s_{1/2}$ and $id_{3/2}$ $(i = 0, 1, 2 \cdots)$ are taken as our model space. Contributions from other continuum channels as the intermediate states of scatterings are taken into account by performing the $\hat{Q}$-box folded-diagram process. In practical computations, the continuum states on the contour L^+ need to be discretized. We use the Gauss-Legendre quadrature method [10, 14, 20] for the discretization.

The intrinsic A-body Hamiltonian has the following form,

$$H = \sum_{i=1}^{A} \frac{p_i^2}{2m} + \sum_{i<j}^{A} v_{ij} - \frac{\mathbf{P^2}}{2Am}, \tag{2}$$

where v_{ij} is the nucleon-nucleon interaction. p_i is the nucleon momentum in the laboratory coordinate, while $\mathbf{P} = \sum_{\mathbf{i=1}}^{\mathbf{A}} \mathbf{p_i}$ is the center-of-mass (CoM) momentum of the system. In shell-model calculations, usually a one-body potential U is written into the Hamiltonian, which makes calculations more convenient,

$$\begin{aligned} H &= \sum_{i=1}^{A} \left(\frac{p_i^2}{2m} + U \right) + \sum_{i<j} \left(v_{ij} - U - \frac{p_i^2}{2Am} - \frac{\mathbf{p_i} \cdot \mathbf{p_j}}{Am} \right) \\ &= H_0 + V, \end{aligned} \tag{3}$$

with $H_0 = \sum_{i=1}^{A} (\frac{p_i^2}{2m} + U)$ having a one-body form. In the present calculations, U takes the WS potential of the ^{16}O core. V is the residual two-body interaction with corrections from the CoM motion. The final goal is to solve the Schrödinger equation $H|\psi_\nu\rangle = E_\nu|\psi_\nu\rangle$, which is done in the framework of the Gamow shell model.

The Hamiltonian has been intrinsic, but the CoM effect in wave functions has not been removed. The CoM motion can bring spurious states. It is difficult to write wave functions in a relative coordinates for many-body systems, though the few-body Jacobi coordinates and the cluster-orbital shell model have been employed to remove the CoM effects. In the present paper, we calculate the low-lying excited states, therefore the CoM effect should be negligible.

3. Calculations and discussions

The CD Bonn potential [21] is considered to derive the realistic effective interaction for the model space of the sd shell with the ^{16}O core. While $0d_{3/2}$ is a resonant orbit, the remaining part of the $d_{3/2}$ channel is continuum. A resonant basis state plays a special role in the CGSM calculation. Therefore, the channel containing resonant state(s) should be included in the model space. In the present calculations, we choose $\{1s_{1/2}, 0d_{5/2}, 0d_{3/2}\text{-resonance}, d_{3/2}\text{-continuum}\}$ as the model space of the CGSM. The bare CD Bonn potential is softened firstly by using the $V_{\text{low-}k}$ renormalization [22]. We have exploited the full $\hat{Q}$-box folded-diagram method to the Berggren complex-k space [12]. The $\hat{Q}$-box perturbation is to include effects from the core polarization

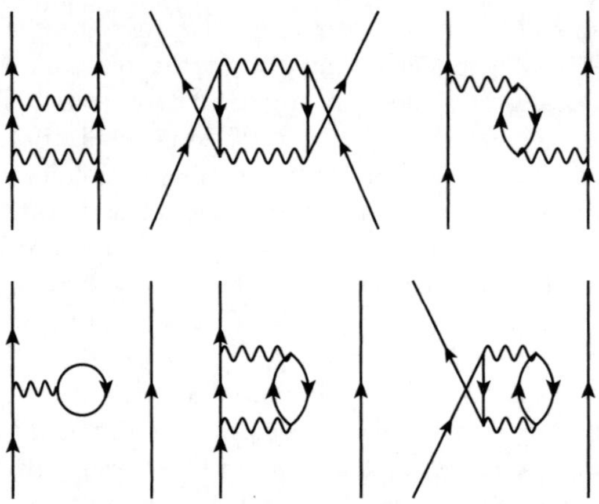

Figure 2. Valence-linked Goldstone diagrams up to the second order of the $\hat{Q}$-box evaluation.

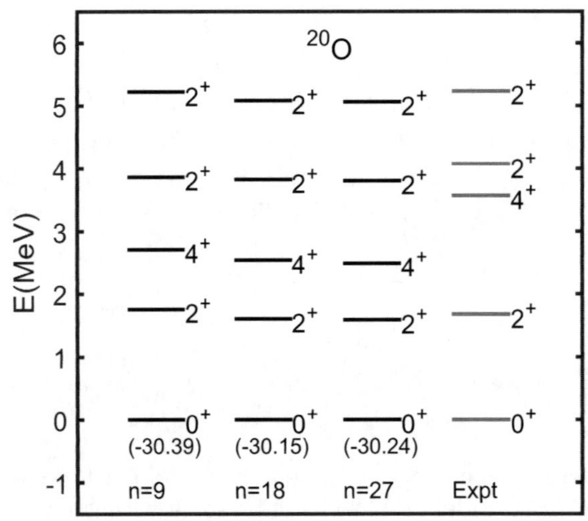

Figure 3. Convergence of spectroscopic calculations against the point number of discretizing the L^+ contour in ^{22}O.

and the excluded space. Fig. 2 displays all the valence-linked diagrams up to the second order in the $\hat{Q}$-box evaluation. The folded-diagram process (i.e., the iteration is used in the extended Krenciglowa-Kuo (EKK) technique [23]) is to consider high-order contributions by summing up the subsets of diagrams to infinite order [24]. The present CGSM space is not degenerate, which includes bound, resonant and continuum orbits. The degenerate $\hat{Q}$-box folded-diagram method should be no longer valid for the nondegenerate Berggren space. At least the continuum channel must cross shells. In the present work, we exploit the EKK method [23] to the complex-k space. The detail can be found in the previous publication [12]. We emphasize that effects from all continuum states are included either by being included as a part of the model space or through the $\hat{Q}$-box folded-diagram process of the effective interaction.

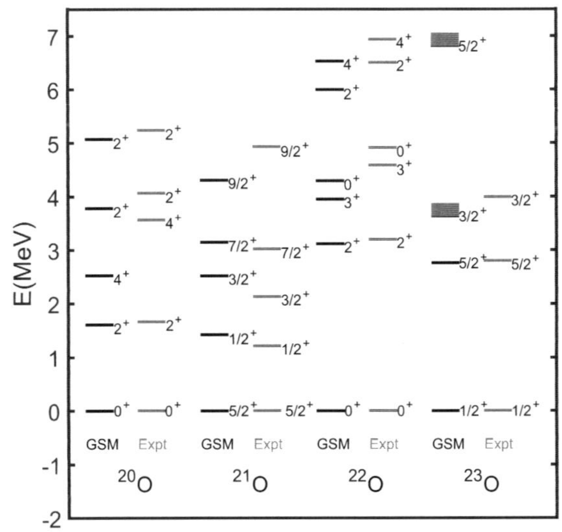

Figure 4. Calculated spectra of $^{20-23}$O, compared with data [30-33]. Resonant states are indicated by shading. $\lambda = 2.6$ fm^{-1} is taken in the $V_{\text{low-}k}$ renormalization.

In practical computations, one has to discretize the continuum scattering states which lie either on the L^+ contour or on the real-k axis, see Fig. 1. We use the Gauss-Legendre method to discretize the contour. An equal number of discretization points is set for each segment of the L^+ contour. Fig. 3 shows the convergence of the CGSM ^{20}O calculations with changing the number of discretization points. We see that 18 discretization points have given a converged result. The convergences as a function of the cutoff Λ of the $V_{\text{low-}k}$ technique and the starting energy of the $\hat{Q}$-box folded diagram have been well tested in our previous calculations [12].

In Fig. 4, the calculated $^{20-23}$O energy spectra are reported, that agree well with experiments. In ^{23}O, the calculations predict two resonant states, $3/2^+_1$ and $5/2^+_2$, at the excitation energies of ~ 3.5 and ~ 6.9 MeV, respectively. The two states are unstable toward neutron emissions with the resonance widths of $\sim 100 - 200$ keV. Oxygen isotopes near the neutron dripline and beyond have been intensely discussed [12]. The loosely-bound properties are well described [12]. However, we should admit that the three-nucleon force (3NF) is missing in the present calculations. Though, from the present calculations, it seems that the 3NF effect on spectra is small and ignorable (at least for the oxygen isotopes), it has been known that 3NF has important effect on binding energy calculations and plays a crucial role in the determination of the dripline [5, 25-29]. In the previous work [12], we have discussed the binding energy calculations and gave improved calculations by taking the experimental single-particle energies. However, the calculations without the inclusion of 3NF cannot give correctly the dripline position of the oxygen chain [12].

4. Summary

In this paper, we present the core Gamow shell model calculations for oxygen isotopes. Nuclear resonance and continuum are treated in a unified framework, named the Berggren complex-k coordinates which describes bound, resonant and scattering states on equal footing. The calculations are based on a realistic nuclear force, the CD-Bonn potential. Due to the nondegenerate property of the Gamow basis, the traditional degenerate $\hat{Q}$-box folded diagram has to be extended to the nondegenerate complex-k space. The bare CD-Bonn potential is renormalized by using $V_{\text{low-}k}$ to soften the hard short-range repulsive core of the strong interaction. ^{16}O is chosen as the inert core for the spectroscopic calculations of oxygen isotopes. The ^{23}O is predicted to have two low-lying resonant states which may be observed in future experiments. The 3NF is not considered in the present GSM calculations, even if we acknowledge that 3NF plays an important role to reproduce correctly for binding energies.

Acknowledgments

This work has been supported by the National Key Basic Research Program of China under Grant No. 2013CB834402; the National Natural Science Foundation of China under Grants No. 11235001, No. 11320101004 and No. 11575007, and the CUSTIPEN (China-U.S. Theory Institute for Physics with Exotic Nuclei) funded by the U.S. Department of Energy, Office of Science under Grant No. DE-SC0009971. We acknowledge the Computer Center of Peking University where partial calculations were done.

References

[1] Michel N, Matsuyanagi K, and Stoitsov M 2008 *Phys. Rev. C* **78** 044319
[2] Bennaceur K, Nowacki F, Okołowicz J, Płoszajczak M 1999 *Nucl. Phys. A* **651** 289
[3] Rotureau J, Okołowicz J, Płoszajczak M 2005 *Phys. Rev. Lett.* **95** 042503
[4] Volya A, Zelevinsky V 2005 *Phys. Rev. Lett.* **94** 052501
[5] Hagen G, Hjorth-Jensen M, Jansen G R, Machleidt R, Papenbrock T 2012 *Phys. Rev. Lett.* **108** 242501
[6] Tsukiyama K, Otsuka T, Fujimoto R 2015 *Prog. Theor. Exp. Phys.* **2015** 093D01
[7] Betan R Id, Liotta R J, Sandulescu N, Vertse T 2002 *Phys. Rev. Lett.* **89** 042501
[8] Michel N, Nazarewicz W, Płoszajczak M, Bennaceur K 2002 *Phys. Rev. Lett.* **89** 042502
[9] Michel N, Nazarewicz W, Płoszajczak M, Okołowicz J 2003 *Phys. Rev. C* **67** 054311
[10] Michel N, Nazarewicz W, Płoszajczak M, Vertse T 2009 *J. Phys. G* **36** 013101
[11] Berggren T 1968 *Nucl. Phys. A* **109** 265
[12] Sun Z H, Wu Q, Zhao Z H, Hu B S, Dai S J, Xu F R 2017 *Phys. Lett. B* **769** 227
[13] Papadimitriou G, Rotureau J, Michel N, Płoszajczak M, Barrett B R 2013 *Phys. Rev. C* **88** 044318
[14] Hagen G, Hjorth-Jensen M, Michel N 2006 *Phys. Rev. C* **73** 064307
[15] Tsukiyama K, Hjorth-Jensen M, Hagen G 2009 *Phys. Rev. C* **80** 051301(R)
[16] Dudek J, Szymański Z, Werner T 1981 *Phys. Rev. C* **23** 920
[17] Ixaru L G, Rizea M, Vertse T 1995 *Comput. Phys. Commun.* **85** 217
[18] Zel′dovich Y B 1961 *Sov. Phys. JETP* **12** 542
[19] Gyarmati B, Vertse T 1971 *Nucl. Phys. A* **160** 523
[20] Liotta R J, Maglione E, Sandulescu N, Vertse T 1996 *Phys. Lett. B* **367** 1
[21] Machleidt R 2001 *Phys. Rev. C* **63** 024001
[22] Bogner S K, Kuo T T S, Schwenk A 2003 *Phys. Rep.* **386** 1
[23] Takayanagi K 2011 *Nucl. Phys. A* **852** 61
[24] Qi C, Xu F R 2008 *Nucl. Phys. A* **800** 47
[25] Otsuka T, Suzuki T, Holt J D, Schwenk A, Akaishi Y 2010 *Phys. Rev. Lett.* **105** 032501
[26] Hergert H, Binder S, Calci A, Langhammer J, Roth R 2013 *Phys. Rev. Lett.* **110** 242501
[27] Bogner S K, Hergert H, Holt J D, Schwenk A, Binder S, Calci A, Langhammer J, Roth R 2014 *Phys. Rev. Lett.* **113** 142501
[28] Jansen G R, Engel J, Hagen G, Navratil P, Signoracci A 2014 *Phys. Rev. Lett.* **113** 142502
[29] Hebeler K, Holt J, Menéndez J, Schwenk A 2015 *Annu. Rev. Nucl. Part. Sci.* **65** 457
[30] Basunia M S 2015 *Nucl. Data Sheets* **127** 69
[31] Elekes Z, et al. 2007 *Phys. Rev. Lett.* **98** 102502
[32] Firestone R B 2015 *Nucl. Data Sheets* **127** 1
[33] Schiller A, et al. 2007 *Phys. Rev. Lett.* **99** 112501

IOP Conf. Series: Journal of Physics: Conf. Series **966** (2018) 012043 doi:10.1088/1742-6596/966/1/012043

Aspects of Shape Coexistence in the Geometric Collective Model of Nuclei

P. E. Georgoudis and A. Leviatan

Racah Institute of Physics, The Hebrew University, 91904, Jerusalem, Israel

E-mail: panos@phys.huji.ac.il, ami@phys.huji.ac.il

Abstract. We examine the coexistence of spherical and γ-unstable deformed nuclear shapes, described by an SO(5)-invariant Bohr Hamiltonian, along the critical-line. Calculations are performed in the Algebraic Collective Model by introducing two separate bases, optimized to accommodate simultaneously different forms of dynamics. We demonstrate the need to modify the β-dependence of the moments of inertia, in order to obtain an adequate description of such shape-coexistence.

1. Introduction

The Geometric Collective Model (GCM) [1] plays a central role in the study of nuclear shapes and, more recently, in providing a convenient framework for incorporating beyond mean-field effects, essential for understanding phase transitions between such shapes. Of particular current experimental and theoretical interest, are first-order transitions involving coexistence of distinct shapes in the same nucleus. In the present contribution, we examine such coexistence of spherical and γ-unstable deformed shapes in the vicinity of the critical-line. In this case, the relevant Bohr Hamiltonian involves a γ-independent potential, supporting two degenerate minima. We employ an algebraic formulation of the GCM, the Algebraic Collective Model (ACM) [2, 3], which makes exact numerical calculations feasible, without recourse to approximations such as β-rigidity and adiabaticty, hence can reveal general capabilities and limitations of the GCM.

2. The Geometric Collective Model with SO(5) symmetry

The Bohr Hamiltonian in the quadrupole variables reads

$$\hat{H} = -\frac{\hbar^2}{2B}\left(\frac{1}{\beta^4}\frac{\partial}{\partial\beta}\beta^4\frac{\partial}{\partial\beta} - \frac{\hat{\Lambda}^2}{\beta^2}\right) + V(\beta,\gamma)\ . \tag{1}$$

Here B is a mass parameter and $\hat{\Lambda}^2 = -\frac{1}{\sin 3\gamma}\frac{\partial}{\partial\gamma}\sin 3\gamma\frac{\partial}{\partial\gamma} + \frac{1}{4}\sum_k \frac{L_k^2}{\sin^2(\gamma - \frac{2}{3}\pi k)}$ is the Casimir operator of SO(5), acting on the γ and three Euler angles Ω. When the potential depends only on β, $V(\beta,\gamma) \mapsto V(\beta)$, the Hamiltonian has SO(5) symmetry and the wave functions can be separated into two parts, $\Psi(\beta,\gamma,\Omega) = f(\beta)\mathcal{Y}_{\tau n_\Delta LM}(\gamma,\Omega)$, satisfying the following equations

$$\hat{\Lambda}^2\mathcal{Y}_{\tau n_\Delta LM}(\gamma,\Omega) = \tau(\tau+3)\mathcal{Y}_{\tau n_\Delta LM}(\gamma,\Omega)\ , \tag{2a}$$

$$\left[-\frac{1}{\beta^4}\frac{\partial}{\partial\beta}\beta^4\frac{\partial}{\partial\beta} + \frac{\tau(\tau+3)}{\beta^2} + v(\beta)\right] f(\beta) = \epsilon f(\beta)\ . \tag{2b}$$

Here $\epsilon = \frac{2B}{\hbar}E$ and $v(\beta) = \frac{2B}{\hbar}V(\beta)$ are the reduced energy and potential, respectively. $\mathcal{Y}_{\tau,n_\Delta,L,M}$ are $SO(5)$ basis states with good $SO(5) \supset SO(3)$ quantum numbers (τ, L), and n_Δ a multiplicity label. By setting $\phi(\beta) = \beta^2 f(\beta)$, Eq. (2b) can be cast in the form of a radial Schrödinger equation

$$-\frac{d^2\phi}{d\beta^2} + \left[\frac{(\tau+1)(\tau+2)}{\beta^2} + v(\beta)\right]\phi = \epsilon\,\phi\ . \tag{3}$$

with an effective τ-dependent potential

$$v^{(\tau)}_{eff}(\beta) = \frac{(\tau+1)(\tau+2)}{\beta^2} + v(\beta)\ . \tag{4}$$

The ACM provides a tractable algebraic scheme for an exact numerical diagonalization of the Bohr Hamiltonian (1), in a basis of SU(1,1) $\times$ SO(5) product wave functions, $R^\lambda_\nu(a\beta)\mathcal{Y}_{\tau n_\Delta LM}(\gamma,\Omega)$. The angular part are the SO(5) spherical harmonics of Eq. (2a), and the radial part are SU(1,1) modified oscillator wave functions given by [2, 3]

$$R^\lambda_\nu(a\beta) = (-1)^\nu\sqrt{\frac{2\nu!a}{\Gamma(\nu+\lambda)}}(a\beta)^{\lambda-1/2}e^{-a^2\beta^2/2}L^{(\lambda-1)}_\nu(a^2\beta^2) \qquad \nu = 0,1,2,\ldots \tag{5}$$

where $L^{(\lambda-1)}_\nu$ is a generalized Laguerre polynomial of oder ν. Any choice of $\lambda > 0$ defines an orthonormal SU(1,1) basis, in which matrix elements of the potential can be evaluated in closed form. For $\lambda = \tau + 5/2$, the above set reduces to the spherical harmonic oscillator basis. In general, the scaling parameter a and λ affect the width and localization of $R^\lambda_\nu(a\beta)$, respectively. An optimal choice of (a,λ), enables a faster convergence as a function of basis size. For γ-independent potentials, with even powers of β, this implies that converged results can be obtained with only a few basis states (small $\nu_{\rm max}$) in the expansion of the radial wave function

$$\phi(\beta; a,\lambda) = \sum_{\nu=0}^{\nu_{max}} c^{(\tau)}_\nu R^\lambda_\nu(a\beta)\ , \tag{6}$$

where the expansion coefficients, $c^{(\tau)}_\nu$, depend on τ. The ACM has so far been tested for potentials with a single minimum [2, 3]. In the present work, we extend this approach to accommodate potentials $v(\beta)$ with multiple minima.

3. Shape coexistence in the GCM with γ-independent potentials

The dynamics associated with γ-independent potentials with a single minimum, has been studied extensively. For a single minimum at $\beta = 0$, the spectrum resembles that of a spherical vibrator, describing quadrupole excitations of a spherical shape. The levels are arranged in n_d-multiplets composed of states with quantum numbers $(n_d = 0,\ \tau = 0,\ L = 0)$, $(n_d = 1,\ \tau = 1,\ L = 2)$, $(n_d = 2,\ \tau = 0,\ L = 0;\ \tau = 2,\ L = 2,4)$ and $(n_d = 3,\ \tau = 3,\ L = 0,3,4,6;\ \tau = 1,\ L = 2)$, in increasing order. For a single minimum at $\beta > 0$, the spectrum resembles that of a γ-unstable deformed roto-vibrator, with β excitations of the deformed equilibrium shape. The ground band is composed of τ-multiplets with quantum numbers $(\tau = 0,\ L = 0)$, $(\tau = 1,\ L = 2)$, $(\tau = 2,\ L = 2,4)$ and $(\tau = 3,\ L = 0,3,4,6)$, in increasing order. The same pattern repeats itself in excited β-bands. The dynamics at the critical-point of a second-order shape-phase transition, where the spherical minimum evolves continuously into a deformed minimum, can be modeled by a flat-bottomed potential. Analytic benchmarks for these three limits of structure are the harmonic spherical vibrator model [4] (similar to the U(5) dynamical symmetry of the interacting boson model (IBM) [5]), the β-rigid Jean-Wilets model [4] (similar to the SO(6) dynamical symmetry of the IBM), and the E(5) critical-point model [6] (an infinite square-well potential), respectively. Characteristic signatures for these solvable limits are listed in Table 1.

IOP Conf. Series: Journal of Physics: Conf. Series **966** (2018) 012043 doi:10.1088/1742-6596/966/1/012043

Table 1. Characteristic properties of yrast states for GCM paradigms: a spherical vibrator [4], E(5) critical-point [6] and γ-unstable deformed rotor [4]. B(E2) values are in units of $B(E2; 2_1^+ \to 0_1^+) = 100$. The E2 operator is proportional to the quadrupole coordinate $\alpha_{2\mu}$.

Observable	Spherical vibrator [U(5)]	E(5)	γ-unstable deformed rotor [SO(6)]
$E(4_1^+)/E(2_1^+)$	2	2.20	2.5
$E(6_1^+)/E(2_1^+)$	3	3.59	4.5
$B(E2; 4_1^+ \to 2_1^+)$	200	168	143
$B(E2; 6_1^+ \to 4_1^+)$	300	221	167

In the present work, we explore the dynamics of shape coexistence in the GCM, with an SO(5)-invariant Bohr Hamiltonian. We focus the discussion to the critical-line of a first-order phase transition involving the coexistence of spherical and γ-unstable deformed shapes. This can modeled by the following sextic potential in the Bohr Hamiltonian

$$v(\beta) \;=\; v_0\beta^2(\beta^2 - \beta_0^2)^2 \;, \tag{7}$$

where $v_0 = \frac{2B}{\hbar}V_0$ is the reduced strength. $v(\beta)$ is an even function of β and is independent of γ, ensuring an SO(5) symmetry for the Hamiltonian. For $v_0 > 0$, it supports two degenerate global minima, spherical ($\beta = 0$) and deformed ($\beta = \beta_0 > 0$), at zero energy. A local maximum at $\beta = \frac{1}{\sqrt{3}}\beta_0$ creates a barrier of height $B_h = \frac{4}{27}v_0\beta_0^6$, separating the two minima. The two wells are asymmetric with different stiffness, $v''(\beta{=}\beta_0) = 4v''(\beta{=}0) = 8v_0\beta_0^4$. The barrier-height and potential stiffness are related, both increasing linearly with v_0 and as a power-law with β_0. The spherical well is finite, while the deformed well has a finite interior boundary and an exterior wall rising as β^6 for $\beta \to \infty$. Both wells become wider for higher energy.

Finite-N aspects of the dynamics in such potentials have been studied [7] in the framework of the IBM, based on a compact U(6) spectrum generating algebra. Here we examine the analogous dynamics in the GCM, which has an inherent non-compact algebraic structure. The sextic potential in question, does not belong to a class of solvable nor quasi-solvable potentials, hence the radial equation (3) necessitates a numerical solution. For that purpose, we employ the ACM approach. To facilitate the identification and convergence of different types of states, we diagonalize the Bohr Hamiltonian with $v(\beta)$, Eq. (7), in two different $SU(1,1) \times SO(5)$ bases

$$\Psi_s \;=\; \beta^{-2}\phi_s(\beta; a_s, \lambda_s)\mathcal{Y}_{\tau n_\Delta LM}(\gamma,\Omega) \;, \tag{8a}$$

$$\Psi_d \;=\; \beta^{-2}\phi_d(\beta; a_d, \lambda_d)\mathcal{Y}_{\tau n_\Delta LM}(\gamma,\Omega) \;. \tag{8b}$$

Here the radial wave functions are defined as in Eq. (6), and are characterized by different parameters $(a_s, \lambda_s) \neq (a_d, \lambda_d)$. In particular, the basis Ψ_s of Eq. (8a) [Ψ_d of Eq. (8b)] is appropriate for spherical [deformed] type of states, associated with the spherical [deformed] minimum, and consequently, $\lambda_s << \lambda_d$. The convergence is verified by examining the energies, B(E2) values and orthogonality of the calculated states, Ψ_s and Ψ_d.

4. Evolution of structure along the critical-line

The spectrum and eigenstates of the Bohr Hamiltonian is governed by the competition between the potential and kinetic-rotational terms. The potential term is attractive and tends to localize the wave functions in the vicinity of its minima. In general, "deepening" a potential lowers the energies of levels confined within it, while "narrowing" a potential raises the level energies. The

IOP Conf. Series: Journal of Physics: Conf. Series **966** (2018) 012043 doi:10.1088/1742-6596/966/1/012043

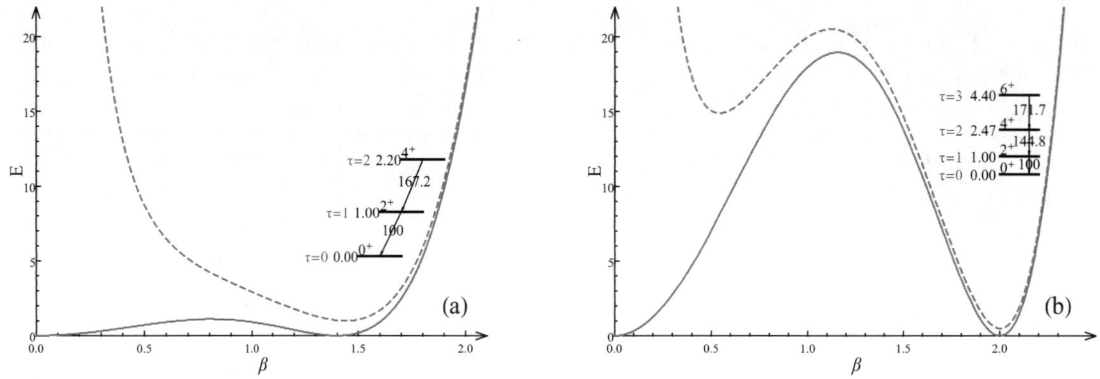

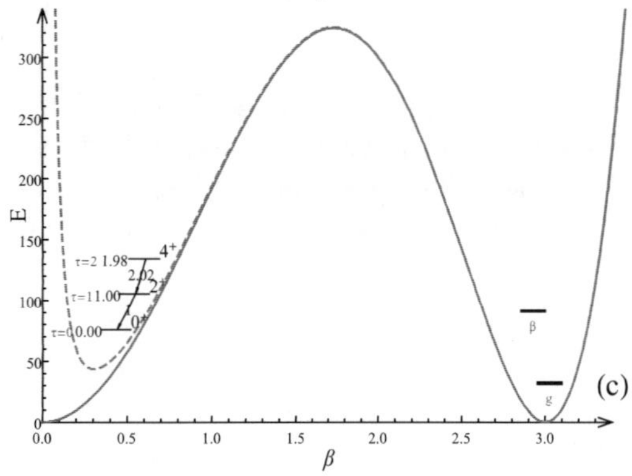

Figure 1. Spectrum of the Bohr Hamiltonian overlayed on the potentials $v(\beta)$, Eq. (7), (blue solid lines) and $v_{eff}^{(\tau=0)}(\beta) = v(\beta) + 2/\beta^2$, Eq. (4), (grey dashed lines), with parameters (a) $(\beta_0, v_0) = (1, 1.4)$, (b) $(\beta_0, v_0) = (2, 2)$, (c) $(\beta_0, v_0) = (3, 3)$, corresponding to low, intermediate and high barriers, respectively. Absolute energies are in units of $\frac{\hbar}{2B} = 1$. Numbers to the left of the drawn levels, indicate relative energies.

kinetic term, $\frac{1}{\beta^2}\hat{\Lambda}^2$, is repulsive and tends to delocalize the wave-functions. Its strength depends on τ and the $1/\beta^2$ dependence energetically penalizes small β values and tends to "push" the wave function towards larger β. This affects strongly the spherical minimum in the effective potential $v_{eff}^{(\tau)}(\beta)$, Eq. (4), shifting its position to larger values of β and distorting its shape. In contrast, the kinetic term has a marginal effect on the shape of the deformed well, yet it governs the rotational splitting of states in the associated ground and excited bands. These effects are seen clearly in Fig. 1, displaying the energy spectrum overlayed on the potentials $v(\beta)$ and $v_{eff}^{(\tau=0)}(\beta)$, for representative values of (v_0, β_0). Fig. 1(a) corresponds to the case of a low-barrier. The energy levels are well above the barrier and experience essentially a flat bottomed potential. The spectrum resembles that of the E(5) critical-point model (see Table 1 for comparison). Higher levels show the spectral character of a β^6 potential. The kinetic term present in the effective potential, completely washes out the spherical minimum. Fig. 1(b) corresponds to the case of an intermediate-barrier. Here a spherical minimum develops in $v_{eff}^{(\tau=0)}(\beta)$ but is shifted to higher energy and is distorted by the kinetic term to such an extent, that it does not support bound states. The spectrum consists only of deformed type of states, forming a ground band with and SO(6)-like structure. Fig. 1(c) corresponds to the case of a high barrier. Here both the spherical and deformed minima are well developed and support bound states localized in their vicinity. The deformed states comprise the ground and β bands. Their rotational splitting, shown in Fig. 2, exhibits an SO(6)-like character. The spherical states are arranged in n_d-multiplets and exhibit a U(5)-like structure.

5. Limitation of the standard GCM Hamiltonian and a possible resolution

Figs. 1(c)-2, demonstrate a limitation of the standard GCM Hamiltonian, namely, very different energy scales for the rotational and vibrational excitations. A pronounced coexistence of many

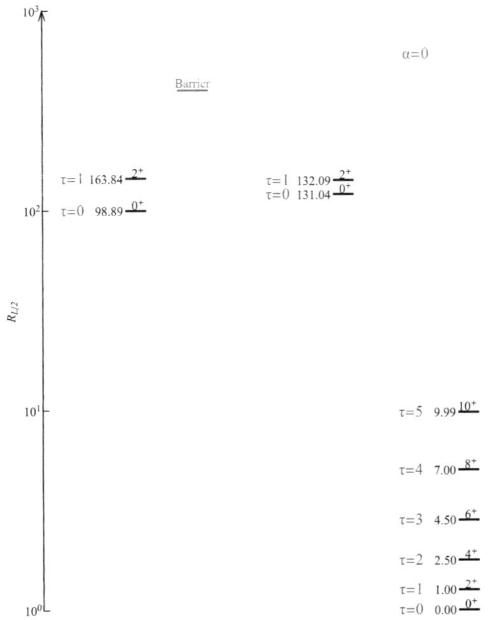

Figure 2. Selected levels in the ground band (right), β-band (middle) and spherical states (left) for the Bohr Hamiltonian with $(\beta_0, v_0, \alpha) = (3, 3, 0)$. Energies are in units of $R_{L/2} = \frac{E(L)-E(0_1^+)}{E(2_1^+)-E(0_1^+)}$.

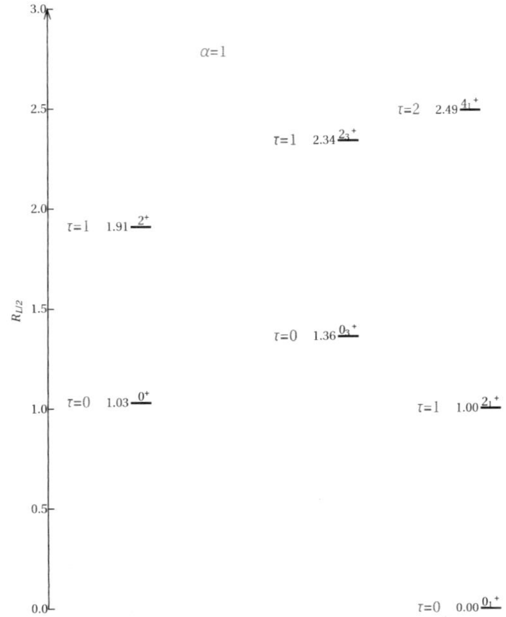

Figure 3. As in Fig. 2, but for a Bohr Hamiltonian with $(\beta_0, v_0, \alpha) = (3, 3, 1)$ in Eqs. (7) and (9). Note the linear energy scale as opposed to the logarithmic energy scale in Fig. 2.

spherical and deformed states requires the two wells to be deep with a high-barrier in-between. For the potential under study, this necessitates a large deformation β_0. The states localized within the deformed potential have an average deformation $\langle\beta^2\rangle$ of order β_0^2, increasing with τ, hence experience a small rotational splitting of order $\tau(\tau+3)/\beta_0^2$, with noticeable centrifugal stretching. For the same β_0, the stiffness of the deformed minimum is large, hence the β-bandhead energy is high, of order $\epsilon_\beta \approx 4\beta_0^2 v_0$. The kinetic term shifts the position of the spherical minimum to a small but non-zero value of β, hence the lowest spherical 0_s^+ state experiences a shift of order $2/\langle\beta^2\rangle$ to higher energy. For the example considered in Figs. 1(c) and 2, $E(2_1^+) - E(0_1^+) = 0.45$, while $\frac{E(0_\beta^+)-E(0_1^+)}{E(2_1^+)-E(0_1^+)} = 131.06$ and $\frac{E(0_s^+)-E(0_1^+)}{E(2_1^+)-E(0_1^+)} = 98.90$, *i.e.*, the resulting rotational and vibrational scales differ by more than two-orders of magnitude. Such a difference is at variance with the experimentally observed patterns of coexistence in nuclei.

The origin of the problem can be traced to the quadratic β-dependence of the irrotational moments of inertia in the kinetic term, $\frac{1}{\beta^2}\hat{\Lambda}^2$. A possible resolution is to allow a departure from such a β^2 behavior. In the present study, we consider the following substitution

$$\frac{1}{\beta^2}\hat{\Lambda}^2 \to \frac{(1+\alpha\beta^2)^2}{\beta^2}\hat{\Lambda}^2 \ . \tag{9}$$

This particular choice is motivated by previous studies within the GCM [8, 9] and the classical limit of the IBM [10]. For $\alpha \neq 0$, this introduces an additional term, $\alpha(2+\alpha\beta^2)\tau(\tau+3)$, in the radial equation, which affects only the levels with $\tau \neq 0$. Typical spectrum and E2 decay pattern, obtained with $\alpha = 1$, are shown in Figs. 3-4. Many states now occur below the barrier, with

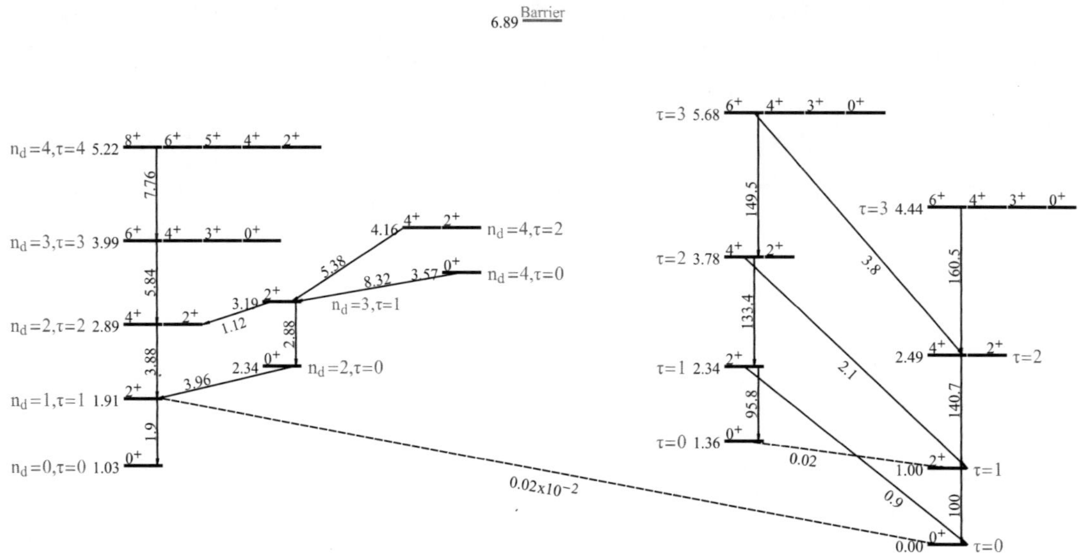

Figure 4. Spectrum of the Bohr Hamiltonian with $(\beta_0, v_0, \alpha) = (3, 3, 1)$ in Eqs. (7) and (9). Converged results for spherical (left side) and deformed (right side) type of states, are obtained by employing $SU(1,1) \times SO(5)$ bases, Eq. (8), with $(a_s, \lambda_s) = (4, 2.5)$ and $(a_d, \lambda_d) = (4.1, 140.5)$, respectively. $L = 0^+_1$, 0^+_3 are the bandhead states of the ground and β bands, with $\sqrt{\langle\beta^2\rangle} =$ 2.97, 2.93, respectively. $L = 0^+_2$ is the spherical ground state with $\sqrt{\langle\beta^2\rangle} = 0.41$.

comparable rotational and vibrational scales. The spherical states exhibit a U(5)-like structure, with strong $\Delta n_d = \pm 1$ E2 decays. The deformed states (comprising the ground and β-bands) maintain an SO(6)-like structure, with strong (weak) intra-band (inter-band) transitions. The E2 decays between the spherical and deformed type of states are extremely weak, reflecting the impact of the high barrier. The ability of a single Hamiltonian to accommodate simultaneously states with different symmetry character, reinforces the view that partial symmetries can play a role in the phenomena of shape coexistence [7, 11, 12].

Acknowledgments

This work is supported by the Israel Science Foundation Grant No. 586/16. P.E.G. acknowledges the Golda Meir Fellowship Fund for partial support. We thank M.A. Caprio for assistance in using his ACM numerical code.

References

[1] Bohr A and Mottelson B R 1975 *Nuclear Structure* (MA: Benjamin Reading)
[2] Rowe D J and Turner P S 2005 *Nucl. Phys.* A **753** 94
[3] Rowe D J, Welsh T A and Caprio M A 2009 *Phys. Rev.* C **79** 054304
[4] Rowe D J and Wood J L 2010 *Fundamentals of Nuclear Models* (Singapore: World Scientific)
[5] Iachello F and Arima A 1987 *The Interacting Boson Model* (Cambridge: Cambridge University Press)
[6] Iachello F 2000, *Phys. Rev. Lett.* **85** 3580
[7] Leviatan A and Gavrielov N 2017 *Phys. Scr.* **92** 114005
[8] Bonatsos D, Georgoudis P E, Lenis D, Minkov N and Quesne C 2011 *Phys. Rev.* C **83** 044321
[9] Georgoudis P E 2014 *Phys. Lett.* B **731** 122
[10] Bonatsos D, Minkov N and Petrellis D 2015 *J. Phys.* G **42** 095104
[11] Leviatan A 2007 *Phys. Rev. Lett.* **98** 242502
[12] Leviatan A and Shapira D 2016 *Phys. Rev.* C **93** 051302(R)

IOP Conf. Series: Journal of Physics: Conf. Series **966** (2018) 012044 doi:10.1088/1742-6596/966/1/012044

Quartetting in even-even and odd-odd N=Z nuclei

M Sambataro[1] and N Sandulescu[2]

[1] Istituto Nazionale di Fisica Nucleare - Sezione di Catania, Via S. Sofia 64, I-95123 Catania, Italy

[2] National Institute of Physics and Nuclear Engineering, P.O. Box MG-6, Magurele, Bucharest, Romania

E-mail: michelangelo.sambataro@ct.infn.it, sandulescu@theory.nipne.ro

Abstract. We report on a microscopic description of even-even $N = Z$ nuclei in a formalism of quartets. Quartets are four-body correlated structures characterized by isospin T and angular momentum J. We show that the ground state correlations induced by a realistic shell model interaction can be well accounted for in terms of a restricted set of $T = 0$ low-J quartets, the $J = 0$ one playing by far a leading role among them. A conceptually similar description of odd-odd self-conjugate nuclei is given in terms of two distinct families of building blocks, one formed by the same $T = 0$ quartets employed for the even-even systems and the other by collective pairs with either $T = 0$ or $T = 1$. Some applications of this formalism are discussed for nuclei in the *sd* shell.

1. Introduction

A distinctive feature of self-conjugate nuclei is that of carrying an equal number of protons and neutrons distributed over the same single particle orbits. In these nuclei, owing to the charge independence of the nuclear interaction, the isovector proton-neutron (*pn*) pairing is expected to come into play on equal footing as the like-particle proton-proton and neutron-neutron pairing of the more common $N > Z$ nuclei. In addition, *pn* pairing is also expected to occur in an isoscalar form. The competition between these two types of *pn* pairing in $N = Z$ nuclei has been matter of great debate in recent years [1].

We have recently carried out an analysis of the *pn* pairing in even-even $N = Z$ nuclei both in the isovector and in the isoscalar channels [2, 3, 4, 5]. This analysis has evidenced, on the one hand, that a description of the ground state correlations induced by this interaction in terms of a condensate of collective pairs (of various form [5]) is not satisfactory and, on the other hand, that these correlations can be accounted for to a high degree of precision by approximating the ground state as a product of identical $T = 0$ quartets. $T = 0$ quartets are four-body correlated structures formed by two protons and two neutrons and, in the case of a spherical mean field, they are also characterized by a total angular momentum $J = 0$. We have also explored a more sophisticated approximation which consists in letting the quartets to be all distinct and we have verified that it leads to basically exact results in the case of *pn* isovector pairing in deformed systems [3]. In all cases the quartets have been constructed variationally for each nucleus.

The *pn* pairing is for sure a key ingredient of the nuclear force for $N = Z$ nuclei but it is nonetheless only a part of it. In the presence of a full Hamiltonian, other quartets are reasonably expected to come into play besides the $T = 0, J = 0$ ones emerging from the analysis of the *pn* pairing. In Ref. [6] we have extended the approach of Refs. [2, 3, 4, 5] to realistic even-even

$N = Z$ nuclei and focused our attention not only on the ground state but also on excited states. The spectra of these nuclei have been constructed by carrying out configuration interaction calculations in spaces built in terms of selected sets of collective quartets of all possible isospins ($T = 0, 1, 2$). It has been evidenced that only the $T = 0$ quartets play an effective role in the structure of the low-lying states of an even-even $N = Z$ nucleus.

Quartetting is espected to play a key role also in the structure of odd-odd $N = Z$ nuclei. These nuclei exhibit a peculiar coexistence of isospin $T = 0$ and $T = 1$ states at very low excitation energies. In Ref. [7] we have shown that the basic features of the spectra of these nuclei can be reproduced by resorting to two distinct families of building blocks, one formed by collective $T = 0$ quartets and the other by collective pairs with either $T = 0$ or $T = 1$ isospin. In this contribution we will summarize the basic results that have been achieved for both even-even and odd-odd nuclei.

2. Quartetting in even-even nuclei

We work in a spherically symmetric mean field and label the single-particle states by $i \equiv \{n_i, l_i, j_i\}$, where the standard notation for the orbital quantum numbers is used. The quartet creation operator is defined as

$$Q^+_{\alpha,JM,TT_z} = \sum_{i_1 j_1 J_1 T_1} \sum_{i_2 j_2 J_2 T_2} C^{(\alpha)}_{i_1 j_1 J_1 T_1, i_2 j_2 J_2 T_2} \times \left[[a^+_{i_1} a^+_{j_1}]^{J_1 T_1} [a^+_{i_2} a^+_{j_2}]^{J_2 T_2} \right]^{JT}_{MT_z}, \quad (1)$$

where a^+_i creates a fermion in the single particle state i and $J(T)$ and $M(T_z)$ denote, respectively, the total angular momentum(isospin) and the relative projections. No restrictions on the intermediate couplings $J_1 T_1$ and $J_2 T_2$ are introduced in the calculations. The criterion adopted for the selection of the quartets has been that of choosing, as representative of the quartets with a given isospin T, those describing the lowest levels with that isospin in nuclei with four active particles outside the inert core of reference [6]. For applications within the *sd* shell, the inert core is represented by ^{16}O and the nuclei which have therefore been considered for the definition of the quartets are $^{20}_{10}\mathrm{Ne}_{10}$, $^{20}_{9}\mathrm{F}_{11}$ and $^{20}_{8}\mathrm{O}_{12}$. The lowest states of these nuclei are characterized by T =0, 1 and 2, respectively. Each of these states therefore identifies a quartet with given T,J (and a projection $T_z = (N - Z)/2$). We have performed configuration interaction calculations in spaces built in terms of these selected sets of quartets. We have also explored reductions of these sets to identify the most relevant quartets in the structure of the nuclei under investigation. The USDB interaction [8] has been employed throughout the calculations.

We start our analysis by examining the self-conjugate nucleus ^{24}Mg. In Fig. 1, we compare the experimental spectrum of this nucleus with that resulting from a shell model (SM) calculation and with the spectrum obtained in the quartet model (QM) when all the selected T=0, 1, 2 quartets are taken into account. The quartet approach is seen to reproduce well both the SM ground state correlation energy and the SM excited states up to an energy of about 9 MeV. The SM is in turn able to fit well the experimental levels.

Having verified that the selected sets of T=0, 1, 2 quartets are sufficient to describe the low-energy spectrum of ^{24}Mg, it becomes of interest to investigate the role of the different quartets. We begin by focusing on isospin. In Fig. 1, on the right hand side, we show the theoretical spectrum obtained in the quartet formalism when only T=0 quartets are retained. On top of the lowest levels we also show the overlaps with the corresponding SM eigenstates. One can see that this spectrum does not exhibit relevant differences with respect to the full QM calculation and that the above overlaps are pretty large, what confirms the good quality of the QM($T = 0$) wave functions. This result provides a clear evidence of the marginal role played by the T=1 and T=2 quartets in the structure of these states.

As a next step, we concentrate on the ground state of ^{24}Mg by employing only $T = 0$ quartets. In Fig. 2, we show how the error in the correlation energy of this state, relative to the SM result,

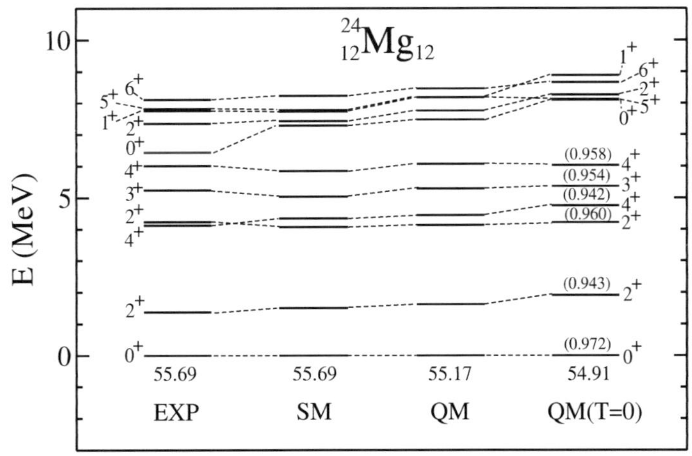

Figure 1. Spectrum of ^{24}Mg obtained in the quartet model (QM) compared to experimental data (EXP) and shell model (SM) results. QM(T=0) denotes the results obtained only with T=0 quartets; the numbers on top of these levels are the overlaps between the QM and SM eigenfunctions. The number below each spectrum gives the ground state correlation energy, namely the difference between the total ground state energy and the energy in the absence of interaction.

varies by reducing, one quartet at a time and starting from the highest one in energy, the set of T=0 quartets. The correlation energy remains basically unchanged up to the point where only the lowest J=0,2,4 quartets are left. From this point on further reductions in the set of quartets lead to significant variations in the energy. The overlaps between the SM and QM ground states which are shown in the same figure exhibit a behavior which is consistent with that of the energies. Thus, these calculations indicate that the T=0 quartets with J=0,2,4 play a major role in the ground state of ^{24}Mg. Among these quartets, the T=0, J=0 quartet is by far the one which contributes most to the correlation energy since, as it can be seen in Fig. 2, an approximation in terms of only this quartet accounts for about 94% of the total energy. A very similar conclusion has emerged also from an analysis of the ground state of the three-quartet system ^{28}Si [6]. It is also worthy noticing that the role of the T=0, J=0 quartet can be further increased in a significant way if this is constructed variationally for each nucleus [9].

3. Quartetting in odd-odd nuclei

The starting point of our analysis of odd-odd $N = Z$ nuclei is the assumption that a state with isospin T can be represented as a superposition of products of one or more $T = 0$ quartets and one extra pair with isospin T. Thus, within this scheme, the total isospin of the state coincides with that of the pair. Based on the conclusions of the previous analysis on even-even nuclei, we involve in the calculations only three $T = 0$ quartets, namely the $J = 0, 2, 4$ quartets describing the lowest three states of ^{20}Ne. In analogy, as collective pairs with isospin T, we assume those describing the lowest three states with that isospin in ^{18}F (one proton and one neutron outside

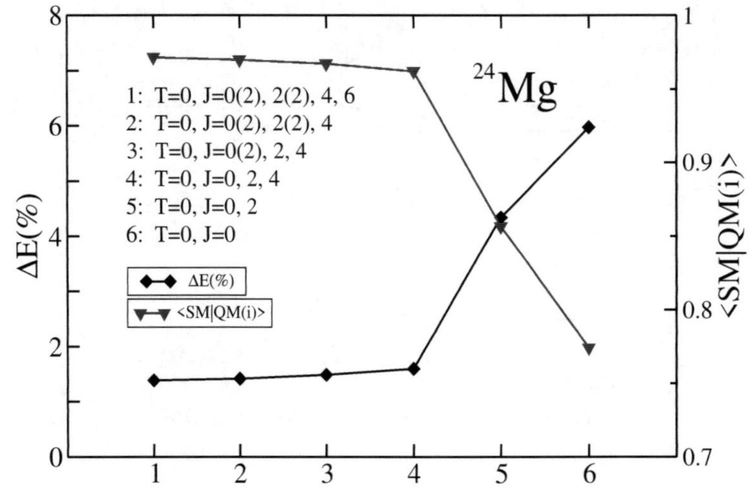

Figure 2. Relative errors (with respect to the SM value) in the ground state correlation energy of ^{24}Mg obtained within the QM with the sets of quartets indicated in the figure. For each set we also show the overlaps between SM and QM wave functions.

the ^{16}O core). These states are characterized by angular momenta $J = 1, 3, 5$ for $T = 0$ and $J = 0, 2, 4$ for $T = 1$. Once fixed, these quartets and pairs are no longer modified. By defining the creation operator for a collective pair as

$$P^{+}_{JM,TT_z} = \sum_{ij} p_{ij} [a^{+}_i a^{+}_j]^{JT}_{MT_z}, \tag{2}$$

where $M(T_z)$ stands for the projection of $J(T)$, we perform configuration interaction calculations in the space (we adopt the m-scheme)

$$\{Q^{+}_{J_1 M_1} P^{+}_{JM,T0}|0\rangle\} \tag{3}$$

for ^{22}Na and in the space

$$\{Q^{+}_{J_1 M_1} Q^{+}_{J_2 M_2} P^{+}_{JM,T0}|0\rangle\} \tag{4}$$

for ^{26}Al, with $|0\rangle$ representing the reference vacuum.

Fig. 3 illustrates the results obtained in the case of ^{26}Al. In this figure we compare the positive-parity experimental spectrum of ^{26}Al up to an energy of about 3 MeV (EXP) with the spectrum obtained with the present quartet formalism (QM) and with the shell model result (SM). The agreement between QM and SM spectra looks satisfactory although one may notice that three $T = 0$ states (with $J = 1, 2, 3$) around 2 MeV are missing in the QM spectrum in comparison with the SM one. This suggests that the inclusion of some extra element either in the set of quartets or in that of $T = 0$ pairs would be welcome. We have also evaluated the overlaps between the lowest QM and SM eigenstates. We find the following values: 0.94 ($J = 5$), 0.94 ($J = 0$), 0.95 ($J = 3$), 0.86 ($J = 1$). These values support the good quality of our approximation scheme.

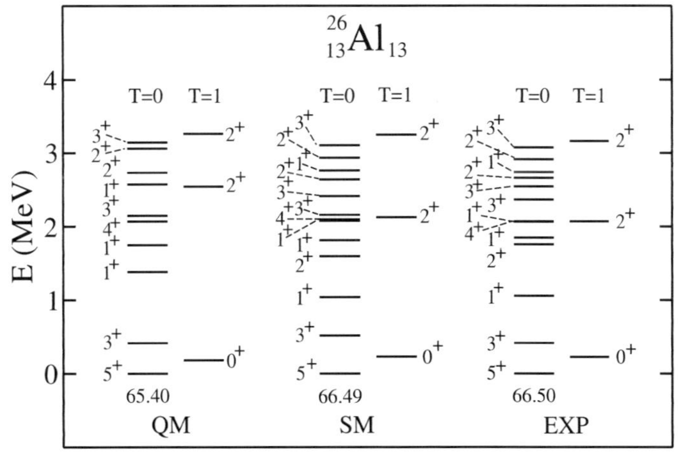

Figure 3. The spectrum of ^{26}Al obtained with the quartet formalism (QM) compared with shell model (SM) and experimental (EXP) spectra. The number below each spectrum is the ground state correlation energy (see Fig. 1).

4. Conclusions

We have reported on a description of both even-even and odd-odd $N = Z$ nuclei in the sd shell in a formalism of $T = 0$ quartets. For all the nuclei that have been analyzed the quartet formalism has provided a description of the low-energy spectra comparable in accuracy with that of shell model calculations. This fact confirms the importance of quartet degrees of freedom in $N = Z$ nuclei and validates the present quartet formalism as the appropriate tool for treating them.

Acknowledgments

This work was supported by the Romanian National Authority for Scientific Research, CNCS UEFISCDI, Project Number PN-II-ID-PCE-2011-3-0596.

References

[1] Frauendorf S and Macchiavelli A 2014 *Progr. Part. Nucl. Phys.* **78** 24
[2] Sandulescu N, Negrea D, Dukelsky J and Johnson C W 2012 *Phys. Rev.* C **85** 061303(R)
[3] Sambataro M and Sandulescu N 2013 *Phys. Rev.* C **88** 061303(R)
[4] Sambataro M, Sandulescu N and Johnson C W 2015 *Phys. Lett.* B **740** 137
[5] Sambataro M and Sandulescu N 2016 *Phys. Rev.* C **93** 054320
[6] Sambataro M and Sandulescu N 2015 *Phys. Rev. Lett.* **115** 112501
[7] Sambataro M and Sandulescu N 2016 *Phys. Lett.* B **763** 151
[8] Brown B A and Richter W A 2006 *Phys. Rev.* C **74** 034315
[9] Sambataro M and Sandulescu N 2013 *J. Phys.* G: *Nucl. Phys.* **40** 055107

IOP Conf. Series: Journal of Physics: Conf. Series **966** (2018) 012045 doi:10.1088/1742-6596/966/1/012045

Interplay of break-up and transfer processes in reactions involving weakly-bound systems [1]

Andrea Vitturi[a,b] **and Laura Moschini**[a,b,c,2]

[a] Dipartimento di Fisica e Astronomia, Via Marzolo 8, 35131 Padova, Italy
[b] INFN, Sezione di Padova, Padova, Italy
[c] Depto. de Física Atómica, Molecular y Nuclear, Universidad de Sevilla, Apartado 1065, E-41080 Sevilla, Spain

E-mail: vitturi@pd.infn.it

Abstract. In this note we illustrate some applications of a simple model which has been devised to clarify the reaction mechanism and the interplay of different reaction channels (elastic, inelastic, transfer, break-up) in heavy-ion collisions. The model involves two potential wells moving in one dimension and few active particles; in spite of its simplicity, it is supposed to maintain the main features, the properties and the physics of the full three-dimensional case. Special attention is given to the role of continuum states in reactions involving weakly-bound systems, and different approximation schemes (as first-order or coupled-channels) as well as different continuum discretization procedures are tested. In the case of two active particles the reaction mechanism associated with two-particle transfer and the effect of pairing intearction are investigated.

1. Introduction

One- and two-particle transfer induced by heavy ions are rather complex processes, difficult to describe from the point of view of the underlying nuclear structure as well as the point of view of the reaction mechanism. This latter is essentially due to the coupling of the transfer channels to the other competing and interplaying channels, starting from the elastic to the inelastic, break-up and fusion ones. For these reasons one typically uses the expedient of resorting to a number of reaction models and approximation schemes (such as coupled-channels, first-order approximation, space truncation, limitation to two-body processes, "effective" optical potentials and form factors, etc).

Particularly critical in the case of weakly-bound systems is the treatment of continuum states and the associated procedures of continuum discretization. For nuclear structure models the coupling to continuum states is popularly included by discretization procedures (box, HO, THO, etc), all aiming at providing a set of confined and square-integrable states with positive energy. These procedures are rather successful as long as one is describing structure properties of isolated systems. In the case of reaction processes that naturally involve asymptotic scattering conditions, the popular choice is provided by the CDCC approach, in which the basis states are obtained by properly "bunching" true scattering states.

[1] Work done in collaboration with Antonio Moro and Kouichi Hagino
[2] Present address: Physique Nuclaire et Physique Quantique, C.P. 229, Universit Libre de Bruxelles, B-1050 Bruxelles, Belgium

The control of all the approximations embedded in these approaches is not an easy task. For this reason we will simplify the problem by assuming particles to move just in a one dimension, and projectile to follow a classical fixed trajectory. In spite of the drastic assumption, the problem may maintain the main features, the properties and the physics of the full three-dimensional case. On the other side, due to its simplicity the scattering problem can be in many cases solved "exactly" and therefore serves as a benchmark for all approximation schemes. In addition this simplified scheme can shed some light on the "reaction mechanism", namely on the description of the process in terms of single or repeated action of the external field in a perturbative expansion. In particular our choice allows us to properly treat in a simple way the action of the pairing correlations and therefore to clarify the long-standing issue of their connection with two-particle transfer (or two-particle break-up processes) and of the competition of sequential vs simultaneus process.

Additional applications of the model to one and two-particle systems can be found in Refs.[1, 3]. A similar line of research, in one and three dimensions, has been also successfully carried out by Samarin and collaborators [2].

2. Systems with one active particle

We first consider processes involving just one active particle, initially sitting on a single-particle level of a one-body potential and feeling the action of a second moving potential, as exemplified in Fig. 1.

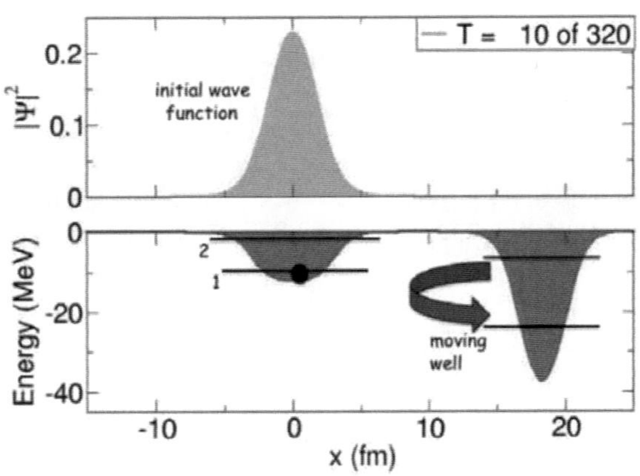

Figure 1. Schematic picture of the process. The upper frame gives the initial wave function probability, while the fixed and moving wells are shown in the lower frame.

Therefore, we have to solve the single-particle time-dependent equation

$$i\hbar\frac{\partial}{\partial t}\Psi(x,t) = \mathcal{H}(x,t)\Psi(x,t) \quad \text{with } \mathcal{H}(x,t) = -\frac{\hbar}{2\mu}\frac{\partial^2}{\partial x^2} + V_T(x) + V_P(x - x_P(t)). \tag{1}$$

The choice of the parameters entering in the calculation will lead to different structural and kinematical conditions, corresponding to rather different physical situations and simulating different bombarding energy regimes, impact parameters, and Q-values for particle transfer: essentially one has to fix the parameters characterizing the two wells V_T and V_P (consequent energies of single-particle states in both potentials), initial condition (selecting one of the single particle state in target potential), and input for the projectile trajectory $x_P(t)$.

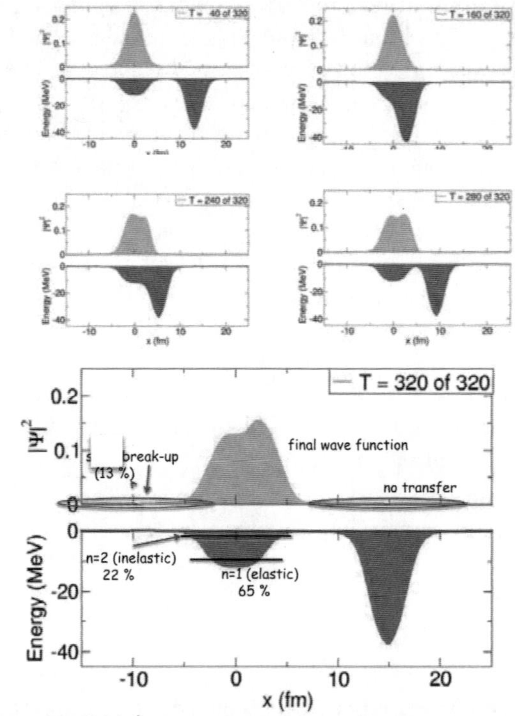

Final population	Exact	First-order	Coupled-channels (only bound states)	Coupled-channels (including Continuum)
n=1 elastic	65 %	71 %	73 %	65 %
n-2 inelastic	22 %	29 %	27 %	22 %
break-up	13 %	≈ 0		13 %

Figure 2. Evolution of the single-particle wave function (and corresponding evolution of the two wells) at different times. The final situation is better shown in the enlarged frame at the bottom of the figure. The parameters for this calculation are given in text. The table at the bottom refers to different models and approximation schemes.

As a first example of the evolution of the wave function we chose a case in which both the target and the projectile potential wells admit two bound levels (with binding energies energies at 2.0 and 9.0 MeV for the fixed target and 27.0 and 9.0 MeV for the moving one). Initially, the particle is sitting in the target ground state and the projectile trajectory is $x_p(t) = x_0 + vt + \frac{1}{2}at^2$; initial energy, reduced mass μ, distance of closest approach x_0 and acceleration[3] are given respectively as 1 MeV, 0.975 amu, 3 fm and $4500/\hbar^2$ fm/ps^2. The evolution of the wave function with time is illustrated in Fig. 2. The different frames refer to different times (the total collision time is divided in 320 steps and the corresponding time is quoted in each frame), and in each frame the upper part gives the square of the one-particle wave function while the lower frame gives the actual position of the two potentials at the same time. At the end of the process, by taking the overlap of the final wave function with the initial wave function and the wave functions of the single-particle states of each well one can determine the probability of elastic,

[3] For this case time is expressed in unit of $\hbar$.

IOP Conf. Series: Journal of Physics: Conf. Series **966** (2018) 012045 doi:10.1088/1742-6596/966/1/012045

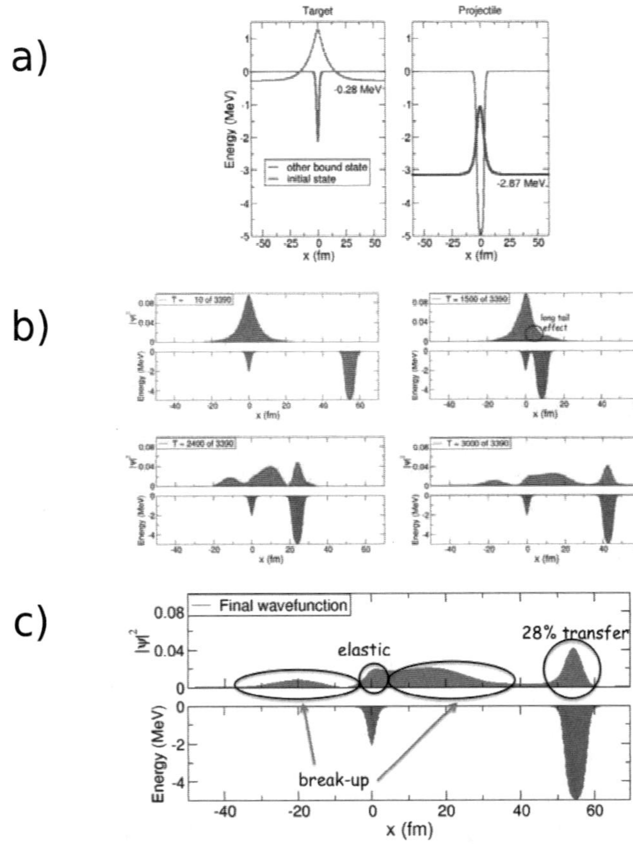

d)

Final population	Exact	CC (with only bound states)	CC (with bound and continuum target states)	CC (with bound and continuum states in both target and projectile)
elastic	20.7 %	95 %	21.4 %	21 %
transfer	28 %	5 %		21 %
break-up	51 %		78.6	57 %

Figure 3. Cf. caption to Fig.2.

inelastic, break-up and of one-particle transfer process. Note that a part of the wave function (although not easily detectable in the figure) appears outside the two wells, indicating a not negligible break-up probability ($\approx 13\%$). On the other side with these kinematical parameters the condition of optimal Q-value seems to inhibit the transfer process.

The table at the bottom of Fig. 2 gives, in comparison with the "exact" values, the values obtained in first order approximation and in a coupled-channels scheme. This is done within the standard time-dependent coupled-channels formalism by constructing the non-diagonal transfer formfactors and expanding the wave function into the dual basis associated with the two wells (cf. Ref. [4]). The continuum is included via a set of pseudostates obtained by discretization procedure in a large box. Results obtained within other procedures are very similar [1].

The "first-order" population of the elastic channel is obtained from the flux conservation as $P_{el} = 1 - \Sigma_i P_i$ in terms of the first-order populations P_i of the different reaction channels. The importance of the full multistep couplings (including continuum, last column) is evidenced by the population of the continuum (break-up) states that, negligible in first order, can only be obtained via the excitation of the weakly-bound excited state.

As a second example we consider a physical situation in which the transfer and excitation to the continuum (breakup) play a more significant role. This is achieved considering as initial state a weakly-bound state of the target well (E_b=-0.28 MeV), as displayed in Fig. 3a. In this case the projectile well follows the trajectory $x_p(t) = x_0 + \sqrt{\rho^2 + (vt)^2} - \rho$ proposed by Ref. [4]; this trajectory differs from the previous one in the fact that the projectile is changing its acceleration over the distance ρ thus simulating the nuclear interaction with the target, in fact at $t = \pm\infty$ the trajectory tends to a uniform motion with zero acceleration. Incident energy, reduced mass μ, ρ and distance of closest approach x_0 are given respectively as 5 MeV, 1.001 amu, 2 fm and 4 fm. The corresponding evolution of the wave function is shown in the different frames of Fig. 3b. The weak-binding situation leads to an initial wave function with a longer tail than in the previous case. As a consequence, part of the wave function is already trasferred to the second well even before the overlap of the two wells (second frame). At the end of the process (cf. Fig. 3c) there is a large transfer probability, but the weak-binding situation has also led to a large fraction of the wave function outside of the two wells, therefore associated to large break-up processes.

The table in Fig. 3d compares the "exact" final probabilities with those obtained within the time-dependent coupled-channels approach. Three different model spaces are compared. In the first calculation we have included only target and projectile bound states. The second includes the target bound levels plus the first 50 continuum pseudostates obtained with the BOX method with a radius equal to 500 fm (this corresponds to an energy cutoff in the continuum of 0.5 MeV). The last (and more complete) calculation uses the target and projectile bound states plus the first 100 pseudostates of target and projectile continuum obtained with the BOX method with a radius of 40 fm (this corresponds to an energy cutoff in the continuum of 300 MeV). It is evident that, even in the simple one-dimensional case, the complexity induced by the different competing channels and the strong role of continuum in the case of weakly-bound systems implies the necessity of a very large model space in order to obtain an acceptable agreement with the exact results.

3. Two-particle transfer processes and pairing interaction

We move now to the case of two-particle sytems and two-particle transfer processes. Starting from the initial two-particle state generated by the fixed well we follow the time evolution of the two-particle wave function due to the action of the moving one-body potential. In addition to the one-body potentials the Hamiltonian includes a residual pairing-like interaction between the two particles, taken as a density-dependent delta interaction (i.e. acting only when the two particles are both inside the same well). Examples of the time evolution are given in Fig. 4, where initial and final wave functions are given as a contour plot as a function of x_1 and x_2. The upper frames refer to the case in which the pairing interaction is switched off (uncorrelated case) and the two particles sit in one single-particle state. From the final wave function we can separate different final states: elastic/inelastic, one-particle transfer, one-particle break-up, two-particle transfer and finally two-particle break-up. In this specific case break-up processes (both one and two-particle) are negligible. The total one-particle probability P_1 amounts to about 40%, while the two-particle transfer probability P_2 amounts to about 4%. Due to the absence of correlations the transfer process is therefore produced by the successive transfer of single particles. In such a situation, in a perturbative approach, we expect a pair transfer probability $P_2 \approx (P_1)^2/4$, which is precisely the value obtained in our calculation.

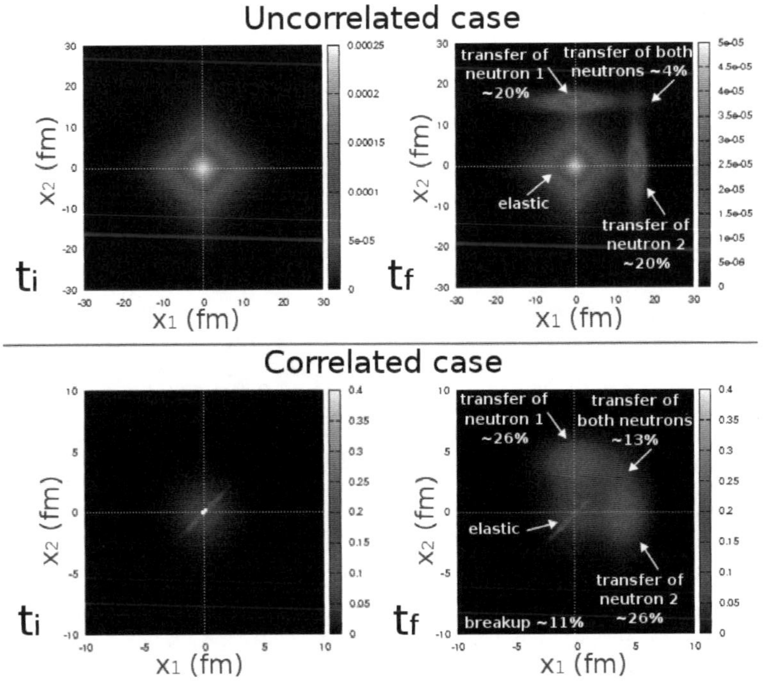

Figure 4. Square of the two-particle wave function shown as a contour plot as a function of x_1 and x_2. The four frames refer to the uncorrelated (upper part) and correlated case (lower part). Initial wave functions in the left column, final ones in the right column.

We switch now to the case with correlations. The corresponding initial and final wave functions are shown in the lower frames of Fig. 4. The initial wave function has been obtained by diagonalizing the residual pairing interaction in the two-particle basis. Continuum states have been included by a discretization procedure (cf. Ref.[5]). Note that due to the correlation the probability of finding both particles on the same side is now clearly favored. The effect of this initial correlation will propagate during the scattering process and affect the final wave function (lower-right frame). At the end of the process one gets a total single-particle probability P_1 equal to 52 % and a pair transfer probability P_2 equal to 13%. This latter value is a factor 2 larger than the uncorrelated estimate $P_2 \approx (P_1)^2/4$. This factor 2 rapresents therefore the enhancement factor due to the pairing correlation. From the point of view of the reaction mechanism, the processes results from the coherent contribution of successive one-particle transfers via the full set of levels in the intermediate one-particle system (continuum states included).

References

[1] Dasso C H and Vitturi A 2007 *Nucl. Phys.* A **787** 476; Dasso C H and Vitturi A 2009 *Phys. Rev.* C **79** 064620; Moschini L 2014 *J. Phys.: Conf. Ser.* **566** 012027; Vitturi A and Moschini L 2015 *J. Phys.: Conf. Ser.* **590** 012007; Hagino K, Vitturi A, Pérez-Bernal F and Sagawa H 2011 *J. Phys. G: Nucl. Part. Phys.* **38** 015105

[2] V.V. Samarin 2015 *Physics of Atomic Nuclei* **78** 128 and references therein

[3] Vitturi A, Moschini L, Hagino K and Moro A 2015 *AIP Conference Proceedings* **1681** 060001; Moschini L 2017, PhD thesis, www.infn.it/thesis/PDF/getfile.php?filename=11604-Moschini-dottorato.pdf

[4] Esbensen H, Broglia R A and Winther A 1983 *Annals of Physics* **146** 149

[5] Moschini L, Perez-Bernal F and Vitturi A 2016 *J. Phys. G: Nucl. Part. Phys.* **43** 045112

Beyond mean-field description of Gamow-Teller resonances and β-decay

Yifei Niu

ELI-NP, Horia Hulubei National Institute for Physics and Nuclear Engineering, 30 Reactorului Street, RO-077125, Bucharest-Magurele, Romania

E-mail: nyfster@gmail.com

Gianluca Colò

Dipartimento di Fisica, Unversità degli Studi di Milano, via Celoria 16, I-20133 Milano, Italy
INFN, Sezione di Milano, via Celoria 16, I-20133 Milano, Italy

Enrico Vigezzi

INFN, Sezione di Milano, via Celoria 16, I-20133 Milano, Italy

Chunlin Bai

School of Physical Science and Technology, Sichuan University, Chengdu 610065, China

Zhongming Niu

School of Physics and Material Science, Anhui University, Hefei 230601, China

Hiroyuki Sagawa

RIKEN, Nishina Center, Wako 351-0198, Japan
Center for Mathematics and Physics, University of Aizu, Aizu-Wakamatsu, Fukushima 965-8580, Japan

Abstract. β-decay half-lives set the time scale of the rapid neutron capture process, and are therefore essential for understanding the origin of heavy elements in the universe. The random-phase approximation (RPA) based on Skyrme energy density functionals is widely used to calculate the properties of Gamow-Teller (GT) transitions, which play a dominant role in β-decay half-lives. However, the RPA model has its limitations in reproducing the resonance width and often overestimates β-decay half-lives. To overcome these problems, effects beyond mean-field can be included on top of the RPA model. In particular, this can be obtained by taking into account the particle-vibration coupling (PVC). Within the RPA+PVC model, we successfully reproduce the experimental GT resonance width and β-decay half-lives in magic nuclei. We then extend the formalism to superfluid nuclei and apply it to the GT resonance in ^{120}Sn, obtaining a good reproduction of the experimental strength distribution. The effect of isoscalar pairing is also discussed.

1. Introduction

The Gamow-Teller (GT) resonance is a prominent collective mode with spin-isospin character [1]. In the GT excitation, the orbital angular momentum of the nucleus is not changed, but the spin and isospin are changed by one unit. GT transitions can be studied in the laboratory by (p,n) [2] or (^{3}He,t)[3] reactions. The low-energy GT transitions lying below the Q_β value also represent the dominant transitions in β-decay. β-decay plays an important role in the rapid neutron capture process (r-process), which is responsible for the production of heavy elements in the universe.

Therefore, much efforts have been made to investigate β-decay half-lives of neutron-rich nuclei. Experimentally, the study of half-lives of neutron-rich nuclei has made much progress due to the development of radioactive beam facilities. Theoretically, self-consistent QRPA model based on Skyrme [4] or relativistic density functionals [5, 6] have been extensively applied in the investigation of β-decay half-lives.

However, the accurate description of β-decay half-lives remains a hard problem. The QRPA model tends to overestimate the half-lives [4, 6]. This problem could be cured to a good extent by including the isoscalar proton-neutron pairing with a free pairing strength parameter. By adjusting this parameter, the half-lives could be reproduced for some open-shell nuclei. However, this does not help at all for closed shell nuclei, like ^{78}Ni and ^{132}Sn [4, 6–8]. Therefore, there must exist other correlations, other than proton-neutron pairing, able to reduce the half-lives. For instance, half-lives can be reduced to some extent by including the effects of the tensor force [9]. However, also the inclusion of tensor terms requires more parameters in the Hamiltonian.

It is well known that the RPA model only considers 1-particle 1-hole (1p-1h) configurations in the model space, so it is interesting to check whether the calculation of lifetime can be improved by including correlations beyond the RPA configuration space (1p-1h states), and considering more complicated states of 2p-2h character. This is also necessary for the description of the experimental spreading width. A direct approach of this kind is represented by the second RPA [10, 11]. However, the associated configuration space, which includes the 2p-2h states, is too large for practical calculations in heavy nuclei. We will instead consider the effects of the particle-vibration coupling (PVC), including 1p-1h states coupled with collective phonons [12]. The self-consistent RPA+PVC model has been developed based on both relativistic and Skyrme density functionals [13–16].

In this paper, we will first present the RPA+PVC model based on Skyrme density functional and its successful application to the GT resonance and to β-decay half-lives in magic nuclei. Then we report on recent progress concerning the extension of RPA+PVC model to the quasiparticle case, i.e., the QRPA + QPVC model, and present some results of GT resonance in superfluid nuclei, including also the effects of isoscalar pairing.

2. RPA+PVC model and its applications

In this section, we will first introduce briefly the formulas of RPA+PVC model, and then we will present the results of GT resonance in ^{208}Pb and β-decay in closed shell nuclei.

2.1. Formulas of RPA+PVC model

The RPA+PVC equation reads

$$\begin{pmatrix} \mathcal{D}+\mathcal{A}_1(\omega) & \mathcal{A}_2(\omega) \\ -\mathcal{A}_3(\omega) & -\mathcal{D}-\mathcal{A}_4(\omega) \end{pmatrix} \begin{pmatrix} F^{(\nu)} \\ \bar{F}^{(\nu)} \end{pmatrix} = (\Omega_\nu - i\frac{\Gamma_\nu}{2}) \begin{pmatrix} F^{(\nu)} \\ \bar{F}^{(\nu)} \end{pmatrix}. \tag{1}$$

We write the RPA+PVC equation in the basis of RPA eigenstates, so $\mathcal{D}$ is a diagonal matrix containing the positive RPA eigenvalues. The $\mathcal{A}_i$ matrices are associated with the coupling to

IOP Conf. Series: Journal of Physics: Conf. Series **966** (2018) 012046 doi:10.1088/1742-6596/966/1/012046

the doorway states. The expressions of $\mathcal{A}_i$ is also written in the RPA basis $|n\rangle$,

$$(\mathcal{A}_1)_{mn} = \sum_{ph,p'h'} W^{\downarrow}_{ph,p'h'}(\omega)X^{(m)}_{ph}X^{(n)}_{p'h'} + W^{\downarrow *}_{ph,p'h'}(-\omega)Y^{(m)}_{ph}Y^{(n)}_{p'h'}, \tag{2}$$

$$(\mathcal{A}_2)_{mn} = \sum_{ph,p'h'} W^{\downarrow}_{ph,p'h'}(\omega)X^{(m)}_{ph}Y^{(n)}_{p'h'} + W^{\downarrow *}_{ph,p'h'}(-\omega)Y^{(m)}_{ph}X^{(n)}_{p'h'}, \tag{3}$$

$$(\mathcal{A}_3)_{mn} = \sum_{ph,p'h'} W^{\downarrow}_{ph,p'h'}(\omega)Y^{(m)}_{ph}X^{(n)}_{p'h'} + W^{\downarrow *}_{ph,p'h'}(-\omega)X^{(m)}_{ph}Y^{(n)}_{p'h'}, \tag{4}$$

$$(\mathcal{A}_4)_{mn} = \sum_{ph,p'h'} W^{\downarrow}_{ph,p'h'}(\omega)Y^{(m)}_{ph}Y^{(n)}_{p'h'} + W^{\downarrow *}_{ph,p'h'}(-\omega)X^{(m)}_{ph}X^{(n)}_{p'h'}, \tag{5}$$

where $W^{\downarrow}$ reads

$$W^{\downarrow}_{ph,p'h'}(\omega) = \sum_N \frac{\langle ph|V|N\rangle\langle N|V|p'h'\rangle}{\omega - \omega_N}. \tag{6}$$

The matrix elements $W^{\downarrow}_{ph,p'h'}$ are given by the sum of the four following terms,

$$\begin{aligned}
W^{\downarrow}_{1php'h'} &= \delta_{hh'}\delta_{j_p j_{p'}} \sum_{p'',nL} \frac{1}{\omega - (\omega_{nL} + \epsilon_{p''} - \epsilon_h) + i\Delta} \frac{\langle p||V||p'',nL\rangle\langle p'||V||p'',nL\rangle}{\hat{j}_p^2}, \\
W^{\downarrow}_{2php'h'} &= \delta_{pp'}\delta_{j_h j_{h'}} \sum_{h'',nL} \frac{1}{\omega - (\omega_{nL} - \epsilon_{h''} + \epsilon_p) + i\Delta} \frac{\langle h''||V||h,nL\rangle\langle h''||V||h',nL\rangle}{\hat{j}_h^2}, \\
W^{\downarrow}_{3php'h'} &= \sum_{nL} \frac{(-)^{j_p - j_{h'} + J + L}}{\omega - (\omega_{nL} + \epsilon_p - \epsilon_{h'}) + i\Delta} \begin{Bmatrix} j_p & j_h & J \\ j_{h'} & j_{p'} & L \end{Bmatrix} \langle p'||V||p,nL\rangle\langle h'||V||h,nL\rangle, \\
W^{\downarrow}_{4php'h'} &= \sum_{nL} \frac{(-)^{j_{p'} - j_h + J + L}}{\omega - (\omega_{nL} + \epsilon_{p'} - \epsilon_h) + i\Delta} \begin{Bmatrix} j_p & j_h & J \\ j_{h'} & j_{p'} & L \end{Bmatrix} \langle p||V||p',nL\rangle\langle h||V||h',nL\rangle.
\end{aligned} \tag{7}$$

In the above formulas, p and h label particle and hole states, respectively. The corresponding angular momentum and single-particle energies are given respectively by j_p, j_h and ϵ_p, ϵ_h. $\hat{j}_i^2$ is a shorthand notation for $2j_i + 1$, while ω_{nL} denotes the energy of the phonon state $|nL\rangle$. The averaging parameter Δ is introduced to avoid singularities in the denominator of Eq. (7).

The GT strength associated with RPA+PVC, is given by

$$S(\omega) = -\frac{1}{\pi}\mathrm{Im}\sum_{\nu}\langle 0|\hat{O}_{\mathrm{GT}^\pm}|\nu\rangle^2 \frac{1}{\omega - \Omega_\nu + i(\frac{\Gamma_\nu}{2} + \Delta)}, \tag{8}$$

where the GT operator is $\hat{O}_{\mathrm{GT}^\pm} = \sum_{i=1}^{A}\sigma(i)\tau_\pm(i)$. In our calculation, we will only focus on the GT^- excitations. More details can be found in Ref. [15, 16].

2.2. GT resonance and β-decay in magic nuclei

Fig. 1 shows the Gamow-Teller strength distributions for ^{208}Pb calculated by RPA and RPA+PVC model with the interaction SkM*, and compared with experimental data [17]. The RPA model could only give discrete peaks, and overestimates the centroid energy of GT resonance. With the inclusion of PVC, the peak energy of GTR is shifted downwards by around 1 MeV, in better agreement with experimental data. At the same time, a spreading width is also developed, and the RPA+PVC result reproduces the experimental lineshape quite well.

We further check if RPA+PVC model could help in reducing the β-decay half-lives. So we present the β-decay half-lives of ^{132}Sn, ^{68}Ni, ^{34}Si, and ^{78}Ni, calculated by RPA and RPA+PVC

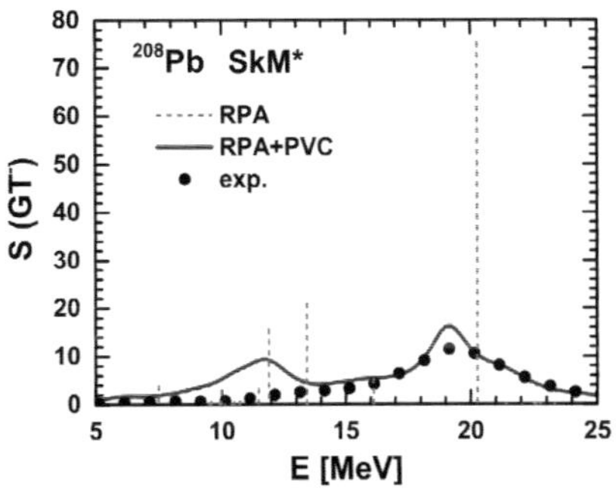

Figure 1. Gamow-Teller strength distributions of ^{208}Pb calculated by RPA and RPA+PVC model with interaction SkM*, in comparison with experimental data [17].

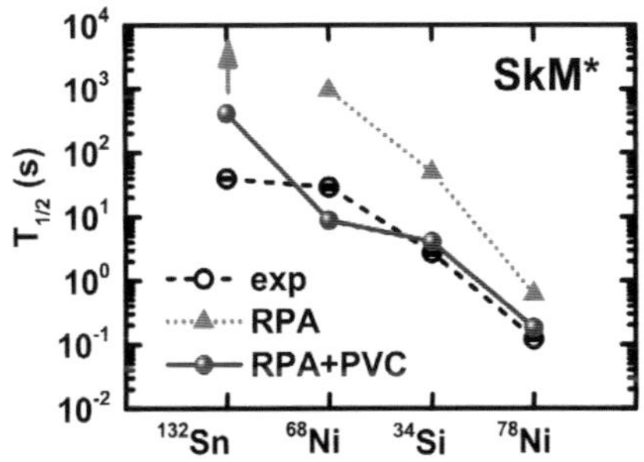

Figure 2. The β-decay half-lives of ^{132}Sn, ^{68}Ni, ^{34}Si, and ^{78}Ni calculated by RPA and RPA+PVC approaches with the interaction SkM*, in comparison with experimental values [18].

approaches with the interaction SkM*, respectively, in comparison with experimental values [18]. The arrow denotes an infinite half-life. The RPA results overestimate the half-lives systematically. In particular, the RPA model predicts an infinite half-life for for ^{132}Sn. With the inclusion of PVC, the half-lives get reduced systematically, and become in reasonable agreement with experimental data. For more results and discussions on half-lives, one can refer to Ref. [19].

In summary, the RPA+PVC model with interaction SkM* gives a good description both of the spreading width and of β-decay half-lives in magic nuclei.

3. QRPA+QPVC model and its applications

In this section, we will first introduce briefly the formulas of QRPA+QPVC model, and then we will present the results of GT resonance in the superfluid nucleus ^{120}Sn.

3.1. Formulas of QRPA+QPVC model

In the case of QRPA+QPVC model, the QRPA+QPVC equation and the $\mathcal{A}$ matrices are the same as Eq. (1) and Eq. (2) in the RPA+PVC model, except that the particle-hole configuration ph is replaced by two-quasiparticle configuration ab. For nuclei not far from the stability line, the BCS quasi-particle states represent a convenient and accurate approximation to the HFB states. So we obtain the following expression for the spreading matrix elements.

$$
\begin{aligned}
W^{\downarrow J}_{1ab,a'b'} &= \delta_{bb'}\delta_{j_aj_{a'}}\frac{1}{\hat{j}_a^2}\sum_{a'',nL}\frac{\langle a||V||a'',nL\rangle\langle a'||V||a'',nL\rangle}{E-[\omega_{nL}+E_{a''}+E_b\pm(\lambda_n-\lambda_p)]+i\Delta},\\
W^{\downarrow J}_{2ab,a'b'} &= \delta_{aa'}\delta_{j_bj_{b'}}\frac{1}{\hat{j}_b^2}\sum_{b'',nL}\frac{\langle b||V||b'',nL\rangle\langle b'||V||b'',nL\rangle}{E-[\omega_{nL}+E_{b''}+E_a\pm(\lambda_n-\lambda_p)]+i\Delta},\\
W^{\downarrow J}_{3ab,a'b'} &= (-)^{j_a+j_b+J}\begin{Bmatrix} j_a & j_b & J\\ j_{b'} & j_{a'} & L\end{Bmatrix}\sum_{nL}\frac{\langle a'||V||a,nL\rangle\langle b||V||b',nL\rangle}{E-[\omega_{nL}+E_a+E_{b'}\pm(\lambda_n-\lambda_p)]+i\Delta},\\
W^{\downarrow J}_{4ab,a'b'} &= (-)^{j_{a'}+j_{b'}+J}\begin{Bmatrix} j_{a'} & j_{b'} & J\\ j_b & j_a & L\end{Bmatrix}\sum_{nL}\frac{\langle a||V||a',nL\rangle\langle b'||V||b,nL\rangle}{E-[\omega_{nL}+E_{a'}+E_b\pm(\lambda_n-\lambda_p)]+i\Delta},
\end{aligned}
\quad (9)
$$

where E_a is the BCS quasi-particle energy, and $\lambda_{n(p)}$ is the chemical potential for neutron (proton). The reduced matrix element has the following form

$$\langle a||V||a'', nL\rangle = \frac{\hat{L}}{\sqrt{1+\delta_{cd}}} \sum_{cd} [\widetilde{V}(cdLa''; a) X_{cd}^{nL} + (-1)^{j_a - j_{a''} + L} \widetilde{V}(cdLa; a'') Y_{cd}^{nL}], \tag{10}$$

where

$$\begin{aligned} \widetilde{V}(cdLa''; a) &= V_{ada''c}^{Lph}(u_a u_{a''} u_c v_d - v_a v_{a''} v_c u_d) + V_{aca''d}^{Lph}(u_a u_{a''} v_c u_d - v_a v_{a''} u_c v_d)(-)^{j_c - j_d + L} \\ &\quad - V_{aa''cd}^{Lpp}(u_a v_{a''} u_c u_d - v_a u_{a''} v_c v_d), \end{aligned} \tag{11}$$

and

$$\begin{aligned} \widetilde{V}(cdLa; a'') &= V_{a''dac}^{Lph}(u_a u_{a''} u_c v_d - v_a v_{a''} v_c u_d) + V_{a''cad}^{Lph}(u_a u_{a''} v_c u_d - v_a v_{a''} u_c v_d)(-)^{j_c - j_d + L} \\ &\quad - V_{a''acd}^{Lpp}(v_a u_{a''} u_c u_d - u_a v_{a''} v_c v_d). \end{aligned} \tag{12}$$

The above matrix elements V are calculated in the canonical basis. v_a is the square root of the occupation probability for the state a in the canonical basis, and $u_a = \sqrt{1 - v_a^2}$. The ph and pp interaction will take the same form as that used for non-charge-exchange QRPA calculation. For more details, see Ref. [20].

3.2. GT resonance in ^{120}Sn

In the QRPA calculation, besides the isovector $T = 1$ pairing both in the ground state and in the residual interaction, the isoscalar $T = 0$ pairing is also included in the residual interaction in the QRPA calculation. We adopt a density-dependent, zero-range surface pairing force in both $T = 1$ and $T = 0$ channel. We multiply the pairing strength of $T = 0$ channel by a constant factor f, and compare the results obtained with the same strength used for $T = 1$ pairing ($f = 1$) or neglecting $T = 0$ pairing ($f = 0$).

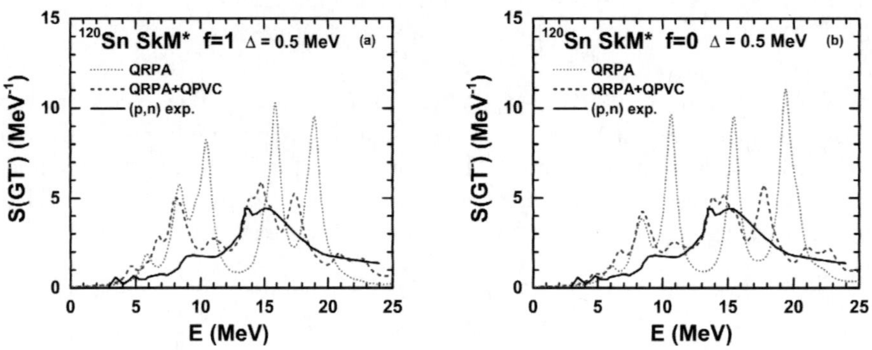

Figure 3. GT strength distributions of ^{120}Sn calculated by QRPA and QRPA+QPVC model using the interaction SkM* with isoscalar pairing ($f = 1$) [panel (a)] and without isoscalar pairing ($f = 0$) [panel (b)], and compared with experimental data [21].

We take ^{120}Sn as an example, and plot the GT strength distributions calculated by QRPA and QRPA+QPVC model using the interaction SkM*. The calculated results with and without isoscalar pairing are shown in panel (a) and (b) respectively. The experimental strength distribution is estimated from the cross section of (p,n) reaction in Ref. [21]. In this work, besides the cross section $\sigma(0^o)$, the unit cross section $\hat{\sigma} = 2.78 \pm 0.16$ mb/sr was also determined. We can obtain an approximate value for the B(GT) strength, using the relation $\sigma(0^o) = \hat{\sigma}F(q,\omega)B(GT)$, and assuming the factor $F(q,\omega)$, that gives the dependence on momentum and energy transfer of cross section, to be constant and equal to 1. Comparing panel (a) and (b), we can see that with the inclusion of the attractive isoscalar pairing the strength in the low-energy region increases; and the strength is redistributed in favour of the lower ones of the two peaks in GR region in the QRPA level. For the QRPA+QPVC results, the profile of the strength function in the giant resonance region is similar in the $f = 0$ and $f = 1$ cases, although the strength of the peaks in the low-energy region is increased and the strength of the highest peak is decreased with the inclusion of isoscalar pairing. With the inclusion of QPVC effect, the spreading width and lineshape of the giant resonance region are very well reproduced. However, the low-lying strength is still overestimated, even neglecting isoscalar pairing.

4. Conclusions

We have reported recent progress made on the study of GT resonance and β-decay within a theoretical framework beyond mean-field. Firstly, the self-consistent RPA+PVC model based on Skyrme density functional has been presented and applied to the study of GT strength distribution and β-decay half-lives in magic nuclei. The model reproduces well the spreading width of GTR in ^{208}Pb with the Skyrme interaction SkM*. β-decay half-lives for ^{132}Sn, ^{68,78}Ni and ^{34}Si are also well described using the same Skyrme interaction. Then we have presented the extension of the model to the calculation of superfluid nuclei, i.e., the self-consistent QRPA+QPVC model, and its application to the GT strength distribution in ^{120}Sn. The QPVC also produces a large spreading width and the model reproduces the experimental data of GTR region, using the same Skyrme interaction SkM*. The effect of isoscalar pairing has also been discussed.

References

[1] Osterfeld F 1992 *Rev. Mod. Phys* **64** 491 – 557
[2] Ichimura M, Sakai H and Wakasa T 2006 *Prog. Part. Nucl. Phys.* **56** 446–531
[3] Fujita Y, Rubio B and Gelletly W 2011 *Prog. Part. Nucl. Phys.* **66** 549–606
[4] Engel J, Bender M, Dobaczewski J, Nazarewicz W and Surman R 1999 *Phys. Rev. C* **60**(1) 014302
[5] Marketin T, Huther L and Martínez-Pinedo G 2016 *Phys. Rev. C* **93**(2) 025805
[6] Niu Z M, Niu Y F, Liang H Z, Long W H, Nikšić T, Vretenar D and Meng J 2013 *Phys. Lett. B* **723** 172 – 176
[7] Nikšić T, Marketin T, Vretenar D, Paar N and Ring P 2005 *Phys. Rev. C* **71**(1) 014308
[8] Marketin T, Vretenar D and Ring P 2007 *Phys. Rev. C* **75**(2) 024304
[9] Minato F and Bai C L 2013 *Phys. Rev. Lett.* **110**(12) 122501
[10] Drożdż S, Nishizaki S, Speth J and Wambach J 1990 *Phys. Rep.* **197** 1 – 65
[11] Gambacurta D, Grasso M and Catara F 2011 *Phys. Rev. C* **84**(3) 034301
[12] Bertsch G F, Bortignon P F and Broglia R A 1983 *Rev. Mod. Phys.* **55**(1) 287–314
[13] Litvinova E, Brown B A, Fang D L, Marketin T and Zegers R G T 2014 *Phys. Lett. B* **730** 307 – 313
[14] Marketin T, Litvinova E, Vretenar D and Ring P 2012 *Phys. Lett. B* **706** 477 – 481
[15] Niu Y F, Colò G, Brenna M, Bortignon P F and Meng J 2012 *Phys. Rev. C* **85**(3) 034314
[16] Niu Y F, Colò G and Vigezzi E 2014 *Phys. Rev. C* **90**(5) 054328
[17] Wakasa T, Okamoto M, Dozono M, Hatanaka K, Ichimura M, Kuroita S, Maeda Y, Miyasako H, Noro T, Saito T, Sakemi Y, Yabe T and Yako K 2012 *Phys. Rev. C* **85**(6) 064606
[18] Audi G, Kondev F G, Wang M, Huang W J and Naimi S 2017 *Chin. Phys. C* **41** 030001
[19] Niu Y F, Niu Z M, Colò G and Vigezzi E 2015 *Phys. Rev. Lett.* **114**(14) 142501
[20] Niu Y F, Colò G, Vigezzi E, Bai C L and Sagawa H 2016 *Phys. Rev. C* **94**(6) 064328
[21] Sasano M, Sakai H, Yako K, Wakasa T, Asaji S, Fujita K, Fujita Y, Greenfield M B, Hagihara Y, Hatanaka K, Kawabata T, Kuboki H, Maeda Y, Okamura H, Saito T, Sakemi Y, Sekiguchi K, Shimizu Y, Takahashi Y, Tameshige Y and Tamii A 2009 *Phys. Rev. C* **79**(2) 024602

IOP Conf. Series: Journal of Physics: Conf. Series **966** (2018) 012047 doi:10.1088/1742-6596/966/1/012047

Quasiparticle formulation of a multiphonon method and its application to the ^{20}O nucleus

P. Veselý[1], G. De Gregorio[1,2,3], F. Knapp[4], N. Lo Iudice[2,3]

[1]Nuclear Physics Institute, Czech Academy of Sciences, 250 68 Řež, Czech Republic
[2]Dipartimento di Fisica, Universitá di Napoli Federico II, Naples, Italy
[3]INFN Sezione di Napoli, Naples, Italy
[4]Faculty of Mathematics and Physics, Charles University, Prague, Czech Republic

E-mail: p.vesely@ujf.cas.cz

Abstract. We outline briefly the quasiparticle formulation of a microscopic multiphonon approach known as equation of motion phonon method and discuss its application to the neutron rich semi-magic ^{20}O. Our calculation emphasizes the crucial role of the two-phonon configurations, responsible for establishing an almost one to one correspondence between theoretical and experimental energy levels and for fragmenting the dipole strength, in qualitative agreement with the data. In particular, the fragmentation yields two dipole levels below the neutron threshold which correspond to the two peaks recently observed.

1. Introduction

The equations of motion phonon method (EMPM) was developed few years ago [1, 2, 3]. It derives and solves iteratively a set of equations of motion to generate an orthonormal basis of multiphonon states built of phonons obtained in Tamm-Dancoff approximation (TDA). The basis so obtained is then adopted to diagonalize the residual Hamiltonian in a multiphonon space. The formalism treats one-phonon as well as multiphonon states on the same footing, takes into account the Pauli principle, and holds for any Hamiltonian.

The method was originally derived in the particle-hole scheme and applied to describe the dipole response in the heavy neutron rich nuclei ^{208}Pb [4] and ^{132}Sn [5]. Recently, it was formulated in terms of Bolgoliubov quasiparticles and applied to the open-shell ^{20}O [6]. Among the neutron rich oxygen isotopes, ^{20}O is one of the most extensively studied.

Several experiments have yielded a rich spectrum [7, 8, 9, 10]. Electromagnetic excitations induced by heavy-ion collisions have determined the dipole cross section over a wide energy range [11], while a virtual-photon scattering experiment have detected two 1^-levels below the neutron decay threshold [12, 13]. These two excitations were further investigated in two recent experiments using (^{20}O $+\alpha$) and (^{20}O + Au) probes [14].

The low-lying spectra of ^{20}O were investigated theoretically within the coupled-cluster (CC) theory [15] and an extended shell-model using chiral two- and three-body interactions up to third order [15, 16, 17, 18]. The dipole strength distribution was studied within the phenomenological Shell-Model [19], relativistic RPA [20], time-dependent HF (TDHF) [21], and RPA plus PVC [22]. The objective of these studies was to establish if the low-lying peaks are a manifestation of the pygmy dipole resonance (PDR) or are simply single particle excitations.

Both spectrum and dipole response are investigated here in a unified fashion within the EMPM.

2. EMPM in the quasiparticle formalism

The main goal of the EMPM is to generate recursively a basis of n-phonon ($n = 0, 1, 2, ..n..$) states $\mid \alpha_n\rangle$, of energy E_{α_n}, having the form

$$\mid \alpha_n\rangle = \sum_{\lambda\alpha} C^{\alpha_n}_{\lambda\alpha_{n-1}} \mid (\lambda \times \alpha_{n-1})^{\alpha_n}\rangle = \sum_{\lambda\alpha} C^{\alpha_n}_{\lambda\alpha_{n-1}} \left\{O^\dagger_\lambda \times \mid \alpha_{n-1}\rangle\right\}^{\beta_n}, \quad (1)$$

where

$$O^\dagger_\lambda = \sum_{r\leq s} c^\lambda_{rs} Z^\lambda_{rs} = -\sum_{r\leq s} c^\lambda_{rs}\zeta_{rs}(\alpha^\dagger_r \times \alpha^\dagger_s)^\lambda \quad (2)$$

is the quasiparticle TDA phonon operator of energy E_λ acting on the $(n-1)$-phonon states $\mid \alpha_{n-1}\rangle$ of energies $E_{\alpha_{n-1}}$, assumed to be known. The normalization factor $\zeta_{rs} = (1 + \delta_{rs})^{-1/2}$ has been used.

We start with constructing the equations of motion

$$< \beta||[H, O^\dagger_\lambda]||\alpha >= (E_\beta - E_\alpha) < \beta||O^\dagger_\lambda||\alpha >, \quad (3)$$

where the subscript n was omitted for simplicity. After expanding the commutator and expressing the quasiparticle operators Z^λ_{rs} in terms of the phonon operators by inverting Eq. (2), we obtain [6] the generalized eigenvalue equations

$$\sum_{\lambda'\alpha'\lambda_1\alpha_1} \left[(E_\lambda + E_\alpha)\delta_{\lambda\lambda_1}\delta_{\alpha\alpha_1} + \mathcal{V}^\beta_{\lambda\alpha,\lambda_1\alpha_1}\right]\mathcal{D}^\beta(\lambda_1\alpha_1, \lambda'\alpha')C^\beta_{\lambda'\alpha'} = E_\beta \sum_{\lambda'\alpha'} \mathcal{D}^\beta(\lambda\alpha, \lambda'\alpha')C^\beta_{\lambda'\alpha'}. \quad (4)$$

Here

$$\mathcal{D}^\beta(\lambda\alpha, \lambda'\alpha') = \langle(\lambda \times \alpha)^\beta \mid (\lambda' \times \alpha')^\beta\rangle \quad (5)$$

is the overlap or metric matrix and $\mathcal{V}^\beta$ is a phonon-phonon potential of the simple form

$$\mathcal{V}^\beta(\lambda\alpha, \lambda_1\alpha_1) = \sum_\sigma W(\beta\lambda_1\alpha\sigma; \alpha_1\lambda)\mathcal{V}^\sigma_{\lambda\alpha,\lambda_1\alpha_1}, \quad (6)$$

where W is a Racah coefficient and

$$\mathcal{V}^\sigma_{\lambda\alpha,\lambda_1\alpha_1} = \frac{1}{2}\sum_{rtsq} \rho_{\lambda\lambda_1}([r \times t]^\sigma)\mathcal{V}^\sigma_{rtsq}\rho^{(n-1)}_{\alpha\alpha_1}([s \times q]^\sigma). \quad (7)$$

The quantities $\rho_{\lambda\lambda_1}([r \times t]^\sigma)$ and $\rho^{(n-1)}_{\alpha\alpha_1}([s \times q]^\sigma)$ denote, respectively, the TDA ($n = 1$) and the $n-$phonon ($n > 1$) density matrices [6], while $\mathcal{V}^\sigma_{rtsq}(22)$ are the quasiparticle matrix elements. Their expressions can be found in Ref. [6].

The generalized eigenvalue problem (4) is formulated in terms of the over-complete set of states $\mid (\lambda \times \alpha_{n-1})^{\alpha_n}\rangle$. A procedure based on the Cholesky decomposition method enables us to extract a basis of linearly independent eigenstates and to turn the singular eigenvalue equation (4) into a regular one [1]. This new equation is solved iteratively starting from the TDA phonons $\mid \alpha_1 = \lambda\rangle$ to yield an orthonormal multiphonon basis $\{\mid 0\rangle, \mid \alpha_1\rangle, ..., \mid \alpha_n\rangle...\}$.

The basis so obtained can be adopted to solve the eigenvalue equations in the multiphonon space

$$\sum_{n'\beta_{n'}} \left[(E_{\alpha_n} - \mathcal{E}_\nu)\delta_{nn'}\delta_{\alpha_n\beta_{n'}} + \mathcal{V}_{\alpha_n\beta_{n'}}\right]\mathcal{C}^{(\nu)}_{\beta_{n'}} = 0, \quad (8)$$

where the potential

$$\mathcal{V}_{\alpha_n \beta_{n'}} = \delta_{n'(n+1)} \mathcal{V}_{\alpha_n \beta_{n'}} + \delta_{n'(n+2)} \mathcal{V}_{\alpha_n \beta_{n'}} \tag{9}$$

couples subspaces differing by one and two phonons for a two-body Hamiltonian.

The solution of the above equations yields the final eigenfunctions

$$| \Psi_\nu \rangle = \sum_{\alpha_n} C^{(\nu)}_{\alpha_n} | \alpha_n \rangle, \tag{10}$$

where $| \alpha_n \rangle$ are the correlated n-phonon states of the form (1).

3. Calculations and results

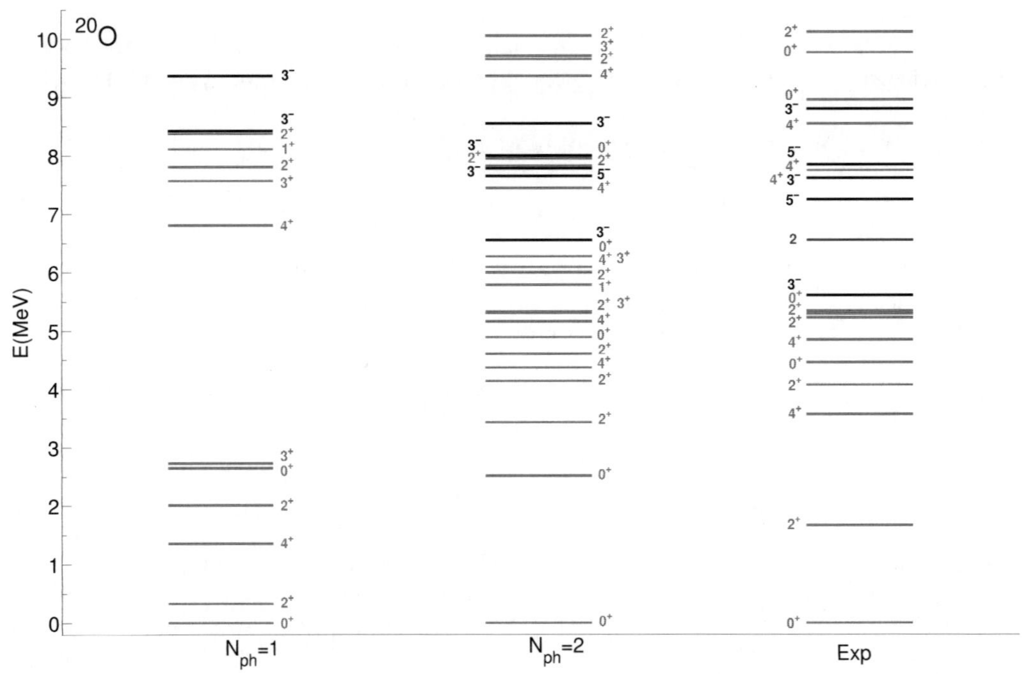

Figure 1. Theoretical versus experimental energy levels in ^{20}O (from [6]).

We used the nucleon-nucleon optimized chiral potential NNLO$_{\text{opt}}$ whose parameters were determined by a new optimization method which minimizes the effects of the three-nucleon force [23].

A canonical HFB basis [24] was generated in the configuration space of 11 harmonic oscillator major shells up to the principal quantum number $N_{\max} = 10$. The TDA phonons were determined in smaller configuration space which included up to the pfh major shell. Their energy and structure did not change when the full space up to $N_{\max} = 10$ was used. A procedure exploiting the Gramm-Schmidt orthogonalization method enabled us to generate phonons free of the spurious admixtures induced by the center of mass motion and the particle-number violation [24].

We built a multiphonon basis up to $n = 2$ and included all two-phonon configurations fulfilling the condition $E_{\lambda_1} + E_{\lambda_2} \leq 30$ MeV. This cut-off was fixed so as to reproduce roughly the first excited 1^- level. The cut-off affects only the energy of the correlated ground state but leaves unchanged the separation energies between excited levels. It would be necessary to include the three-phonon states in a space large enough to restore the correct separation between excited and ground state energies. These configurations are known to couple strongly to the one-phonon states and to push them down in energy [3]. Though feasible in principle, the inclusion of a large number of three-phonon states would have required unbearably lengthy calculations unless we resorted to some efficient approximations. We plan to do this in the future.

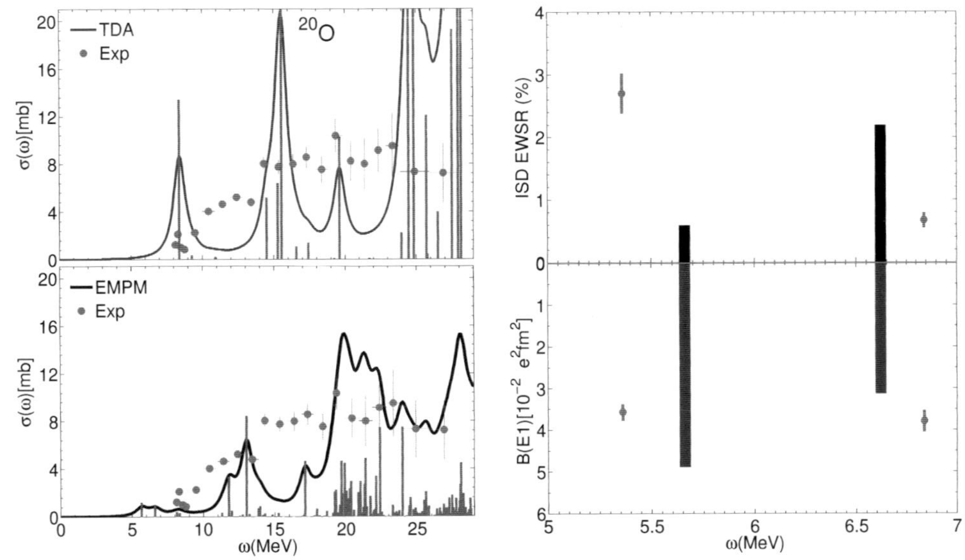

Figure 2. Theoretical and experimental dipole cross sections are plotted on the left. The isoscalar and isovector dipole transition strengths collected by the lowest two 1^- states are compared with recent data on the right.

Theoretical versus experimental [7, 8, 9, 10] spectra are shown in Figure 1. A satisfactory agreement between theory and experiments is achieved only once the two phonons are included. Indeed, they produce a much richer spectrum and redistribute the energies so that an almost one to one correspondence between theoretical and experimental energy levels is achieved.

The two-phonon configurations play a crucial role also in shaping the dipole cross section. As shown in Figure 2, the phonon coupling induces a severe fragmentation of the strength and a consequent quenching and spreading of the single peaks, consistently with the data. Appreciable discrepancies are noticeable, however in the low-energy region, where the experimental cross section is seriously underestimated. In fact, the strength collected by all the states up to ~ 15 MeV accounts only for $\sim 6\%$ of the TRK sum rule, just half the $\sim 12\%$ fraction estimated experimentally [11, 12, 13]. The cross section integrated up to ~ 27 MeV accounts for $\sim 38\%$ of the TRK sum rule to be compared with the $\sim 45\%$ determined experimentally.

Another intriguing effect of the phonon coupling is observed at low-energy. The lowest TDA 1^- peak gets fragmented into several smaller peaks with the appearance of two levels below the neutron decay threshold in agreement with the data obtained in a virtual-photon scattering experiment performed at MSU [12, 13] at around 100 MeV/nucleon. The isovector transition strengths measured by those experiments were reproduced with great accuracy by our calculation[6].

Very recently, a new measurement [14] using the ($^{20}O + \alpha$) and (^{20}O + Au) probes produced for these two levels the isoscalar dipole (ISD) strengths and, in addition, new isovector transition probabilities which differ appreciably from the previous measurements [12, 13]. The agreement with the theoretical isovector strengths is slightly spoiled but still satisfactory. Concerning the ISD transitions, the calculation reproduces the overall strength. In fact, the two levels are estimated to exhaust 2.78 % of the energy weighted sum rule (EWSR), almost within the experimental errors of the measured value. However, it is distributed among the two levels just in the opposite way.

These two low-lying 1^- states have a dominant two-phonon character. The one-phonon components, responsible for the transition, account for 10% of the lower state and for 20% of the one at higher energy. A level crossing would be desirable. This may be induced for instance by the inclusion of the three-phonon configurations.

4. Conclusions

We have seen that the quasiparticle formulation keeps the same simplicity of EMPM in the particle-hole scheme and is well suited for clarifying the role of complex configurations on the properties of open shell nuclei, including those far from the stability line. The neutron-rich ^{20}O isotope is an important illustrative example. For the study of this nucleus, we have performed a calculation in a truncated space including up to a fraction of two-phonon states. The truncation of the two-phonon configurations is the only free parameter of our calculation. It was dictated by the necessity of keeping a reasonable energy gap between the excited and correlated ground states.

A calculation in the two-phonon space, while affecting little the excited states, would have depressed very significantly the ground state. It would have been, therefore, necessary to restore the correct energy gap by including a sufficiently large number of three-phonon states, which are known to couple strongly to the one-phonon configurations and to push the excited states down in energy. While feasible, the inclusion of the three phonon states would require a more efficient numerical implementation of EMPM. We plan to cope with this task in the future.

Although limited to a truncated space, the calculation has shown that the two-phonon configurations have a very strong impact on the spectra and the dipole response. They improve greatly the agreement between theoretical and experimental spectra by enhancing the levels density and redistributing the TDA energies, provide a fairly satisfactory description of the cross section in the region of the giant dipole resonance in virtue of the strong fragmentation of the dipole strength induced by the phonon coupling. They are also responsible for the appearance of two 1^- levels below the neutron threshold, detected in recent experiments, and reproduce the overall strength measured for these two levels.

Acknowledgments

This work was partly supported by the Czech Science Foundation (P203-13-07117S). Two of the authors (F.K. and P.V.) thank the Istituto Nazionale di Fisica Nucleare (Italy) for financial support. Highly appreciated was the access to computing and storage facilities provided by the Meta Centrum under Program No. LM2010005 and the CERITSC under the program Centre CERIT Scientific Cloud, part of the Operational Program Research and Development for Innovations, Register No. CZ.1.05/3.2.00/08.0144.

References

[1] Andreozzi F, Knapp F, Lo Iudice N, Porrino A and Kvasil J 2007 *Phys. Rev.* C **75** 044312
[2] Andreozzi F, Knapp F, Lo Iudice N, Porrino A and Kvasil J 2008 *Phys. Rev.* C **78** 054308
[3] Bianco D, Knapp F, Lo Iudice N, Andreozzi F, and Porrino A 2012 *Phys. Rev.* C **85** 014313
[4] Knapp F, Lo Iudice N, Veselý P, Andreozzi F, De Gregorio G and Porrino A 2014 *Phys. Rev.* C **90** 014310
[5] Knapp F, Lo Iudice N, Veselý P, Andreozzi F, De Gregorio G and Porrino A 2015 *Phys. Rev.* C **92** 054315
[6] De Gregorio G, Knapp F, Lo Iudice N and Veselý P 2016 *Phys. Rev.* C **93** 044314
[7] Tilley D R, Cheves C M, Kelley J H, Raman S and Weller H R 1998 *Nucl. Phys.* A **636** 249

[8] Jewell J K, Riley L A, Cottle P D, Kemper K W, Glasmacher T, Ibbotson R W, Scheit H, Chromik M, Blumenfeld Y, Hirzebruch S E, Marechal F and Suomijarvi T 1999 *Phys. Lett.* B **454** 181
[9] Khan E, Blumenfeld Y, Van Giai N, Suomijarvi T, Alamanos N, Auger F, Colo G, Frascaria N, Gillibert A, Glasmacher T, Godwin M, Kemper K W, Lapoux V, Lhenry I, Marechal F, Morrissey D J, Musumarra A, Orr N A, Ottini-Hustache S, Piattelli P, Pollacco E C, Roussel-Chomaz P, Roynette J C, Santonocito D, Sauvestre J E, Scarpaci J A and Volpe C 2000 *Phys. Lett.* B **490** 45
[10] Stanoiu M et al 2004 *Phys. Rev.* C **69** 034312
[11] Leistenschneider A et al 2001 *Phys. Rev. Lett.* **86** 5442
[12] Tryggestad E, Aumann T, Baumann T, Bazin D, Beene J R, Blumenfeld Y, Brown B A, Chartier M, Halbert M L, Heckman P, Liang J F, Radford D C, Shapira D, Thoennessen M and Varner R L 2002 *Phys. Lett.* B **541** 52
[13] Tryggestad E, Baumann T, Heckman P, Thoennessen M, Aumann T, Bazin D, Blumenfeld Y, Beene J R, Lewis T A, Radford D C, Shapira D, Varner R L, Chartier M, Halbert M L and Liang J F 2003 *Phys. Rev.* C **67** 064309
[14] Nakatsuka N et al 2017 *Phys. Lett.* B **768** 387
[15] Jansen G R, Engel J, Hagen G, Navrátil P and Signoracci A 2014 *Phys. Rev. Lett.* **113** 142502
[16] Jensen O, Hagen G, Hjorth-Jensen M, Brown B A and Gade A 2011 *Phys. Rev. Lett.* **107** 032501
[17] Hagen G, Hjorth-Jensen M, Jansen G R, Machleidt R and Papenbrock T 2012 *Phys. Rev. Lett.* **108** 242501
[18] Holt J D, Menendez J and Schwenk A 2013 *Eur. Phys. J.* A **49** 39
[19] Sagawa H and Suzuki T 1999 *Phys. Rev.* C **59** 3116
[20] Vretenar D, Paar N, Ring P and Lalazissis G A 2001 *Nucl. Phys.* A **692** 496
[21] Tohyama M and Umar A S 2001 *Phys. Lett.* B **516** 415
[22] Colò G and Bortignon P F 2001 *Nucl. Phys.* A **696** 427
[23] Ekstroem A, Baardsen G, Forssen C, Hagen G, Hjorth-Jensen M, Jansen G R, Machleidt R, Nazarewicz W, Papenbrock T, Sarich J and Wild S M 2013 *Phys. Rev. Lett.* **110** 192502
[24] Bianco D, F Knapp F, Lo Iudice N, Veselý P, Andreozzi F, De Gregorio G and Porrino 2014 *J. Phys. G: Nucl. Part. Phys.* **41** 025109

IOP Conf. Series: Journal of Physics: Conf. Series **966** (2018) 012048 doi:10.1088/1742-6596/966/1/012048

Neutron knockout from ^{68,70}Ni ground and isomeric states.

F. Recchia[1,2], D. Weisshaar[1], A. Gade[1,3], J. A. Tostevin[4], R. V. F. Janssens[5], M. Albers[5], V. M. Bader[1,3], T. Baugher[1,3], D. Bazin[1], J. S. Berryman[1], B. A. Brown[1,3], C. M. Campbell[6], M. P. Carpenter[5], J. Chen[7], C. J. Chiara[5,7], H. L. Crawford[6], C. R. Hoffman[5], F. G. Kondev[8], A. Korichi[5,9], C. Langer[1], T. Lauritsen[5], S. N. Liddick[1,10], E. Lunderberg[1,3], S. Noji[1], C. Prokop[1,10], S. R. Stroberg[1,3], S. Suchyta[1,10], K. Wimmer[1,11], S. Zhu[5]

[1]National Superconducting Cyclotron Laboratory, Michigan State University, East Lansing, Michigan 48824, USA.
[2]Dipartimento di Fisica e Astronomia “Galileo Galilei”, Università degli Studi di Padova and INFN Padova, I-35131 Padova, Italy.
[3]Department of Physics and Astronomy, Michigan State University, East Lansing, Michigan 48824, USA.
[4]Department of Physics, University of Surrey, Guildford, Surrey GU2 7XH, United Kingdom.
[5]Physics Division, Argonne National Laboratory, Argonne, Illinois 60439, USA.
[6]Nuclear Science Division, Lawrence Berkeley National Laboratory, Berkeley, California 94720, USA.
[7]Department of Chemistry and Biochemistry, University of Maryland, College Park, Maryland 20742, USA
[8]Nuclear Engineering Division, Argonne National Laboratory, Argonne, Illinois 60439, USA.
[9]CSNSM-IN2P3/CNRS, F-91405 Orsay Campus, France.
[10]Department of Chemistry, Michigan State University, East Lansing, Michigan 48824, USA.
[11]Department of Physics, Central Michigan University, Mt. Pleasant, Michigan 48859, USA.

E-mail: `francesco.recchia@unipd.it`

Abstract.

Neutron-rich isotopes are an important source of new information on nuclear physics. Specifically, the spin-isospin components in the nucleon-nucleon (NN) interaction, e.g., the proton-neutron tensor force, are expected to modify shell structure in exotic nuclei. These potential changes in the intrinsic shell structure are of fundamental interest. The study of the excitation energy of states corresponding to specific configurations in even-even isotopes, together with the single-particle character of the first excited states of odd-A, neutron-rich Ni isotopes, probes the evolution of the neutron orbitals around the Fermi surface as a function of the neutron number a step forward in the understanding of the region and the nature of the NN interaction at large N/Z ratios.

In an experiment carried out at the National Superconducting Cyclotron Laboratory [1], new spectroscopic information was obtained for ^{68}Ni and the distribution of single-particle strengths in ^{67,69}Ni was characterized by means of single-neutron knockout from ^{68,70}Ni secondary beams. The spectroscopic strengths, deduced from the measured partial cross sections to the individual states tagged by their de-exciting gamma rays, is used to identify and quantify configurations that involve neutron excitations across the $N = 40$ harmonic oscillator shell closure. The de-excitation γ rays were measured with the GRETINA tracking array [2].

The results challenge the validity of the most current shell-model Hamiltonians and effective interactions, highlighting shortcomings that cannot yet be explained. These results suggest that our understanding of the low-energy states in such nuclei is not complete and requires further investigation.

1. Introduction

Nuclear physics aims to understand the properties of nuclei, complex highly-correlated quantum systems, in terms of theoretical models of their structure. The most important degrees of freedom need to be sensed and highlighted by means of experimental measurements. Among these are the trends and notable changes in between nuclei in the same mass-isospin region that shed light on the relevant aspects of the underlying nuclear interactions.

Sophisticated nuclear models have been developed in order to account for the large variety of nuclear phenomena observed in nuclear systems under different conditions of excitation and isospin. The shell model is indeed one of the most successful efforts to reproduce basic features in a wide range of nuclei: many nuclear properties can be explained by means of bunching of nuclear orbitals, separated by energy gaps corresponding to stabilizing number of protons and neutrons. As nucleons are added to a closed-shell configuration, strong correlations in between them drive the shape of the nucleus from a spherical shape to either an oblate or a prolate one, with a substantial modification of the structure that is better understood by means of collective models. The exploration of the Segré chart by means of rare isotopes has made possible to track several changes in the shell-model structure of nuclei having an unbalanced number of protons and neutrons.

Spin-isospin parts in the nucleon-nucleon interaction, e.g., the proton-neutron tensor force, are expected to modify shell structure in exotic nuclei [3–5]. Potential changes in the intrinsic shell structure are of fundamental interest, and also have implications for nucleosynthesis. For example, they could directly affect the role of ^{78}Ni as a waiting point nucleus in the r-process [6, 7].

Shell evolution has been shown to imply that nuclei expected to show closed-shell features instead display an open-shell behavior as a consequence of the breakdown of magic numbers [8, 9]. And in some cases, new unexpected magic numbers appear. New regions of deformation appear typically to be related with the promotion of pairs of quasi-particles to high-j orbitals located near the shell closure [10]. Recently, the way this mechanism can be at play within a single nucleus was highlighted. Deformation driving particle-hole pair configurations can compete with spherical ones, resulting in two different structures coexisting at low excitation in the same nucleus.

Of particular interest is the ^{68}Ni region where a shell closure occurs at Z=28 and the harmonic-oscillator shell gap separates the fp and $g_{9/2}$ neutron orbitals. A systematic study of the region shows that, in ^{68}Ni, there is a maximum of the excitation energy of the 2_1^+ state [11] and a minimum of the $B(E2; 0_1^+ \rightarrow 2_1^+)$ [12] transition probability.

For decades, the excitation energy E_x = 1770(30) keV of the 0_2^+ isomer, obtained by Bernas et al. [13] using particle-spectroscopy techniques with the ^{70}Zn(^{14}C, ^{16}O) transfer reaction, was the only direct measurement for this state. Although the uncertainty was quoted to be 30 keV, comparisons of early transfer-reaction data for ^{67,68}Ni [13, 14] with subsequent, higher-precision γ-ray spectroscopy (e.g., Refs. [11, 15, 16]) reveal a systematic 100- to 200-keV offset in the excitation spectrum from which one may infer the 0_2^+ isomer to be lower in energy than originally reported. However, only recently did Suchyta et al. [17] directly measure the electrons produced in the E0 decay. A parallel work by Recchia et al. [18] provided not only an improved value for the energy of the 0_2^+ state [1603.5(3) keV], but also indicated the presence of prompt

IOP Conf. Series: Journal of Physics: Conf. Series **966** (2018) 012048 doi:10.1088/1742-6596/966/1/012048

1138- and 2420-keV transitions preceding the $E0$ decay [17]. In addition to the 0_1^+ and 0_2^+ states, a 2511-keV 0_3^+ level was tentatively proposed following β decay [19]. This assignment was later firmly established in a deep-inelastic-scattering (DIS) experiment [20].

The investigation of the excitation energy of low spin states of nuclei in the ^{68}Ni region is particularly interesting because they are often associated with rather simple configurations interpreted in terms of excitations of protons and neutrons across the gaps at Z=28 and N=40. A number of recent examples has been reported [17, 21–32] together with several theoretical works [33–38] .

2. Experiment, results and discussion

How sharply does the neutron occupancy change when the Fermi surface moves from the fp to the $g_{9/2}$ orbitals near ^{68}Ni? To what extent does the ground state of ^{68}Ni have a configuration corresponding to a spherical shape dominated by a completely filled fp shell? What is the configuration of the ^{70}Ni ground state (g.s.)?

In order to answer these questions, an experiment was performed in which we used exclusive one-neutron knockout cross sections, measured using γ-ray tagged neutron removal reactions from ^{68,70}Ni [39]. This technique allowed us to identify and quantify configurations that involve neutron excitations across the N=40 gap. Specifically, the partial cross sections to the lowest-lying $1/2^-$, $5/2^-$, $3/2^-$ and $9/2^+$ levels are measured and compared to calculations using shell-model spectroscopic strengths and eikonal reaction theory for both ^{68,70}Ni. Partial cross sections were measured as well for higher-lying states populated by neutron knockout from a long-lived isomer present along with the g.s. of ^{68}Ni.

The experiment was performed at the Coupled Cyclotron Facility (CCF) at the National Superconducting Cyclotron Laboratory (NSCL) [1]. Secondary beams of ^{68,70}Ni were produced by fragmentation of a 140 MeV/u ^{82}Se primary beam on a 423-mg/cm^2 ^{9}Be production target located at the entrance of the A1900 separator [40]. The secondary beams, selected and purified with the separator, were delivered at momentum acceptances of 1% for ^{68}Ni and 3% for ^{70}Ni, respectively. The ^{68}Ni and ^{70}Ni beams impinged on ^{9}Be reaction targets (100- and 281-mg/cm^2 thick, respectively) with mid-target energies of 85 and 74 MeV/u. The measurements described below were carried out with GRETINA [2] surrounding the target and the S800 spectrograph [41].

The projectile-like reaction residues, after traversing the trigger plastic scintillator at the back of the S800 focal plane, were implanted into an aluminum plate placed in front of a CsI(Na) detector array [42]. The delayed γ-ray detection with the scintillator array in this IsoTagger configuration [43] enabled measurements of transitions from isomers in the knockout residues.

With the ~50-ns flight time for ^{68}Ni ions through the S800 spectrograph, it was possible to correlate isomeric decays measured using the CsI(Na) detectors at the focal plane (see below) with prompt γ rays at the target position. The coincidence relationships observed between prompt 1139- and delayed 511-keV γ rays (see fig.1), as well as the relative intensities of the 663- and 1139-keV lines, indicate that the latter should be placed above the 0_2^+ isomer, likely depopulating the known 2743-keV, 2_2^+ level and, thus, fixing the 0_2^+ energy at 1604(1) keV, in agreement with Suchyta et al. [17]. The placement described above is further supported by determination of the half-life of the 511-keV isomeric decay line. The measured $t_{1/2}$ values are reasonably consistent with the 0.270(5)-s half-life reported for the 0_2^+ state in Ref. [13].

The position of the 0_2^+ isomer at 1603.5(3) keV was measured and the comparisons with shell-model (SM) calculations reveal the importance of mixing to account for the observed decay patterns, as reported in [18], but quantification of the amount of mixing requires further experimental data.

The ^{68}Ni nucleus has two known isomeric states, the 0_2^+ (τ = 390 ns) and the 5^- (τ = 1.24 ms) levels that, if produced in the fragmentation process, survive sufficiently long to be transmitted

IOP Conf. Series: Journal of Physics: Conf. Series **966** (2018) 012048 doi:10.1088/1742-6596/966/1/012048

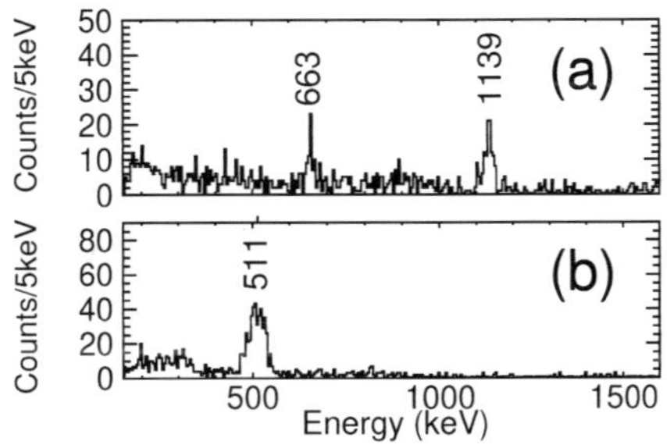

Figure 1. (a) Prompt GRETINA spectrum coincident with the identification of a ^{68}Ni recoil and the detection of a delayed 511-keV γ ray in the CsI(Na) focal plane detectors. (b) Delayed γ-ray spectrum recorded in the CsI(Na) scintillators in coincidence with implanted ^{68}Ni ions [18].

to the reaction target. ^{70}Ni has only one known isomeric state, the 2861-keV 8^+ ($\tau = 335$ ns) level that, if populated, will survive to the experimental end station as well. Accordingly, to interpret the cross sections for the population of individual final states, the isomeric ratio had to be determined for each projectile. For this purpose, a stopper was placed in the center of the GRETINA array and the γ-decay radiation from the implanted nuclei was measured over several microseconds. To determine the ^{68}Ni 5^- and ^{70}Ni 8^+ isomeric contents, the number of implanted ions was counted with a plastic scintillator upstream of the stopper. The efficiency-corrected yields of the γ-ray cascades depopulating the two isomers were then used to determine the isomeric ratios. These were 39(4)% for the 5^- level in ^{68}Ni and 8(1)% for the 8^+ one in ^{70}Ni, respectively. For the ^{68}Ni 0_2^+ state, the isomeric ratio was determined by the measurement of the 511-keV γ rays following positron annihilation associated with the internal pair formation, the main decay mode of this level. Including corrections for internal conversion, this ^{68}Ni 0_2^+ isomeric ratio was determined to be less than 1%.

The ^{67}Ni and ^{69}Ni one-neutron knockout residues, produced upon collision with the ^{9}Be reaction target, were detected and identified on an event-by-event basis using the time-of-flight and energy-loss information measured with the beam line timing detectors and the S800 spectrograph focal-plane detector system [41]. Prompt γ rays, emitted in-flight from the de-excitation of the knockout residues, were detected with the GRETINA array [2] surrounding the target at the entrance of the S800 spectrograph. The γ-ray energies were Doppler corrected on an event-by-event basis using the reconstructed momentum vector provided by the S800 spectrograph for each reaction residue [41].

The measured inclusive cross sections for the ^{9}Be(^{68}Ni,^{67}Ni)X and ^{9}Be(^{70}Ni,^{69}Ni)X one-neutron removal reactions, $\sigma_{inc}^{-1n} = 133(10)$ and 168(13) mb, respectively, were derived from the yield of the ^{67,69}Ni reaction residues relative to the number of incoming ^{68,70}Ni projectiles and the number density of the reaction target [39].

The theoretical partial and inclusive cross section calculations follow Refs. [44, 45]. The cross section for neutron removal from an initial state i of the A-body projectile – here its ground state or isomer – to a given residue final state f is given by

$$\sigma_{\alpha}^{fi} = \left(\frac{A}{A-1}\right)^{N} C^2S\ \sigma_{\alpha}^{sp}, \tag{1}$$

where C^2S is the shell-model spectroscopic factor and σ_{α}^{sp} the single-particle cross section. Here, α labels the quantum numbers n, ℓ, j of the removed nucleon. The A-dependent multiplicative term is a center-of-mass correction factor to the shell-model C^2S value for removal from an orbital with N oscillator quanta [46]. The single-particle cross section σ_α^{sp} is obtained with the eikonal approach [44], assuming a normalized single-particle overlap form factor for the removed nucleon and complex residue- and nucleon-target optical potentials.

Usually, the shell- plus eikonal models overestimate the cross section (and the total spectroscopic strength) leading to bound final states. This reduction is not observed here. However, this might be expected considering the shell-model orbitals used here, which, so far, neglect any possible neutron $f_{7/2}$ single-particle strength. Furthermore, for the ^{68}Ni→^{67}Ni reaction, the following applies as reported in [39]: (i) a systematic over-prediction of the cross sections to the low-spin, negative-parity states ($1/2^-$ - $5/2^-$), populated in removal from the ground state of ^{68}Ni, and (ii) a systematic underestimation of the cross sections to high-spin states ($9/2^+$ - $15/2^+$) populated by the removal from the 5^- isomer in ^{68}Ni. One may speculate that the discrepancy reported here points to an opportunity to improve the neutron configurations in the shell-model framework. However, any improvement to the Hamiltonians in this region would have to preserve the good description of the spectroscopic strength distribution that is found in the knockout to ^{69}Ni [39].

3. Conclusion

With this experiment, it was possible to obtain the precise excitation energy for the 0_2^+ state in ^{68}Ni, that was correctly predicted by the state-of-the-art shell-model calculations. Moreover, the new data enabled the identification of a number of new states in both ^{67}Ni and ^{69}Ni, in particular, levels that carry the largest part of the single-particle strength of the neutron orbitals lying at the Fermi surface. In comparison to shell-model calculations, the high measured inclusive cross sections hint at significant bound $f_{7/2}$ neutron strength, which is outside the shell-model space considered here. In contrast, the relative single-particle strength distribution for the knockout to ^{69}Ni is described well by the shell-model calculations if we consider the tentative ^{69}Ni level scheme proposed from this work. The results for the neutron removal to ^{67}Ni challenge the validity of one of the most current shell-model Hamiltonians, highlighting its shortcomings and providing benchmarks for future interactions developed for this neutron-rich region of the nuclear chart. The results suggest that our understanding of the low-energy states in these interesting nuclei is not yet complete and requires further investigation.

This work was supported in part by the the National Science Foundation (NSF) under Contract No. PHY-1102511, by the US Department of Energy (DOE), Office of Nuclear Physics, under Grants No. DE-FG02-08ER41556 (Michigan State University) and No. DE-FG02-94-ER40834 (University of Maryland), and Contract No. DE-AC02-06CH11357 (Argonne National Laboratory), and by the DOE, National Nuclear Security Administration, under Award No. DE-NA0000979. GRETINA was funded by the DOE, Office of Science. Operation of the array at NSCL was supported by the NSF under Cooperative Agreement No. PHY-1102511 (NSCL) and DOE under Grant No. DE-AC02-05CH11231 (LBNL). B.A.B. acknowledges support from NSF Grant No. PHY-1404442 and J.A.T. from the United Kingdom Science and Technology Facilities Council (STFC) Grant No. ST/L005743/1. We are grateful for Augusto Macchiavelli's support of the GRETINA campaign at NSCL.

References

[1] Gade A and Sherrill B M 2016 *Physica Scripta* **91** 053003
[2] Paschalis S *et al.* 2013 *Nucl. Instrum. Methods Phys. Res., Sect. A* **709** 44 – 55
[3] Otsuka T *et al.* 2001 *Phys. Rev. Lett.* **87**(8) 082502
[4] Otsuka T *et al.* 2005 *Phys. Rev. Lett.* **95**(23) 232502
[5] Otsuka T *et al.* 2010 *Phys. Rev. Lett.* **105**(3) 032501
[6] Hosmer P T *et al.* 2005 *Phys. Rev. Lett.* **94**(11) 112501
[7] Hosmer P *et al.* 2010 *Phys. Rev. C* **82**(2) 025806
[8] Gade A and Glasmacher T 2008 *Prog. Part. Nucl. Phys.* **60** 161 – 224

[9] Sorlin O and Porquet M G 2008 *Prog. Part. Nucl. Phys.* **61** 602 – 673
[10] Poves A 2016 *J. Phys. G.* **43** 024010
[11] Broda R *et al.* 1995 *Phys. Rev. Lett.* **74**(6) 868–871
[12] Sorlin O *et al.* 2002 *Phys. Rev. Lett.* **88**(9) 092501
[13] Bernas M *et al.* 1982 *Physics Letters B* **113** 279 – 282
[14] Girod M *et al.* 1988 *Phys. Rev. C* **37**(6) 2600–2612
[15] Zhu S *et al.* 2012 *Phys. Rev. C* **85**(3) 034336
[16] Broda R *et al.* 2012 *Phys. Rev. C* **86**(6) 064312
[17] Suchyta S *et al.* 2014 *Phys. Rev. C* **89**(2) 021301
[18] Recchia F *et al.* 2013 *Phys. Rev. C* **88**(4) 041302
[19] Mueller W F *et al.* 2000 *Phys. Rev. C* **61**(5) 054308
[20] Chiara C J *et al.* 2012 *Phys. Rev. C* **86**(4) 041304
[21] Olaizola B *et al.* 2017 *Phys. Rev. C* **95**(6) 061303
[22] Campo L C *et al.* 2017 *Phys. Rev. C* **96**(1) 014312
[23] Klintefjord M *et al.* 2017 *Phys. Rev. C* **95**(2) 024312
[24] Crider B *et al.* 2016 *Physics Letters B* **763** 108 – 113
[25] Albers M *et al.* 2016 *Phys. Rev. C* **94**(3) 034301
[26] Morfouace P *et al.* 2016 *Phys. Rev. C* **93**(6) 064308
[27] Prokop C J *et al.* 2015 *Phys. Rev. C* **92**(6) 061302
[28] Santamaria C *et al.* 2015 *Phys. Rev. Lett.* **115**(19) 192501
[29] Liddick S N *et al.* 2015 *Phys. Rev. C* **92**(2) 024319
[30] Ayangeakaa A D *et al.* 2015 *Phys. Rev. C* **91**(4) 044327
[31] Chiara C J *et al.* 2015 *Phys. Rev. C* **91**(4) 044309
[32] Flavigny F *et al.* 2015 *Phys. Rev. C* **91**(3) 034310
[33] Nowacki F *et al.* 2016 *Phys. Rev. Lett.* **117**(27) 272501
[34] Rodríguez T R *et al.* 2016 *Phys. Rev. C* **93**(5) 054316
[35] Hergert H *et al.* 2014 *Phys. Rev. C* **90**(4) 041302
[36] Lay J A *et al.* 2014 *Phys. Rev. C* **89**(3) 034618
[37] Tsunoda Y *et al.* 2014 *Phys. Rev. C* **89**(3) 031301
[38] Péru S and Martini M 2014 *Eur. Phys. J. A* **50** 88
[39] Recchia F *et al.* 2016 *Phys. Rev. C* **94**(5) 054324
[40] Morrissey D *et al.* 2003 *Nucl. Instrum. Methods Phys. Res., Sect. B* **204** 90 – 96
[41] Bazin D *et al.* 2003 *Nucl. Instrum. Methods Phys. Res., Sect. B* **204** 629 – 633
[42] Meierbachtol K *et al.* 2011 *Nucl. Instrum. Methods Phys. Res., Sect. A* **652** 668 – 670
[43] Wimmer K *et al.* 2015 *Nucl. Instrum. Methods Phys. Res., Sect. A* **769** 65 – 71
[44] Hansen P and Tostevin J 2003 *Annu. Rev Nucl. Part. S.* **53** 219–261
[45] Stroberg S R *et al.* 2014 *Phys. Rev. C* **90**(3) 034301
[46] Dieperink A E L and Forest T d 1974 *Phys. Rev. C* **10**(2) 543–549

Pre-equilibrium GDR in a heavy composite system

C Parascandolo[1], D Pierroutsakou[1], R Alba[2], A Del Zoppo[2], C Maiolino[2], D Santonocito[2], C Agodi[2], V Baran[3,2], A Boiano[1], M Colonna[2], R Coniglione[2], E De Filippo[4], M Di Toro[2,5], U Emanuele[6], F Farinon[7], A Guglielmetti[8], M La Commara[9,1], B Martin[9,1], C Mazzocchi[8], M Mazzocco[10], C Rizzo[2,5], M Romoli[1], C Signorini[10], R Silvestri[9,1], F Soramel[10], E Strano[10], D Torresi[10], A Trifirò[6], M Trimarchi[6]

[1]INFN - Napoli, Napoli, Italy [2]INFN - LNS, Catania, Italy [3]University of Bucharest, Bucharest and NIPNE-HH, Magurele, Romania [4]INFN - Catania, Catania, Italy [5]Dip. di Fisica e Astronomia, Università di Catania, Catania, Italy [6]INFN - Gr. Coll. di Messina and Università di Messina, Messina, Italy [7]GSI, Darmstadt, Germany [8]Dip. di Fisica, Università di Milano and INFN - Milano, Milano, Italy [9]Dip. di Fisica, Università di Napoli, Napoli, Italy [10]Dip. di Fisica e Astronomia, Università di Padova and INFN - Padova, Padova, Italy

E-mail: concetta.parascandolo@na.infn.it

Abstract. The excitation of the Dynamical Dipole mode along the fusion path was investigated for the first time in the formation of a heavy composite system in the $A{\sim}190$ mass region, in fusion–evaporation and fission events. The composite system was formed at identical conditions of excitation energy and spin from two reactions with different charge asymmetry: the charge asymmetric ^{40}Ca + ^{152}Sm and the nearly charge symmetric ^{48}Ca + ^{144}Sm at E_{lab}=11 and 10.1 MeV/nucleon, respectively. In this paper, we report the results on fusion evaporation events and the preliminary analysis on fission channel.

1. Introduction

During the charge equilibration mechanism taking place in the first stages of dissipative reactions between colliding ions with different N/Z ratios, a large amplitude dipole collective motion develops in the composite dinuclear system, the so-called "Dynamical Dipole mode" (DD), or pre-equilibrium Giant Dipole Resonance (GDR). The DD gives rise to a prompt γ-ray emission [1, 2, 3, 4, 5, 6], characterized by: i) a centroid energy lower than that of the compound nucleus (CN) GDR in the same mass region indicating a high deformation of the emitting source [3, 4], ii) an anisotropic angular distribution with respect to the beam axis since the oscillation is confined in the reaction plane [7] and iii) a γ yield that should depend on both the reaction dynamics and the symmetry term of the nuclear matter Equation Of State (EOS), that is acting as a restoring force [4].

Experimentally, the DD excitation has been observed in deep inelastic [5, 8, 9, 10] and fusion-evaporation heavy-ion collisions [10, 11, 12, 13, 14, 15, 16, 17], while the first systematic study of its incident energy dependence was performed in an experimental campaign [12, 13, 14, 15] for the ^{132}Ce CN. In this case, the DD was evidenced with a model independent method, the so-called "difference technique", consisting in: (a) forming the same CN at identical conditions of excitation energy and spin from two entrance channels, a nearly charge symmetric and a

IOP Conf. Series: Journal of Physics: Conf. Series **966** (2018) 012049 doi:10.1088/1742-6596/966/1/012049

charge asymmetric one; (b) obtaining the difference between the γ-ray spectra and angular distributions of the two channels for fusion-evaporation events. The DD was evidenced through the observation of an excess of yield in the γ-ray energy spectrum of the more charge asymmetric system at a centroid energy lower than that of the CN GDR and with a higly anisotropic angular distribution, presenting a maximum around 90° with respect to the beam direction.

At the moment very few data exist on the DD γ multiplicity and on its angular distribution that can be directly compared with theoretical calculations. Furthermore, calculations are not able to simultaneously reproduce all the existing experimental findings. The emission of DD γ-rays decreases the excitation energy and the initial temperature of the nucleus reaching the statistical phase. This cooling mechanism could be suitable to favour the CN survival against statistical fission and thus, the formation of a super heavy element in hot fusion processes [4, 19]. In order to investigate the behavior of the DD in heavier systems than those studied before and to test its usefulness in superheavy element production, we decided to study the DD in a composite system in the mass region A=190 [18].

2. The experiment: ^{40,48}Ca+^{152,144}Sm at 11 MeV/nucleon

We performed the experiment by using the ^{40}Ca (^{48}Ca) pulsed beam provided by the Cyclotron of the Laboratori Nazionali del Sud (INFN-LNS, Italy), impinging on a 1 mg/cm^2 thick self-supporting ^{152}Sm (^{144}Sm) target enriched to 98.4%(93.8%) in ^{152}Sm (^{144}Sm) at E_{lab} = 440 (485) MeV. Both reactions populate the same composite system through a quite different initial dipole moment, 30.6 fm for the ^{40}Ca + ^{152}Sm charge asymmetric reaction and 5.3 fm for the ^{48}Ca + ^{144}Sm more charge symmetric one, while the mass asymmetry of the two entrance channels is very similar, namely 0.22 (0.18) for the ^{40}Ca + ^{152}Sm (^{48}Ca + ^{144}Sm) system. The formed system had identical excitation energy in both reactions, E^* = (220±7) MeV, and identical spin distribution: $L_{max} = 74\hbar$ for fusion and $L_{max} = 42\hbar$ for fusion-evaporation, based on PACE2 calculations [20] with a level density parameter a = A/9.5 MeV^{-1}, A being the CN mass.

The γ-rays and the light charged particles were detected by using the BaF_2 detectors of the MEDEA setup [26], covering the polar angular range $\theta_{lab} = 42.4° - 170°$ and the full range in the azimuthal angle. It operates in vacuum to allow a simultaneous detection of γ-rays and charged particles. The discrimination between γ-rays and light particles was performed by combining a shape analysis of the BaF_2 signal with a time of flight (TOF) measurement, between each BaF_2 and the radiofrequency signal of the Cyclotron.

The fusion-evaporation residues (ER) were detected by four Parallel Plate Avalanche Counters (PPACs) located symmetrically around the beam direction at 70 cm from the target. The PPACs were centered at θ_{lab} = 7° with respect to the beam direction, subtending 7° in θ and covering a total solid angle of 0.089 sr. The PPACs provided the energy loss ΔE and the TOF of the reaction products. The fission events were selected by detecting the two kinematically coincident fission fragments (FF) with position sensitive PPACs, centered at θ_{lab} = 52.5° symmetrically around the beam axis, at 16 cm from the target covering 22° in both θ and ϕ. The PPACs gave information on the energy loss ΔE, the TOF and the position (x,y) of the FF that allowed to reconstruct angles, masses and velocity vectors of the fragments in the laboratory and the center-of-mass reference frame. Both ER and FF were selected by applying appropriate contours in the relative bi-dimensional plots (ΔE, TOF) of the PPACs.

Down-scaled single PPAC events together with coincidence ones between a PPAC and at least one fired BaF_2 were collected during the experiment. The energy threshold in a BaF_2 detector was set at ~5.5 MeV. The coincidence request eliminated any cosmic ray contamination of the γ-ray spectra. By using the above trigger there are no normalization factors in the γ-ray (charged particle) spectra as the double differential γ (charged particle) multiplicity is obtained from the ratio of the number of coincidences between γ-rays (charged particles) and ER (FF) and the

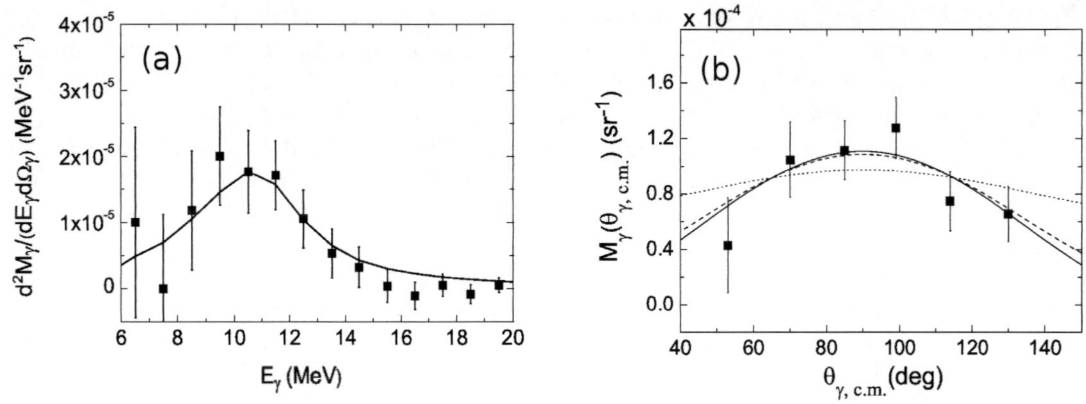

Figure 1. (a) The symbols represent the DD γ-ray spectrum for fusion-evaporation events. The line is a Lorentzian curve folded by the experimental set up response function. (b) 9 MeV$\leq E_\gamma \leq$15 MeV DD γ-ray angular distribution, corrected for the detection efficiency. The lines represent different Legendre polynomial fits, obtained by using different a_2 values. See the text for details.

number of single events of ER (FF).

3. Analysis and Results

The experiment was designed in such a way to form the same composite system at identical excitation energy in both reactions by taking into account the pre-equilibrium particle emission. That reduces the CN average mass, average charge and average excitation energy and cannot be discarded in the TOF spectrum of the reaction products because they have overlapping velocity distributions with those of the complete fusion events [21, 22]. Therefore, in the present work, the average excitation energy, the average mass and the average charge of the composite system, after pre-equilibrium particle emission, were evaluated by studying the energy spectra of the light charged particles (α-particles and protons) detected in coincidence with ER and FF, while the pre-equilibrium neutron emission was estimated from our proton data and from existing neutron emission studies (see [18]).

3.1. Fusion–evaporation results

The light particle spectra detected by BaF_2 scintillators in coincidence with the ER were analyzed assuming that particles have been emitted isotropically from two moving sources: a *slow* one describing the statistical evaporation from the hot CN and an *intermediate-velocity* (between the CN and the projectile velocity) one due to the pre-equilibrium particles emitted by the composite system before thermalization (see [18]). To evaluate the average energy taken away by pre-equilibrium neutrons, not detected in this experiment, we assumed that their energy spectra were very similar to the proton ones, apart from the Coulomb barrier. Then, the average kinetic energy of a pre-equilibrium neutron was taken to be that of a pre-equilibrium proton minus the Coulomb barrier while the pre-equilibrium neutron multiplicity was deduced by that of pre-equilibrium protons multiplied with the N/Z ratio of the CN. The adopted pre-equilibrium neutron multiplicity is in agreement within errors with neutron emission studies performed at similar center-of-mass incident energy above the Coulomb barrier [23]. We found that two units of charge and three units of mass were carried away from the initial composite system leading to the ^{189}Hg nucleus in both reactions, with the same average excitation energy E^*=(220±7)

MeV, giving us confidence that any difference between the γ-ray spectra and γ-ray angular distributions of the two reactions is related to an entrance channel effect.

After the evaluation of the excitation energy of both reactions, the incoherent bremsstrahlung component considered to originate primarily in neutron-proton (np) collisions and dominant for $E_\gamma > 30$ MeV, must be evaluated and subtracted from the data. An equal bremsstrahlung component is expected for the two reactions because of their very similar beam energy and size of the reaction partners and of the same temperature of the composite system [24, 25]. The np bremsstrahlung component was deduced by fitting simultaneously the center-of-mass fusion-evaporation γ-ray spectra of the two reactions at different polar angles in the energy range 30 MeV$<E_\gamma<$40 MeV. The fit was performed assuming an exponentially decreasing behavior.

The difference between the center-of-mass bremsstrahlung-subtracted γ-ray spectra of the two reactions for fusion-evaporation events (symbols in panel (a) of Figure 1) shows an excess of γ-rays in the more charge asymmetric channel, between 8 and 15 MeV. This can only be related to the DD excitation in the composite system of the ^{40}Ca+^{152}Sm reaction because of its larger charge asymmetry. This spectrum is reproduced well by means of a Lorentzian curve folded with the response function of the apparatus [27] (solid line), with the DD centroid energy $E_{DD} = 11$ MeV and the width $\Gamma_{DD} = 3.5$ MeV. E_{DD} is 3 MeV lower than E_{GDR} (obtained by using Cascade code [28]), indicating a high deformation of the emitting source during the DD γ emission. This result is in excellent agreement with expectations for a dipole oscillation along the symmetry axis of a deformed dinucleus during the early moments of the reaction [4, 3, 29, 5, 30, 4, 19] and with previous experimental works on lighter systems [11, 19, 12, 14, 15].

The DD γ-ray angular distribution is a sensitive probe of the fusion dynamics and of the DD lifetime. Indeed, the amount of anisotropy, if present, is related to the interplay of the rotation angular velocity of the dinuclear system during the prompt DD emission and the instant at which this emission occurs [7, 14, 15]. Panel (b) of Figure 1 displays the fusion-evaporation γ-rays angular distribution with respect to the beam axis for the difference between the reactions, integrated over energy from 9 to 15 MeV and corrected for the detection efficiency, i.e. the DD γ-ray angular distribution. The lines describe the angular distribution of the emitted γ-rays given by the Legendre polynomial expansion $M_\gamma(\theta_\gamma) = M_0[1 + Q_2 a_2 P_2 cos(\theta_\gamma)]$, where a_2 is the anisotropy coefficient and Q_2 is an attenuation factor for the finite γ-ray counter [31] (0.98 in our case). The DD γ-ray angular distribution is very anisotropic around 90° and can be reproduced well with a_2 = -1 (solid line) that describes an emission from a dipole oscillation along the beam axis. The dashed line corresponds to a value of a_2 = -0.84 obtained within BNV calculations [4] for evaporation events, while the dotted one shows a more isotropic angular distribution (a_2 = -0.25). The above a_2 values indicate a preferential oscillation axis of the DD along an axis that has not rotated much on the reaction plane during the DD lifetime, confirming the pre-equibrium nature of this emission. This is in agreement with our previous results for evaporation events [14, 15] and with theoretical expectations [7].

By taking into account the DD γ-ray angular distribution (a_2 = -1) for evaporation events and the response function of the experimental setup, the DD yield, integrated over energy and over angle, is $(1.2 \pm 0.2)*10^{-3}$ [18]. The experimental results were also compared with BNV calculations. These calculations reproduced all the DD features, except for the yield that is overestimated. This discrepancy suggests that BNV calculations do not take into account some aspects of the reaction dynamics which could inhibit the DD γ-ray emission. An ingredient neglected is the deformation of the ^{152}Sm target ground state that could influence the DD excitation mechanism. Calculations are under way to evaluate this point.

3.2. Fission: preliminary results

As mentioned before, FF were detected by two position-sensitive PPACs placed at $\theta_{lab} = 52.5°$. The information obtained from these detectors were the ΔE, TOF and x, y positions of the

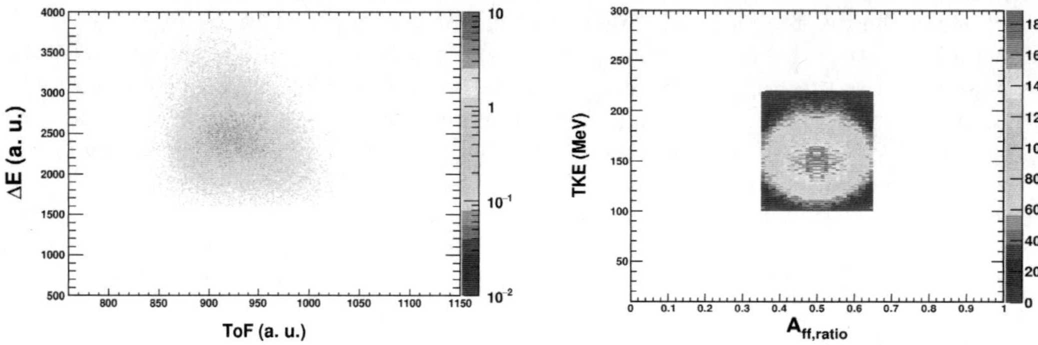

Figure 2. ΔE vs TOF (left-hand side) and TKE vs composite system mass ratio (right-hand side) bidimensional plot for a PPAC at $\theta_{lab} = 52.5°$ in the ^{40}Ca + ^{152}Sm fission reaction. Here, only coincidence events between the two PPACs at $\theta_{lab} = 52.5°$, after the selection procedure, are shown. See the text for details.

fragments. The position signals from the PPACs were transformed event by event to polar and azimuthal angles (θ, ϕ). From TOF and angles of the FF, velocity vectors, masses and Total Kinetic Energy (TKE) were obtained by using kinematical considerations. Moreover, an iterative procedure was done to compensate for the energy losses in the target, assuming that the reaction takes place at the middle of the target.

By applying appropriate conditions in the bidimensional plot (ΔE, TOF) of the PPACs and selecting the same mass and TKE distributions for two coincident FF, the observables of the two reactions can be compared properly. Figure 2 shows the resulting ΔE vs TOF (left-hand side) and TKE vs composite system mass ratio (right-hand side) bidimensional plots for a PPAC in the charge asymmetric reaction, where only coincidence events between the two PPACs are considered.

As done in fusion-evaporation, we evaluated the average excitation energy and the mass of the composite system in fission events (selected as mentioned above) in both reactions, by studying the pre-equilibrium particle emission. Protons and α-particles detected in coincidence with FF were extracted for several polar angles with respect to the beam direction and analyzed by means of a multiple-source least-squares fit where the particles are assumed to be emitted isotropically (in the respective center of mass reference frame) from four moving sources. In this case, the light particle spectra were supposed to originate from: 1) the emission of fast particles of non-statistical origin (the *intermediate-velocity* source) from the composite system before reaching the equilibration, 2) the statistical particle emission from the CN (the *slow* source) before fission and 3-4) the statistical particle emission of the two excited FF. From this analysis, the pre-equilibrium particle emission was found to be lower than in evaporation events, showing a dependence from the impact parameter of the selected channel [32]. Furthermore, we demonstrated that the composite system was formed in both reactions with the same average excitation energy and mass.

After applied the selection procedure of fission events in the two reactions, γ-ray - fission events were determined with a triple coincidence between γ-rays detected by MEDEA detetector and two FF detected by the PPACs. Since the evaluation of the pre-equilibrium particle emission in fission events proved that the composite system was formed in both reactions with the same average excitation energy and average mass, we are entitled to compare the γ-ray spectra of both reactions. This comparison showed that there is also in this channel a γ-ray excess between 8

IOP Conf. Series: Journal of Physics: Conf. Series **966** (2018) 012049 doi:10.1088/1742-6596/966/1/012049

and 15 MeV, in the more charge asymmetric reaction. This excess should be ascribed to the only different parameter between the two reactions, namely the initial charge asymmetry. A more accurate analysis is under way to obtain the γ-ray angular distribution in fission events and to understand the DD dependence from the impact parameter.

4. Conclusion

The present results on the pre-equilibrium GDR radiation in ^{40}Ca + ^{152}Sm fusion-evaporation reaction allows to take a step forward in the study of superheavy element formation, demonstrating that the DD γ radiation, a possible cooling mechanism of the composite system along the fusion path, survives in heavy composite systems in the fusion-evaporation channel.

Furthermore, we proved the DD excitation also in fission events. This observation provides inedited information on the DD excitation at higher partial waves.

By using the prompt DD radiation as a probe and with the advent of more intense radioactive beams, new possibilities for the investigation of the EOS symmetry energy at sub-saturation density are foreseen [7]. Indeed, radioactive beams are expected to maximize the difference of the DD yield between the different prescriptions of the symmetry energy dependence on density and to allow a clear experimental discrimination [7, 33].

References

[1] Brink D 1990 *Nucl. Phys.* A **519** 3c
[2] Chomaz P, Di Toro M and Smerzi A 1993 *Nuc. Phys.* A **563** 509
[3] Simenel C, Chomaz P and de France G 2001 *Phys. Rev. Lett.* **86** 2971
[4] Baran V *et al* 2001 *Phys. Rev. Lett.* **86** 182501 and 2001 *Nuc. Phys.* A **679** 373
[5] Papa M *et al* 2005 *Phys. Rev.* C **72** 064608
[6] Zheng H *et al* 2017 *Phys. Lett.* B **769** 424
[7] Baran V *et al* 2009 *Phys. Rev.* C **79** 021603 (R)
[8] Pierroutsakou D *et al* 2001 *Nucl. Phys.* A **687** 245c
[9] Pierroutsakou D *et al* 2003 *Eur. Phys. Jour.* A **16** 423
[10] Amorini F *et al* 2004*Phys. Rev.* C **69** 014608
[11] Flibotte S *et al* 1996 *Phys. Rev. Lett.* **77** 1448
[12] Pierroutsakou D *et al* 2003 *Eur. Phys. Jour.* A **17** 71
[13] Pierroutsakou D *et al* 2005 *Phys. Rev.* C **71** 054605
[14] Martin B *et al* 2008 *Phys. Lett.* B **664** 47
[15] Pierroutsakou D *et al* 2009 *Phys. Rev.* C **80** 024612
[16] Corsi A *et al* 2009 *Phys. Lett.* B **679** 197
[17] Giaz A *et al* 2014 *Phys. Rev.* C **90** 014609
[18] Parascandolo C *et al* 2016 *Phys. Rev.* C **93** 044619
[19] Simenel C, Chomaz P and de France G 2007 *Phys. Rev.* C **76** 024609
[20] Gavron A 1980 *Phys. Rev.* C **21** 230
[21] Morgenstern H *et al* 1984 *Phys. Rev. Lett.* **52** 1104
[22] Kelly M P *et al* 1997 *Phys. Rev.* C **56** 3201
[23] Holub E *et al* 1983 *Phys. Rev.* C **28** 252
[24] Migneco E *et al* 1993 *Phys. Lett.* B **298** 46
[25] d'Enterria D G *et al* 2002 *Phys. Lett.* B **538** 27
[26] Migneco E *et al* 1992 *Nuc. Inst. Meth.* A **314** 31
[27] Bellia G *et al* 1993 *Nuc. Inst. Meth.* A **329** 173
[28] Puhlhofer F 1977 *Nucl. Phys.* A **280** 27 and Harakeh M N extended version (private communication)
[29] Umar A S and Oberacker V E 2007 *Phys. Rev.* C **76** 014614
[30] Wu H L *et al* 2010 *Phys. Rev.* C **81** 047602; Ma Y G *et al* 2012 *Phys. Rev.* C **85** 024618; Ye S Q *et al* 2013 *Phys. Rev.* C **88** 047602
[31] Rose M E 1953 *Phys. Rev.* **91** 610
[32] Alba R *et al* 1994 *Phys. Lett.* B **322** 38
[33] Di Toro M *et al* 2009 *Int. Jour. Mod. Phys.* E **17** 110

IOP Conf. Series: Journal of Physics: Conf. Series **966** (2018) 012050 doi:10.1088/1742-6596/966/1/012050

Subtraction method in the Second Random Phase Approximation

Danilo Gambacurta

Extreme Light Infrastructure - Nuclear Physics (ELI-NP), Horia Hulubei National Institute for Physics and Nuclear Engineering, 30 Reactorului Street, RO-077125 Magurele, Jud. Ilfov, Romania

E-mail: danilo.gambacurta@eli-np.ro

Abstract. We discuss the subtraction method applied to the Second Random Phase Approximation (SRPA). This method has been proposed to overcome double counting and stability issues appearing in beyond mean-field calculations. We show that the subtraction procedure leads to a considerable reduction of the SRPA downwards shift with respect to the random phase approximation (RPA) spectra and to results that are weakly cutoff dependent. Applications to the isoscalar monopole and quadrupole response in ^{16}O and to the low-lying dipole response in ^{48}Ca are shown and discussed.

1. Introduction

The random-phase approximation (RPA) is a microscopic theory applied very successful for the study of the main properties of collective states. In RPA, collective excitations are described as a superposition of 1 particle-1 hole (1p1h). This method, especially when applied within the Energy Density Functional (EDF) framework, allows to describe fairly good global properties of giant resonances, such as the centroid energy and the total strength distribution.

On the other hand, the RPA model has some very well known limits. Among them we recall that the width and fine structure of the nuclear excitations cannot be reproduced except for the single-particle Landau damping and for the escape width (if continuum states are taken into account). A well-known extension of the RPA scheme is the second RPA (SRPA) model which amounts to enlarge the space of basic elementary excitations by including 2 particle-2 hole (2p2h) configurations. This leads to a richer treatment of the excitation modes.

Only in the last years, large-scale SRPA calculations have been performed, showing merits and limits of this approach. Performing such calculations has allowed to show some features of the SRPA that could not be seen in previous applications, because of strong truncations in the model space or approximations. In particular, the SRPA spectrum is systematically lowered by several MeV with respect to that obtained in the ordinary RPA.

The origin of this SRPA can be traced back to the implicit inclusion of correlations in the ground state that have been already taken into account at mean-field level [1, 2]. This issue leads also to a violation of the Thouless theorem within the SRPA and to stability problems [3]. A solution to both drawbacks was recently applied [4, 5], based on a subtraction procedure [1, 2] suited for beyond RPA approaches.

In this paper, we first briefly discuss the subtraction method in the SRPA framework and then some applications are presented. In particular, the case of the monopole and quadrupole response in the nucleus ^{16}O and of the low-lying dipole response in ^{48}Ca, will be shown and discussed.

2. The subtraction procedure in SRPA

In the RPA approximation the excitations operators are assumed to be a linear superposition of 1p1h operators:

$$Q_\nu^\dagger = \sum_{ph} X_{ph}^\nu a_p^\dagger a_h - \sum_{ph} Y_{ph}^\nu a_h^\dagger a_p. \tag{1}$$

where for notation simplicity, the coupling to total quantum numbers is not indicated.

In the case of the SRPA, the 2p2h configurations are explicitly considered in the description of the excitations operators, having the following structure:

$$Q_\nu^\dagger = \sum_{ph} X_{ph}^\nu a_p^\dagger a_h - \sum_{ph} Y_{ph}^\nu a_h^\dagger a_p + \sum_{p<p',h<h'} (X_{php'h'}^\nu a_{p'}^\dagger a_p^\dagger a_{h'} a_h - Y_{php'h'}^\nu a_h^\dagger a_{h'}^\dagger a_p a_{p'}). \tag{2}$$

In both cases, the energies ω_ν of the excited states and their wave function (e.g. the $X's$ and $Y's$ amplitudes) are obtained by solving an eigenvalue problem of this form:

$$\begin{pmatrix} \mathcal{A} & \mathcal{B} \\ -\mathcal{B}^* & -\mathcal{A}^* \end{pmatrix} \begin{pmatrix} \mathcal{X}^\nu \\ \mathcal{Y}^\nu \end{pmatrix} = \omega_\nu \begin{pmatrix} \mathcal{X}^\nu \\ \mathcal{Y}^\nu \end{pmatrix}. \tag{3}$$

In the RPA case the $\mathcal{A}$ and $\mathcal{B}$ matrices describe the coupling among the 1p1h configurations, while in SRPA, more general block matrices appear, describing the coupling of the 1p1h configurations with the 2p2h configurations and of the 2p2h configurations among themselves, (see for example Ref. [6]).

The subtraction procedure consists in subtracting in the A_{11} block of the SRPA matrix the quantity

$$E_{11'} = -\sum_{2,2'} A_{12}(A_{22'})^{-1} A_{2'1'} - \sum_{2,2'} B_{12}(A_{22'})^{-1} B_{2'1'} \tag{4}$$

where the indices 1 and 2 stand for the 1p1h and 2p2h configurations, respectively. This subtraction guarantees that the subtracted SRPA (SSRPA) response reduces to the RPA one in the zero–frequency limit. When density-dependent interactions are employed, the subtraction procedure requires also the modification of the B_{11} matrix [4].

As discussed above, large scale SRPA calculations [6, 7, 8, 9] show that the SRPA spectrum is systematically lowered by several MeV with respect to that obtained in RPA. On the other hand, RPA results based on the EDF framework are typically in very good agreement with experimental data, at least concerning the centroid energies and total strength of the giant resonances. Therefore, the inclusion of the 2p2h configurations often worsens the RPA description. In the next section, we show the effect of the subtraction procedure in strongly reducing this strong shift.

From a numerical point of view, the main cost in calculating the correction (4) consists in inverting the A_{22} matrix. This inversion becomes trivial if the matrix is assumed to be diagonal. Therefore the correction can be calculated with a reasonable extra–numerical effort. Hereafter, we will use the acronyms SSRPA$_F$ to denote the subtracted SRPA in the full scheme, e.g. full inversion of the A_{22} matrix, and SSRPA$_D$ to denote the subtracted SRPA assuming diagonal the A_{22} matrix in the correction term (4).

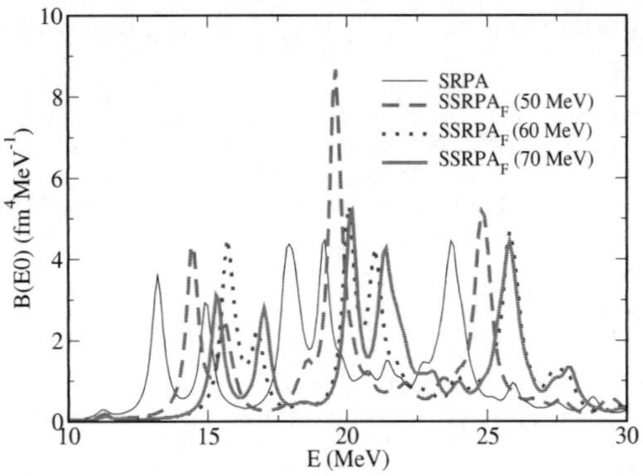

Figure 1. (Color online) Isoscalar monopole response in ^{16}O calculated in the standard SRPA (solid-thin line), and with the SSRPA$_F$, with a cutoff for the correction terms at 50 (dashed line), 60 (dotted line), and 70 (solid-thick line) MeV.

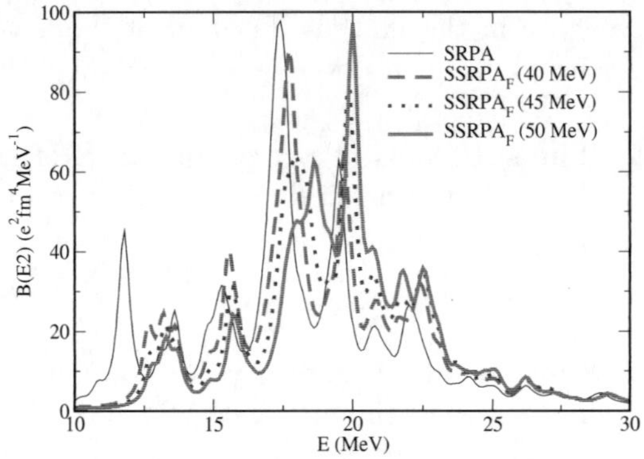

Figure 2. (Color online) Isoscalar quadrupole response in ^{16}O calculated in the standard SRPA (solid-thin line), and with the SSRPA$_F$, with a cutoff for the correction terms at 45 (dashed line), 45 (dotted line), and 50 (solid-thick line) MeV.

3. Subtracted SRPA results

3.1. Results in ^{16}O: isoscalar monopole and quadrupole response

In this section we show the effect of the subtraction method, in the case of the isoscalar monopole and quadrupole response in ^{16}O. We perform Hartree-Fock plus RPA calculations. The SGII [10] interaction is used. We include all the 1p1h configurations with energy less than 100 MeV. For the 2p2h space in the SRPA calculations, we take the cutoff to be at 70 MeV and 50 MeV for the monopole and the quadrupole cases, respectively. Those values lead to about 5000 $2p2h$ configurations in each of the two cases. This number is small enough so that we can still fully invert the matrix $A_{22'}$ to perform the subtraction without approximation. In all the figures that follow, we fold the calculated response with a Lorentzian of width 0.5 MeV.

Fig. 1 shows the isoscalar monopole strength distribution, calculated with the unmodified SRPA and with the SSRPA$_F$, using a cutoff on the 2p2h configurations in the correction term of Eq. (4) equal to 50, 60, and 70 MeV. In the last of these cases, all the SRPA $2p2h$ configurations are included in the correction. The effect of the subtraction, as we expected, is to shift the SRPA spectrum upwards, by amounts that increase with the cutoff in the correction terms. The important differences between the three subtracted strength functions indicate that it is crucial to include all the $2p2h$ states in the correction terms containing $(A_{22'})^{-1}$ in Eq. (4). Figs. 2 displays the same results as Fig. 1 but in the isoscalar quadrupole channel (where the $2p2h$ cutoff is at 50 MeV in the SRPA calculation). The cutoffs in the correction terms are at 40, 45, and 50 MeV. The remarks made about the first figure apply here as well.

Figs 3 and 4 compare the full and diagonal subtracted SRPA results with the 70 MeV and 50 MeV cutoff in the correction terms, respectively. We observe that the SSRPA$_F$ and SSRPA$_D$ results are very similar, the difference being a small shift to larger excitation energies in the SSRPA$_D$.

Another very positive feature of the SSRPA results is the very weakly dependence on the $2p2h$ cutoff. This can been seen in Fig. 5 , where we show the isoscalar monopole responses, with cutoffs for the correction terms at 70, 80, and 90 MeV. The three strength functions are

IOP Conf. Series: Journal of Physics: Conf. Series **966** (2018) 012050 doi:10.1088/1742-6596/966/1/012050

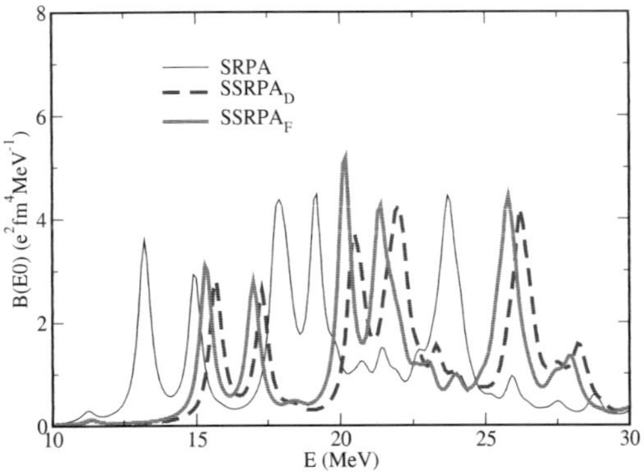

Figure 3. (Color online) Isoscalar monopole response in ^{16}O calculated in the SSRPA$_F$ and SSRPA$_D$ approximation. (See the text for more details).

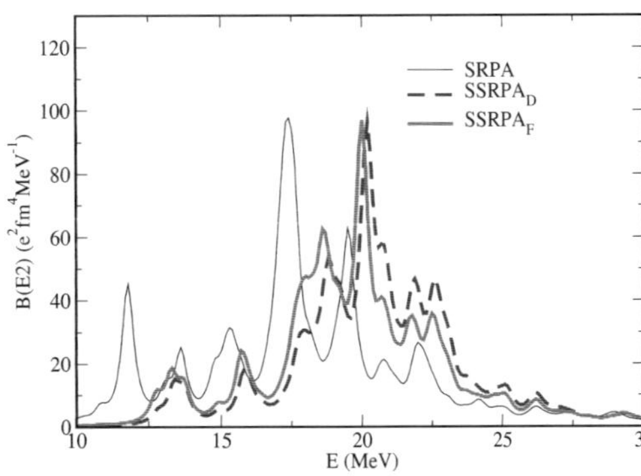

Figure 4. (Color online) Isoscalar quadrupole response in ^{16}O calculated in the SSRPA$_F$ and SSRPA$_D$ approximation. (See the text for more details).

very similar showing that the subtraction procedure provides much more robust, e.g. cutoff–insensitive, results. A similar behaviour has been found also for the quadrupole case.

3.2. Results in ^{48}Ca: low-lying dipole response

The dipole response below the neutron threshold in ^{48}Ca has been measured with the (γ, γ') reaction [11]. Systematic future studies of the nuclear dipole response are planned at ELI-NP in Bucharest, both below [12] and above [13] the neutron separation threshold.

A cutoff of 100 MeV is used in building the 1p1h configurations, ensuring a full preservation of the isoscalar and isovector Energy Weighted Sum Rules (EWSRs). Deviations of less than 1% are found in both cases in RPA. In the SRPA calculations, a cutoff of 60 MeV in the 2p2h configurations is used. The EWSRs are satisfied as in RPA, with deviations less than 1%. On the other side, the SSRPA model provides by construction the same inverse moment m_{-1} as in RPA, but not the same moment m_1 and deviations of $\approx 5 - 7$ % are found for the EWSRs.

We recall that RPA predictions fail in reproducing such a low-lying strength either because the lowest RPA energies are larger than 10 MeV or because the RPA model cannot provide the observed fragmentation. Only beyond mean-field models including correlations, such as the SRPA or the extended theory of finite Fermi systems (ETFFS) [11, 14], based on the quasiparticle–phonon coupling, could account for this low–lying strength in ^{48}Ca. A fragmented SRPA dipole strength below 10 MeV was indeed illustrated in Ref. [8] for ^{48}Ca. However, the SRPA $B(E1)$ transition probability, integrated up to 10 MeV, was found to be definitely much larger than the experimental value. By using the implemented SRPA model augmented via the subtraction method, we show here that problem of overestimation of the transition probability is strongly reduced, providing thus a rather good description of the experimental data. Some states having strong 1p1h components, which are too much shifted downwards by the standard SRPA, are pushed to higher energies by the subtraction, reducing in this way the $B(E1)$ value below 10 MeV. Moreover, previous applications of the SRPA and SSRPA were done without including the spin-orbit and Coulomb terms in the residual interaction. In this work, all the terms of the interaction consistently with the mean-field Hartree-Fock description are included. Due to the inclusion of all terms of the residual interaction, the present SRPA results are different compared to those illustrated in Ref. [8], where the spin–orbit and the Coulomb contributions were not taken into account. In Fig. 6(a) we show the experimental transition probabilities

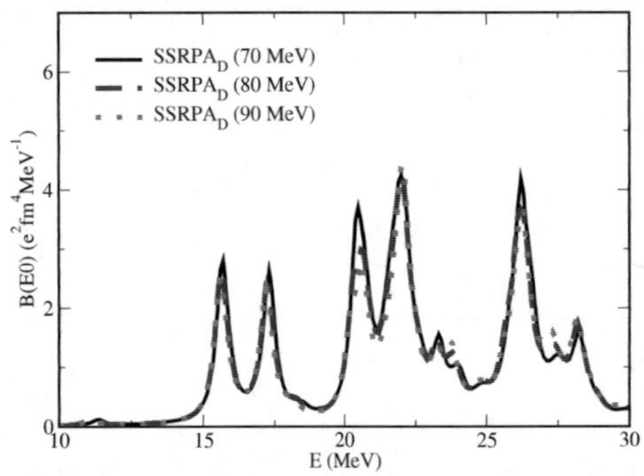

Figure 5. (Color online) Isoscalar monopole response in ^{16}O in the diagonal approximation with cutoff for the correction terms at 70 (black line and magenta area), 80 (green dashed line), and 90 (blue dotted line) MeV.

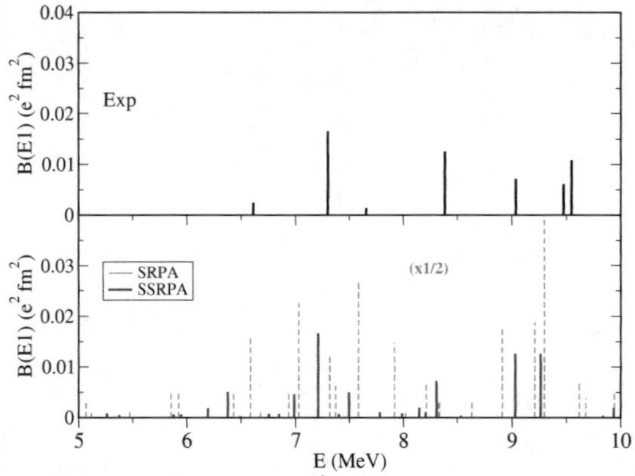

Figure 6. Low-lying dipole strength in ^{48}Ca :(a) Experimental $B(E1)$ values [11]; (b) Theoretical predictions for the transision probabilities $B(E1)$ calculated with the standard SRPA (dashed red bars; the values have been devided by 2) and with the SSRPA (blue thick bars).

$B(E1)$ [11]. The corresponding theoretical results are displayed in Fig. 6(b). One observes that the total strength provided by the standard SRPA is much higher than the experimental one as already found and discussed in Ref. [8]. The theoretical prediction given by the SSRPA model, on the other side, is in a very satisfactory agreement with respect to the experimental data. The fragmentation of the states follows quite well the experimental distribution in the energy position of the main peaks. The total strength integrated between 5 and 10 MeV is shown in Table I. The first column shows the experimental result and the second and third columns the corresponding values obtained with the SRPA and the SSRPA models, respectively. Whereas the SRPA value is eight times larger than the experimental result, the SSRPA summed $B(E1)$ is very close to the measured value. A similar behaviour is found also for the EWSR integrated up to 10 MeV, shown in the second line of Table I.

An interesting information that can be analysed in connection with the strength distribution is the composition of the excitation modes in terms of 1p1h and 2p2h configurations. By extracting the expression of N_1 from the normalization condition,

$$\sum_{ph}(\mid X^{\nu}_{ph}\mid^2 - \mid Y^{\nu}_{ph}\mid^2) + \sum_{p<p',h<h'}(\mid X^{\nu}_{php'h'}\mid^2 - \mid Y^{\nu}_{php'h'}\mid^2) = N_1 + N_2 = 1, \tag{5}$$

we plot in Figs 7 and 8 the $B(E1)$ values corresponding and the quantity N_1 (lower panel) for each excited state in SRPA and SSRPA, respectively. One observes that, in the SRPA case, a strong mixing of 1p1h and 2p2h components is present. On the contrary, in the SSRPA case, states having mainly a 2p2h nature remain as a consequence of the subtraction procedure, pushing up state with a strong 1p1h component.

4. Conclusions

In this paper we have shown and discussed recent applications of the SRPA framework where the subtraction method has been employed. The implemented subtraction method is able to

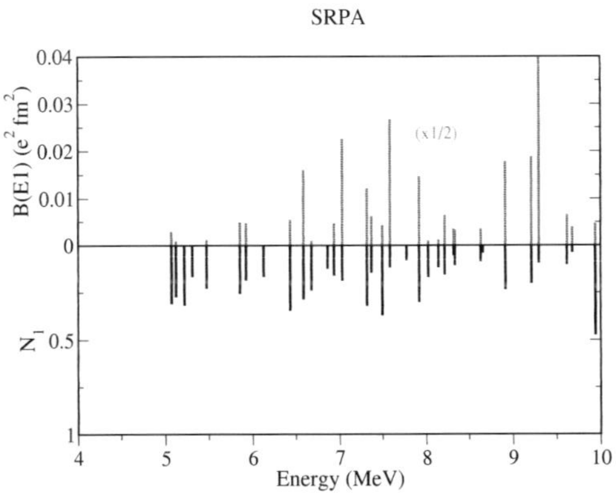

Figure 7. (Color online) For each state the SRPA $B(E1)$ value and the total 1p1h contribution N_1 to the norm of the state defined in Eq. (5) (lower panel), are shown.

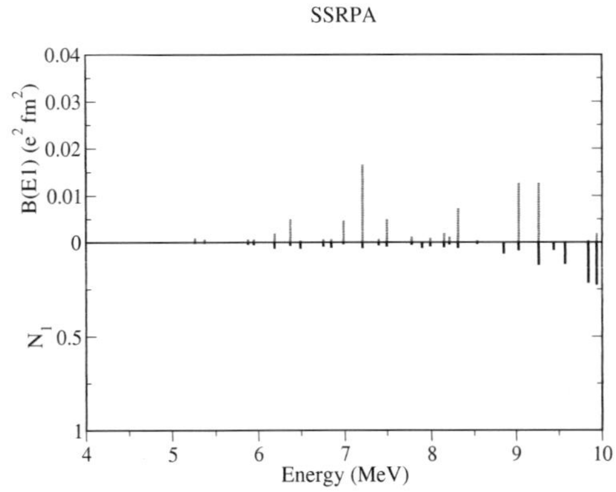

Figure 8. (Color online) For each state the SSRPA $B(E1)$ value and the total 1p1h contribution N_1 to the norm of the state defined in Eq. (5) (lower panel), are shown.

	Exp	SRPA	SSRPA
$\sum B(E1)$	0.068 $\pm$ 0.008	0.563	0.078
$\sum_i E_i B_i(E1)$	0.570 $\pm$ 0.062	4.618	0.621

Table 1. Experimental [11] and theoretical $\sum B(E1)$ in (e^2 fm^2) and $\sum_i E_i B_i(E1)$ in (MeV e^2 fm^2) summed between 5 and 10 MeV.

strongly reduce the shift of the strength found in SRPA. and to provide robust and weakly cutoff dependent predictions. Applications for the description of the monopole and quadrupole response in ^{16}O are shown. After that, the low-lying dipole response in ^{48}Ca, is studied, showing that the subtracted SRPA results improve to a great extent the agreement with experimental data.

References

[1] Tselyaev V I 2007 *Phys. Rev. C* **75** 024306
[2] Tselyaev V I 2013 *Phys. Rev. C* **88** 054301
[3] Papakonstantinou P 2014, *Phys. Rev. C* **90**, 024305
[4] Gambacurta D, Grasso M and Engel J 2015 *Phys. Rev. C* **92** 034303
[5] Gambacurta D and Grasso M 2016 *Eur. Phys. J. A* **52** 198
[6] Gambacurta D, Grasso M, and Catara F 2010 *Phys. Rev. C* **81** 054312
[7] Gambacurta D, Grasso M, and Catara F 2011 *J. Phys. G* **38** 035103
[8] Gambacurta D, Grasso M, and Catara F 2011 *Phys. Rev. C* **84** 034301
[9] Gambacurta D, Grasso M, De Donno V, Co G and Catara F 2012 *Phys. Rev. C* **86** 021304(R)
[10] Giai N V and Sagawa H 1981, *Phys. Lett. B* **106**, 379
[11] Hartmann T *et al.* 2004, *Phys. Rev. Lett.* **93** 192501
[12] Ur C A *et al.* 2016 *Romanian Reports in Physics*, Vol. **68**, Supplement, P. S483S538,
[13] Camera F *et al.* 2016 *Romanian Reports in Physics*, Vol. **68**, Supplement, P. S539S619
[14] Tertychny G 2007 *et al. Nucl. Phys. A* **788**, 159c

IOP Conf. Series: Journal of Physics: Conf. Series **966** (2018) 012051 doi:10.1088/1742-6596/966/1/012051

Surface Flows of Soft Monopole Modes of ^{40}Mg

J.C. Pei

State Key Laboratory of Nuclear Physics and Technology, School of Physics, Peking University, Beijing 100871, China

E-mail: peij@pku.edu.cn

Kai Wang

State Key Laboratory of Nuclear Physics and Technology, School of Physics, Peking University, Beijing 100871, China

M. Kortelainen

Department of Physics, P.O. Box 35 (YFL), University of Jyvaskyla, FI-40014 Jyvaskyla, Finland

Helsinki Institute of Physics, P.O. Box 64, FI-00014 University of Helsinki, Finland

Abstract. In order to explore the nature of collective modes in weakly bound nuclei, we have investigated surface flow patterns of isoscalar monopole modes in a shape-coexisting nucleus ^{40}Mg. The calculations were done in a fully self-consistent continuum finite-amplitude-method quasiparticle-random-phase-approximation (FAM-QRPA) in a large deformed spatial mesh. Our results demonstrated that the transition current densities at surfaces are nontrivial and dependent on deformations. The toroidal mode is favorable in the oblate shape, demonstrating the geometric deformation effects on nuclear surface flows. The surface flow patterns become more complicated as excitation energies increase.

1. Introduction

The low-energy collective modes associated with neutron skin or halo are of great interests in the area of radioactive nuclear beams. Nuclei close to drip lines are weakly bound nuclei and can have significant extended nuclear surfaces. The dilute surfaces are loosely interacting with the cores, and nuclei become a quantum liquid-gases coexisted system. The spatial extended surface allows low energy surface-core relative motions which is novel in contrast to the giant resonances. The resulting soft or pygmy modes are related to a variety of physics aspects [1] such as the equation of state, symmetry energy, size of neutron skin/halo, the incompressibility. The low energy resonances can also enhance the neutron capture rates in astrophysical r-process.

The strengthes of the low-energy modes have been extensively studied theoretically and experimentally. However, the detailed nature of the low-energy modes is still of great of interests. In the framework of QRPA, the transition current density can be a useful tool to directly understand the collective modes. In the superfluid quantum systems, the current flows are irrotational and can have various self-organized patterns. In the finite nuclei, the superfluid flows can also dependent on the excitation energies, deformations, the multipole operators and the weakly binding effects. For the pygmy dipole modes, the superfluid flow can be compressional

IOP Conf. Series: Journal of Physics: Conf. Series **966** (2018) 012051 doi:10.1088/1742-6596/966/1/012051

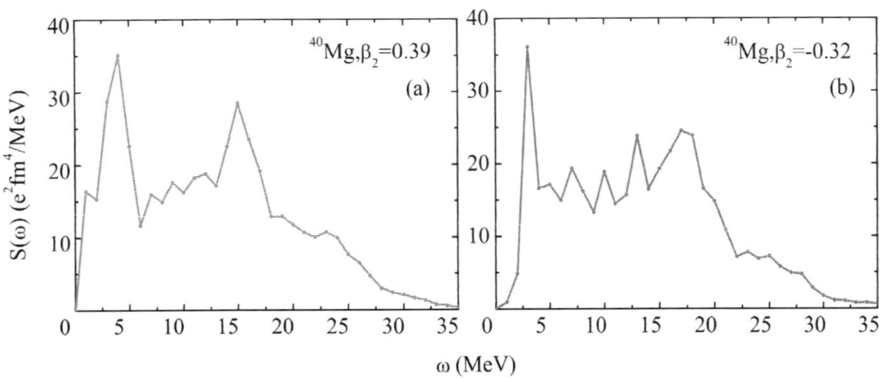

Figure 1. (Color online) Transition strength functions of isoscalar monopole resonances in the shape-coexisting nucleus ^{40}Mg as a function of excitation energies ω. (a) prolate case; (b) oblate case.

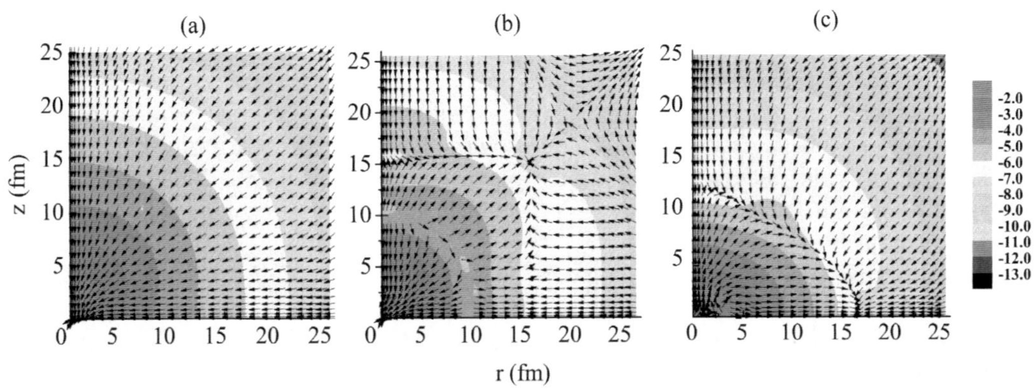

Figure 2. (Color online) Neutron transition current densities $\vec{j}(r, z, \phi = 0)$ of isoscalar monopole soft modes in ^{40}Mg. The color scales denote the logarithm of the current density and the arrows denote the flow directions. (a) ω=1 MeV of prolate shape; (b) ω=4 MeV of prolate shape; (c) ω=3 MeV of oblate shape.

or toroidoal modes [2, 3, 4]. In this work, we are interested in the soft monopole modes. The monopole modes are also called breathing modes. The dilute surfaces can have significant soft monopole modes since the incompressibility is very small [5]. It is interesting to explore the superfluid flows of the soft monopole modes.

The ideal tool is the fully self-consistent continuum QRPA [6]. However, the QRPA calculations are very expensive in deformed cases with continuum configurations [7, 8, 9]. Recently, FAM-QRPA has been proposed [10] and it is very efficient compared to the conventional QRPA method [11, 12, 13, 14]. The FAM-QRPA solves the non-linear QRPA equations iteratively and avoids the computations of huge QRPA matrix elements. We recently implement the FAM-QRPA approach in the large deformed spatial mesh with high precision so that the details of the multipole collective modes can be revealed [15].

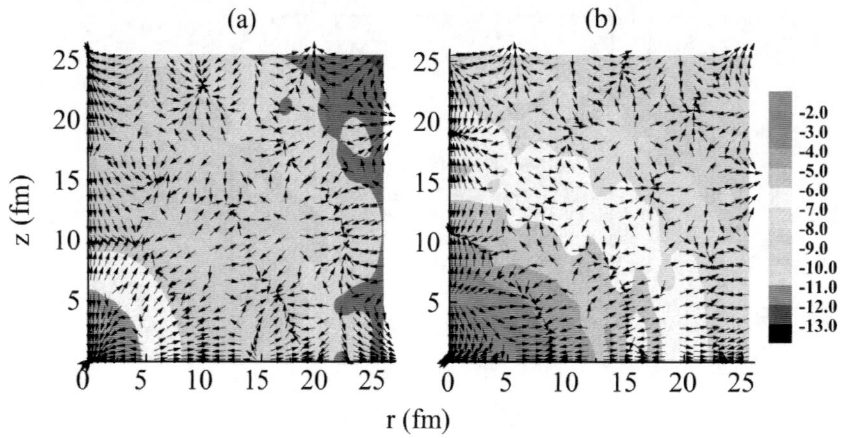

Figure 3. (Color online) Transition currents similar to Fig.2 but for giant resonances. (a) ω=15 MeV of prolate shape; (b) ω=18 MeV of oblate shape.

2. Calculations of transition strengths

In this work, we studied the monopole modes of a weakly-bound deformed ^{40}Mg, which has N=28 magic neutron number but has a well-established prolate-oblate shape coexistence [16, 17]. The shape coexistence provides a good opportunity to study the deformation dependent surface flows. Recently there are strong experimental interests in the spectroscopy of ^{40}Mg as studied in RIKEN.

Firstly, the Hartree-Fock-Bogoliubov equation is solved in the axial-deformed coordinate-spaces of 27.6 fm, using the B-spline techniques [18]. The precise HFB solutions are essential for describing deformed halo structures and continuum discretizations. In our HFB calculations, the ground state of ^{40}Mg is prolate with β_2=0.4, and its energy is 1.9 MeV lower than the prolate shape with β_2=-0.31. Then obtained HFB wavefunctions and quasiparticle energies are inputs for the solutions of the non-linear FAM-QRPA equations which is implemented in a Gauss-Legendre lattice. Our goal is to iteratively solve the FAM-QRPA non-linear equation [10],

$$\begin{aligned}(E_\mu + E_\nu - \omega)\, X_{\mu\nu}(\omega) + \delta H^{20}_{\mu\nu}(\omega) &= -F^{20}_{\mu\nu},\\ (E_\mu + E_\nu + \omega)\, Y_{\mu\nu}(\omega) + \delta H^{02}_{\mu\nu}(\omega) &= -F^{02}_{\mu\nu},\end{aligned} \tag{1}$$

where $X_{\mu\nu}(\omega)$ and $Y_{\mu\nu}(\omega)$ are the FAM amplitudes; $\delta H^{20}_{\mu\nu}$, $\delta H^{02}_{\mu\nu}$ are the induced oscillations in Hamiltonian; F is the external time-dependent field. Then the transition strength can be obtained with X and Y. The expressions of H^{20} and H^{02} in the HFB approach can be found in Ref. [10], which are not needed in the ground state calculations. Note that the time-odd terms [19] of the Skyrme functional are very important in QRPA calculations and have been included. The non-linear equation is solved iteratively using the modified Broyden method. In the solution of the FAM-QRPA equation, a smooth factor of 0.5 MeV has been adopted. The FAM-QRPA solver is implemented with hybrid OpenMP+MPI parallel scheme. The calculations adopt the Skyrme force, new-fitting extended SLy4 force [20] and the density dependent pairing interactions as described in Refs. [15, 21]. The parameters of the extended SLy4 force are given in Ref. [20].

The monopole transition strengths of the prolate and oblate energy minima of ^{40}Mg are displayed in Fig. 1. We see that there are remarkable soft monopole modes in ^{40}Mg. In the prolate case, the peak of a soft monopole mode is at 4 MeV. The soft monopole mode is very

IOP Conf. Series: Journal of Physics: Conf. Series **966** (2018) 012051 doi:10.1088/1742-6596/966/1/012051

broad and is close to the threshold. There is another possible peak at 1 MeV. In the oblate case, the peak of a soft monopole mode appears at 3 MeV. In contrast to the prolate case, the soft monopole mode is narrow and evident. The features of the strengthes of monopole modes can be understand as the influences of deformation effects, which has been demonstrated in isovector dipole modes [15]. We like to point out that the monopole strength of the prolate shape is almost close to zero energy, which is special and very different from other multipole modes of ^{40}Mg and it is interesting to probe it by experimental cross sections near zero degrees.

3. Transition Current Densities

The transition current densities $\vec{j}(r, z, \phi = 0)$ of monopole modes of ^{40}Mg have been studied to reveal the surface flow patterns. Fig.2 displays the neutron current flows of the soft monopole modes at E=1 and 4 MeV of the prolate shape and at E=3 MeV of the oblate shape, respectively. In Fig.2(a), the in-phase compressional flow pattern is very simple at E=1 MeV. In Fig.2(b), at E=4 MeV of the prolate shape, the compressional flow pattern has two boundary lines at 10 fm and 15 fm respectively. This is similar to the flow patterns of dipole modes around 4 MeV [15]. In Fig.2(c), with E=3 MeV of the oblate shape, the compressional flow forms a distinct circulation, or a toroidal mode. In the oblate K=1 dipole mode, there is also an ambiguous flow circulation. Thus we say the toroidal mode is more favorable in the oblate shape in contrast to the prolate shape, demonstrating the geometric deformation effects on nuclear surface flows.

Fig.3 displays the neutron current flows of the giant monopole modes at E=15 MeV of the prolate shape and at E=18 MeV of the oblate shape. The flows are compressional but are more complicated compared to the flows of soft modes in Fig.2. This is similar to the flows of giant dipole modes. Both the soft and giant monopole modes have an in-phase pattern of transition densities. It is difficult to identify the geometric deformation effects in flows of giant monopole modes. In both soft and giant resonances, the flows of protons involve a small region with a distance less than 12 fm. Thus the novel surface flow patterns we demonstrated in Fig.2 and Fig.3 are unique phenomena in weakly bound nuclei with largely-extended neutron surfaces.

4. Summary

We have developed the deformed continuum FAM-QRPA solver for multipole collective excitations in a large spatial mesh. This enable us to reveal the detailed nature of soft collective excitations. In the previous work, we studied the surface flows of the pygmy dipole modes. In this work, we studied the surface flows of the soft monopole modes in the shape coexisting weakly-bound ^{40}Mg. We see the monopole modes are compressional and the flow patterns are dependent on the geometric deformations and excitation energies. The lowest-energy mode has the simplest flow patterns. The oblate shape is favorable for the toroidal mode. The flow patterns certainly become complex as excitation energies increase.

References

[1] Paar N, Vretenar D, Khan E and Colò G 2007 *Rept. Prog. Phys.* **70** 691
[2] Vretenar D, Paar N and Ring P 2002 *Phys. Rev.*C **65** 021301
[3] Repko A, Reinhard P -G, Nesterenko V O and Kvasil J 2013 *Phys. Rev.* C **87** 024305
[4] Papakonstantinou P, Wambach J, Mavrommatis E and Ponomarev V Yu 2004 *Phys. Lett.* B **604** 157
[5] Pei J C, Kortelainen M, Zhang Y N and Xu F R 2014 *Phys. Rev.* C **90** 051304(R)
[6] Ring P and Schuck P 1980 *The nuclear many-body problem*(Springer, Berlin)
[7] Matsuo M 2001 *Nucl. Phys.* A **696** 371
[8] Pei J C, Zhang Y N and Xu F R 2013 *Phys. Rev.* C **87** 051302(R)
[9] Zhang Y N, Pei J C and Xu F R 2013 *Phys. Rev.* C **88** 054305
[10] Avogadro P and Nakatsukasa T 2011 *Phys. Rev.* C **84** 014314
[11] Stoitsov M, Kortelainen M, Nakatsukasa T, Losa C and Nazarewicz W 2011 *Phys. Rev.* C **84** 041305(R); Liang H, Nakatsukasa T, Niu Z and Meng J 2013 *Phys. Rev.* C **87** 054310; Hinohara N, Kortelainen M, and Nazarewicz W 2013 *Phys. Rev.* C **87** 064309

[12] Kortelainen M, Hinohara N and Nazarewicz W 2015 *Phys. Rev.* C **92** 051302(R)
[13] Oishi T, Kortelainen M and Hinohara N 2016 *Phys. Rev.* C **93** 034329
[14] Washiyama K and Nakatsukasa T, arXiv:1708.06878v1[nucl-th]
[15] Wang K, Kortelainen M and Pei J C 2017 *Phys. Rev.* C **96** 031301(R)
[16] Terasaki J, Flocard H, Heenen P-H. and Bonche P 1997 *Nucl. Phys.* A **621** 706
[17] Rodríguez-Guzmán R, Egido J L and Robledo L M 2002 *Phys. Rev.* C **65** 024304
[18] Pei J C, Stoitsov M V, Fann G I, Nazarewicz W, Schunck N and Xu F R 2008 *Phys. Rev.* C **78** 064306
[19] Bender M, Dobaczewski J, Engel J and Nazarewicz W 2002 *Phys. Rev.* C **65** 054322
[20] Xiong X Y, Pei J C and Chen W J 2016 *Phys. Rev.* C **93** 024311
[21] Pastore A, Barranco F, Broglia R A and Vigezzi E 2008 *Phys. Rev.* C **78** 024315

IOP Conf. Series: Journal of Physics: Conf. Series **966** (2018) 012052 doi:10.1088/1742-6596/966/1/012052

Opportunities for nuclear astrophysics at FAIR

K. Langanke[1,2]

[1]GSI Helmholtzzentrum für Schwerionenforschung, Planckstr. 1, 64291 Darmstadt, Germany
[2]Institut für Kernphysik, TU Darmstadt, Schlossgartenstr. 9, 64291 Darmstadt, Germany

E-mail: `k.langanke@gsi.de`

Abstract. The Facility for Antiproton and Ion Research FAIR is a new accelerator complex currently under construction in Darmstadt, Germany. FAIR will offer unprecedented research opportunities in nuclear, hadron and atomic physics and in several applied sciences. This manuscript discusses how FAIR will advance our understanding of the origin of the elements in the universe and of the astrophysical objects which produce them.

1. Introduction

Our understanding of the origin of the elements in the universe [1] and of the astrophysical objects which produce them has significantly advanced since Radioactive Ion-Beam (RIB) facilities allow to produce and study the short-lived nuclei which play often crucial roles in these astrophysical scenarios. A further advance is expected from FAIR, with its unique combination of high-intensity, high-energy accelerators for all ion species (and antiprotons), high- and low-energy storage rings, a superconducting Fragment Separator and a large variety of fore-front detector systems. In particular the use of the storage rings will allow precision measurements of masses, half-lives and reaction cross sections of many short-lived nuclei which until now have not been experimentally accessible. Furthermore the high-intensity beams will be exploited in relativistic heavy-ion collisions which will shed light on the nuclear Equation of State (EoS) and will progress our knowledge about the properties of matter inside neutron stars.

There has been significant advances in the understanding of explosive astrophysical objects due to progress in nuclear (and astrophysical) modelling and due to some key measurements performed at existing RIB facilities or in selected stable-beam experiments [2, 3, 4]. In this manuscript we will briefly review two recent developments: the improved description of nuclear physics input in supernova simulations and the discussion of neutron-star mergers as a potential production site of heavy elements in the Universe via the astrophysical r-process. Despite the important progress achieved in the combined efforts of experimental and theoretical nuclear physics and of astrophysical modelling and astronomical observation many open questions, however, still remain which hopefully can be addressed once next-generation facilities like FAIR are operational.

2. FAIR and supernovae

A massive star ends its life in a supernova explosion triggered by the gravitational collapse of its inner core. The collapse comes to a halt when the inner region of the core has reached nuclear matter densities. A shock wave is created which, with the help of energy transport by neutrinos and by convective motion, ejects the outer shells of the star [5, 6]. Nuclear physics plays an important role at various stages of the explosion and FAIR can contribute to all of them, as we briefly discuss in the following.

2.1. Electron capture during collapse

Electron captures onto nuclei play crucial roles for the dynamics of the collapse [7, 8]. These captures are made possible as the Fermi energy ϵ_F of the relativistic electron gas exceeds typical nuclear Q-values at densities of 10^9 g/cm^3 and beyond. Continuous electron captures reduce the pressure which the degenerate electron gas can exert against the collapse and cools the core due to the emission of neutrinos associated with the captures. Finally the electron capture drives the matter composition in the core, which is in nuclear statistical equilibrium, to heavier and more neutron-rich nuclei. Due to the effective cooling, the entropy in the core stays low and heavy nuclei survive the collapse [7].

During the early collapse, where the core composition is dominated by nuclei from the iron-nickel mass range (pf shell nuclei), the capture process is dominated by Gamow-Teller (GT) transitions. As under these conditions $\epsilon_F \sim Q$, the capture rates are quite sensitive to a detailed and accurate description of the GT distribution. The pioneering work to measure GT strength functions has been performed at TRIUMF using (n,p) reactions, achieving, however, only modest energy resolutions [9]. This situation has been dramatically improved by the development of the (d,^{2}He) technique at the KVI Groningen which allowed to determine GT distributions for many nuclei in the iron-nickel mass range with an energy resolution of about 150 keV [10]. Furthermore, modern diagonalization shell model calculations have been proven to reproduce the measured distributions quite well [11] and are hence the method of choice to calculate the relevant stellar electron capture rates for the early collapse phase [12, 13]. In a detailed comparison to experiment Cole *et* [14] confirmed that the diagonalization shell model indeed gives a reliable desription of early stellar electron capture. We note that in the stellar environment electron capture is reduced (approximately by a factor of order 2) due to screening effects [15]. As beta decays are enhanced by screening, stellar environment effects can alter the conditions at which URCA pairs operate [43]. This is particularly important for the core collapse of $8-12M_\odot$ stars, (electron capture supernovae) [17], where several URCA pairs influence the dynamics of the core evolution.

In the later stage of the collapse of massive stars, the nuclei present in the core composition get heavier and more neutron-rich. At these conditions the electron Fermi energy is noticeably larger than the nuclear Q values and the electron capture rates are mainly sensitive to the total GT strength and its centroid. These quantities are well described by an hybrid approach proposed in [18]. The calculations are then based on a statistical shell model approach (Shell Model Monte Carlo (SMMC) [19]) which allows to determine nuclear properties at finite temperature and in large multi-shell model spaces taking the relevant nuclear correlations into account. Such correlations are particularly important for nuclei with proton number below and neutron number above an oscillator shell closure (like N=40). In such situations GT transitions would be completely blocked by the Pauli principle in the Independent Particle Model (IPM) [20] suppressing electron capture on nuclei drastically. However, it has been shown in [18] that nuclear correlations induced by the residual interaction move nucleons across the shell gap enabling GT transitions and making electron capture on nuclei the dominating weak interaction process during collapse [18, 21]. This unblocking of the GT strength across the $N = 40$ shell closure has been experimentally confirmed for ^{76}Se (with 34 protons and 42 neutrons) [22], in agreement with shell model studies [23]. Furthermore, shell model studies clearly show that the description of cross-shell correlations is a rather slowly converging process requiring the consideration of multi particle-multi hole configurations [24, 25, 26, 23].

As most of the relevant neutron-rich nuclei, encountered in the later stage of the collapse, are unstable, the measurement of the respective GT distributions requires charge-exchange experiments in inverse kinematics. Due to the mass difference between the (unstable) heavy projectile and the lighter target such experiments are only possible at RIB facilities with high-energy secondary beams. In a landmark experiment the GT strength has been measured for the unstable nucleus ^{56}Ni [27]. At FAIR the EXL collaboration will perform such experiments in storage rings with stored and cooled ions which increases the effectiveness drastically. FAIR will decisively contribute to the fundamental issue of cross-shell correlations by spectroscopy and reaction experiments with neutron-rich nuclei, in this way deepening our understanding of GT unquenching and guiding improved models.

The shell model weak-interaction rates have significant impact on collapse simulations. In the presupernova phase ($\rho < 10^{10}$ g/cm^3) the captures proceed slower than assumed before and for a short period during silicon burning β-decays can compete [28, 29]. As a consequence, the core is cooler, more massive and less neutron rich before the final collapse. Based on the SMMC calculations it has been shown in [18] that capture on nuclei dominates over capture on free protons. Also changes of the supernova trajectories compared to previous simulations with less accurate capture rates are significant [6, 18, 30].

The shell model electron capture rates [12, 13], which are noticeably slower than the pioneering rates of Fuller *et al.* (FFN) for *pf* shell nuclei [31], have also important consequences in nucleosynthesis studies for thermonuclear (type Ia) supernovae by yielding larger electron-to-nucleon ratio behind the burning front than obtained with the FFN rates [32].

2.2. Shock wave and Nuclear Equation of State

When the interior of the collapsing core reaches nuclear matter densities (or slightly beyond), a shock wave is created at the surface of the so-called homologous core of about 0.5 $M_\odot$. The shock looses energy by dissociation of nuclei into free nucleons so that its energy is insufficient to traverse the infalling matter of the iron core. The shock gets stalled and becomes revived by energy transport due to neutrinos from the hot interior and to convective motion. Details of this picture depend strongly on the nuclear Equation of State (EoS) which is subject of intensive experimental studies at FAIR exploiting relativistic collisions in which a high-intensity, high-energy beam of heavy ions is impinged on a fixed target and creates, for a short time, a hot fireball of nuclear matter, where energy is transferred into the generation of pairs of short-lived particles. Due to their relatively long mean free paths, di-lepton pairs are well-suited probes to study the properties of the fireball matter. This is achieved by the FAIR detectors HADES and CBM, where CBM due to its unprecedented event rate capabilities will also observe rare probes like particles with charm quarks created in the fireball [33]. Recent supernova simulations [34], which also considered muons as substitute of high-energy electrons in the EoS, indicate that the density/temperature regime reached in the collapse correspond very well to the parameter space which is exploited by the HADES experiment [35].

Two properties of the EoS are subject of other experimental approaches at FAIR. The compression modulus can be studied by excitation of the giant monopole resonance, which is planned as an experimental storage ring program by the EXL collaboration. Using inverse kinematics, such experiments can be performed for different (stable and unstable) nuclei within a long isotope chain to explore additionally the isospin dependence of this quantity. The isospin dependence enters also via the symmetry energy into the EoS. It has been shown that the symmetry energy scales with the dipole polarizability [36] and the difference between nuclear mass and charge radii [37]. An intensive program to study these quantities is envisioned by the R^{3}B collaboration at FAIR by means of Coulomb excitation.

IOP Conf. Series: Journal of Physics: Conf. Series **966** (2018) 012052 doi:10.1088/1742-6596/966/1/012052

2.3. Explosive and neutrino nucleosynthesis

During the explosion matter is ejected from the surface of the nascent neutron star, triggered by neutrino interactions. Due to the high temperatures involved, the matter exists originally as free nucleons, which upon reaching cooler regions can combine to nuclei. The abundances produced in this 'neutrino-driven wind' nucleosynthesis depends crucially on the proton-to-neutron ratio of the ejected matter which is determined by the competition of neutrino and antineutrino absorptions on neutrons and protons, respectively, and depends strongly on the neutrino luminosities and spectra [38]. Simulations show that the proton-to-neutron ratio can vary with time so that the ejected matter can be either proton rich (leading to νp-process nucleosynthesis [39, 40, 41]) or slightly neutron rich (resulting in a weak r-process which can produce nuclides up to mass number $A \sim 130$, but not the elements in the third r-process peak ($A \sim 195$) [43, 44]). Finally, when the neutrinos pass through the outer stellar shells they can contribute to the production of selective nuclides by neutrino-induced spallation reactions on more abundant nuclei [42]. Calculations show that this neutrino-nucleosynthesis process is important for production of ^{11}B, ^{19}F, ^{138}La and ^{180}Ta [45], but also for isotopes ^{22}Na and ^{26}Al [46], which both play an important role in radioastronomy.

3. Neutron-star mergers as a site for the r-process

The r-process is the astrophysical origin of about half of the elements heavier than iron. It occurs in an astrophysical environment with extreme neutron densities [47]. As a consequence the r-process path in the nuclear chart runs through nuclei with such large neutron excesses that most of them have yet not been made in the laboratory and their properties have to be modelled. Of particular importance are the nuclei with magic neutron numbers. With their relatively long half-lives, as predicted by theory, they act as obstacles in the r-process flow (so-called waiting points).

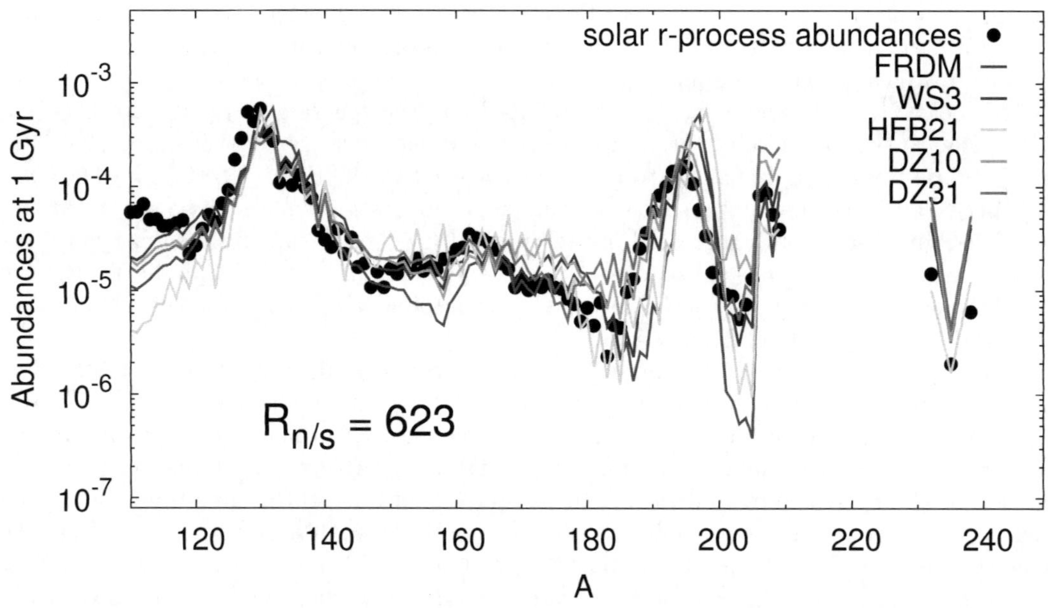

Figure 1. Final r-process abundances at a time of 1 Gyr for various mass models and a selected neutron-star merger trajectory with initial neutron-to-seed, $R_{n/s} = 623$. The peaks at around $A = 130$ and 195 are refered to as second and third r-process peaks. (from [53])

The astrophysical site of the r-process is still an open question. Recent observations of the abundances of r-process elements in old metal-poor stars might even point to (at least) two different sites: the abundances of the elements with mass numbers $A > 130$ show the same relative pattern as found in the solar system (i.e. accumulated over the history of the galaxy until the solar system was formed), while the abundances of the lighter elements in these stars are underproduced relative to solar (e.g. [48]). Neutron-star mergers are discussed as the site for the robust production of the heavy r-process nuclides (with $A > 130$) [49, 50], while the neutrino-driven wind from the nascent neutron star in a core-collapse supernova might contribute additionally to the production of the lighter nuclides [43, 44].

R-process simulations in a neutron-star merger environment show several noticeable features. At first, the matter ejected during the merger expands and cools and by beta-decay of very neutron-rich nuclei is reheated to establish an $(n,\gamma) \rightleftarrows (\gamma,n)$ equilibrium [51]. Another characteristic of r-process simulations for neutron star mergers is the occurence of fission cycling. Here the very extreme neutron densities support mass flow up to the mass range $A \sim 280 - 300$ where the nuclei decay by fission, producing nuclides around ^{132}Sn as fission fragments. Simulations thus have the interesting feature that the third peak is generated by neutron captures already before freeze-out, while the second peak at $A \sim 130$ and the lead peak develop as products of fission and alpha decays, respectively, at late stages of the process after freeze out. In fact, using neutron-star trajectories from Ref. [52], Ref. [53] succeeded to achieve a rather robust r-process pattern between the second and third peaks in close resemblance with the solar r-process distribution (see Fig. 1).

It is intriguing to speculate that neutron star mergers might be responsible for the robust r-process patterns observed in old metal-poor stars [48]. This requires, however, that the frequency of neutron star mergers is sufficiently large during the early time of our galaxy which is yet an open question. A proof that neutron-star mergers are indeed sites of r-process nucleosynthesis might come from the direct observation of the associated electromagnetic transient, called kilonova [54]. The respective lightcurve is produced by the radioactive decay of freshly synthesized nuclei, in particular, by alpha decays of actinides. Hence the lightcurves are very sensitive to the amout of actinides produced in the r-process which in turn depends on the mass flow through the third r-process peak.

Despite these intruiging results, the simulation of Ref. [53] shows a sensitivity to the properties of nuclei around the third r-process peak. These nuclei are the domain of FAIR, which with its high-energy primary beams, isotopic-pure secondary beams and storage rings offer unique opportunities to produce and study, for the first time, the masses, half-lives and beta-delayed neutron emission probabilities of the nuclei in the $A \sim 195$ r-process peak. In particular, these measurements will test the recent predictions [55, 56] that forbidden transitions contribute significantly to the half-lives of the $N = 126$ r-process nuclei resulting in a faster mass flow through the third r-process peak.

The work has benefitted from many discussions with Hans-Thomas Janka, Gabriel Martinez-Pinedo and Friedel Thielemann.

References

[1] Burbidge E M, Burbidge G R, Fowler W A and Hoyle F 1957 *Rev. Mod. Phys.* **29** 547
[2] Wiescher M, Käppeler F and Langanke K 2012 *Ann. Rev. Astron. Astrophys.* **50** 165
[3] Thielemann F -K *et al.* 2011 *Prog. Part. Nucl. Phys.* **66** 346
[4] Langanke K and Schatz H 2013 *Physica Scripta T* **152** 014011
[5] Bethe H A 1990 *Rev. Mod. Phys.* **62** 801
[6] Janka H-Th *et al.* 2007 *Phys. Rep.* **442** 38
[7] Bethe H A, Brown, G E, Applegate J and Lattimer J M 1979 *Nucl. Phys. A* **324** 487
[8] Langanke K, Martínez-Pinedo G 2003 *Rev. Mod. Phys.* **75** 819
[9] Vetterli M C *et al.* 1992 *Phys. Rev. C* **45** 997
[10] Frekers D 2006 *Prog. Part. Nucl. Phys.* **57**

[11] Caurier E, Langanke K, Martínez-Pinedo G and Nowacki F 1999 *Nucl. Phys. A* **653** 439
[12] Langanke K and Martínez-Pinedo G 2000 *Nucl. Phys. A* **673** 481
[13] Langanke K and Martínez-Pinedo G 2001 *At. Data Nucl. Data Tables* **79** 1
[14] Cole A L *et al.* 2012 *Phys. Rev. C* **86** 015809
[15] Juodagalvis A *et al* 2010 *Nucl. Phys. A* **848** 454
[16] Martínez-Pinedo G *et al.*, 2014 *Phys. Rev. C* **89** 045806
[17] Nomoto K 1987 *Astrophys. J.* **322** 206
[18] Langanke K *et al.* 2003 *Phys. Rev. Lett.* **90** 241102
[19] Koonin S E, Dean D J and Langanke K 1997 *Phys. Rep.* **278** 2
[20] Fuller G M 1982 *Astr. J.* **252** 741
[21] Fischer T, Langanke K, and Martínez-Pinedo G 2013 *Phys. Rev. C* **88** 065804.
[22] Grewe E W *et al.* 2008 *Phys. Rev. C* **78** 044301
[23] Zhi Q *et al.*, 2011 *Nucl. Phys. A* **859** 172
[24] Dean D J *et al.* 1999 *Phys. Rev. C* **59** 2474
[25] Dean D J *et al.* 1997 *Phys. Lett. B* **399** 1
[26] Caurier E *et al.* 2001 *Phys. Lett. B* **522** 69
[27] Sasano M *et al.* 2011 *Phys. Rev. Lett.* **107** 202501
[28] Heger A *et al.* 2001 *Phys. Rev. Lett.* **86** 1678
[29] Heger A *et al.* 2001 *Astr. J.* **560** 307
[30] Hix R W *et al.* 2003 *Phys. Rev. Lett.* **91** 210102
[31] Fuller G M, Fowler W A and Newman M J 1980 *ApJS* **42** 447; 1982 *ApJS* **48** 279
[32] Brachwitz F *et al.* 2000 *Astrophys. J.* **536** 934
[33] Ablyazimov T *et al.* 2017 *Eur. J. Phys. A* **53** 60
[34] Martínez-Pinedo G *et al.*, to be published
[35] Galatyuk T *et al.* 2016 *Eur. J. Phys. A* **52** 131
[36] Reinhard P G and Nazarewicz W 2010 *Phys. Rev. C* **81** 051303
[37] Brown B A 2000 *Phys. Rev. Lett.* **85** 5296
[38] Qian Y -Z and Fuller G M 1995 *Phys. Rev. D* **52** 656
[39] Fröhlich C*et al.* 2005 *Astrophys. J.* **637** 415
[40] Fröhlich C, Martínez-Pinedo G *et al.* 2006 *Phys. Rev. Lett.* **96** 142502
[41] Pruet J *et al.* 2005 *Astrophys. J.* **623** 325
[42] Woosley S E *et al.* 1990 *Astrophys. J.* **356** 272
[43] Martínez-Pinedo G *et al.* 2014 *J. Phys. G* **41** 044008
[44] Wanajo S, Müller B, Janka H -T and Heger A, in print
[45] Heger A *et al.* 2005 *Phys. Lett. B* **606** 258
[46] Sieverding A *et al.* 2016 *EPJ Web of Conferences* **109** 06004
[47] Cowan J J, Thielemann F -K and Truran J W 1991 *Phys. Rep.* **208** 267
[48] Sneden C, Cowan J J and Gallino R 2008 *Annu. Rev. Astron. Astrophys.* **46** 241
[49] Goriely S, Bauswein A and Janka H -Th 2011 *Astrophys. J.* **738** L32
[50] Korobkin O *et al.* 2012 *Mon. Not. Roy. Ast. Soc.* **426** 1940
[51] Arnould M, Goriely S and Takahashi K 2007 *Phys. Rept.* **450** 97
[52] Bauswein A, Goriely S and Janka H -T 2013 *Astr. J.* **773** 78
[53] de Jesus Mendoza-Temis J *et al.* 2015 *Phys. Rev. C* **92** 055805
[54] Metzger B D *et al.*, 2010 *Mon. Nat. R. Astr. Soc.* **406** 2650
[55] Suzuki T *et al.* 2012 *Phys. Rev. C* **85** 015802
[56] Zhi Q *et al.* 2013 *Phys. Rev. C* **87** 025803

Hard two-photon contribution to elastic lepton-proton scattering determined by the OLYMPUS experiment

D K Hasell for the OLYMPUS Collaboration

Massachusetts Institute of Technology, 77 Massachusetts Avenue, Cambridge, MA 02139 USA

E-mail: hasell@mit.edu

Abstract. The OLYMPUS collaboration has recently made a precise measurement of the positron-proton to electron-proton elastic scattering cross section ratio, $R_{2\gamma}$, over a wide range of the virtual photon polarization, $0.456 < \epsilon < 0.978$. This provides a direct measure of hard two-photon exchange in elastic lepton-proton scattering widely thought to explain the discrepancy observed between unpolarized and polarized measurements of the proton form factor ratio, $\mu_p G_E^p / G_M^p$. The OLYMPUS results are small, within 1% on unity, over the range of momentum transfers measured and significantly lower than theoretical calculations that can explain part of the observed discrepancy in terms of two-photon exchange at higher momentum transfers. However, the results are in reasonable agreement with predictions based on phenomenological fits to the available form factor data.

The motivation for measuring $R_{2\gamma}$ will be presented followed by a description of the OLYMPUS experiment. The importance of radiative corrections in the analysis will be shown also. Then we will present the OLYMPUS results and compare with results from two similar experiments and theoretical calculations.

1. Introduction

It has been about 100 years since Ernst Rutherford named the hydrogen nucleus the proton; later discovered to be a fundamental component in all nuclei. Yet many fundamental parameters of the proton are still not completely understood and still excite both theoretical and experimental research. The proton radius [1], the proton spin [2], and how the proton mass arises from the energy of the constituent and current quarks in lattice QCD [3] are all still topical subjects in nuclear physics. The OLYMPUS experiment addressed yet another "proton puzzle" [4, 5] concerning the ratio of the charge and magnetic form factors.

Electron scattering has long been a standard technique for studying nucleons and nuclei. The electromagnetic interaction is well understood and the point-like nature of electrons make them ideal for probing electric and magnetic charge distributions. Historically, unpolarized electron-proton scattering has been analysed in terms of one-photon exchange (Born approximation) to determine the electric, G_E^p, and magnetic, G_M^p, form factors for the proton. But recent experiments with polarized electrons, polarized targets, and measurements of the polarization transferred to the proton are in striking disagreement with the unpolarized results for the proton form factor ratio, $\mu_p G_E^p / G_M^p$ (see Fig. 1). The unpolarized results [6–13] obtained using the Rosenbluth technique, are known to be insensitive to the electric form factor, G_E^p, at high

IOP Conf. Series: Journal of Physics: Conf. Series **966** (2018) 012053 doi:10.1088/1742-6596/966/1/012053

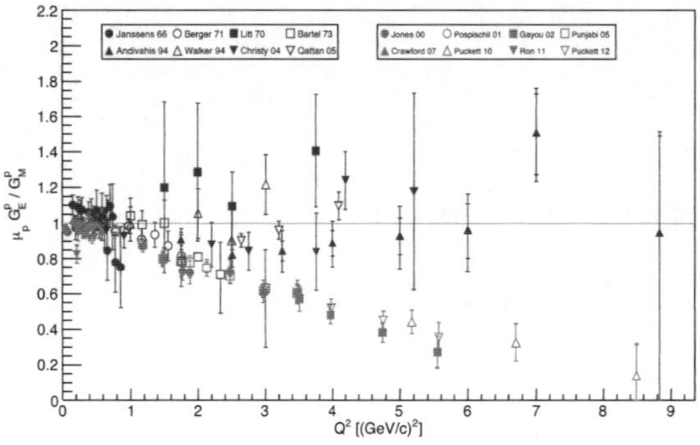

Figure 1. Ratio of proton form factors measured using the unpolarized Rosenbluth technique and using polarization transfer.

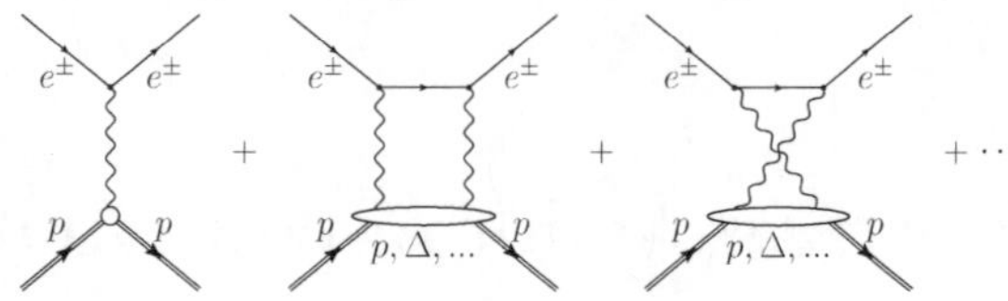

Figure 2. Some of the Feynman diagrams for lepton-proton scattering showing one- and two-photon exchange. Diagrams for self-energy, vertex corrections, vacuum polarization, and bremsstrahlung (not shown) must also be included in any calculation.

momentum transfer while the polarization measurements [14–21] make a direct measurement of the form factor ratio, by measuring the ratio of transverse to longitudinal nuclear polarization.

It has been suggested that two-photon exchange (see Fig. 2) might be able to explain the observed discrepancy in the measured form factor ratio. Radiative corrections must be applied to the measured cross sections to extract the equivalent one-photon exchange value so results from different experiments and theoretical calculations can be compared. These radiative corrections can be significant and are complicated by details of the experimental acceptance, efficiency, and resolution. Two-photon exchange is generally included in the standard radiative corrections but only in the "soft" limit where one of the photons imparts negligible energy to the proton. These calculations generally do not include models for the proton structure. A more complete handling of two-photon exchange contributions might be able to resolve the discrepancy. A proper calculation of "hard" two-photon exchange is more difficult because details of the proton ground state and nucleon resonances for the intermediate state must also be considered.

To determine the contribution of "hard" two-photon exchange, the OLYMPUS experiment proposed to measure the ratio of positron-proton to electron-proton elastic scattering. If two-photon exchange is a significant factor in lepton-proton scattering the ratio will deviate from unity because the interference between one- and two-photon exchange changes sign between electron and positron scattering. Naively, one would expect a small effect, of order $\alpha \approx 1/137$, but that wouldn't explain the striking discrepancy observed.

2. OLYMPUS Experiment

The OLYMPUS experiment [22] ran on the DORIS storage ring at the DESY Laboratory in Hamburg, Germany. DESY undertook significant modifications to the DORIS storage ring to accommodate the OLYMPUS experiment. RF cavities and quadrupoles had to be relocated from the straight section of the storage ring where OLYMPUS was to be located. Services for cooling water and power for the OLYMPUS toroidal magnet had to be installed and the shielding walls extended to make room for the detector. The power supplies for the DORIS ring were also modified so their polarity could be changed quickly when switching between positron and electron running. This capability was crucial for the OLYMPUS experiment, which switched daily between beams of electrons and positrons. A large transport frame was also produced to support the OLYMPUS detector on rails. This allowed the detector to be assembled outside the ring and then rolled into the ring for the experiment.

A hydrogen gas target [23] was installed internal to the storage ring. The target consisted of a thin-walled, elliptical tube 600 mm long without entrance or exit windows. Hydrogen gas was

IOP Conf. Series: Journal of Physics: Conf. Series **966** (2018) 012053 doi:10.1088/1742-6596/966/1/012053

injected into the centre of the tube and allowed to diffuse to either end where series of vacuum pumps were used to maintain the high vacuum required by the storage ring. The nominal target areal density was approximately 3×10^{15} atoms/cm^2. Additionally, the target region required collimators to shield against synchrotron radiation and specially designed transition pieces to minimize wakefield effects.

In 2010, the former BLAST detector [24] from MIT-Bates was disassembled and shipped to DESY where it was reassembled. The detector, shown schematically in Fig. 3, consisted of an eight-sector toroidal magnetic spectrometer with the two horizontal sectors instrumented with large acceptance drift chambers covering polar angles $20° \leq \theta \leq 80°$ and azimuthal angles $-15° \leq \phi \leq 15°$ for 3D particle tracking and walls of time-of-flight scintillator bars for triggering and particle identification. The detector was left-right symmetric and this was used as a cross-check during the analysis.

Two new detector systems were built to monitor the luminosity. These were symmetric Møller/Bhabha calorimeters at $\theta = \pm 1.29°$ and two telescopes of three triple GEM (gas electron multiplier) detectors interleaved with three multi-wire proportional chambers mounted at $\theta = \pm 12°$ relative to the beam axis.

The timeline for the OLYMPUS experiment was very tight. OLYMPUS received approval and funding in December, 2009, and faced a fixed deadline of December, 2012, when DORIS was scheduled to be shut down. The detector rolled into the DORIS ring in July, 2011. After a few commissioning tests, it ran for one month in February, 2012, and then for two months at the end of 2012, alternating daily between electrons and positrons at 2.01 GeV with a typical current around 65 mA. In total OLYMPUS collected approximately 4.5 fb^{-1} of data, 25% more than the original proposal.

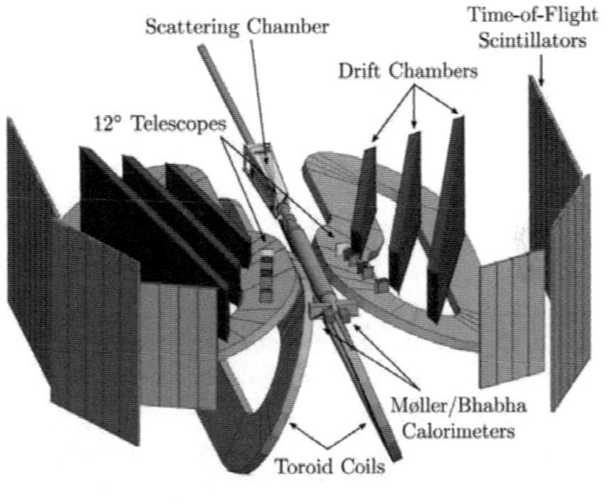

Figure 3. Schematic of OLYMPUS detector. The top four toroid coils are not shown to reveal the horizontal sectors. The drift chambers are shown as three separate chambers in each sector but are actually combined into a single gas volume.

3. Analysis and Results

The analysis of the OLYMPUS experiment was complicated by an inhomogeneous magnetic field and drift chamber inefficiencies due to the high rate of Møller and Bhabha electrons bent into the innermost drift chambers. It was planned to change the toroid magnet polarity each day to reduce tracking systematics but the background with negative polarity prevented operation at high currents so the current analysis is with positive polarity only. To properly analyse the OLYMPUS data a detailed Monte Carlo simulation of the experiment was written using GEANT4. This allowed the Monte Carlo simulation to account for the differences between electrons and positrons with respect to radiative effects, changing beam position and energy, the spectrometer acceptance, track reconstruction efficiency, luminosity, and elastic event selection. The resulting ratio for the positron-proton to electron-proton cross sections was then determined by calculating:

$$R_{2\gamma} = \frac{\sigma_{e^+p}}{\sigma_{e^-p}} = \frac{N^{Data}_{e^+p}}{N^{Data}_{e^-p}} \times \frac{N^{MC}_{e^-p}}{N^{MC}_{e^+p}}$$

IOP Conf. Series: Journal of Physics: Conf. Series **966** (2018) 012053 doi:10.1088/1742-6596/966/1/012053

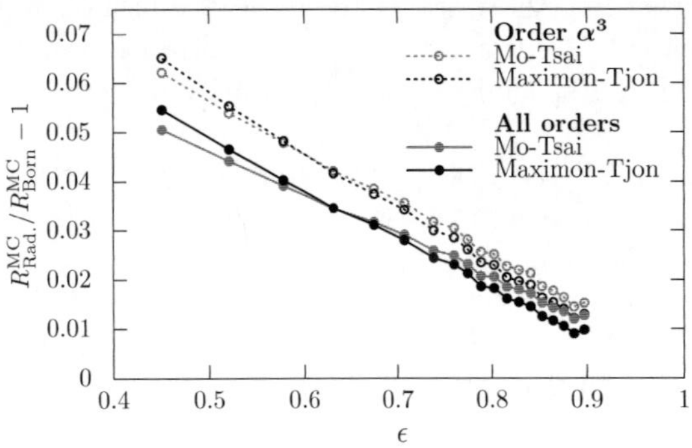

Figure 4. Magnitude of radiative corrections as a function of ϵ for two standard "soft" photon prescriptions for two-photon exchange with exponentiation (all orders) or just to α^3.

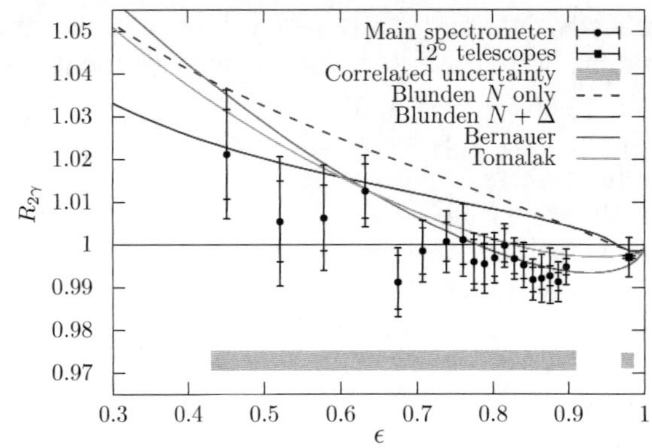

Figure 5. OLYMPUS results for $R_{2\gamma}$ as a function of ϵ. Inner error bars are statistical while the outer error bars include uncorrelated systematic uncertainties added in quadrature. The gray band is correlated systematic uncertainty.

where the N_i are luminosity normalized experimental and Monte Carlo yields.

A custom event generator was developed to convolve the standard radiative corrections with the detector acceptance and resolutions. Since lepton-proton bremsstrahlung interference also changes sign between electrons and positrons, care was taken to calculate contributions as accurately as possible, without resorting to peaking or soft-photon approximations. The calculations for "soft" two-photon exchange employed two standard prescriptions: Mo-Tsai [25] or Maximon-Tjon [26]. The generator could calculate corrections to order α^3, or all orders using exponentiation. The approximate size of the radiative corrections to OLYMPUS is shown in Fig. 4 relative to the Born approximation.

The OLYMPUS results [27] are shown in Fig. 5 together with various calculations. The results are below unity at high ϵ but tend to rise with decreasing ϵ. The dispersive calculations of Blunden [28], which can account for part of the discrepancy observed in the form factor ratio at higher Q^2, are systematically above the OLYMPUS results in this energy regime. The phenomenological prediction from Bernauer [29] and the subtractive dispersion calculation from Tomalak [30], using Bernauer's fit, are in reasonable agreement with the OLYMPUS results.

Two other recent experiments, VEPP-3 [31] and CLAS [32], also measured the ratio of positron-proton to electron-proton scattering. However, it is difficult to compare these results directly with OLYMPUS since their measurements were performed at different energies yielding results at different points in the (ϵ, Q^2) plane. To partially account for this, we can compare all the two-photon exchange results by taking the difference with respect to a theoretical calculation (in this case Blunden's N+Δ calculation) evaluated at the correct (ϵ, Q^2) for each data point. This is shown in Fig. 6 plotted versus ϵ. In this view, the results from the three experiments are seen to be in reasonable agreement with each other over the range in ϵ but are systematically below the theoretical calculation. This supports the previous assertion that the theoretical calculation does not reproduce the results in this energy regime. However, the ϵ dependence of both the results and calculations appears to be in agreement.

4. Conclusions

At the momentum transferred range measured by OLYMPUS the effect of "hard" two-photon exchange is small, on the order of 1%. This is good news for historical electron scattering measurements made at low energies but does not explain the observed discrepancy in the form factor ratio at higher energies. The rising trend in the ratio with increasing Q^2 (decreasing ϵ) may indicate that two-photon exchange is present and may become significant at higher energies. However, to prove this will require measurements at higher energies that will be difficult due to the rapid decrease in the cross section.

Current theoretical calculations that explain part of the observed discrepancy at higher energies overestimate the effect at the energies measured by the three recent experiments. Possibly higher order radiative corrections are required or nucleon states beyond N+Δ need to be considered.

The discrepancy in the form factor ratio measured using unpolarized and polarized techniques and the possible role played by two-photon exchange continues to be topical within the nuclear physics community. A parallel session at the NSTAR 2017 Workshop [33] will be devoted to two-photon exchange. Also, the need for future experiments at higher energy have stimulated discussions at JLab [34] as well as other laboratories. Hopefully, more theoretical and experimental work will bring a better understanding of the proton's structure in the near future.

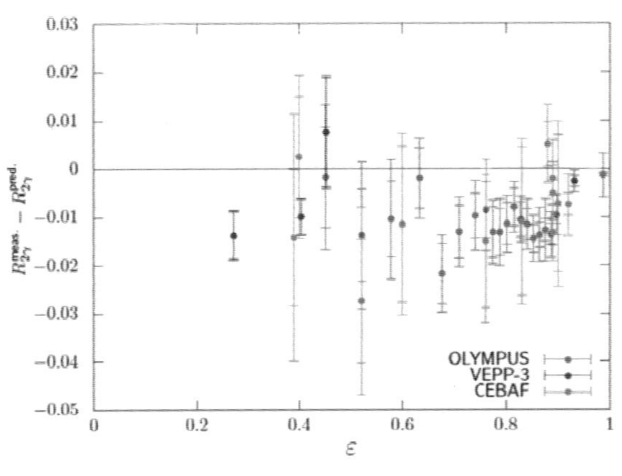

Figure 6. Difference between experimental results and Blunden's N+Δ calculation.

Acknowledgments

The OLYMPUS experiment was an international collaboration of 55 physicists from 13 institutions plus significant support from engineering and technical staff. We gratefully acknowledge the support from: the Ministry of Education and Science of Armenia, the Deutsche Forschungsgemeinschaft, the European Community-Research Infrastructure Activity, the United Kingdom Science and Technology Facilities Council and the Scottish Universities Physics Alliance, the United States Department of Energy and the National Science Foundation, and the Ministry of Education and Science of the Russian Federation.

References

[1] Pohl R *et al* 2013 *Ann. Rev. Nucl. Part. Sci.* **63** 175–204
[2] Aidala C A *et al* 2013 *Rev. Mod. Phys.* **85** 655–691
[3] Durr S *et al* 2008 *Science* **322** 1224–1227
[4] Afanasev A *et al* 2017 *Prog. Part. Nucl. Phys* **95** 245–278
[5] Blau S K 2017 *Physics Today* **70** 14–15
[6] Janssens T *et al* 1966 *Phys. Rev.* **142** 922–931
[7] Berger C *et al* 1971 *Phys. Lett.* **B35** 87
[8] Litt J *et al* 1970 *Phys. Lett.* **B31** 40–44
[9] Bartel W *et al* 1973 *Nucl. Phys.* **B58** 429–475
[10] Andivahis L *et al* 1994 *Phys. Rev.* **D50** 5491–5517
[11] Walker R C *et al* 1994 *Phys. Rev.* **D49** 5671–5689
[12] Christy M E *et al* 2004 *Phys. Rev.* **C70** 015206

IOP Conf. Series: Journal of Physics: Conf. Series **966** (2018) 012053 doi:10.1088/1742-6596/966/1/012053

[13] Qattan I A *et al* 2005 *Phys. Rev. Lett.* **94** 142301
[14] Jones M K *et al* 2000 *Phys. Rev. Lett.* **84** 1398–1402
[15] Pospischil T *et al* 2001 *EPJ Web of Conferences* **A12** 125–127
[16] Gayou O *et al* 2001 *Phys. Rev.* **C64** 038202
[17] Punjabi V *et al* 2005 *Phys. Rev.* **C71** 055202
[18] Crawford C *et al* 2007 *Phys. Rev. Lett.* **98** 052301
[19] Puckett A J R *et al* 2010 *Phys. Rev. Lett.* **104** 242301
[20] Ron G *et al* 2011 *Phys. Rev.* **C84** 055204
[21] Puckett A J R *et al* 2012 *Phys. Rev.* **85** 045203
[22] Milner R *et al* 2014 *Nucl. Instrum. Methods* **A741** 1–17
[23] Bernauer J C *et al* 2014 *Nucl. Instrum. Methods* **A755** 20–27
[24] Hasell D *et al* 2009 *Nucl. Instrum. Methods* **A603** 247–262
[25] Mo L W and Tsai Y S 1969 *Rev. Mod. Phys.* **41** 205–235
[26] Maximon L C and Tjon J A 2000 *Phys. Rev.* **C62** 054320
[27] Henderson B S *et al* (OLYMPUS) 2017 *Phys. Rev. Lett.* **118** 092501
[28] Blunden P G and Melnitchouk W 2017 *Phys. Rev.* **nucl-th**
[29] Bernauer J C *et al* (A1) 2014 *Phys. Rev.* **C90** 015206
[30] Tomalak O and Vanderhaeghen M 2015 *EPJ Web of Conferences* **A51** 24
[31] Rachek I A *et al* 2015 *Phys. Rev. Lett.* **114** 062005
[32] Adikaram D *et al* 2015 *Phys. Rev. Lett.* **114** 062003
[33] NSTAR 2017, The 11th International Workshop on the Physics of Excited Nucleons URL `http://nstar2017.physics.sc.edu/`
[34] JPos17 URL `https://www.jlab.org/conferences/JPos2017/index.html`

IOP Conf. Series: Journal of Physics: Conf. Series **966** (2018) 012054 doi:10.1088/1742-6596/966/1/012054

Role of exact pairing in the description of nuclear level density and radiative strength function

N. Dinh Dang
Quantum Hadron Physics Laboratory, RIKEN Nishina Center for Accelerator-Based Science, 2-1 Hirosawa, Wako City, 351-0198 Saitama, Japan
E-mail: dang@riken.jp

N. Quang Hung
Institute of Research and Development, Duy Tan University, 3 Quang Trung, Danang City, Vietnam
E-mail: nqhungdtu@gmail.com

L.T. Quynh Huong
Department of Natural Science and Technology, University of Khanh Hoa, Nha Trang City, Khanh Hoa Province, Vietnam and Faculty of Physics and Engineering Physics, Ho Chi Minh City University of Science, Ho Chi Minh City, Vietnam
E-mail: lethiquynhhuong@yahoo.com

Abstract. An approach utilizing the microscopic pairing Hamiltonian is proposed to simultaneously describe the nuclear level density and radiative strength function, taking into account the thermal effects of the exact pairing as well as the giant resonances within the phonon-damping model. The good agreement between the results of calculations and experimental data extracted by the Oslo group for ^{170,171,172}Yb isotopes demonstrates the importance of exact thermal pairing in the description of nuclear level densities at low and intermediate excitation energies and invalidates the assumption based on the Brink-Axel hypothesis in the description of the radiative strength functions.

1. Introduction

The level spacing in medium and heavy nuclei decreases rapidly with increasing the excitation energy so that the consideration of average properties of nuclear excitations becomes convenient and meaningful. The nuclear level density (NLD) and radiative γ-ray strength function (RSF) are two main quantities, which are often employed to describe these properties. The former is the number of excited levels per unit of excitation energy E^*, whereas the latter is the average transition probability per γ-ray energy E_γ. The NLD offers the information on the pairing correlations and nuclear thermodynamic properties such as temperature, entropy, heat capacity, etc. [1]. The RSF is a tool to study the average nuclear electromagnetic properties [2]. The study of NLD and RSF has been one of the most important topics in nuclear structure physics owing to their important contributions in the study of low-energy nuclear reactions and nuclear astrophysics [3]. The recent developments of the experimental technique proposed by Oslo's group, which simultaneously extracts both NLD and RSF from the primary γ-decay spectra of highly-excited nuclei created in compound nuclear reactions [4, 5, 6], have made the study of NLD and RSF particularly attractive. Despite the concepts of NLD and RSF have been proposed long time ago [2, 7], no unified theory, which can describe simultaneously and microscopically both the NLD and RSF, has been developed so far. The finite-temperature shell model quantum Monte-Carlo method [8] descibes quite well the NLD, but it is time consuming in its application to heavy nuclei. Moreover, it cannot handle the γ-strength functions, which involve giant resonances, and the related RSF. The Hartree-Fock BCS [9] and Hartree-Fock-Bogolyubov plus combinatorial method (HFBC) [10] offer a global description of NLD but they violate the particle number. As a result, to match the experimental data, the NLD has to be renormalized by using two parameters, whose values are extracted from the experimental analysis of the cumulative number of discrete levels and s-wave neutron resonance spacing at the neutron binding energy [11]. The predictive power of these theories is therefore limited. In the present talk I will report on a unified theoretical approach, proposed for the first time in Ref. [12], which can simultaneously describe both the NLD and RSF.

2. Formalism

The present approach includes both the effects of exact thermal pairing and temperature-dependent resonance width. Thermal pairing is treated by using the eigenvalues $\mathcal{E}_S$, obtained by diagonalizing the pairing Hamiltonian [13] $H = \sum_k \epsilon_k(a^\dagger_{+k}a_{+k} + a^\dagger_{-k}a_{-k}) - G\sum_{kk'} a^\dagger_{+k}a^\dagger_{-k}a_{-k'}a_{+k'}$ at zero temperature and different numbers of unpaired particles (seniorities) S. The notations $a^\dagger_{\pm k}(a_\pm)$ denote the creation (annihilation) operators of a nucleon with angular momentum k, projection $m_{\pm k}$, and energy ϵ_k. The total seniorities S are equal to $0, 2, ..., \Omega$ (number of single-particle levels) for a system with an even number of particles and $1, 3, ..., \Omega$ for a system with an odd number of particles. These exact eigenvalues are used to construct the partition function of the canonical ensemble (CE) [14], from which all the thermodynamic quantities such as free energy $\mathcal{F}$, total energy $\mathcal{E}$, entropy $\mathcal{S}$, heat capacity $\mathcal{C}$, and thermal pairing gap Δ [15, 16] are calculated. Because of the limitation by the size of the matrix to be diagonalized, the exact solutions of the pairing Hamiltonian are limited to the levels around the Fermi surface (truncated levels). The total partition function of the whole system is obtained by combining the exact CE partition function of the truncated levels with those obtained within the independent-particle model (IPM) [17] for the levels beyond the truncated space, where the independent motion of nucleons is assumed.

The density of state $\omega(E^*)$ at excitation energy E^* as $\omega(E^*) = e^{\mathcal{S}}/(T\sqrt{2\pi\mathcal{C}})$ [18] is obtained from the inverse Laplace transformation of the partition function [1]. The total NLD $\rho(E^*)$ is calculated from the state density $\omega(E^*)$ as $\rho(E^*) = \omega(E^*)/(\sigma\sqrt{2\pi})$ [19], where σ is the spin cut-off parameter. The latter, in axially deformed nuclei, has two components, the perpendicular $\sigma_\perp = \mathcal{I}_\perp T/\hbar$ and parallel $\sigma_\parallel = \mathcal{I}_\parallel T/\hbar$ ones, where $\mathcal{I}_\perp$ and $\mathcal{I}_\parallel$ are the perpendicular and parallel moment of inertia of the nucleus, respectively. Based on the limit of rigid body with the same density distribution as of the nucleus, $\sigma_\perp$ is empirically described as $\sigma_\perp^2 \approx 0.015A^{5/3}T$ [20], whereas $\sigma_\parallel$ is given as $\sigma_\parallel = \sigma_\perp\sqrt{(3-2\beta_2)/(3+\beta_2)}$ [21], where β_2 and A are the quadrupole deformation parameter and mass number, respectively. The collective vibrational and rotational excitations, not included in the pairing Hamiltonian, are expressed in terms of the vibrational k_{vib} and rotational k_{rot} enhancement factors [21, 22, 23], whose explicit forms are empirically given as $k_{vib} = \exp[0.0555A^{2/3}T^{4/3}]$ [23] and $k_{rot} = (\sigma_\perp^2 - 1)/[1 + e^{(E^*-U_C)/D_C}] + 1$, where E^* is the excitation energy obtained within the exact CE of the pairing Hamiltonian plus the IPM (EP+IPM), whereas D_C and U_C are given as $D_C = 1400\beta_2^2A^{-2/3}, U_C = 120\beta_2^2A^{1/3}$ [21]. The final total NLD, including the effects of vibrational and rotational enhancements, is given as

$\rho(E^*) = k_{rot}k_{vib}\omega(E^*)/(\sigma_{\|}\sqrt{2\pi})$ [21, 24]. Methods to obtain exact angular-momentum projected NLD and collective effects from the microscopic random-phase approximation within the static path approximation have been developed in Refs. [25] and [26], respectively, whose application is beyond the scope of this work.

The RSF $f_{X\lambda}(E_\gamma)$ for the electric ($X = E$) or magnetic ($X = M$) excitations with multipolarity λ is calculated via the $X\lambda$ strength function $S_{X\lambda}(E_\gamma)$. In the phenomenological models, such as the Kadmenskij-Markushev-Furman (KMF) one [27], a Lorentzian is used for the strength function $S_{X\lambda}(E_\gamma)$ with an approximated resonance width for $E1$ excitations as a function of T^2, whereas the widths for $M1$ and $E2$ excitations take their values at $T = 0$ as T varies (See Eqs. (14) – (17) in Ref. [5] or Eqs. (9) – (11) in Ref. [28]). In the present work, we calculate the strength function $S_{X\lambda}(E_\gamma)$ within the Phonon Damping Model (PDM), where the temperature-dependent resonance width $\Gamma_{X\lambda}(T)$ is obtained microscopically, including the effect of non-vanishing thermal pairing [29] by using the exact CE pairing mentioned above [30]. The resonance width in the PDM is the sum of the quantal width Γ_Q caused by coupling the collective giant excitations to the non-collective ph configurations at zero and finite T, and the thermal width Γ_T caused by coupling of giant resonances to pp and hh configurations at $T \neq 0$. The model has two parameters $F_1^{(\lambda)}$ and $F_2^{(\lambda)}$ for the couplings to ph, and pp (hh) configurations, respectively. The value of $F_1^{(\lambda)}$ is chosen to reproduce the resonance width $\Gamma_{X\lambda}(T = 0)$, whereas $F_2^{(\lambda)}$ is selected at $T = 0$ so that the resonance energy $E_{X\lambda}$ does not changes significantly as T varies. In numerical calculations within the present work, the small fluctuation of the resonance peak is neglected by setting the resonance energies $E_{X\lambda}$ for $E1, M1$, and $E2$ excitations at their corresponding experimental values extracted at $T = 0$. The numerical calculations are carried out for ^{170,171,172}Yb isotopes, whose single-particle spectra are taken from the axially deformed Woods-Saxon potential [31]. The quadrupole deformation parameters β_2 are 0.295 for ^{170,171}Yb and 0.296 for ^{172}Yb. The other parameters of the Woods-Saxon potential are the same as those in Refs. [15, 32]. The values of the pairing interaction parameter G for neutrons and protons are chosen so that the exact neutron and proton pairing gaps obtained at $T = 0$ reproduce the corresponding experimental values extracted from the odd-even mass formulas. The diagonalization of the pairing Hamiltonian is carried out for 12 doubly degenerate single-particle levels with 6 levels above and 6 levels below the Fermi surface. The exact CE chemical potential and pairing gap are calculated, from which one obtains the quantities that mimic the "exact" quasiparticle energy E_k, the coefficients u_k and v_k of the Bogolyubov transformation between particles and quasiparticles, as well as the quasiparticle occupation numbers n_k based on their conventional definitions. These quantities are used as inputs in the RSF calculations within the PDM for the levels with pairing around the Fermi surface, whereas for the remaining spectrum, where $u_k = 1$ (0) and $v_k = 0$ (1) for $k = p$ (h) according to the IPM, one has $E_k = |\epsilon_k - \epsilon_F|$, $n_p = f_p$ and $n_h = 1 - f_h$ with ϵ_F and f_k being the Fermi energy and the single-particle occupation number described by the Fermi-Dirac distribution at finite T, respectively.

3. Analysis of numerical results

The results of the exact neutron (solid lines) and proton (dotted lines) gaps as functions of T are plotted in Figs. 1(a) – 1(c). They decrease with increasing T and remain finite up to $T \sim 3$ MeV, that is much larger than the critical temperature $T_c \sim 0.57\Delta(T = 0)$, where the BCS gap collapses. A slight increase in the exact neutron gap at low $T < 0.5$ MeV takes place in ^{171}Yb due to the blocking effect from the odd neutron [33]. Because of this non-vanishing pairing gaps, the NLDs obtained within the EP+IPM (solid lines) agree well with the experimental data for all nuclei considered in the present paper as seen in Figs. 1(d) – 1(f). The NLDs obtained within the EP+IPM almost coincide with results of the global microscopic calculations within the HFBC for both negative (dashed lines) and positive (dotted lines) parities [34].

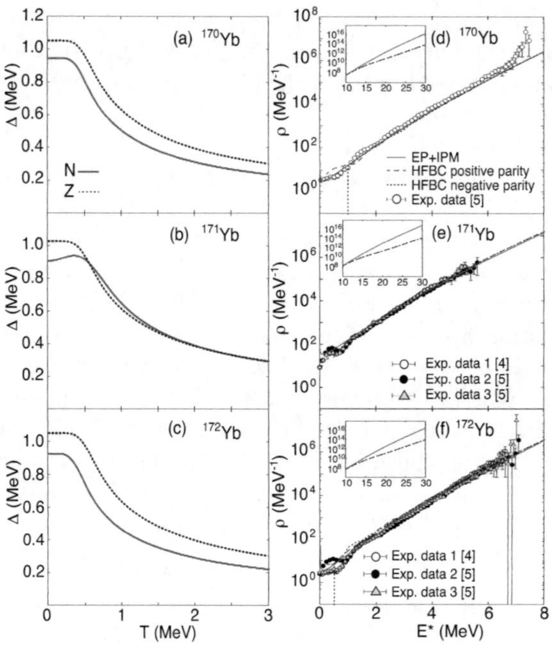

Figure 1. (Color online) Neutron and proton pairing gaps Δ [(a) – (c)] as functions of T and total level densities ρ [(d) – (f)] as functions of E^* obtained within the EP+IPM in comparison with predictions of HFBC calculations for the positive and negative parities and the experimental data for ^{170,171,172}Yb nuclei.

However, to have a good description of the experimental data the NLDs obtained within the HFBC have to be renormalized based on two phenomenological parameters. Moreover, since the HFBC was derived based on the partition function of the incoherent ph states built on top of the HFB single-particle spectra, it fails to predict the NLD in the region of high excitation energy, where the contributions of the pp, hh, as well as of higher states like $2p2h, 3p3h$, etc. become significant. Within the EP+IPM, the exact CE partition function is obtained from the direct diagonalization of the matrix elements of the Hamiltonian, which consist of all possible couplings between the ph, pp and hh states. Therefore, this exact CE partition in combination with that of the IPM is capable to describe the NLD up to high E^* region. The insets of Figs. 1(d) – 1(f), where the NLDs obtained within the EP+IPM are compared with those obtained within the HFBC in the region $10 \leq E^* \leq 30$ MeV, clearly show that the former are significantly higher than the latter. Beyond the Woods-Saxon mean field, the EP+IPM uses only two temperature-independent parameters, namely the monopole pairing strengths G for protons and neutrons. Shown in Fig. 2 are the RSF [(a) – (c)] and the sum $\mathcal{S}_{PDM}(E_\gamma)$ of the strength functions $S_{X\lambda}(E_\gamma)$ calculated within the PDM for $E1$, $M1$, and $E2$ resonances at several values of $T \leq 0.7$ MeV [(d) – (f)]. They have been multiplied by the corresponding cross sections $\sigma(X\lambda)$ at their maxima and normalized by $(2\lambda+1)$, namely $\mathcal{S}_{PDM}(E_\gamma) = \sigma(E1(I))S_{E1(I)}(E_\gamma)/3 + \sigma(E1(II))S_{E1(II)}(E_\gamma)/3 + \sigma(M1)S_{M1}(E_\gamma)/3 + \sigma(E2)S_{E2}(E_\gamma)/5$, where $E1(I)$ and $E1(II)$ correspond to the two components of the GDR determined from the photoabsorption experiments [5]. The values of resonance energies $E_{X\lambda}$, their FWHM $\Gamma_{X\lambda}$, and cross sections $\sigma(X\lambda)$ at $T=0$ for ^{170,171,172}Yb are taken from Table I of Ref. [5]. The GDR with the largest values of $\sigma(X\lambda)$ $(X\lambda = E1(I), E1(II))$ gives the largest contribution the total

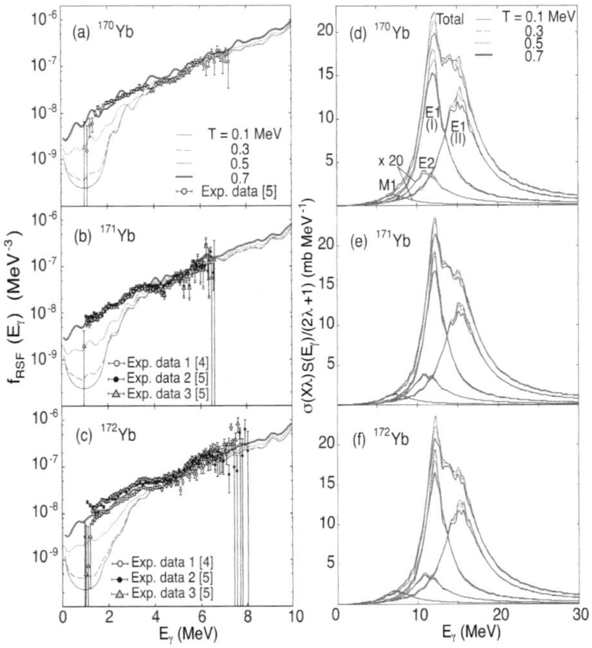

Figure 2. (Color online) Radiative strength functions [(a) – (c)] obtained within the EP+PDM in comparison with experimental data for ^{170,171,172}Yb nuclei, and the corresponding total strength functions [(d) – (f)] together with their components for $E1$, $E2$, and $M1$ excitations as functions of E_γ at different temperatures.

strength function [Figs. 2(d) – 2(f)]. The widths of its two components remain nearly constant at $T \leq 0.4$ MeV and increase with T at $T > 0.4$ MeV, resulting in a significant increase in the total RSF at low $E_\gamma < 4$ MeV as seen in Figs. 2(a) – 2(c). The RSFs obtained within the PDM at $T = 0.7$ MeV agree well with the experimental data for all nuclei under consideration. This value of T is higher than that obtained from the fitting by using the KMF model in Ref. [5], which is below 0.4 MeV. This result invalidates the assumption of the Brink-Axel hypothesis [35], which states that the GDR built on an excited state should be the same as that built on the ground state, and based on which the experimental RSFs were extracted. Based on the fitting by using the KMF model, Ref. [5] has also suggested that there should appear a two-component pygmy dipole resonance (PDR) in the region $2.1 < E_\gamma < 3.5$ MeV in ^{171}Yb and ^{172}Yb. This two-component PDR was added on top of the GDR in fitting the experimental RSF in Ref. [5]. Within the PDM, it has been shown in Ref. [30] that exact pairing enhances the $E1$ strength function in the region $E_\gamma < 5$ MeV. By including this exact pairing, the RSFs, predicted by the PDM, agree well with the experimental data [thick solid lines in Figs. 2(e) and 2(f)]. Hence, the enhancement of the experimental RSF at low E_γ, which was suggested to be caused by the PDR, is explained microscopically by the effect of exact thermal pairing within the PDM.

4. Conclusions

An approach utilizing the same microscopic pairing Hamiltonian to describe simultaneously both the nuclear level density (NLD) and radiative γ-ray strength function (RSF) is proposed for the very first time. It employs the exact solutions of the pairing problem to construct the partition function, based on which the NLD and thermal pairing gap are calculated. The latter is included

IOP Conf. Series: Journal of Physics: Conf. Series **966** (2018) 012054 doi:10.1088/1742-6596/966/1/012054

in the phonon-damping model to calculate the RSF. The good agreement between the results obtained within this approach and the experimental data for NLD and RSF in ^{170,171,172}Yb demonstrates the crucial role of exact thermal pairing in the description of both NLD and RSF in the low and intermediate region of excitation and γ-ray energies. The results obtained also show that, to have a good description of the RSF, the microscopic strength function with the temperature-dependent width for the giant resonances should be used instead of the Brink-Axel hypothesis. The merits of this approach are its microscopic nature and the use of only the parameters taken over from previous calculations as well as the very short computing time, which takes less than five minutes even for a heavy nucleus.

Acknowledgment
The numerical calculations were carried out using the FORTRAN IMSL Library by Visual Numerics on the RIKEN supercomputer HOKUSAI-GreatWave System. This work is funded by the National Foundation for Science and Technology Development (NAFOSTED) of Vietnam through Grant No. 103.04-2017-69.

References
[1] Ericson T 1961 *Adv. Phys.* **9** 425
[2] Blatt J M and Weisskopf V F 1952 *Theoretical Nuclear Physics* (New York: Willey)
[3] Rauscher T and Thielemann F K 2000 *At. Data Nucl. Data Tables* **75** 1
[4] Schiller A *et al.* 2001 *Phys. Rev. C* **63** 021306(R)
[5] Agvaanluvsan U *et al.* 2004 *Phys. Rev. C* **70** 054611
[6] Melby E *et al.* 2001 *Phys. Rev. C* **63** 044309; Siem S *et al.* 2002 *Phys. Rev. C* **65** 044318; Guttormsen M *et al.* 2003 *Phys. Rev. C* **68** 064306; Guttormsen M *et al.* 2015 *Eur. Phys. J. A* **51** 170
[7] Bethe H A 1936 *Phys. Rev.* **50** 332; 1937 *Rev. Mod. Phys.* **9** 69
[8] Alhassid Y, Liu S and Nakada H 1999 *Phys. Rev. Lett.* **83** 4265; Alhassid Y, Liu S and Nakada H 2007 *Phys. Rev. Lett.* **99** 162504
[9] Demetriou P and Goriely S 2001 *Nucl. Phys. A* **695** 95
[10] Hilaire S and Goriely S 2006 *Nucl. Phys. A* **779** 63
[11] Goriely S, Hilaire S and Koning A J 2008 *Phys. Rev. C* **78** 064307
[12] Quang Hung N, Dinh Dang N and Quynh Huong L T 2017 *Phys. Rev. Lett.* **118** 022502
[13] Volya A, Brown B A and Zelevinsky V 2001 *Phys. Lett. B* **509** 37
[14] Hung N Q and Dang N D 2009 *Phys. Rev. C* **79** 054328
[15] Hung N Q and Dang N D 2010 *Phys. Rev. C* **81** 057302
[16] Dang N D and Hung N Q 2012 *Phys. Rev. C* **86** 044333
[17] Alhassid Y, Bertsch G F and Fang L 2003 *Phys. Rev. C* **68** 044322
[18] Nakada H and Alhassid Y 1997 *Phys. Rev. Lett.* **79** 2939
[19] Behkami A N and Huizenga J R 1973 *Nucl. Phys. A* **217** 78
[20] Dilg W, Schantl W, Vonach H and Uhl M 1973 *Nucl. Phys. A* **217** 269
[21] Junghans A R *et al.* 1998 *Nucl. Phys.* A **629** 635; Nasrabadi M N and Sepiani M 2014 *Act. Phys. Pol. B* **45** 1865
[22] Ignatyuk A V 1983 *The statistical properties of the excited atomic nuclei* (Moscow: Energoatomizdat)
[23] Iljinov A S 1992 *Nucl. Phys. A* **543** 517
[24] Capote R *et al.*, 2009 *Nucl. Data Sheets* **110** 3107
[25] Rossignoli R, Ansari A and Ring P 1993 *Phys. Rev. Lett.* **70** 1061
[26] Rossignoli R and Ring P 1998 *Nucl. Phys. A* **633** 613
[27] Kadmenskij S G, Markushev V P and Furman V I 1983 *Sov. J. Nucl. Phys.* **37** 165
[28] Voinov A, Guttormsen M, Melby E, Rekstad J, Schiller A and Siem S 2001 *Phys. Rev. C* **63** 044313
[29] Dang N D and Arima A 2003 *Phys. Rev. C* **68** 044303
[30] Dinh Dang N and Quang Hung N 2013 *J. Phys. G* **40** 015103; Balaram Dey *et. al.* 2014 *Phys. Lett. B* **731** 92
[31] Cwiok S *et al.* 1987 *Comput. Phys. Commun.* **46** 379
[32] Hung N Q and Dang N D 2010 *Phys. Rev. C* **82** 044316
[33] Hung N Q, Dang N D and Quynh Huong L T 2016 *Phys. Rev. C* **94** 024341
[34] https://www-nds.iaea.org/RIPL-3/.
[35] Brink D M 1955 *PhD thesis* (Oxford University); Axel P 1962 *Phys. Rev.* **126** 617

Fundamental Physics with Electroweak Probes of Nuclei

Saori Pastore

Theoretical Division, Los Alamos National Laboratory, Los Alamos, NM 87545, USA

E-mail: saori@lanl.gov

Abstract. The past decade has witnessed tremendous progress in the theoretical and computational tools that produce our understanding of nuclei. A number of microscopic calculations of nuclear electroweak structure and reactions have successfully explained the available experimental data, yielding a complex picture of the way nuclei interact with electroweak probes. This achievement is of great interest from the pure nuclear-physics point of view. But it is of much broader interest too, because the level of accuracy and confidence reached by these calculations opens up the concrete possibility of using nuclei to address open questions in other sub-fields of physics, such as, understanding the fundamental properties of neutrinos, or the particle nature of dark matter.

In this talk, I will review recent progress in microscopic calculations of electroweak properties of light nuclei, including electromagnetic moments, form factors and transitions in between low-lying nuclear states along with preliminary studies for single- and double-beta decay rates. I will illustrate the key dynamical features required to explain the available experimental data, and, if time permits, present a novel framework to calculate neutrino-nucleus cross sections for $A > 12$ nuclei.

The nuclear *ab initio* approach aims at describing the widest range of nuclear phenomena through interactions occurring between nucleons inside the nucleus. In this microscopic picture, nucleons interact with each other via two- and three-body interactions, and with external electroweak probes via couplings to individual nucleons and to nucleon-pairs (a contribution described by two-nucleon currents). *Albeit* limited to light nuclei ($A \leq 12$), Quantum Monte Carlo calculations based on the AV18 [1] two-body and IL7 [2] three-body interactions successfully explain available experimental data in a broad energy range, from the keV regime relevant to astrophysics studies to the GeV regime where short-range correlations become predominant [3, 4, 5]. These studies yield a rather complex picture of the nucleus with many-body correlations in both the nuclear wave functions and electroweak currents playing an important role in reaching agreement with the data. For example, corrections from two-body electromagnetic currents are as large as 40% in the calculated magnetic moment of ^{9}C [6], while electron scattering experiments have demonstrated the requirement of two-body currents in quasi-elastic scattering from nuclei, where they enhance the transverse response by up to $\sim 40\%$ [4, 5].

The success of the microscopic picture in explaining the data both qualitatively and quantitatively is an important achievement from the nuclear physics point of view. But it is of much broader interest too, because the level of accuracy and confidence reached by these *ab inito* Quantum Monte Carlo calculations opens up the concrete possibility of using nuclei to address prominent and pressing open questions in nuclear physics and their connection to fundamental physics quests.

Recently, we addressed the "g_A problem", that is the systematic overprediction ($\sim 20\%$ in $A \leq 18$ nuclei) of Gamow-Teller matrix elements in simplified nuclear calculations [7, 8]. The overprediction of the calculated matrix elements is possibly attributable to the fact that for larger nuclear systems, in order for the calculations to be computationally feasible, one has to approximate the *ab initio* framework, by, *e.g.*, leaving out correlations and/or truncate the model space. Another approximation that can contribute to the manifestation of the "g_A problem" is in the adopted model for the nuclear axial current, which typically neglects many-body terms. In order resolve this long-standing problem, we performed numerically exact Quantum Monte Carlo calculations of Gamow-Teller matrix elements in $A = 6$–10 nuclei [9], accounting systematically for many-body effects in nuclear interactions and coupling to the axial one-, two- and three-body currents derived in chiral effective field theory [10, 11].

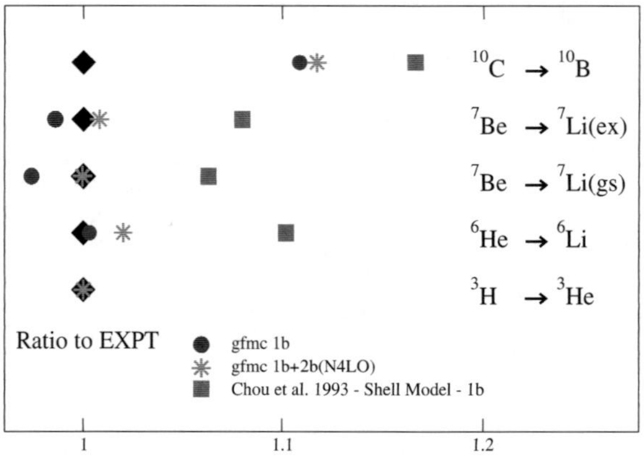

Figure 1. (Color online) Ratios of GFMC to experimental values of the Gamow-Teller reduced matrix elements in the ^{3}H, ^{6}He, ^{7}Be, and ^{10}C weak transitions. Theory predictions (from Ref. [9]) correspond to the χEFT axial current in LO (blue circles) and up to N4LO (magenta stars). Green squares indicate 'unquenched' shell model calculations from Ref. [7] based on the LO axial current.

Our results are summarized in Fig. 1, where we show Green's Function Monte Carlo (GFMC) [3] calculations of the Gamow-Teller reduced matrix elements normalized to the corresponding experimental values in ^{3}H, ^{6}He, ^{7}Be, and ^{10}C weak transitions. The effect of many-body components in the axial current can be appreciated by comparing blue dots, based on the one-body axial current, with the magenta stars which include in addition to the one-body also two- and three-body terms in the axial current [10]. We find that the effect of many-body currents is negligible, in fact they increase by only $\sim$ 2–3% the one-body Gamow-Teller contributions. On the other hand, correlations in the wave functions significantly reduce the matrix elements, a fact that can be appreciated by comparing the GFMC (blue circles in Fig. 1) and the shell model calculations (green squares in the same figure) from Ref. [7], both based

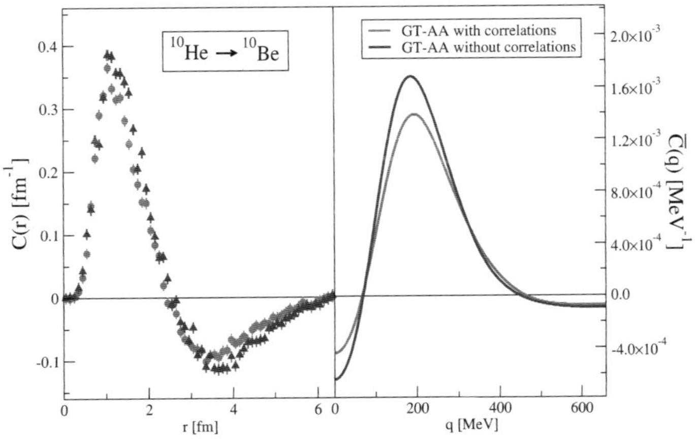

Figure 2. The left (right) panel shows the light Majorana neutrino exchange mechanism (GT-AA) distribution in r-space (q-space) for the ^{10}He→^{10}Be transition, with and without "one-pion-exchange-like" correlations in the nuclear wave functions. See text and Ref. [14] for explanation.

on the one-body axial current. The reduced matrix elements in the shell model calculations are enhanced by $\sim$ 8% ($\sim$ 17%) in $A = 6$–7 ($A = 10$) transitions with respect to the experimental data, while inclusion of correlations in the nuclear wave functions leads to an agreement with the data that is excellent in $A = 6$–7 transitions and at the 10% level in the ^{10}C weak transition. The discrepancy in this last case may be attributable to deficiencies in the AV18+IL7 wave functions of $A = 10$ nuclei. In fact, GFMC calculations based on the Norfolk two- and three-nucleon chiral potentials [12, 13] and the one-body axial current, bring the ^{10}C prediction only $\sim$ 4% above the experimental datum. These findings suggest that the longstanding "g_A-problem" may be resolved primarily by correlation effects.

While the "g_A-problem" in single-beta decays, limited to the transitions we have studied, seems to be largely attributable to missing correlations in the nuclear wave functions, little it is know about how it propagates at moderate values of momentum transfer. This lack of knowledge impacts ongoing world-wide experimental enterprises aimed at potentially observe neutrinoless double beta decay, a process in which two neutrons decay into two protons with the emission of two electrons but no neutrinos, thus violating the lepton number conservation by two units. The average momentum transfer in these transitions is of the order of 100 MeV [8], a scale that is set by the average distance between the two decaying neutrons. Observation of this decay would be a clear signature of physics beyond the standard model and it will have tremendous consequences on our understanding of the Majorana nature of the neutrino and the neutrino mass hierarchy, and potentially the observed matter-antimatter asymmetry in the universe. The interpretation of either a positive or null experimental result heavily relies on accurate evaluations of nuclear matrix elements. The latter are at present characterized by large theoretical uncertainties [8], primarily attributable to the fact that for nuclei relevant to experimental purposes ($A \geq 48$), one has to approximate the *ab initio* framework by leaving out some correlations.

In order to address how the "g_A-problem" propagates at moderate values of momentum transfer, we recently performed Quantum Monte Carlo calculations of neutrinoless double beta decay matrix elements in A = 6–12 nuclei [14]. In particular, we have studied the effect of artificially turning off correlations in the nuclear wave functions, finding a $\sim$ 10% increase in

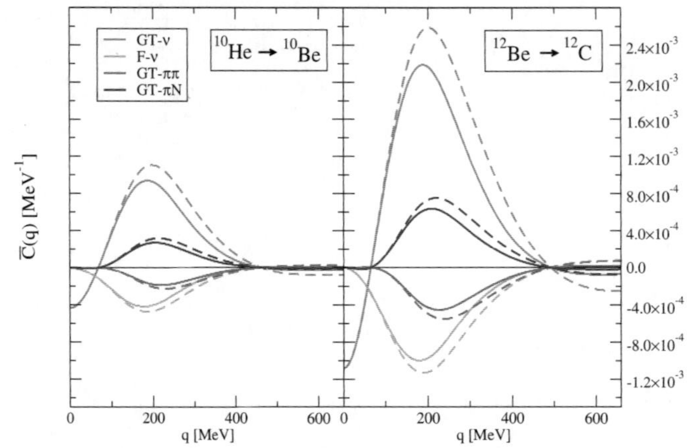

Figure 3. The GT-ν, F-ν, GT-$\pi\pi$, and GT-πN distributions in momentum space for the ^{10}He$\to$^{10}Be and ^{12}Be$\to$^{12}C decays. Solid and dashed lines are obtained, respectively, with and without the inclusion of the momentum dependence in nucleonic form factors. See text and Ref. [14] for explanation.

the calculated nuclear matrix elements for the light Majorana neutrino exchange mechanism. This corresponds to having to "quench" g_A by ~ 0.95 to accommodate for correlation effects. On the other hand, we saw that in single-beta decay, a zero momentum transfer observable, the required "quenching" of g_A in say, the ^{10}C weak transition evaluated in the shell model, is ~ 0.83 as one can read off Fig. 1 but comparing the $A = 10$ green square with the experimental datum (black diamond). These findings may indicate that the g_A "quenching" required in calculations based on more approximated nuclear models (for $A > 12$ nuclei) is larger in single beta decay than in neutrinoless double beta decays. In Fig. 2 we show the effect of artificially turning off correlations in the nuclear wave functions when calculating the neutrinoless double beta decay matrix elements induced by the light Majorana neutrino exchange potential. In particular, in Fig. 2 the blue triangles (solid line) in the left (right) panel represent the r-space (q-space) transition distribution obtained by turning off the correlations to be compared with the red dots (solid line) obtained with the correlated wave function. The associated matrix element, obtained by integrating the distribution in dr where r is the inter-particle distance, undergoes a 10% increase when correlations are turned off.

We emphasize that the nuclear systems we studied in Ref. [14] are not relevant from the experimental point of view, nevertheless they are interesting and provide us with an extremely useful set of test cases to benchmark theoretical nuclear models and/or computational methods. We studied the matrix elements of light Majorana neutrino exchange potentials [16, 15] (denoted by ν in Fig. 3) as well as short-distance sources (denoted by $\pi\pi$ and πN in Fig. 3) of lepton number violation encoded in dimension-seven and -nine operators (see Ref. [17] and references therein). These generically lead to Gamow-Teller and Fermi spin-isospin structures, in Fig. 3 denoted with GT and F, respectively. In the left and right panels of Fig. 3 we show the q-space transition distributions of the ^{10}He$\to$^{10}Be and ^{12}Be$\to$^{12}C decays, respectively. The different height of the peaks is due to the different spatial overlaps between an initial diffuse neutron distribution and a final compact proton distribution in the case of the $A = 10$ transition, and between two compact initial neutron and final proton distributions in the $A = 12$ transition [14].

IOP Conf. Series: Journal of Physics: Conf. Series **966** (2018) 012055 doi:10.1088/1742-6596/966/1/012055

These findings may have implications in our understanding of the dynamics entering the neutrinoless double beta decay matrix elements in nuclei of experimental interest.

This work has a Los Alamos National Laboratory preprint number LA-UR-17-29533

References

[1] Wiringa R B, Stoks V G J and Schiavilla R 1995 *Phys. Rev. C* **51** (1995) 38
[2] Pieper Steven C 2008 *AIP Conference Proceedings* **1011** 143
[3] Carlson J *et al.* 2015 *Rev. Mod. Phys.* **87** 1067
[4] Carlson J and Schiavilla R 1998 *Rev. Mod. Phys.* **70** 743
[5] Bacca S and Pastore S 2014 *Journal of Physics G: Nuclear and Particle Physics* **41** 123002
[6] Pastore S *et al.* 2013 *Phys. Rev. C* **87** 035503
[7] Chou W T *et al.* 1993 *Phys. Rev. C* **47** 163
[8] Engel J and Menendez J 2016 arXiv:1610.06548
[9] Pastore S *et al.* 2017 arXiv:1709.03592
[10] Baroni A *et al.* 2016 *Phys. Rev. C* **93** 015501
[11] Krebs H *et al.* 2017 *Annals Phys.* **387** 317
[12] Piarulli M *et al.* 2015 *Phys. Rev. C* **91** 024003
[13] Piarulli M *et al.* 2017 arXiv:1707.02883
[14] Pastore S, Carlson J, Cirigliano V, Dekens W, Mereghetti E and Wiringa R B 2017 arXiv:1710.05026
[15] Bilenky Samoil M and Petcov S T 1987 *Rev. Mod. Phys.* **59** 671
[16] Bilenky S M and Giunti C 2015 *Int. J. Mod. Phys. A* **30** 1530001
[17] Cirigliano V, Dekens W, de Vries J, Graesser M L and Mereghetti E 2017 arXiv:1708.09390

IOP Conf. Series: Journal of Physics: Conf. Series **966** (2018) 012056 doi:10.1088/1742-6596/966/1/012056

Merging first principle structure studies and few-body reaction formalism

R. Crespo

Instituto Superior Técnico, Universidade de Lisboa, Av. Rovisco Pais 1, 1049-001, Lisboa, Portugal
Centro de Ciências e Tecnologias Nucleares, Universidade de Lisboa, Estrada Nacional 10, 2695-066 Bobadela, Portugal

E-mail: raquel.crespo@tecnico.ulisboa.pt

E. Cravo, A. Arriaga

Departamento de Física, Faculdade de Ciências, Universidade de Lisboa, Edifício C8, Campo Grande, 1749-016 Lisboa, Portugal

R. Wiringa

Physics Division, Argonne National Laboratory, Argonne, Illinois, 60439, USA

A. Deltuva

Institute of Theoretical Physics and Astronomy, Vilnius University, Saulėtekio al. 3, LT-10222 Vilnius, Lithuania

R. Diego

Centro de Ciências e Tecnologias Nucleares, Universidade de Lisboa, Estrada Nacional 10, 2695-066 Bobadela, Portugal

Abstract.

Calculations for nucleon knockout from a ^{7}Li beam due to the collision with a proton target at 400 Mev/u are shown based on *ab initio* Quantum Monte Carlo (QMC) and conventional shell-model nuclear structure approaches to describe the relative motion between the knockout particle and the heavy fragment of the projectile. Structure effects on the total cross section are shown.

1. Introduction

At the very earlier days of Nuclear Physics, nuclear experimental data have shown evidence that at the stability line, certain number of nucleons result in a greater stability, leading to the concept of magic numbers. Closely related to this idea, the independent shell model was developed, which assumes that the interactions within the many-nucleon system can be conveniently represented by a mean field potential (with a central and spin-orbit parts) for each single particle. In the independent shell model the nucleons move independently in this potential. Shell model presupposes that residual interactions to this independent picture can be handled perturbatively.

The development of the Radioactive Ion Beam (RIB) facilities allowing the study away from the stability line unveiled the concept of shell evolution. It was verified that along the nuclear landscape some magic numbers are either disappearing or emerging. This novel behaviour has trigger numerous theoretical structure studies, that attempt to trace back shell evolution effects to the NN components of the interaction or 3-body nuclear forces. Nevertheless, despite considerable progress, a clear picture has not emerged yet.

Extracting structure information from nuclear reactions, in particular nuclear correlations or NNN force effects, faces a severe handicap. In fact, the current state-of-the-art reaction formalism relies on the fundamental assumption that essentially only a limted number of particles participate dynamically in the reaction process. Attempts have been made to incorporate the degrees of the excitation of one of the clusters into the reaction mechanisms. Clearly, *ab initio* many body structure information is required to be merged into the reaction framework.

Quantum Monte Carlo methods (QMC) allow the incorporation of sophisticated NN and NNN interactions and correlations in microscopic calculations of light nuclei bound states [1, 2]. In this work, first steps are made for merging state-of-the-art QMC wave functions into Faddeev reaction approach [3, 4]. Results of calculated total cross sections for nucleon knockout from a ^{7}Li beam due to the collision with a proton target at 400 Mev/u are shown.

2. The Faddeev/AGS multiple scattering framework

Let us consider the scattering of a many body projectile of A nucleons with a proton target. A key ingredient of standard reaction approaches to the scattering is the reduction of the overwhelming many body problem, into an effective 3-body system composed by an inert (A-1) core denoted here as C, plus a valence nucleon and a target proton. The exact dynamics of the collision in this truncated Hilber space can then be described by standard Faddeev/Alt, Grassberger and Sandhas (Faddeev/AGS) [5, 6]. This consists of a three-body non-relativistic reaction formalism treating all open channels (elastic, breakup and transfer) simultaneously.

In the assumed truncated Hilbert space $\mathcal{H}_{C+N+p}$ for $C+N+p$ free relative motion the heavy fragment can only be either in the ground state or in a low-lying excited state. Therefore, core dynamical excitations during the collision process, highly excited states above evaporation or multiple particle knockout (p,pNN' ..) are not taken into account.

In this reaction approach, the transition amplitudes leading to the observables are the on-shell matrix elements of the operators $U^{\beta\alpha}$, calculated from the solution of the three-body AGS integral equations

$$U^{\beta\alpha} = \bar{\delta}_{\beta\alpha} G_0^{-1} + \sum_{\gamma} \bar{\delta}_{\beta\gamma} t_{\gamma} G_0 U^{\gamma\alpha} , \tag{1}$$

with $\alpha, \beta, \gamma = (1, 2, 3)$. Here, $\bar{\delta}_{\beta\alpha} = 1 - \delta_{\beta\alpha}$ and the two-body transition operator is

$$t_{\gamma} = v_{\gamma} + v_{\gamma} G_0 t_{\gamma} , \tag{2}$$

with the free resolvent $G_0 = (E + i0 - H_0)^{-1}$, with E the total energy of the three-particle system in the center of mass (c.m.) frame. The solution of the Faddeev/AGS equations at higher energies can be found by iteration

$$U^{\beta\alpha} = \bar{\delta}_{\beta\alpha} G_0^{-1} + \sum_{\gamma} \bar{\delta}_{\beta\gamma} t_{\gamma} \bar{\delta}_{\gamma\alpha} + \sum_{\gamma} \bar{\delta}_{\beta\gamma} t_{\gamma} \sum_{\xi} G_0 \bar{\delta}_{\gamma\xi} t_{\xi} \bar{\delta}_{\xi\alpha} + \cdots . \tag{3}$$

The successive terms of this series can be considered as terms of zero order (which contribute only for rearrangement transitions), first order (single scattering), second order (double scattering)

IOP Conf. Series: Journal of Physics: Conf. Series **966** (2018) 012056 doi:10.1088/1742-6596/966/1/012056

and so on in the transition operators t_γ. The $\beta = 0$ partition corresponds to three free particles in the continuum.

The breakup/nucleon knockout observables are calculated from the on-shell matrix elements of the AGS operators, as described in [7].

3. Interactions

In this work, two different structure approaches were used to obtain the projectile and residue overlap wave functions: (i) a two-body model with a Woods-saxon potential with depth constrained by the nucleon separation energy S_N, standard geometric factors (WS) and spectroscopic factors taken from [12] (ii) Quantum Monte Carlo (QMC) *ab initio* calculations based on the Argonne (AV18) NN interaction [8] with additional 3-body interactions [9]. In the work of [9] an effective interaction was found to reproduce the same overlap functions. This interaction can then be fed in the reaction formalism to calculate the Nucleon-Core $(N - C)$ transition amplitude, Eq. 2.

The comparison between the effect of these two structure approaches on the calculated observables will provide insight about the importance of structure effects due to modifications of *both* geometry and spectroscopic factors.

The central effective interactions and the radial overlaps for the ^{6}Li+n and ^{6}He+p partitions are represented in Fig. 1 and Fig. 2 respectively. As shown in the figures, the *ab initio* VMC effective interactions have a larger diffuseness and a smaller depth than the WS representation.

As in earlier works, we take the realistic nucleon-nucleon CD Bonn potential [10] for the proton-nucleon particle pair. The results are essentially independent on the choice of the NN interaction. To describe the interaction between the target proton and the core (not well known) we take a phenomenological nuclear optical potential where the parameters are taken from the Koning-Delaroche (KD)[11] global optical potential parametrization evaluated at 200 MeV. In addition, we also arbitrarily take this parametrization for the knockout nucleon - Core interaction, in all the partial waves other than partial wave for the ground state of the projectile. The KD parametrization is not really designed for nuclei $A < 24$, and there is an associated uncertainty with the choice of this parametrization. Nevertheless, one might anticipate that the results presented here, with respect to the effects of the the description of the overlap functions, will be independent of this choice.

4. Results

The theoretical cross sections for the knockout of a nucleon considering a Core in its ground state, are evaluated from the transition amplitudes Eq. 3.

In the calculations, all multiple scattering terms were included to achieve convergence. In addition, all few-body kinematical contributions are evaluated accurately. For p-knockout antu-symmetrization is taken into account in the multiple scattering expansion.

Taken the spectroscopic factor unity, these cross sections denoted as single particle σ_{sp} calculated with the two structure approaches and reproduced in Table 1 differ from about 10%. Upon introducing the spectroscopic factors taken from the *ab initio* [9] and the mean field Shell model framework [12], the difference between the theoretical cross sections σ_{th} remains of the same order of magnitude for the case of neutron- knockout, but increases significantly for the case of proton- knockout. This might be related to related to the fact that ^{6}He is a diffuse Core.

Our results show the same trend than those obtained from nucleon knockout due to the collision with a light target using the Glauber eikonal with the adiabatic reaction approach of [12].

IOP Conf. Series: Journal of Physics: Conf. Series **966** (2018) 012056 doi:10.1088/1742-6596/966/1/012056

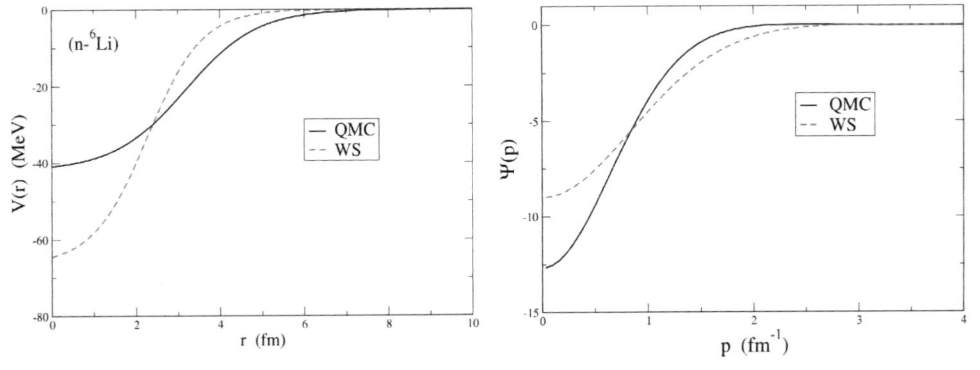

Figure 1. (Color online) Interactions and respective wave functions for describing the ^{6}Li+n

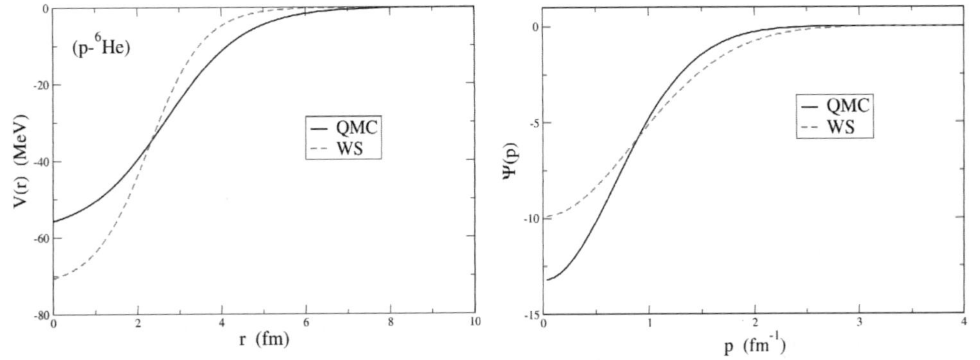

Figure 2. (Color online) Interactions and respective wave functions for describing the ^{6}He+p partition as described in the text.

Table 1. Total cross sections for N-knockout

Reaction	σ_{sp}^{QMC} (mb)	σ_{th}^{QMC} (mb)	σ_{sp}^{IPM} (mb)	σ_{th}^{IPM} (mb)	$\Delta_{sp}(\%)$	$\Delta_{th}(\%)$
^{7}Li(p,pn)^{6}Li	18.8	15.4	16.3	16.9	13	9
^{7}Li(p,2p)^{6}He	12.7	5.08	11.2	7.84	12	54

5. Conclusions and outlook

We have calculated nucleon knockout from a ^{7}Li projectile due to the collision with a proton target at 400 MeV/u. We have used the standard Faddeev/AGS few-body reaction framework considering that the heavy fragment is found in its ground state and remains inert in the collision process.

We have shown that the calculated single particle cross sections using an *ab initio* VMC approach to evaluate the radial overlaps between the projectile and the heavy fragment differ moderately of about 10% from the results obtained from a simplified Woods Saxon interaction, standard diffuseness and depth adjusted to reproduce the nucleon separation energy. On the other hand, when multiplied by the microscopic spectroscopic factors the results using the *ab initio* structure approach differ significantly from that taken form a mean field Shell model framework, for the case of proton- knockout. This might be related to related that to the fact

IOP Conf. Series: Journal of Physics: Conf. Series **966** (2018) 012056 doi:10.1088/1742-6596/966/1/012056

that a two-body model of an ^{6}He core coupled with a proton is a poor description for ^{7}Li.

Future developments to consistently include the projectile ground state and continuum from the *ab initio* structure approach into the reaction formalism will be performed.

Acknowledgments: The work of R.C. and E.C. is supported by Fundação para a Ciência e Tecnologia (FCT) of Portugal under the contract No. PTD/FIS-NUC/2240/2014. The work of A.D. is supported by Lietuvos Mokslo Taryba (Research Council of Lithuania) under contract No. MIP-094/2015.

References

[1] Arriaga A, Pandharipande V R and Wiringa R B 1995, *Phys. Rev. C* **52**, 2362.
[2] Wiringa R B, Schiavilla R, Pieper S C and Carlson J 2014, *Phys. Rev. C* **89**, 024305.
[3] Deltuva A 2013, *Phys. Rev. C* **88**, 011601R.
[4] Crespo R, Cravo E and Deltuva A 2014, *Phys. Rev. C* **90**, 044606.
[5] Faddeev L D 1960 *Zh. Eksp. Teor. Fiz.* **39** 1459 [Sov. Phys. JETP **12**, 1014 (1961)]
[6] Alt E O, Grassberger P and Sandhas W 1967 *Nucl. Phys.* **B2** 167
[7] Crespo R, Deltuva Rodríguez-Gallardo A M, Cravo E, and Fonseca A C 2009, *Phys. Rev. C* **79**, 014609 .
[8] Wiringa R B, Stocks V G J and Schiavilla R 1995, , *Phys. Rev. C* **51**, 38.
[9] Brida I, Pieper S C, and Wiringa R B 2011, *Phys. Rev. C* **84**, 024319.
[10] Machleidt R 2001, *Phys. Rev. C***63**, 024001.
[11] Koning A J and Delaroche J P 2003, Nucl. Phys. **A713**, 231.
[12] Grinyer G F et al 2012, *Phys. Rev. C* **86**, 024315.

IOP Conf. Series: Journal of Physics: Conf. Series **966** (2018) 012057 doi:10.1088/1742-6596/966/1/012057

Study of the $^{249\text{-}251}$Cf + ^{48}Ca reactions: recent results and outlook

A A Voinov[1], Yu Ts Oganessian[1], F Sh Abdullin[1], N T Brewer[2], S N Dmitriev[1], R K Grzywacz[2, 3], J H Hamilton[4], M G Itkis[1], K Miernik[2, 5], A N Polyakov[1], J B Roberto[6], K P Rykaczewski[2], A V Sabelnikov[1], R N Sagaidak[1], I V Shirokovsky[1], M V Shumeiko[1], M A Stoyer[7], V G Subbotin[1], A M Sukhov[1], Yu S Tsyganov[1], V K Utyonkov[1], G K Vostokin[1]

[1]FLNR, JINR, 141980 Dubna, Russian Federation
[2]Dep. of Phys. and Astron., ORNL, Oak Ridge, TN 37831, USA
[3]Dep. of Phys. and Astron., UT, Knoxville, TN 37996, USA
[4]Dep. of Phys. and Astron., VU, Nashville, TN 37235, USA
[5]Faculty of Physics, UW, Warsaw, Poland
[6]Sci. and Tech. Partnerships Directorate, ORNL, Oak Ridge, TN 37831, USA
[7]Nucl. and Chem. Sci. Division, LLNL, Livermore, CA 94551, USA

Author e-mail: voinov@jinr.ru

Abstract. Experiment aiming at the synthesis of heavy isotopes of Z=118 (Og) using beam of ^{48}Ca and a target of $^{249\text{-}251}$Cf was undertaken in October 2015 – April 2016 employing the Dubna Gas-Filled Recoil Separator (FLNR JINR). The target of mixed isotopes of $^{249\text{-}251}$Cf (50.7% of ^{249}Cf, 12.9% of ^{250}Cf, and 36.4% of ^{251}Cf) was irradiated by ^{48}Ca ions at two beam energies of 252 and 258 MeV with the corresponding accumulated beam doses of 1.6×10^{19} and 1.1×10^{19}. A single event observed at lower beam energy was assigned to the isotope ^{294}Og, the product of the reaction ^{249}Cf(^{48}Ca, 3n); its decay pattern and the observed radioactive properties of the nuclides in the decay chain reproduce in full those observed for ^{294}Og in our earlier experiments of 2002-2005 and 2012. At higher beam energy we observed no decay chains that could be attributed to the isotopes of Og. The possibility of renewal of this experiment in the future is discussed.

1. Introduction

More than 50 new isotopes of the nuclei with Z = 104-118 from the predicted "Island of stability" of superheavy elements (SHE) were synthesized for the first time at the Flerov Laboratory of Nuclear Reactions (JINR, Dubna) using Dubna Gas-Filled Recoil Separator (DGFRS) and heavy ions cyclotron U-400 during last 15 years (Fig. 1). The complete-fusion reactions of ^{238}U, ^{237}Np, $^{242,\,244}$Pu, ^{243}Am, $^{245,\,248}$Cm, ^{249}Bk and ^{249}Cf targets with ^{48}Ca ions were studied and cross sections of these reactions were measured [1]. Recently, these results were approved also by the other researches performed by IVO [2], SHIP [3], BGS [4] and TASCA [5-7] experimental setups. Decay properties of these nuclei that belong to the domain of superheavy nuclei, point to a considerable stabilizing effect when approaching the predicted neutron closure shell at N=184.

Few attempts have been also undertaken to expand the region of SHN by synthesis of elements 119 and 120 with target nuclei ranging from ^{238}U to ^{249}Cf and projectiles heavier than ^{48}Ca. Five reactions ^{64}Ni + ^{238}U (SHIP [8]), ^{58}Fe + ^{244}Pu (DGFRS [9]), ^{54}Cr + ^{248}Cm (SHIP [10]), ^{50}Ti + ^{249}Bk (TASCA

[11]) and ^{50}Ti + ^{249}Cf (TASCA [12]) were employed, however, no decay chains of SHE were observed in these experiments. The upper cross section limits were set at 0.07 – 1.1 pb range depending on the reaction.

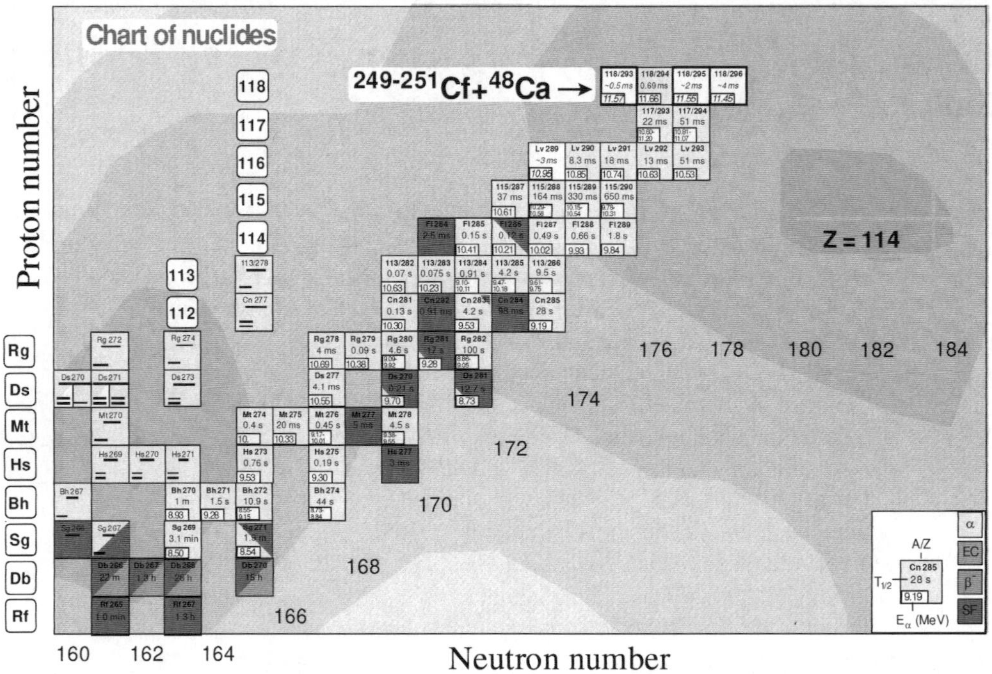

Fig. 1. The top of the nuclear landscape with the heaviest identified nuclei and their measured half-lives and decay modes that were obtained in ^{48}Ca-induced reactions. White squares show the isotopes ^{289}Lv, ^{293}Og, ^{295}Og and ^{296}Og which can be produced in $^{249\text{-}251}$Cf+Ca reactions.

Recently we continued exploring the island of stability of the SHN employing the DGFRS in searching for the new isotopes at the edge of the new SHE region. In irradiation of $^{249\text{-}251}$Cf targets with ^{48}Ca beam three new isotopes of the heaviest 118 element can be produced (Fig. 1). New nuclide ^{289}Lv can be observed as daughter of ^{293}118. Two new isotopes ^{284}Fl and ^{285}Fl have been already synthesized and studied in the reactions ^{239}Pu(^{48}Ca, *4-3*n)^{283,284}Fl and ^{240}Pu(^{48}Ca, *4-3*n)^{284,285}Fl [13].

In this work, we present the preliminary results of the new campaign aimed at the synthesis of new Og isotopes in the $^{249\text{-}251}$Cf + ^{48}Ca fusion-evaporation reactions.

2. Experiment

The new series of the experiments was undertaken at the DGFRS during October, 2015 – April, 2016. The target of mixed Cf isotopes (50.7% of ^{249}Cf, 12.9% of ^{250}Cf, 36.4% of ^{251}Cf) was provided by Oak Ridge National Laboratory (ORNL, TN). Twelve identical sectors were delivered to FLNR in February 2015 and were mounted on a disk that was rotated at 1700 rpm so that the target was perpendicular to the direction of the incoming beam. In the course of the bombardment with the ^{48}Ca beam, the target layers were systematically monitored by counting α-particles from the decay of the target isotopes. The lab-frame beam energies in the middle of the target layers, excitation energy ranges and beam doses for the experiments studied are summarized in Table I.

The array of Silicon detectors at the DGFRS final focus has been modified to increase the position resolution of recorded signals and subsequently reduce the probability of observing sequences of random events that mimic decay chains of implanted nuclei. The new detection system includes a 0.3-mm thick Double-sided Silicon Strip Detector (DSSD) manufactured by Micron Semiconductor Ltd. This large DSSD has 1-mm wide strips, 48 at the front side and 128 at the back side, creating over

IOP Conf. Series: Journal of Physics: Conf. Series **966** (2018) 012057 doi:10.1088/1742-6596/966/1/012057

6000 1-mm^2 pixels in one Silicon wafer. Such high pixilation helps to achieve superior position resolution for recoil-correlated decay sequences reducing potential random events. This new Si-detector array was designed, assembled, commissioned off-line and provided by ORNL. The signals from all detectors were processed using MESYTEC linear preamplifiers [17]. Further, these analog signals were split into two independent measurement branches by special spectroscopic splitter-amplifier PA32-64 [18] designed by DGFRS group. Thus, all detectors' signals were processed simultaneously by analog electronics similar to those used in previous DGFRS experiments [19], and by digital electronics system based on XIA Pixie-16 modules provided by ORNL [20]. This new DSSD assemblage and two independent registering systems were successfully applied recently in ^{239}Pu + ^{48}Ca and ^{240}Pu + ^{48}Ca experiments, see [13].

Table I. The ^{249}Cf target thickness, lab-frame beam energies in the middle of the target layers, resulting excitation-energy intervals, total beam doses, and numbers of observed decay chains assigned to the parent nuclei ^{294}Og (3*n*) are listed.

Reaction	Target thickness (mg/cm^2)	E_{lab} (MeV)	E_{exc} (MeV)	Beam dose $\times 10^{19}$	Number of chains 2*n* / 3*n* /4*n* / 5*n*	Ref.
^{249}Cf + ^{48}Ca	0.23	245	26.6–31.7	2.5	- / 1 / - / -	[14]
	0.34	251	32.1–36.6	1.6	- / 2 / - / -	[15]
^{249}Cf + ^{48}Ca*	0.30	252	26.6–37.7	1.6	- / 1 / - / -	[16]
^{249}Cf + ^{48}Ca	0.35	252	33.0–37.4	1.6	- / 1 / - / -	This work
	0.35	258	37.8–42.4	1.1	- / - / - / -	This work
^{251}Cf + ^{48}Ca	0.35	252	33.9–38.3	1.6	- / - / - / -	This work
	0.35	258	39.0–43.6	1.1	- / - / - / -	This work

* ^{249}Cf as buildup result of ^{249}Bk β-decay ($T_{1/2}$ = 320 d).

The FWHM energy resolution of the implantation detector was 34 to 78 keV depending on the strip, while the summed signals recorded by the side and implantation detectors had an energy resolution of 147 to 263 keV. Other experimental conditions, including the method of calibration of the detectors, were the same as in previous DGFRS experiments (see [1] and references therein).

3. Results

One of the heaviest nuclei from the "Island of Stability", the isotope ^{294}Og, was produced for the first time in the ^{249}Cf+^{48}Ca reaction in 2002 [14]. Three years later in the same reaction we observed two similar decay chains more [15]. Additionally, the fourth event of ^{294}Og formation was observed in the experiment on Z=117 isotopes synthesis as a result of the buildup of ^{249}Cf in the ^{249}Bk target material due to its β-activity [16].

Present experiment with mixed $^{249-251}$Cf target was carried out from October 1, 2015 till April 6, 2016 at two ^{48}Ca-beam energies, See Table I. At the beam energy E = 252 MeV with accumulated beam dose of 1.6×10^{19} particles just one correlated decay chain was observed which we assign to ^{294}Og, the product of 3*n*-evaporation channel of the ^{249}Cf(^{48}Ca, 3*n*)^{294}Og reaction (Fig. 2). Its decay properties reproduce in full those

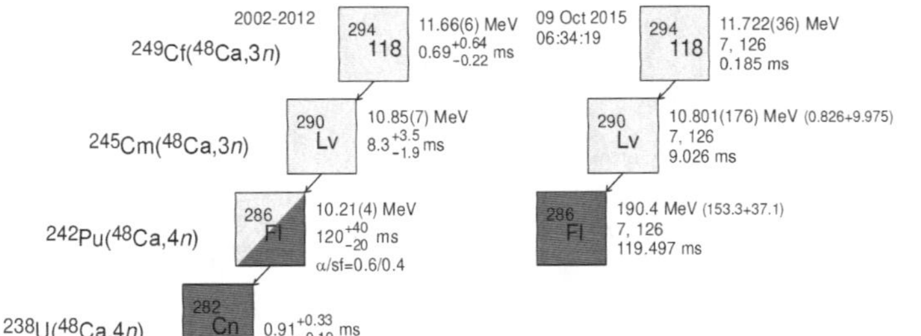

Fig. 2. a). Average decay properties ($T_{1/2}$ and E_α) of ^{294}Og-^{282}Cn determined in the reactions ^{48}Ca + ^{249}Cf, ^{245}Cm, ^{242}Pu, ^{238}U [1]. b). Properties of ^{294}Og-^{286}Fl decay measured in this work. The left rows provide energies, time intervals between events and their DSSD strip numbers for α-decay and SF. Energies of summed signals are given in parentheses.

IOP Conf. Series: Journal of Physics: Conf. Series **966** (2018) 012057 doi:10.1088/1742-6596/966/1/012057

observed in [14-16]. The cross section of the ^{249}Cf(^{48}Ca,3*n*)^{294}Og reaction for 252-MeV projectiles was measured to be about 0.9 pb (Fig. 3).

At 258 MeV beam energy the bombardment was started on February 10, 2016 with the aim to produce ^{295}Og or ^{296}Og isotopes, the products of the ^{251}Cf(^{48}Ca,x*n*)$^{294-x}$Og reaction with evaporation of 4 and 3 neutrons. The beam dose of 1.1×10^{19} particles was accumulated during 56 days, no decay chains were observed at this energy which could be attributed to SHN.

Finally, this experiment was stopped due to significant pollution of the target material caused by melted glue from sector frames which covered all the target sectors and prevented ERs escaping from the target (the target surface was monitored during the experiment by registering of the α–particle count from the target). The target was sent back to ORNL for the regenerating. We are looking forward for continuation of this experiment with $^{249-251}$Cf target in 2017-2018.

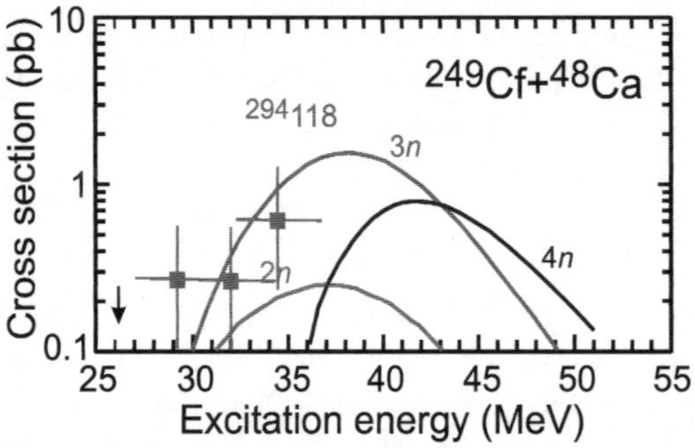

Fig. 3. The calculated excitation functions [21] of the ^{249}Cf(^{48}Ca,x*n*)$^{294-x}$Og reactions are compared with the experimental data [14-16] and this work. Vertical error bars correspond to total uncertainties.

Acknowledgments
We acknowledge the support of the RFBR Grant No. 13-02-12052 and 16-52-55002. Research at ORNL was supported by the U.S. DOE Office of Nuclear Physics under DOE Contract No. DE-AC05-00OR22725 with UT-Battelle, LLC. Research at LLNL was supported by LDRD Program Project No. 08-ERD-030, under DOE Contract No. DEAC52-07NA27344 with Lawrence Livermore National Security, LLC. This work was also supported by the U.S. DOE through a Grant No. DE-FG-05-88ER40407 (Vanderbilt University).

References
[1] Oganessian Yu Ts and Utyonkov V K 2015 *Rep. Prog. Phys.* **78**, 036301.
[2] Eichler R *et al.* 2004 *Nature* **447**, 72-5; Eichler R *et al.* 2010 *Radiochim. Acta* **98**, 133.
[3] Hofmann S *et al.* 2007 *Eur. Phys. J A* **32** 251-60; Hofmann S *et al.* 2012 *Eur. Phys. J A* **48** 62.
[4] Stavsetra L *et al.* 2009 *Phys. Rev. Lett.* **103**, 132502;
Ellison P A *et al.* 2010 *Phys. Rev. Lett.* **105**, 182701.
[5] Gates J M *et al.* 2011 *Phys. Rev. C* **83**, 054618.
[6] Rudolph D *et al.* 2013 *Phys. Rev. Lett.* **111**, 112502.
[7] Khuyagbaatar J *et al.* 2014 *Phys. Rev. Lett.* **112**, 172501.
[8] Hofmann S *et al.* 2009 *GSI Sci. Rep. 2008*, p.131.
[9] Oganessian Yu Ts *et al.* 2009 *Phys. Rev. C* **79**, 024603.
[10] Hofmann S *et al.* 2012 *GSI Sci. Rep. 2011*, p. 205.
[11] Dullmann Ch E *et al.* 2012 *GSI Sci. Rep. 2011*, p.206.
[12] Khuyagbaatar J *et al.* 2012 *GSI Sci. Rep. 2011*, p. 131.
[13] Utyonkov V K *et al.* 2015 *Phys. Rev. C* **92**, 034609.
[14] Oganessian Yu Ts *et al.* 2002 *JINR Communication* D7-2002-287.
[15] Oganessian Yu Ts *et al.* 2006 *Phys. Rev. C* **74**, 044602.
[16] Oganessian Yu Ts *et al.* 2012 *Phys. Rev. Letters* **109**, 162501.
[17] Rajabali M M *et al.* 2012 *Phys. Rev. C* **85,** 034326, http://www.mesytec.com.
[18] Subbotin V G *et al.* (to be published).
[19] Oganessian Yu Ts *et al.* 2013 *Phys. Rev. C* **87**, 054621.
[20] Grzywacz R *et al.* 2003 *Nucl. Instrum. Methods Phys. Res. B* **204**, 649.
[21] Zagrebaev V I *et al.* 2012 *Phys. Rev. C* **85**, 014608.

IOP Conf. Series: Journal of Physics: Conf. Series **966** (2018) 012058 doi:10.1088/1742-6596/966/1/012058

Shell evolution and $E2$ collectivity: new spectroscopic information

N Pietralla, C Kremer, M Lettmann, S Aslanidou, A Krugmann, C Lizarazo, P von Neumann-Cosel, V Werner and W Witt

Institut für Kernphysik, TU Darmstadt, Schlossgartenstraße 9, 64289 Darmstadt, Germany

E-mail: pietralla@ikp.tu-darmstadt.de

Abstract. We report here on the first evidence for shape coexistence caused by Type II Shell Evolution which is firmly based on the measurement of absolute $E2$ transition rates. The data have been obtained in high-resolution inelastic electron scattering spectroscopy of the nucleus ^{96}Zr at the Superconducting Darmstadt Linear electron Accelerator (S-DALINAC). The data are presented and discussed. The neutron sub-shell closure at neutron number $N = 56$ plays a crucial role for the occurence of shape coexistence in ^{96,98}Zr. We have studied the structure of exotic neutron-rich isotopes near the $N = 56$ neutron sub-shell closure using γ-ray spectroscopy and radioactive beams. We also present and discuss spectroscopic data on ^{84,86,88}Ge taken recently at RIBF, RIKEN. The new data suggest the hitherto unknown existence of a region of pronounced nuclear triaxiality in neutron-rich Ge isotopes.

1. Introduction

The way how the complex nuclear forces form atomic nuclei from their constituent protons and neutrons represents the core topic of nuclear structure research. The evolution of nuclear structure across the nuclear chart is dominated by the evolution of effective single-particle energies as a function of the numbers of protons and neutrons that make up the nuclei. Shell evolution has been realized as the reason for the disappearance of shell closures at magic numbers known from nuclei near the valley of stability and for the occurrence of new ones for exotic nuclides, in particular those with large neutron excess. The understanding of shell evolution has been pushed especially by the Tokyo group around T. Otsuka [1]. It is supported by vast experimental evidence in particular from the structure of neutron-rich nuclei that have become accessible due to new facilities for intense beams of radioactive ions.

Recently, the concept of *Type II Shell Evolution* has been introduced [2, 3]. It emphasizes, that the effective single-particle energies of nucleon orbitals systematically depend on the nucleonic configuration in each individual nuclear state due to the mutual interaction between the nucleons. This general mechanism can be particularly pronounced in key-nuclei with certain numbers of protons and neutrons, for which Type II Shell Evolution may lead to drastic nuclear structure effects such as even shape coexistence and first-order shape phase transitions.

The most rapid change of the shape of nuclear ground states in the nuclear chart occurs for the chain of Zirconium isotopes. While ^{96}Zr exhibits features of a doubly-closed shell nucleus, ^{100}Zr is already strongly deformed. In particular, the first excited state of the even-even nucleus ^{96}Zr is a $J^\pi = 0^+$ state instead of a 2^+ state. The 2^+_1 state has a high excitation energy of

1750 keV and the $B(E2; 2_1^+ \to 0_1^+) = 2.3(3)$ W.u. is non-collective. In contrast, in ^{100}Zr the 2_1^+ state lies at 213 keV and decays collectively to the ground state with an $E2$ strength of $B(E2; 2_1^+ \to 0_1^+) = 75(4)$ W.u. The corresponding shape phase transition has recently been shown to be due to shell evolution [4].

Therefore, the Zirconium isotopes provide a unique opportunity for investigating the mechanisms that dominate the formation of nuclear structure. In particular, the following general questions are intriguing: Is the shape-phase transition in the Zirconium isotopic chain of first or second order as a function of neutron number? If it is of first order, can we observe a region of shape coexistence in spherical Zirconium isotopes corresponding to the region between the spinodal point and the critical point of the quantum phase transition? Are the conditions for the occurrence of the quantum phase transition in Zirconium isotopes specific to this isotopic chain or can we observe traces of them in other nuclei in this mass region? We have performed nuclear spectroscopy experiments for contributing crucial data to answer these questions. In particular, we have studied inelastic electron scattering on ^{96}Zr and γ-ray spectroscopy on neutron-rich nuclides in the $A \approx 80$ - 100 mass region.

2. Electron-scattering experiment on ^{96}Zr

Inelastic electron scattering on the nuclide ^{96}Zr have been studied [5] at the Superconducting DArmstadt LINear electron ACcelerator (S-DALINAC) at TU Darmstadt using the Lintott high-resolution magnetic spectrometer. The electron-beam energies and observation angles have been set such that momentum transfers from 0.08 to 0.35 fm^{-1} were realized. Fig. 1 shows the relevant part of the level scheme of ^{96}Zr and corresponding electromagnetic transition strengths. The $E2$ strength in the deformed structure has been measured here for the first time.

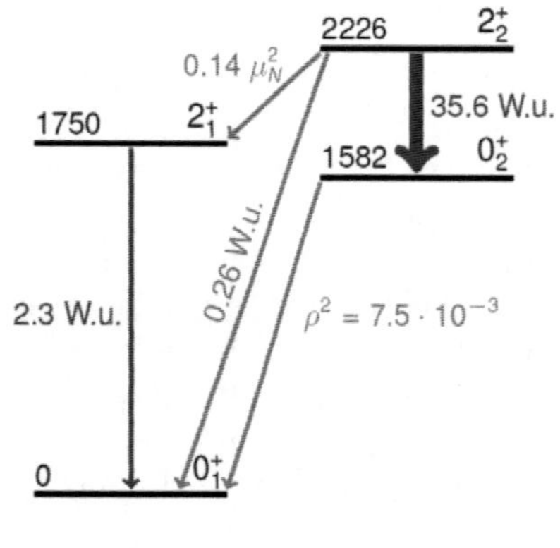

Figure 1. Partial level scheme of ^{96}Zr. The electromagnetic decay rates of the 2_2^+ state were measured [5] in this work. The $E2$ transition strength between the excited 2_2^+ and 0_2^+ states is an order of magnitude larger than the one from the 2_1^+ state to the ground state and two orders of magnitude larger than the one from the 2_2^+ state to the ground state.

A typical part of the data is shown in Fig. 2. In particular, the 2_2^+ state of ^{96}Zr at an excitation energy of 2.226 MeV has been observed. Its electromagnetic excitation strength from the ground state has been measured relative to the excitation of the 2_1^+ state with a known excitation strength of $B(E2; 0_1^+ \to 2_1^+) = 0.030(4)$ e^2b^2 [6] with the method described in Refs. [7, 8]. Together with the pre-existing information on the γ-decay intensities this measurement provides first information on the $E2$ strength within the deformed sequence of states built on top of the first excited 0^+ state of ^{96}Zr. Moreover, the $E2$ decay strength from the 2_2^+ state to the ground state is small. It amounts to less than 1% of the intra-band transition strength in the deformed structure and indicates very weak mixing between the spherical ground state and the coexisting deformed first-excited state and hence a high potential barrier between spherical and deformed configurations at low energies.

The coexistence of spherical and deformed structures of ^{96}Zr at low excitation energies can be attributed to the mechanism of type II shell evolution [4]. The ground state exhibits a

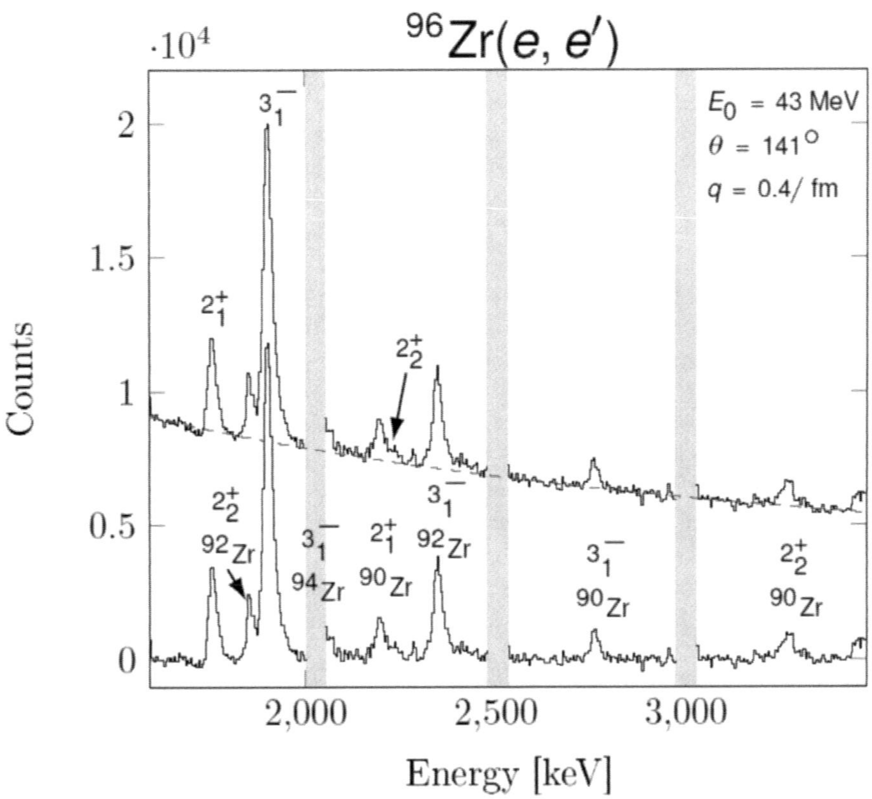

Figure 2. Summed and efficiency corrected energy-loss spectrum of the ^{96}Zr(e, e') electron-scattering reaction with a beam energy of 43 MeV on a sample of Zirconium metal enriched in the isotope ^{96}Zr to 57% observed at a polar angle of $\vartheta = 141°$. The radiative tail of the elastic line has been subtracted in the spectrum at the bottom. Gray areas correspond to inactive segments of the detector system. At 2.226 MeV a signal from the 2_2^+ state of ^{96}Zr is observed. See [5].

doubly closed-shell configuration with almost full occupancies of the proton $\pi(fp)$-shell and the neutron $\nu(d_{5/2})$ orbital and pronounced shell gaps for Z=40 and N=56. In contrast to that, the occupancy of the $\nu(d_{5/2})$ orbital is considerably reduced in the excited deformed structure. This leads to a modification of the effective proton single-particle energies due to the effects of the tensor force of the nucleon-nucleon interaction such that the Z=40 shell gap is significantly reduced and the $\pi(g_{9/2})$ proton orbital is occupied by about 4 protons. This, in turn, modifies the effective neutron single-particle energies due to the tensor force such that the $\nu(g_{7/2}, d_{3/2}, h_{11/2})$ orbitals are lowered in energy, the $\nu(d_{5/2})$ orbital is lifted, and the shell gap for N=56 is reduced facilitating the excitation of neutrons from the $\nu(d_{5/2})$ orbital to other orbitals. This is a self-reinforcing mechanism and the nuclear wave functions of the deformed states are spread out over several single-particle orbitals indicative of collectivity and deformation. The coexisting spherical and deformed structures of ^{96}Zr represent the first case of type II shell evolution, the observation of which is firmly based on the measurement of absolute transition rates [5].

3. Coulomb excitation of ^{98}Zr and evolution of $E2$ strength in the Zirconium isotopic chain

Information on the $E2$ excitation strength of the neighboring neutron-rich radioactive isotope ^{98}Zr does not exist to date. We have attempted a Coulomb-excitation experiment with a beam of radioactive ^{98}Zr ions delivered by the CARIBU facility at the ANL. The experiment took advantage of the powerful combination of the position-sensitive GRETINA array [9] consisting at the time of 4 quad modules in combination with the CHICO2 detector [10], to identify the beam-like recoils and fix the kinematics. Unfortunately, the yield of ^{98}Zr ions was too small and the contamination of the beam with A=98 isobars was too strong at the time of the experiment for us being able to see an in-beam excitation of ^{98}Zr. However, the superior Doppler correction making use of GRETINA and CHICO2 allowed to separate the expected transition energy of the 2_1^+ state of ^{98}Zr at 1223 keV from a contaminant transition ($3_1^- \rightarrow 2_1^+$) at 1230 keV from the main component of the beam, ^{98}Mo. Analysis of the background count rate at the position of the expected peak for the $2_1^+ \rightarrow 0_1^+$ transition in ^{98}Zr provided an upper limit for its intensity. This converts into an upper limit for the $B(E2; 2_1^+ \rightarrow 0_1^+)$ value with a strength smaller than 16 W.u. The new experimental information is included in Fig. 3 showing the evolution of $E2$ transition rates in heavy Zirconium isotopes and corresponding large-scale shell model calculations as taken from Ref. [4]. In particular, the new data points on ^{96,98}Zr support the interpretation of a smooth evolution of deformed structures in Zirconium isotopes, a region of shape coexistence for $^{94-98}$Zr and a first-order ground state shape phase transition between ^{98}Zr and ^{100}Zr.

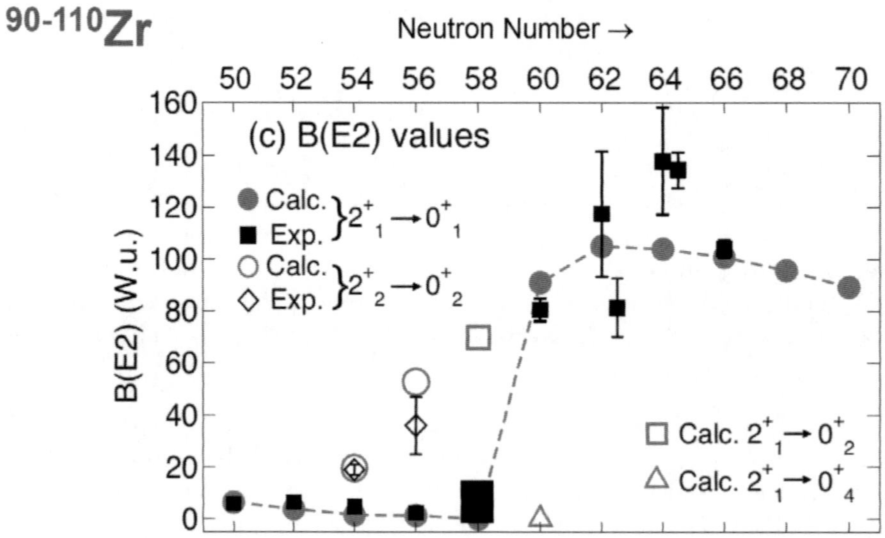

Figure 3. Evolution of $E2$ transition rates in Zirconium isotopes over the shape-phase transition at ^{98,100}Zr as a function of neutron number. See [4, 5]. The new data on ^{96,98}Zr support the conclusion that a coexisting deformed structure in the lighter-mass Zirconium isotopes evolve gradually into the deformed ground-state structure of the heavier mass isotopes.

4. Spectroscopy at the $N = 56$ sub-shell closure at the neutron-rich extremes

The occupancy of the $\nu(d_{5/2})$ neutron orbital has been recognized as the driving ingredient for the type II shell evolution in the Zirconium chain of isotopes. It is interesting to study the structure near N=56 for even more neutron-rich isotopes. However, spectroscopic information is scarce. We have recently studied [11] neutron-rich Germanium isotopes in γ-ray spectroscopy

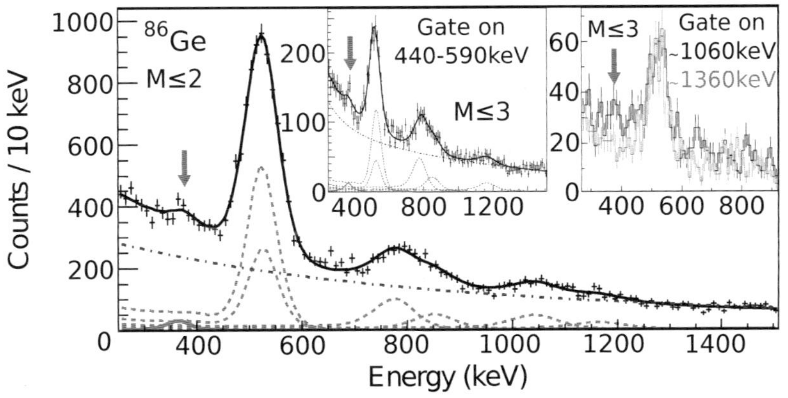

Figure 4. Single and coincidence spectra of ^{86}Ge from the ^{87}As$(p, 2p)$ reaction observed with an array of NaI detectors superimposed with the fit of the whole spectrum (black solid line). The respective background (blue dash-dotted line) and simulated response functions for each transition (red dotted and solid lines) are shown. The insets show spectra after gating on the energy region for the $2^+_{1,2}$ transition and a neighboring region (red). The multiplicity cutoff M is chosen to optimize the peak-to-background ratio and available statistics. The red arrows emphasize the 380-keV tentatively assigned $(3^+_1) \rightarrow (2^+_2)$ transition [11].

experiments at the Radioactive-Ion Beam Facility (RIBF) at RIKEN, Tokyo, Japan. Excited states of ^{84,86,88}Ge were populated by (p, pn) or $(p, 2p)$ reactions of beams of odd-mass Ge or As ions, from in-flight fission of a primary ^{238}U beam, on a liquid-hydrogen target. It was the first time that γ-ray spectroscopy of the exotic isotope ^{88}Ge at neutron number N=56 has been successfully performed and first information on γ transitions in ^{88}Ge has been obtained [11].

For ^{86}Ge six new γ transitions were identified [11] and placed in the level scheme together with the only previously known 527-keV $2^+_1 \rightarrow 0^+_1$ transition. Fig. 4 shows a γ-ray spectrum from ^{86}Ge. The data suggest pronounced triaxiality of ^{86}Ge. Namely, a 380-keV transition, observed with better than 95% confidence [11], is in agreement with predictions from shell model and SCCM calculations, and would correspond to a transition from a relatively low-lying 3^+ state to the 2^+_2 state. A corresponding staggering parameter, which is a measure of the triaxial rigidity [12, 13], of $S(4) = 0.20(4)$ can be deduced. As such, data support the model predictions of a triaxial potential minimum about 30°, with a degree of rigidity larger than that in ^{76}Ge with $S(4) = 0.091(2)$ which was recently established [14]. This pronounced triaxial shape together with its comparatively high rigidity makes ^{86}Ge an intriguing test ground for further studies of triaxial nuclear deformation.

Acknowledgments

We thank all those who have contributed to the experiments, in particular, the accelerator crew, students, and doctoral researchers in accelerator science at the S-DALINAC and in the spectrometer group at TU Darmstadt. We also thank our collaborators in the DALI - MINOS experiment at RIKEN, in particular, M.L. Cortes, P. Doornenbal, and A. Obertelli and in the ^{98}Zr experiment at Argonne Nat'l Laboratory, in particular, W. Korten, R.V.F. Janssens, and G. Savard. We thank T. Otsuka for numerous discussions. We gratefully acknowledge the financial support by the German Research Council (DFG) under grant No. SFB 1245, by the Helmholtz International Center for FAIR (HIC for FAIR) funded by the State of Hesse, and by the German Ministry for Education and Research (BMBF) under grant No. 05P15RDFN1.

References

[1] Otsuka T, Suzuki T, Fujimoto R, Grawe H and Akaishi Y 2005 *Phys. Rev. Lett.* **95** 232502
[2] Tsunoda Y, Otsuka T, Shimizu N, Honma M and Utsuno Y 2014 *Phys. Rev. C* **89** 031301
[3] Otsuka T and Tsunoda Y 2016 *J. Phys. G* **43** 024009
[4] Togashi T, Tsunoda Y, Otsuka T and N. Shimizu N 2016 *Phys. Rev. Lett.* **117** 172502
[5] Kremer K, Aslanidou S, Bassauer S, Hilcker M, Krugmann A, von Neumann-Cosel P, Otsuka T, Pietralla N, Ponomarev V Yu, Shimizu N, Singer M, Steinhilber G, Togashi T, Tsunoda Y, Werner V and Zweidinger M 2016 First measurement of collectivity of coexisting shapes based on type II shell evolution: The case of 96Zr *Phys. Rev. Lett.* **117** 172503
[6] Kumbartzki G, Benczer-Koller N, Holden J, Jakob G, Mertzimekis T J, Taylor M J, Speidel K-H, Ernst R, Stuchbery A E, Beausang C W and Krücken R 2003 *Phys. Lett. B* **562** 193
[7] Scheikh Obeid A, Burda O, Chernykh M, Krugmann A, von Neumann-Cosel P, Pietralla N, Poltoratska I, Ponomarev V Yu and Walz C 2013 *Phys. Rev. C* **87** 014337
[8] Scheikh Obeid A, Aslanidou A, Birkhan J, Krugmann A, von Neumann-Cosel P, Pietralla N, Poltoratska I and Ponomarev V Yu 2014 *Phys. Rev. C* **89** 037301
[9] Paschalis S, Lee I Y, Macchiavelli A O, Campbell C M, Cromaz M, Gros S, Pavan J, Qian J, Clark R M, Crawford H L, Doering D, Fallon P, Lionberger C, Loew T, Petri M, Stezelberger T, Zimmermann S, Radford D C, Lagergren K, Weisshaar D, Winkler R, Glasmacher T, Anderson J T and Beausang C W 2013 *Nucl. Instr. and Meth. in Phys. Res. A* **709** 44 - 55
[10] Wu C Y, Cline D, Hayes A, Flight R S, Melchionna A M, Zhou C, Lee I Y, Swan D, Fox R and Anderson J T 2016 *Nucl. Instr. and Meth. in Phys. Res. A* **814** 6 - 11
[11] Lettmann M, Werner V, Pietralla N, Doornenbal P, Obertelli A, Rodrguez T R, Sieja K *et al.* 2017 Triaxiality of neutron-rich ^{84,86,88}Ge from low-energy nuclear spectra *Phys. Rev. C* **96** 011301(R)
[12] Zamfir N V and Casten R F 1991 *Phys. Lett. B* **260** 265
[13] McCutchan E A, Bonatsos D, Zamfir N V and Casten R F 2007 *Phys. Rev. C* **76** 024306(R)
[14] Toh Y, Chiara C J, McCutchan E A, Walters W B, Janssens R V F, Carpenter M P, Zhu S, Broda R, Fornal B, Kay B P et al. 2013 *Phys. Rev. C* **87** 041304(R)

IOP Conf. Series: Journal of Physics: Conf. Series **966** (2018) 012059 doi:10.1088/1742-6596/966/1/012059

Nuclear deformation in the configuration-interaction shell model

Y. Alhassid,[1] G.F. Bertsch[2,3] C.N. Gilbreth,[2] and M.T. Mustonen[1]
[1]Center for Theoretical Physics, Sloane Physics Laboratory, Yale University, New Haven, Connecticut 06520, USA
[2]Institute of Nuclear Theory, Box 351550, University of Washington, Seattle, WA 98915
[3]Department of Physics, Box 351560, University of Washington, Seattle, WA 98195

E-mail: `yoram.alhassid@yale.edu`

Abstract. We review a method that we recently introduced to calculate the finite-temperature distribution of the axial quadrupole operator in the laboratory frame using the auxiliary-field Monte Carlo technique in the framework of the configuration-interaction shell model. We also discuss recent work to determine the probability distribution of the quadrupole shape tensor as a function of intrinsic deformation β, γ by expanding its logarithm in quadrupole invariants. We demonstrate our method for an isotope chain of samarium nuclei whose ground states describe a crossover from spherical to deformed shapes.

1. Introduction

Deformation is a central concept in understanding the physics of heavy nuclei [1]. However, since intrinsic deformation is introduced by invoking a mean-field approximation that breaks rotational symmetry, it is a challenge to determine the probability density of the intrinsic deformation in the configuration-interaction (CI) shell model, a framework that preserves rotational symmetry.

Here we review a recent technique we introduced to calculate the axial quadrupole distribution in the laboratory frame using the auxiliary-field Monte Carlo (AFMC) method [2, 3]. We found that this lab-frame distribution exhibits a model-independent signature of deformation. We then discuss recent work in which we used quadrupole invariants [4, 5] to model the quadrupole shape distribution in the intrinsic frame [6]. We demonstrate our method for an isotope chain of samarium nuclei, using the model space and interaction of Refs. [7, 8]. Quadrupole invariants were used to extract the effective intrinsic deformation within the framework of the CI shell model in lighter nuclei; see the recent examples in Refs. [9, 10] and references therein.

2. Auxiliary-field Monte Carlo method

AFMC, also known in the context of the nuclear shell model as the shell model Monte Carlo (SMMC) method [11, 12, 13], is based on the Hubbard-Stratonovich (HS) transformation [14]. The Gibbs operator $e^{-\hat{H}/T}$ of a nucleus described by the Hamiltonian $\hat{H}$ at temperature T is represented as a superposition of non-interacting propagators $\hat{U}_\sigma$ of nucleons moving in auxiliary fields $\sigma = \sigma(\tau)$ that depend on imaginary time τ

$$e^{-\hat{H}/T} = \int \mathcal{D}[\sigma] \, G_\sigma \hat{U}_\sigma \,, \tag{1}$$

where G_σ is a Gaussian weight. The thermal expectation value of an observable $\hat{O}$ can then be written as

$$\langle \hat{O} \rangle = \frac{\mathrm{Tr}\left(\hat{O}e^{-\hat{H}/T}\right)}{\mathrm{Tr}\, e^{-\hat{H}/T}} = \frac{\int \mathcal{D}[\sigma] G_\sigma \langle \hat{O} \rangle_\sigma \mathrm{Tr}\, \hat{U}_\sigma}{\int \mathcal{D}[\sigma] G_\sigma \mathrm{Tr}\, \hat{U}_\sigma}, \tag{2}$$

where $\langle \hat{O} \rangle_\sigma \equiv \mathrm{Tr}\,(\hat{O}\hat{U}_\sigma)/\mathrm{Tr}\,\hat{U}_\sigma$. The integrands in Eq. (2) can be calculated using matrix algebra in the single-particle space, and the integration over the large number of auxiliary fields $\sigma(\tau)$ is carried out by Monte Carlo methods. Canonical expectation values at fixed number of protons and neutrons are calculated using a discrete Fourier representation of the particle-number projection [15, 16].

3. Quadrupole distribution in the laboratory frame

The lab-frame distribution $P(q)$ of the axial quadrupole $\hat{Q}_{20}$ at temperature T is defined by

$$P(q) = \mathrm{Tr}\left[\delta(\hat{Q}_{20} - q)e^{-\hat{H}/T}\right] / \mathrm{Tr}\, e^{-\hat{H}/T}. \tag{3}$$

Using compete sets of many-particle eigenstates $|e_m\rangle$ and $|q_n\rangle$ of $\hat{H}$ and $\hat{Q}_{20}$, respectively, we have (note that $[H, \hat{Q}_{20}] \neq 0$)

$$P(q) = \sum_n \delta(q - q_n) \sum_m \langle q_n | e_m \rangle^2 e^{-e_m/T}. \tag{4}$$

In the CI shell model, the spectrum of $\hat{Q}_{20}$ is discrete, but for a heavy nucleus it becomes a quasi-continuum.

3.1. Projection on the axial quadrupole

To carry out the projection in AFMC, we represent the δ function as a Fourier integral

$$\delta(\hat{Q}_{20} - q) = \frac{1}{2\pi} \int_{-\infty}^{\infty} d\varphi\, e^{-i\varphi q}\, e^{i\varphi \hat{Q}_{20}}, \tag{5}$$

and use this in (3) together with the HS transformation (1) for $e^{-\hat{H}/T}$. For each configuration σ of the auxiliary fields, we replace the Fourier integral by a discrete Fourier transform. Choosing an interval $[-q_{\max}, q_{\max}]$, dividing it into $2M+1$ intervals of equal length $\Delta q = 2q_{\max}/(2M+1)$, and defining $q_m = m\Delta q$, we have

$$\mathrm{Tr}\left[\delta(\hat{Q}_{20} - q_m)\hat{U}_\sigma\right] \approx \frac{1}{2q_{\max}} \sum_{k=-M}^{M} e^{-i\varphi_k q_m}\, \mathrm{Tr}(e^{i\varphi_k \hat{Q}_{20}} \hat{U}_\sigma), \tag{6}$$

where $\varphi_k = \pi k / q_{\max}$ $(k = -M, \ldots, M)$. Since $\hat{Q}_{20}$ is a one-body operator, we can calculate the grand-canonical traces on the r.h.s. of (6) in terms of the matrices $\mathbf{Q}_{20}$ and $\mathbf{U}_\sigma$ representing, respectively, $\hat{Q}_{20}$ and $\hat{U}_\sigma$ in the single-particle space, i.e., $\mathrm{Tr}\left(e^{i\varphi_k \hat{Q}_{20}} \hat{U}_\sigma\right) = \det\left(1 + e^{i\varphi_k \mathbf{Q}_{20}} \mathbf{U}_\sigma\right)$.

3.2. Angle averaging

When using the usual Metropolis algorithm, we find that for a deformed nucleus, the distribution $P(q)$ and its moments are slow to thermalize and have a large decorrelation length. We resolved this problem by averaging over a specific set of rotation angles Ω_j

$$\langle e^{i\varphi \hat{Q}_{20}} \rangle_\sigma \to \frac{1}{N_\Omega} \sum_{j=1}^{N_\Omega} \langle e^{i\varphi \hat{Q}_{20}} \rangle_{\sigma, \Omega_j}, \tag{7}$$

where $\langle e^{i\varphi\hat{Q}_{20}}\rangle_{\sigma,\Omega} = \mathrm{Tr}\left(e^{i\varphi\hat{Q}_{20}}\hat{R}\hat{U}_\sigma\hat{R}^\dagger\right)/\mathrm{Tr}\left(\hat{R}\hat{U}_\sigma\hat{R}^\dagger\right)$, with $\hat{R} = \hat{R}(\Omega)$ being the rotation operator with angles Ω. We note that any rotation of $\hat{Q}_{20}$ in (5) does not affect the distribution $P(q)$ since the Hamiltonian $\hat{H}$ is invariant under rotations. The angles Ω_j are chosen such that $\hat{Q}_{20}^m$ is proportional to the invariant of order m up to a given order n. We have determined a set of 6 angles for $n = 2$ and a set of 21 angles for $n = 3$ [3]. All calculations shown here are based on a 21-angle average. We used a time slice of $\Delta\beta = 1/64$ MeV^{-1} in a discretized version of the HS transformation (1) and $\sim$ 5000 auxiliary-field configurations for each temperature.

3.3. Application to samarium isotopes

In Fig. 1 we show AFMC distributions $P(q)$ for an isotope chain of samarium nuclei $^{148-154}$Sm at low, intermediate and high temperatures. We observed that the low-temperature distribution for ^{154}Sm, whose Hartree-Fock-Bogoliubov (HFB) ground state is deformed, is skewed and in qualitative agreement with the distribution for a prolate rigid rotor (dashed line). In contrast, the low-temperature distribution for the spherical nucleus ^{148}Sm is close to a Gaussian. We conclude that the axial quadrupole distribution in the lab frame is a model-independent signature of deformation. At low temperatures, we observe a crossover from a spherical to a prolate shape as we increase the number of neutrons. In the isotopes that are deformed at low temperature ($^{150-154}$Sm), we observe a crossover to a spherical shape as we increase T.

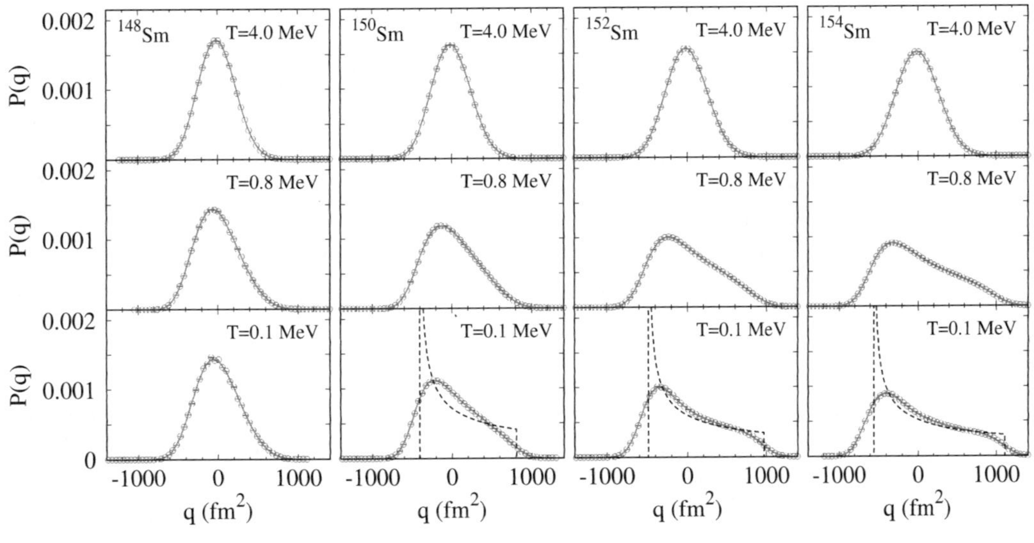

Figure 1. AFMC distributions $P(q)$ vs. q (blue circles) for an isotope chain of samarium nuclei at high, intermediate and low temperatures. The dashed lines are rigid-rotor distributions. The red solid lines are the distributions obtained from (12) (see Sec. 4.2). Adapted from Ref. [3].

4. Quadrupole distribution in the intrinsic frame

In physical applications, we are interested in the intrinsic deformation of the nucleus. Information on intrinsic deformation can be extracted without invoking a mean-field approximation by using quadrupole invariants which are frame-independent.

4.1. Quadrupole invariants and their relation to moments of $\hat{Q}_{20}$.

A quadrupole invariant is a linear combination of products of the components $\hat{Q}_{2\mu}$ that is invariant under rotations. These invariants can be constructed from tensor products of the second-rank quadrupole tensor $\hat{Q}_{2\mu}$ [4, 5]. For any given order $2 \le n \le 4$, these invariants are unique and their expectation values can be expressed in terms of the corresponding moments of $\hat{Q}_{20}$:

$$\langle \hat{Q} \cdot \hat{Q} \rangle = 5\langle \hat{Q}_{20}^2 \rangle \, , \quad \langle (\hat{Q} \times \hat{Q})^{(2)} \cdot \hat{Q} \rangle = -5\sqrt{\frac{7}{2}} \langle \hat{Q}_{20}^3 \rangle \, , \quad \langle (\hat{Q} \cdot \hat{Q})^2 \rangle = \frac{35}{3} \langle \hat{Q}_{20}^4 \rangle \, . \tag{8}$$

For given values $q_{2\mu}$ of the quadrupole tensor, we define dimensionless quadrupole deformation parameters $\alpha_{2\mu}$ as in the liquid drop model, i.e., $q_{2\mu} = \frac{3}{\sqrt{5\pi}} 3r_0^2 A^{5/3} \alpha_{2\mu}$, where $r_0 = 1.2$ fm and A is the mass number of the nucleus. For each set of deformation parameters $\alpha_{2\mu}$, we define an intrinsic frame whose orientation is characterized by Euler angles Ω, and in which the deformation parameters $\tilde{\alpha}_{2\mu}$ are given by

$$\tilde{\alpha}_{20} = \beta \cos\gamma, \quad \tilde{\alpha}_{21} = \tilde{\alpha}_{2,-1} = 0, \quad \tilde{\alpha}_{22} = \tilde{\alpha}_{2,-2} = \text{real} = \frac{1}{\sqrt{2}} \beta \sin\gamma \, . \tag{9}$$

The parameters β, γ are known as the Hill-Wheeler parameters. The metric of the transformation from the lab-frame variables $\alpha_{2\mu}$ to the intrinsic-frame variables β, γ, Ω is given by

$$\prod_\mu d\alpha_{2\mu} = \frac{1}{2} \beta^4 |\sin(3\gamma)| \, d\beta \, d\gamma \, d\Omega \, . \tag{10}$$

Quadrupole invariants can also be constructed from $\alpha_{2\mu}$, and up to fourth order, they are given by

$$\alpha \cdot \alpha = \beta^2 \, , \quad [\alpha \times \alpha]_2 \cdot \alpha = -\sqrt{\frac{2}{7}} \beta^3 \cos(3\gamma) \, , \quad (\alpha \cdot \alpha)^2 = \beta^4 \, . \tag{11}$$

4.2. Landau-like expansion

The distribution $P(T, \alpha_{2\mu})$ of the quadrupole deformation $\alpha_{2\mu}$ at temperature T is a rotational invariant and therefore it depends only on the intrinsic parameters β, γ. Using a Landau-like expansion [17], we expand the logarithm of P in the quadrupole invariants up to fourth order

$$P(T, \beta, \gamma) = \mathcal{N}(T) e^{-a(T)\beta^2 - b(T)\beta^3 \cos(3\gamma) - c(T)\beta^4} \, , \tag{12}$$

where a, b, c are temperature-dependent coefficients and $\mathcal{N}$ is a normalization constant determined from $4\pi^2 \int d\beta \, d\gamma \, \beta^4 |\sin(3\gamma)| P(T, \beta, \gamma) = 1$. The parameters a, b, c are determined by matching the expectation values [calculated with the distribution (12)] of the three quadrupole invariants as expressed in (11) with their AFMC values, which can be computed from the corresponding moments of $P(q)$ using Eqs. (8).

4.3. Validation of the Landau-like expansion

To test the validity of (12), we construct the lab-frame distribution $P(T, \alpha_{2\mu})$ by expressing the quadrupole invariants in terms of the lab-frame deformation $\alpha_{2\mu}$ [see Eq. (11)]. We then integrate over all $\alpha_{2\mu}$ with $\mu \neq 0$ to find the lab-frame distribution of α_{20}, or equivalently $P(q)$, and compare it with the AFMC distribution. The distributions $P(q)$ calculated from the model (12) are shown by the solid red lines in Fig. 1 and are in excellent agreement with the AFMC distributions (open blue circles).

4.4. Application to samarium isotopes

Figure 2 shows the calculated shape distributions $P(T, \beta, \gamma)$ defined in (12) vs. β, γ for the samarium isotopes at the same temperatures as in Fig. 1. The maxima of the distributions (12) mimic the shape transition observed in the HFB mean-field approximation but in the

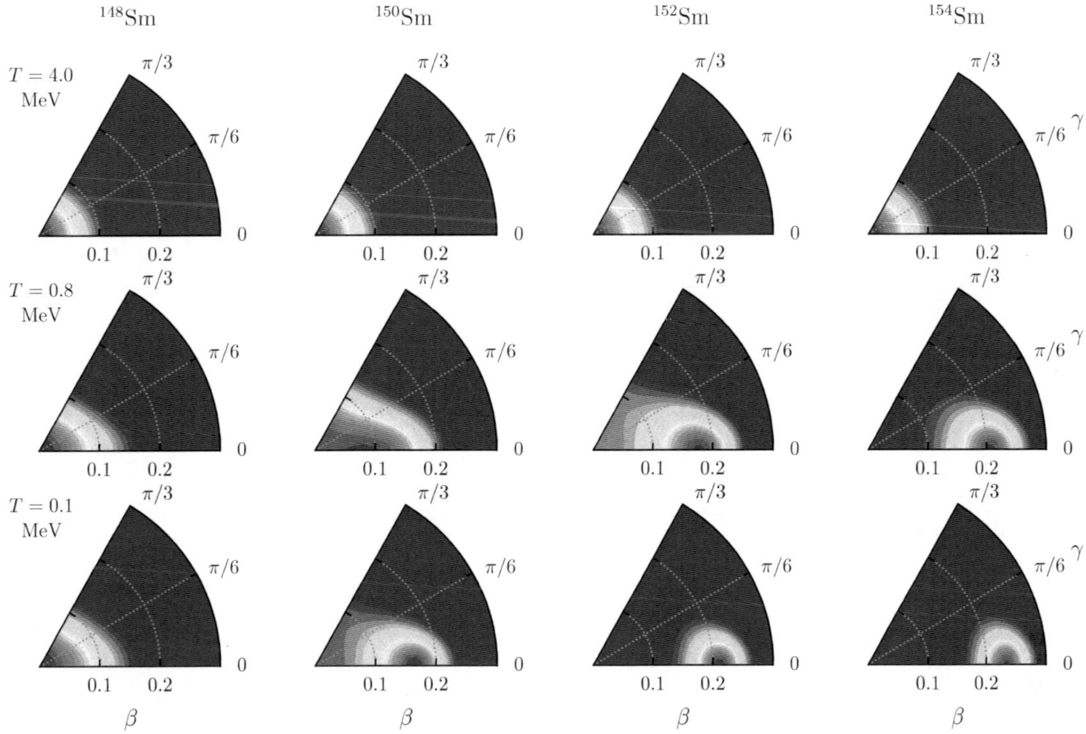

Figure 2. Intrinsic shape distributions $P(T, \beta, \gamma)$ at low, intermediate and high temperatures for the even-mass samarium isotopes $^{148-154}$Sm. Adapted from Ref. [6].

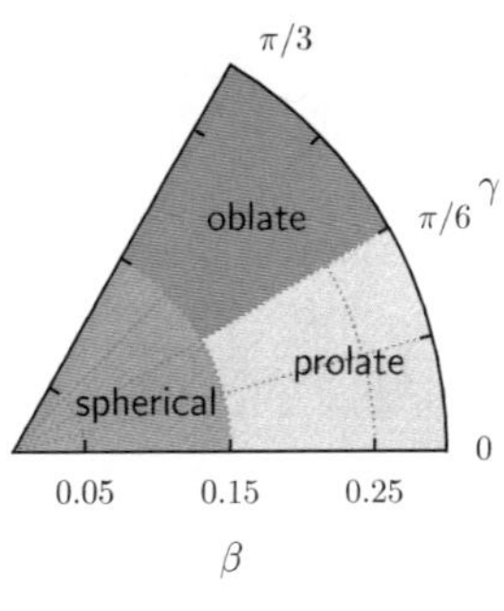

Figure 3. Definition of a spherical, prolate and oblate shape regions in the $\beta - \gamma$ plane. These regions are used in presenting the results of Fig. 4. Taken from Ref. [6].

framework of the CI shell model [18]. As a function of neutron number we observe a transition from a spherical to prolate shape, while nuclei that are deformed in their ground state make a transition from deformed to spherical shape as a function of temperature.

To simplify the presentation of our results we divide the β-γ plane into the three regions as shown in Fig. 3, which we choose to represent spherical, prolate and oblate shapes. For each region, we then define $P_{\rm shape}(T)$ to be the probability to find the nucleus in the corresponding region, i.e., $P_{\rm shape}(T) = 4\pi^2 \int_{\rm shape} d\beta\, d\gamma\, \beta^4 |\sin 3\gamma| P(T, \beta, \gamma)$. In Fig. 4 we show these probabilities as a function of temperature T for the four even-mass samarium isotopes. In the spherical ^{148}Sm, the spherical region dominates at all temperatures, while in the deformed ^{152,154}Sm isotopes, the prolate region has a probability close to 1 at low temperatures and the spherical region becomes the most probable above a certain temperature. The transitional nucleus ^{150}Sm exhibits an intermediate behavior.

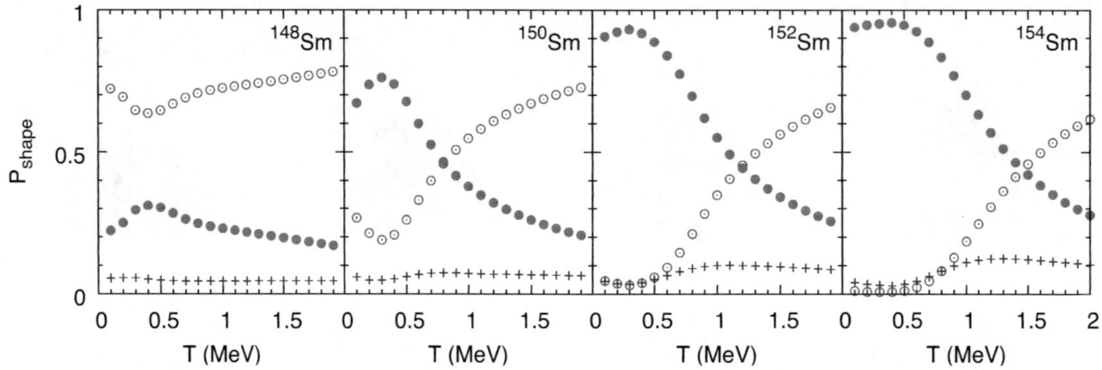

Figure 4. Probabilities of spherical (red open circles), prolate (green solid circles), and oblate (blue pluses) regions as a function of T for $^{148-154}$Sm isotopes. Adapted from Ref. [6].

5. Conclusion and outlook

We discussed a method we recently introduced to calculate lab-frame and intrinsic shape distributions within the CI shell model without invoking a mean-field approximation. Using the saddle-point approximation, it is also possible to convert the finite-temperature intrinsic shape distribution (12) to level densities $\rho(E_x, \beta, \gamma)$ as a function of excitation energy E_x and intrinsic deformation β, γ [6]. Deformation-dependent level densities are useful in the modeling of nuclear shape dynamics, such as fission.

Acknowledgments

This work was supported in part by the U.S. DOE grant Nos. DE-FG02-91ER40608 and DE-FG02-00ER41132. The research presented here used resources of the National Energy Research Scientific Computing Center, which is supported by the Office of Science of the U.S. Department of Energy under Contract No. DE-AC02-05CH11231. This work was also supported by the HPC facilities operated by, and the staff of, the Yale Center for Research Computing.

References

[1] Bohr A and Mottelson B R 1975 *Nuclear Structure* vol II (Reading, MA: Benjamin)
[2] Alhassid Y, Gilbreth C N and Bertsch G F 2014 *Phys. Rev. Lett.* **113** 262503
[3] Gilbreth C N, Alhassid Y and Bertsch G F arXiv:1710.00072
[4] Kumar K 1972 *Phys. Rev. Lett.* **28** 249
[5] Cline D 1986 *Ann. Rev. Nucl. Part. Sci.* **36** 683
[6] Mustonen M T, Gilbreth C N, Alhassid Y and Bertsch G F, to be published.
[7] Alhassid Y, Fang L and Nakada H 2008 *Phys. Rev. Lett.* **101** 082501
[8] Özen C, Alhassid Y and Nakada H 2013 *Phys. Rev. Lett.* **110** 042502
[9] Hadyńska-Klek K *et al.* 2016 *Phys. Rev. Lett.* **117** 062501
[10] Schmidt T, Heyde K L G, Blazhev A and Jolie J 2017 *Phys. Rev.* C **96** 014302
[11] Lang G H, Johnson C W, Koonin S E and Ormand W E 1993 *Phys. Rev.* C **48** 1518
[12] Alhassid Y, Dean D J, Koonin S E, Lang G, and Ormand W E 1994 *Phys. Rev. Lett.* **72** 613
[13] For a recent review, see Alhassid Y in 2017 *Emergent Phenomena in Atomic Nuclei from Large-Scale Modeling: a Symmetry-Guided Perspective*, ed K D Launey (Singapore: World Scientific)
[14] Hubbard J 1959 *Phys. Rev. Lett.* **3** 77; Stratonovich R L 1957 *Sov. Phys. - Dokl.* **115** 1097
[15] Ormand W E, Dean D J, Johnson C W, Lang G H and Koonin S E 1994 *Phys. Rev.* C **49** 1422
[16] Alhassid Y, Liu S and Nakada H 1999 *Phys. Rev. Lett.* **83** 4265
[17] Alhassid Y, Levit S and Zingman J 1986 *Phys. Rev. Lett.* **57** 539
[18] Note, however, that the maxima of the probability density distribution in β, γ, given by $4\pi^2\beta^4|\sin 3\gamma|P(T, \beta, \gamma)$, do not exhibit sharp shape transitions.

IOP Conf. Series: Journal of Physics: Conf. Series **966** (2018) 012060 doi:10.1088/1742-6596/966/1/012060

Experimental evidence of chaos in the bound states of ^{208}Pb

L. Muñoz[1], R. A. Molina[2], J. M. G. Gómez[1], A. Heusler[3]

[1] Grupo de Física Nuclear, Facultad de Ciencias Físicas, Universidad Complutense, E-28040 Madrid, Spain

[2] Instituto de Estructura de la Materia, IEM-CSIC, Serrano 123, Madrid, E-28006, Spain

[3] Gustav-Kirchhoff-Str. 7/1, D-69120 Heidelberg, Germany

E-mail: jmgomezg@ucm.es

Abstract. Theoretical calculations, especially shell-model calculations, have shown a strongly chaotic behavior of bound states at higher excitation energy, in regions of high level density. However, it had not been possible up to now to observe chaos in the experimental bound energy levels of any single nucleus. In this paper we study the spectral fluctuations of the ^{208}Pb nucleus using the complete experimental spectrum of 151 states up to excitation energies of 6.20 MeV. For natural parity states the results are very close to the predictions of Random Matrix Theory (RMT) for the nearest-neighbor spacing distribution. By contrast, the results for unnatural parity states are far from RMT behavior. We interpret these results as a consequence of the strength of the residual interaction in ^{208}Pb, which, according to experimental data, is much stronger for natural than for unnatural parity states. In addition our results show that chaotic and non-chaotic nuclear states coexist in the same energy region of the spectrum.

1. Introduction

The atomic nucleus is generally considered a paradigmatic case of quantum chaos. Intuitively one can expect that fast moving nucleons interacting with the strong nuclear force and bound in the small nuclear volume should give rise to a chaotic motion. The quest for chaos in nuclei has been quite intensive, both with theoretical calculations using nuclear models and with detailed analyses of experimental data. Statistical spectroscopy studies in nuclei have been also motivated by a desire to understand the implications of chaotic behavior in many-body quantum systems. Theoretical calculations, especially shell-model calculations, have shown a strongly chaotic behavior of bound states at higher excitation energy, in regions of high level density. However, as we discuss below, it has not been possible up to now to observe chaos in the experimental bound energy levels of any single nucleus. For a comprehensive review of chaos in nuclei see for example Gómez *et al.* [1] and Weidenmüller and Mitchell [2].

In this paper we first present in Section 2 a brief outline of the spectral properties that characterize quantum chaos using two different approaches, the Random Matrix Theory (RMT) and the time series approach. Then we look at the theoretical and experimental evidence for the existence of chaos in nuclei in Section 3, and we study in Section 4 the spectral fluctuations of the lowest 151 states in ^{208}Pb, which have been recently identified with spin and parity assignements [3]. It turns out that nuclear motion in ^{208}Pb is strongly chaotic at least for the natural parity states [4]. For unnatural parity states the evidence is less clear, but it seems to be intermediate

IOP Conf. Series: Journal of Physics: Conf. Series **966** (2018) 012060 doi:10.1088/1742-6596/966/1/012060

between chaotic and regular motion. Finally we compare our results with those of another recent study of chaos in ^{208}Pb [5].

2. Quantum chaos and spectral fluctuations

Quantum chaos has been the subject of many investigations during the last decades. The pioneering work of Berry, Bohigas *et al.*, and others [6, 7] leads to an important and concise statement: the spectral properties of simple systems known to be ergodic in the classical limit follow very closely those of the Gaussian orthogonal ensemble (GOE) of random matrices. On the contrary, integrable systems lead to level fluctuations that are well described by the Poisson distribution, i.e., levels behave as if they were uncorrelated.

Fluctuations are the departure of the actual level density from a local uniform density. Therefore it is essential to eliminate the smooth part of the exponential increase of the nuclear density, mapping the actual spectrum onto a quasiuniform spectrum with mean spacing $\langle s\rangle = 1$. This step, called unfolding, is delicate and of utmost importance, because some of the unfolding procedures used in the literature can lead to completely wrong results on the behavior of level fluctuations [8].

Short-range correlations between energy states are usually measured by the nearest neighbor spacing (NNS) distribution $P(s)$. The Poisson distribution is given by $P_P(s) = \exp(-s)$. The GOE distribution has no analytical formula, but is well approximated by the Wigner surmise, $P_W(s) = (\pi s/2)\exp(-\pi s^2/4)$. Note that $P_P(0) = 1$ and $P_W(0) = 0$. Quantum chaos is characterized by strong state correlations and level repulsion. An assesment of chaotic or regular behavior of a system is given by comparison of the NNS distribution to Poisson and Wigner. Real complex systems are often not fully chaotic or fully regular. These intermediate situations are often described in terms of a single parameter ω by means of the Brody distribution $P_B(s,\omega)$, which is just an interpolation formula between the Poisson ($\omega = 0$) and Wigner ($\omega = 1$) distributions. Although it has no obvious physical meaning, the Brody parameter ω is a simple and widely used measure of chaoticity of a system. When the number of spacings is not very large, it is preferable to fit the cumulative distribution, $I(s) = \int_0^s P(x)dx$.

Long-range correlations are usually studied with the Dyson-Mehta statistic $\Delta_3(L)$. For Poisson $\langle\Delta_3(L)\rangle = \dfrac{L}{15}$ and for GOE $\langle\Delta_3(L)\rangle = \dfrac{1}{\pi^2}\log(L) + b + \mathcal{O}(L^{-1}),\quad L \gg 1$.

There is another approach to quantum chaos based on the analogy between a discrete time series and a quantum energy spectrum, if time t is replaced by the energy E of the quantum states [9]. The spacing between two consecutive levels is $s_i = \varepsilon_{i+1} - \varepsilon_i$. We define the statistic δ_n as a signal,

$$\delta_n = \sum_{i=1}^{n}(s_i - \langle s\rangle) = \varepsilon_{n+1} - \varepsilon_1 - n \tag{1}$$

and the discrete power spectrum is $P(k) = \left|\widehat{\delta_k}\right|^2$, where $\widehat{\delta_k}$ is the Fourier transform of δ_n,

$$\widehat{\delta_k} = \frac{1}{\sqrt{N}}\sum_n \delta_n \exp\left(\frac{-2\pi i k n}{N}\right) \tag{2}$$

and N is the size of the series. It turns out that the energy spectra of chaotic quantum systems are characterized by 1/f noise, and on the contrary, the energy spectra of regular quantum systems are characterized by 1/f^2 noise (Brown noise) [9].

3. Chaos in nuclei

3.1. Spectral fluctuations in shell-model spectra

Theoretical calculations provide long sequences of J^π or $J^\pi T$ levels suitable for statistical analysis of fluctuations (no missing levels, no uncertain spin and parity assignments). Calculations performed with the spherical shell model, the cranking model, the interacting boson model and other models have shown examples of highly regular energy spectra in deformed nuclei and examples of highly chaotic spectra in spherical nuclei, although there are exceptions in some nuclei.

In spherical nuclei, where the shell model is most appropriate, many calculations have shown that for large configuration spaces, with $J^\pi T$ level sequences up to several thousand, the usual fluctuation measures $P(s)$ and $\Delta_3(L)$ agree very well with GOE predictions. To mention just some examples, Fig. 1 illustrates the good agreement of $P(s)$ and $\Delta_3(L)$ with GOE for the $J^\pi = 2^+$ states of ^{28}Si calculated with the shell model [10]. Fig. 2 illustrates the 1/f noise behavior of shell-model states in ^{34}Na ($\alpha = 1.11$) and ^{24}Mg ($\alpha = 1.06$). The straight line is the best fit of the power spectrum assuming $P(k) = 1/k^\alpha$.

In the $2p1f$ shell the configuration space and the level density are much larger than in sd-shell nuclei. Shell-model calculations with a realistic interaction have been performed to investigate the degree of chaos in different isotopes as a function of excitation energy by Molina *et al.* [11]. For example, in ^{46}Sc a total of 25,498 spacings are included in the calculations, ensuring excellent statistics. Fig 3 shows the behavior of $\Delta_3(L)$ for the $J^\pi = 0^+, T = T_z$ states of ^{46}Ca, ^{46}Sc and ^{46}Ti. Clearly, there is an isospin dependence in the chaoticity of these isobars. For ^{46}Ti the agreement with GOE (dotted line in Fig. 3) is excellent. For ^{46}Sc there is GOE behavior up to $L \sim 40$, and ^{46}Ca clearly deviates from GOE towards Poisson (dashed line) for $L \geq 10$. Furthermore, at low energies the fluctuations in Ca isotopes are more regular than chaotic, for instance $\omega = 0.25$ for the levels up to 5 MeV above yrast in ^{52}Ca. Similar results were obtained for Pb isotopes, with only valence neutrons outside the ^{208}Pb core [12].

We interpret the observed isospin dependence of chaoticity as a result of the strength of the residual interaction. The shell-model residual nn interaction is much weaker than the residual pn interaction. Ca isotopes have only neutrons in the pf shell, but if just one neutron is replaced by a proton, the pn interactions destroy the mean-field order. Therefore Sc or Ti isotopes, having both protons and neutrons in the valence space of the pf shell, exhibit strong chaotic characteristics even in the ground-state region.

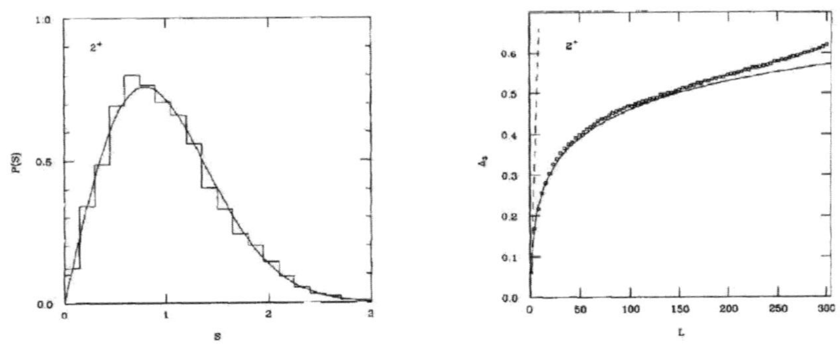

Figure 1. Shell-model $P(s)$ and $\langle\Delta_3(L)\rangle$ values for the $J^\pi = 2^+$, $T = 0$ states of ^{28}Si. Taken from [10].

3.2. Spectral fluctuations in experimental nuclear bound states

Haq, Pandey and Bohigas [13] analyzed the spectral fluctuations of a very large number of experimentally identified neutron and proton $J^\pi = 1/2^+$ resonances just above the one-nucleon emission threshold and showed that they agree very well with the spectral fluctuations of GOE.

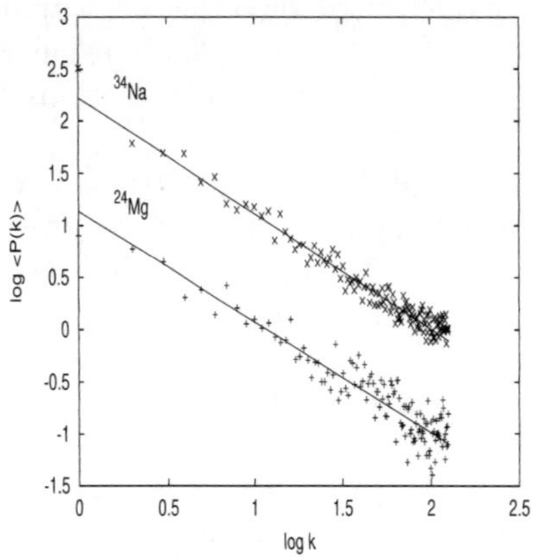

Figure 2. Power spectrum of δ_n statistic for nuclear shell-model spectra exhibiting 1/f noise. Adapted from [9].

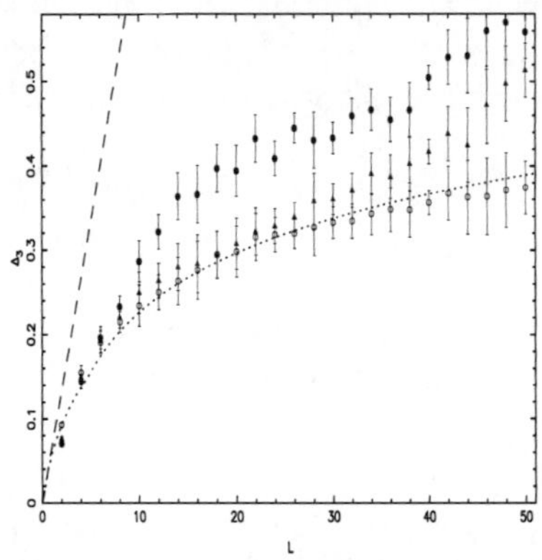

Figure 3. Comparison of $\langle\Delta_3(L)\rangle$ for $J^\pi = 0^+$ states in ^{46}Ca (full circles), ^{46}Sc (triangles) and ^{46}Ti (open circles). Adapted from [11].

Thus, in this sense, it is clear that nuclei are very chaotic in the energy region just above the one-nucleon emission threshold. But for bound states, the situation is not so clear, because a good analysis of fluctuations in experimental energy spectra requires the knowledge of sufficiently long, pure and complete sequences, i.e. with the same $J^\pi T$ values and without missing levels or $J^\pi T$ misassignments. But this ideal situation is rarely found in nuclei. For very light nuclei the number of bound levels is not sufficient for statistical purposes. For medium and heavy nuclei the identified levels are limited to the ground state region, because at higher energy the level density becomes very high and the experimental identification of the energy and J^π values becomes generally impossible.

In order to improve statistics, level spacings from different nuclei can be combined into a single set to analyze the behavior of the NNS distribution $P(s)$. An extensive analysis of low-lying energy levels was performed by Shriner *et al.* [14] using experimental data along the whole nuclear chart. A total of 988 spacings from 60 different nuclei were included in the analysis, giving $\omega = 0.43 \pm 0.05$, which is closer to Poisson than to GOE. Separating the data in six different mass regions a clear trend from GOE to Poisson is observed as A increases. For $A \leq 50$, the fit gives $\omega = 0.72 \pm 0.16$, and for $A > 230$ it gives $\omega = 0.24 \pm 0.11$. Generally spherical nuclei are closer to GOE and deformed nuclei are closer to Poisson. The latter is not necessarily a manifestation of regular behavior for the low-lying states of these nuclei, because a deviation towards Poisson may be also due to the omission of some symmetry. In the present case of deformed nuclei it may be due to omission of the K quantum number.

4. Chaos in the experimental bound states of ^{208}Pb

Recently the complete experimental spectrum of the lowest 151 states in ^{208}Pb, up to 6.20 MeV excitation energy, has been identified with spin and parity assignments. This represents the largest ensemble known up to now which can be used for a statistical investigation of the chaotic behavior among bound states in an atomic nucleus. We consider that these accurate data on ^{208}Pb enable a meaningful statistical analysis of level fluctuations, with pure, complete, and reasonably long sequences. There are 14 J^π sequences with a minimum of 5 and up to 19 known consecutive states.

Separate unfolding has been performed for all J^π sequences using the constant temperature

IOP Conf. Series: Journal of Physics: Conf. Series **966** (2018) 012060 doi:10.1088/1742-6596/966/1/012060

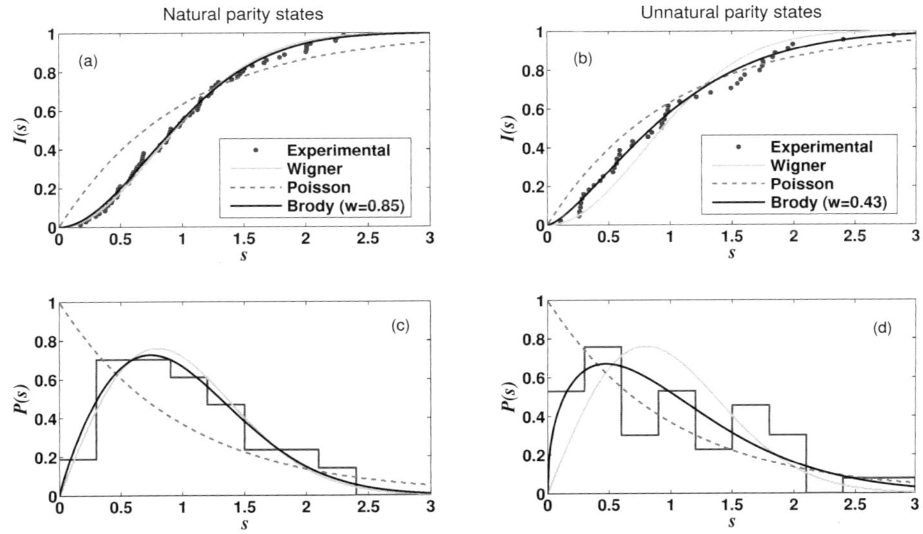

Figure 4. Comparison of spectral fluctuation statistics for natural parity states (left panels) and unnatural parity states (right panels) in ^{208}Pb. Panels (a) and (b) show the cumulative nearest neighbor spacing (cumulative NNS) distribution for experimental levels (dots). Panels (c) and (d) show the NNS distribution $P(s)$ for experimental levels (histogram). In all panels the green (light gray) line is the Wigner surmise, the red dashed line is the Poisson distribution, and the black solid line is the best fit Brody distribution. Adapted from [4].

formula [14],

$$\overline{\rho}(E) = \frac{1}{T}\exp[(E - E_0)/T], \tag{3}$$

where T and E_0 are taken as parameters. Gathering the unfolded spacings for all J^π into a single set, there are 115 spacings. The $P(s)$ distribution and the cumulative $I(s)$ distribution for the full set of spacings exhibit shapes much closer to GOE than to Poisson, but still the difference with GOE is substantial. The Brody parameter for the cumulative distribution is $\omega = 0.68$.

The relevant question with these results is how to interpret them regarding chaotic motion in ^{208}Pb. The result $\omega = 0.68$ is similar to the highest value $\omega = 0.72 \pm 0.16$ obtained for experimental states of nuclei with $A < 50$ [14]. Therefore we may wonder whether it represents more or less a practical limit of possible chaos in nuclear bound states.

Heusler *et al.* [3] have shown that the number of identified states at $E_x < 6.20$ MeV nearly agrees with the number of states in this energy interval predicted by a simplified shell-model consisting of 1p-1h mean field configurations. However, 70 unnatural parity states agree with this model within ~ 0.2 MeV, and 20 natural parity states differ by > 0.5 MeV. Hence they conclude that configuration mixing is stronger for natural parity states and therefore the residual interaction is much larger than for unnatural parity states. To check if this effect can be observed in the fluctuation measures we have analyzed separately the spectral fluctuations of those two sets of states and have found that they behave very differently.

Fig. 4 shows the experimental $P(s)$ and $I(s)$ distributions for natural and for unnatural parity states, compared to Wigner and Poisson. Clearly the agreement with GOE is good for natural parity states ($\omega = 0.85$), but for unnatural parity states the results are somewhat closer to Poisson ($\omega = 0.43$). Table 1 summarizes the number of spacings and the value of ω for

Table 1. Number of spacings, Brody parameter ω and rms deviation from Wigner and Poisson distributions for different combinations of parity in the experimental states of ^{208}Pb at $E_x < 6.20$ MeV.

	Number of spacings		
Parity	all	natural	unnatural
even	45	29	16
odd	70	42	28
all	115	71	44
Brody ω	0.68	0.85	0.43
$(\mathrm{RMSD})_W$	0.040	0.025	0.077
$(\mathrm{RMSD})_P$	0.115	0.129	0.088

the different combinations of parity and natural/unnatural parity. Independently of the Brody parameter, the table gives as well the root mean square deviation (RMSD) of the experimental $I(s)$ from the Wigner and Poisson limits, showing the strong chaoticity of natural parity states.

These results agree with the observation of a stronger residual interaction for natural parity states pointed out in Ref. [3]. In the shell model the mean field gives rise to a regular motion and the residual interaction produces the mixing of basis states in the eigenstates, destroying the regular mean-field motion. This effect has been observed in shell-model calculations introducing a strength parameter to modulate the residual interaction. As the strength parameter increases, the fluctuations measures of energy levels approach GOE behavior and thus the motion becomes chaotic [10].

Finally, let us point out that another analysis of spectral fluctuations has been recently published [5]. They use two chaoticity measures different from ours and do not observe a significant difference between the experimental natural and unnatural parity states. Possibly this difference with our results is due to the different unfolding procedures used. The influence of the unfolding method is important when the number of levels is relatively small, as is the case for the unnatural parity states of ^{208}Pb. On the other hand, the spectral fluctuations have been analyzed by the same authors for two different shell-model calculations in ^{208}Pb. It should be noticed that the shell-model J^π sequences are much larger than the experimental ones and therefore the unfolding is more reliable. The shell-model values of the chaoticity parameters are different for natural and unnatural parity, the natural parity states being more chaotic.

References

[1] Gomez JMG, Kar K, Kota VKB, Molina RA, Relaño A and Retamosa J 2011 *Phys. Rep.* **499** 103.
[2] Weidenmüller HA and Mitchell GE 2009 *Rev. Mod. Phys.* **81** 539.
[3] Heusler A, Jolos RV, Faestermann T, Hertenberger R, Wirth H.-F. and von Brentano P 2016 *Phys. Rev. C* **93** 054321.
[4] Muñoz L, Molina RA, Gómez JMG and Heusler A 2017 *Phys. Rev. C* **95** 014317.
[5] Dietz B, Heusler A, Maier KH, Richter A and Brown BA 2017 *Phys. Rev. Lett.* **118** 012501.
[6] Berry MV 1981 *Ann. Phys. (N.Y.)* **131** 163.
[7] Bohigas O, Giannoni MJ and Schmit C 1984 *Phys. Rev. Lett.* **52** 1.
[8] Gómez JMG, Molina RA, Relaño A and Retamosa J 2002 *Phys. Rev. E* **66** 036209.
[9] Relaño A, Gómez JMG, Molina RA, Retamosa J and Faleiro E 2002 *Phys. Rev. Lett.* **89** 244102.
[10] Zelevinsky V, Brown BA, Frazier N and Horoi M 1996 *Phys. Rep.* **276** 85.
[11] Molina RA, Gómez JMG and Retamosa J 2000 *Phys. Rev. C* **63** 014311.
[12] Molina RA 2006 *Eur. Phys. J. A* **28** 125.
[13] Haq RU, Pandey A and Bohigas O 1982 *Phys. Rev. Lett.* **48** 1086.
[14] Shriner Jr. JF, Mitchell GE and von Egidy T, *Z. Phys. A* **338** 309.

IOP Conf. Series: Journal of Physics: Conf. Series **966** (2018) 012061 doi:10.1088/1742-6596/966/1/012061

Shell-model investigation of odd-mass nuclei in the ^{132}Sn region

H. Naïdja[†,⋆], F. Nowacki[†]

† Université de Strasbourg, CNRS, IPHC UMR 7178, F-67000 Strasbourg, France
⋆ Université Constantine 1, LPMS, 1 route Ain El Bey, 25000 Constantine, Algeria

E-mail: nhouda@gmail.com

Abstract. New shell-model calculations are performed in the mass region around ^{132}Sn robust core, where the low-lying states energies and the magnetic dipole moments of even-odd (^{135,137}Te, ^{137,139}Xe, ^{139,141}Ba, ^{141,143}Ce), and odd-even (^{135}Sb, ^{135,137}I, ^{135,139}Cs, ^{139,141}La) nuclei are obtained with the N3LOP effective interaction. The position of $5/2^+$ state in ^{135}Sb is discussed with the variation of the proton gap between $0g_{7/2}$ and $1d_{5/2}$ orbits.

1. Introduction

The study of odd nuclei with few neutrons and protons above ^{132}Sn doubly magic nucleus constitute a challenging topic for experimentalists and theorists, and provide them an important source of information. Experimentally [1, 2, 3, 5, 4, 6], this class of nuclei can yield information about β-decay rates, half-lives $T_{1/2}$ and single particle excitation energies. For nuclear theorists [7, 8, 9] they remain a severe test of their Hamiltonians, and the observed single particle energies can serve in the calculations of the two-body matrix elements.

In this context, before starting the study of odd-mass nuclei, the analysis of even ones has been necessary. The first investigations were carried out for ^{134,136,138,140}Sn isotopes, where the energy levels, the isomeric transitions and the masses were widely exposed in [10, 11, 12]. The non-sub-shell closure at $N = 90$ in ^{140}Sn was discussed and proved in [12], where increasing the neutron gap $1f_{7/2} - 2p_{3/2}$ was incompatible with the observed energy levels and the transition rates in tin isotopes.

Another important step was the survey of the spectroscopic properties of open neutron and proton systems with $52 \leq Z \leq 60$ and $82 \leq N \leq 88$ [13, 14, 15, 16]. One of the most interesting content in this study was to demonstrate the first signs of collectivity in $N = 86$ and 88 isotones, with the signature of soft γ bands supported by the quadrupole properties and the deformation parameters. It is worth noticing that the calculations performed by the way of new effective interaction N3LOP exhibit an overall excellent agreement with the available experimental data.

The main task in this part of work is to extend our shell-model study of ^{132}Sn mass region, by the calculation of the energy levels and the magnetic moments of odd-mass nuclei with $52 \leq Z \leq 58$ and $82 \leq N \leq 85$, using the *same* effective interaction N3LOP applied for even nuclei. It was important for us to return to the question about the position of $5/2^+$ state in ^{135}Sb, where its sharply lowered position was not fully understood.

All these applications constitute the best test for different matrix elements components: neutron-neutron, proton-proton, and neutron-proton, of our N3LOP effective interaction.

This report is organized as follows: after a brief insight into the model space and the employed effective interaction N3LOP (section 2), the results of the energy levels and the magnetic moments are presented in detail in sections 3 and 4 respectively. The conclusions are collected in section 5.

2. Model space and interaction

The calculations are performed in the model space spanned by the $1f_{7/2}$, $0h_{9/2}$, $1f_{5/2}$, $2p_{3/2}$, $2p_{1/2}$, $0i_{13/2}$ neutron orbitals and the $0g_{7/2}$, $1d_{5/2}$, $1d_{3/2}$, $2s_{1/2}$, $0h_{11/2}$ proton orbitals referred to $r4h - r5i$, taken above the robust core ^{132}Sn. The corresponding neutrons and protons single particle energies are borrowed from ^{133}Sn and ^{133}Sb experimental data [17]. However the energies of $0i_{13/2}$ neutron and $2s_{1/2}$ proton orbital energies are empirical values taken from [18] and [19] respectively.

The diagonalizations are achieved using ANTOINE and NATHAN shell-model codes [20, 21], where the full configuration space is included for all the nuclei of interest, except for ^{143}Ce where we adopt a truncation scheme by limiting the occupation numbers of $0h_{11/2}$ proton orbital.

The realistic V_{NN} interaction is derived from the Chiral effective field theory potentials N3LO [22] employing the low momentum approach V_{low-k}, by integrating V_{NN} down to a cutoff momentum $\wedge$=2.2fm^{-1}. The renormalized interaction is adapted to the model space by many body perturbation theory techniques, including all the $\hat{Q}$-box folded-diagrams up to the second order [23]. After, some monopole and multipole modifications to the realistic interaction were necessary: first of all, we reduced slightly (by about 120 keV) the $1f_{7/2}$ neutron-neutron pairing matrix elements, to obtain the isomeric transitions in the ^{134,136,138}Sn isotopes, where the initial realistic interaction failed to reproduce the isomeric transition in the mid-shell nucleus ^{136}Sn. This quenching is due to seniority mixing effects discussed extensively in our previous works [10, 11, 12]. After, we did some adjustments to the proton-proton and proton-neutron monopole matrix elements, in order to reproduce the energy of the single particle states of $N = 82$ and 83 isotones.

The *same* effective interaction dubbed hereafter N3LOP already applied in the survey of the spectroscopic properties and collectivity of even-even chains of $N = 86$ and 88 isotones with $52 \leq Z \leq 60$ [13, 14, 15, 16], is also employed in this part, to study odd-mass nuclei above ^{132}Sn core.

3. Energy levels

3.1. Odd-even nuclei

The energy levels of ^{135}Sb, ^{135,137}I, ^{135,139}Cs and ^{139,141}La displayed in figures 1-4 and calculated using N3LOP interaction are fairly well reproduced compared to the data [17], and some tentative spin-parity assignment are confirmed by our calculations (i.e ^{141}La and ^{139}Cs). However a discrepancy of order 200 keV appeared in some high excited states like $11/2^+$ and $15/2^+$ states in ^{137}Cs. The $7/2^+$ ground state in ^{135}Sb and ^{135,137}I is interpreted as the occupation of $\nu(1f_{7/2})^n \times \pi(0g_{7/2})^m$ (n and m are the neutron and the proton valence numbers), while in ^{137}Cs and ^{139,141}La it is characterized by $\nu(1f_{7/2})^n \times \pi(1d_{5/2})^5(1d_{3/2})^{(m-5)}$.

One of the striking remarks in this investigation, is the good position of $5/2^+$ state in all nuclei in particular in ^{135}Sb, which was unexpected behavior discussed before in many theoretical and experimental works. According to the experimentalists in [5], the fall down in energy of the first excited $5/2^+$ state in ^{135}Sb was viewed as a downship of $1d_{5/2}$ proton orbit relative to $0g_{7/2}$, which was justified by logft values, and was interpreted to the formation of the neutron-skin. In this context many shell-model calculations were performed to explain this experimental surprising information. In [7] using an empirical interaction the $5/2^+$ energy state was overestimated by about 300 keV. To overcome this difficulty A. Brown et *al.* proposed in [24] a downship of

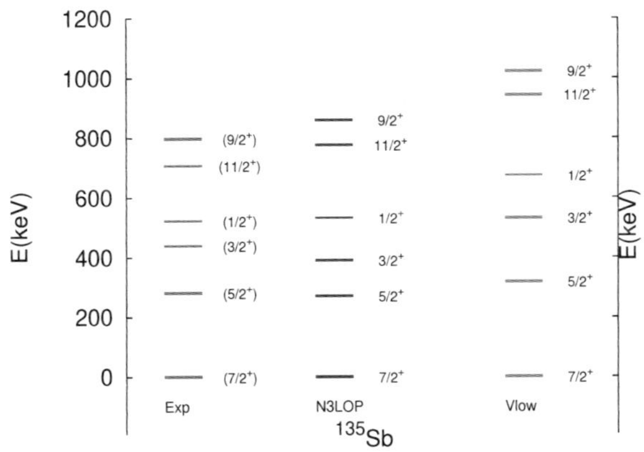

Figure 1. The energy levels of ^{135}Sb

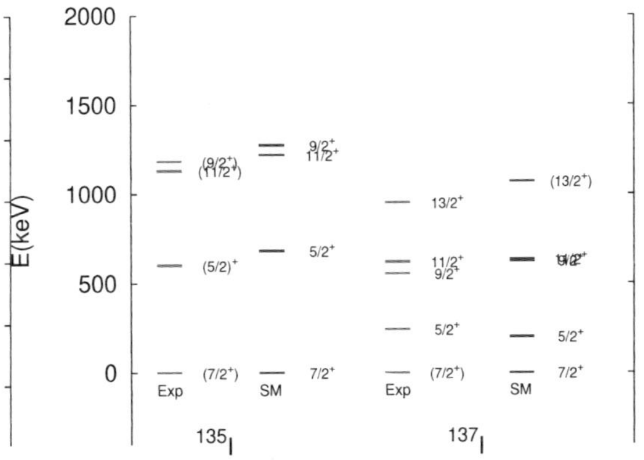

Figure 2. The energy levels of ^{135,137}I

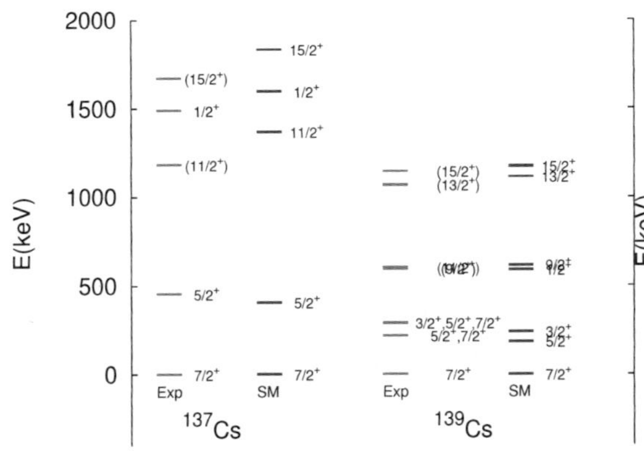

Figure 3. The energy levels of ^{137,139}Cs

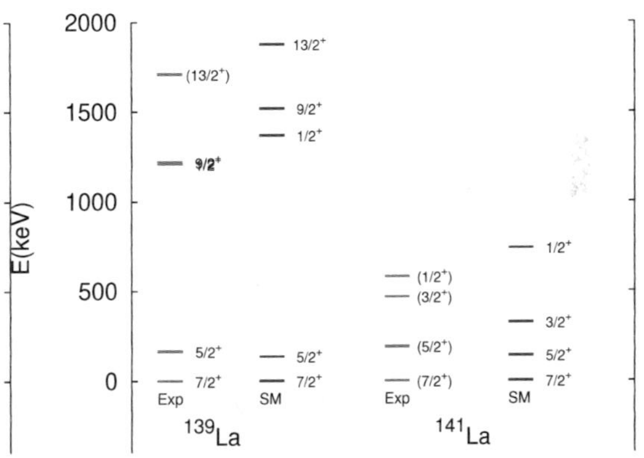

Figure 4. The energy levels of ^{139,141}La

the proton $1d_{5/2}$ single particle energy by about 300 keV using an effective interaction derived from CD-Bonn, to obtain a $5/2^+$ energy state consistent with the data. The Naples shell-model calculations [8] overestimated this state by about 100 keV, without any modification to their CD-Bonn effective interaction.

After this briefly review of some shell-model calculations, it was important for us to check this unusual behavior of $5/2^+$ state in ^{135}Sb. As it is shown in figure [1], its position calculated using N3LOP interaction is consistent with the data, characterized by the large weight (43%) of $\nu(1f_{7/2})^2 \times \pi 1d_{5/2}$ configuration with 18% of $\nu(1f_{7/2})^2 \times \pi 0g_{7/2}$. We turn attention that this coherence is not originated from the monopole corrections made in our interaction, where the same result is obtained using the initial realistic interaction V_{low-k}.

Since the position of $5/2^+$ state is sensitive to the variation of the proton gap between $0g_{7/2}$ and $1d_{5/2}$, we report in Fig. [5] the splitting between the $0g_{7/2}$ and $1d_{5/2}$ effective single particle energies (espe). As it is displayed, the gap decreases by increasing the neutron number, until the inversion at N=90, where we expect that $5/2^+$ state becomes the ground state in ^{141}Sb. As we have proposed and detailed in [6], this behavior could be explained by the fact that the neutron-proton monopole part $V^0_{1f_{7/2}0g_{7/2}}$ is more attractive than $V^0_{1f_{7/2}1d_{5/2}}$, which leads to the reduction of the $0g_{7/2} - 1d_{5/2}$ proton gap by increasing the neutron numbers.

IOP Conf. Series: Journal of Physics: Conf. Series **966** (2018) 012061 doi:10.1088/1742-6596/966/1/012061

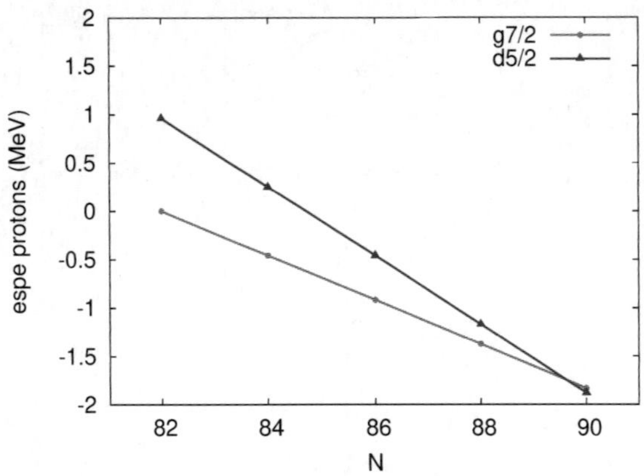

Figure 5. The variation of the $0g_{7/2}$ and $1d_{5/2}$ protons effective single particle energies

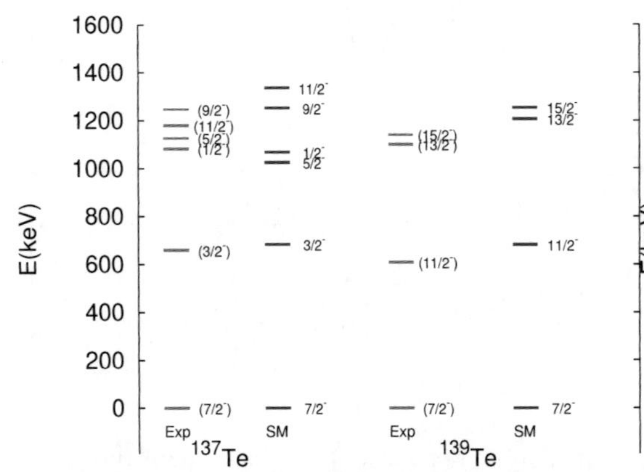

Figure 6. The energy levels of ^{135,137}Te

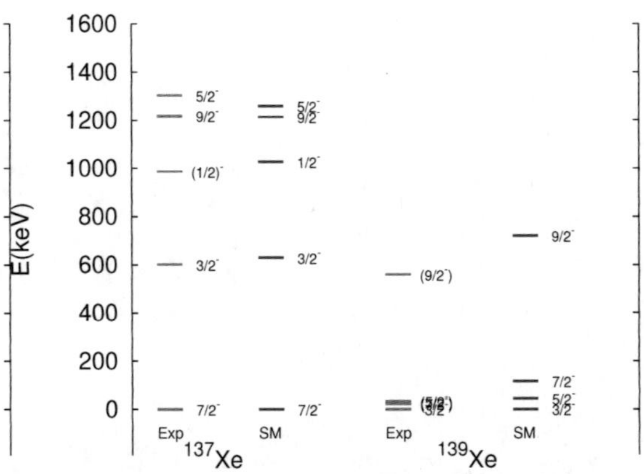

Figure 7. The energy levels of ^{137,139}Xe

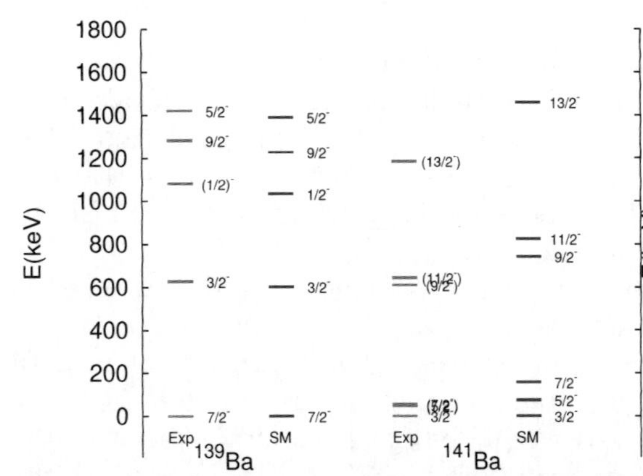

Figure 8. The energy levels of ^{139,141}Ba

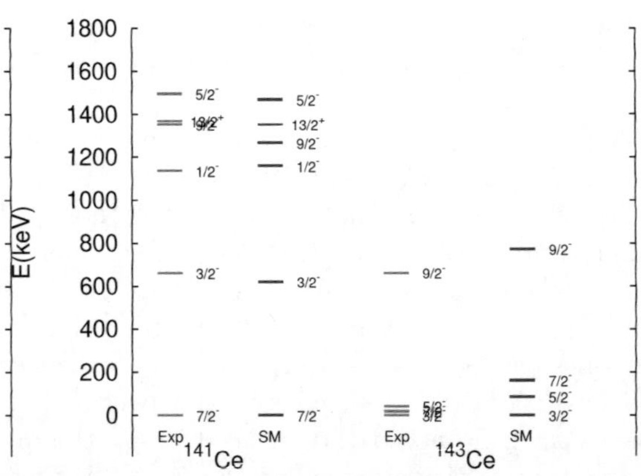

Figure 9. The energy levels of ^{141,143}Ce

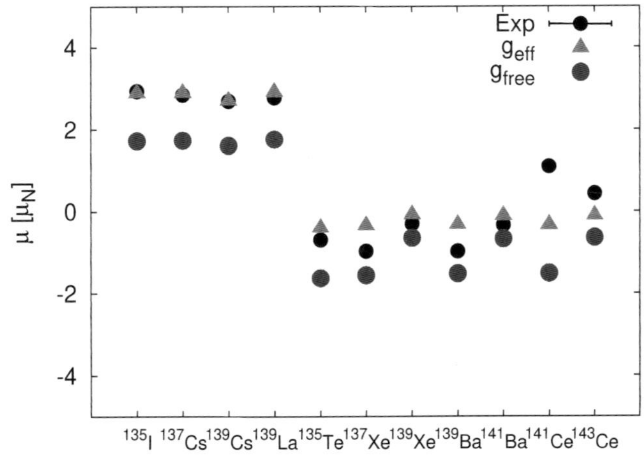

Figure 10. Ground states magnetic dipole moments of some odd-even and even-odd nuclei.

3.2. Even-odd nuclei

Similar to the previous section, the energy levels of even-odd nuclei: ^{135,137}Te, ^{137,139}Xe, ^{139,141}Ba and ^{141,143}Ce are gathered in figures .6-9. A close agreement within 100keV with the measured data is visible. Moreover, some spin with negative parity attribution are confirmed by our calculations (i.e ^{135,137}Te). The $7/2^-$ ground state in ^{135,137}Te and ^{137}Xe is characterized by the occupation of $\nu(1f_{7/2})^n \times \pi(0g_{7/2})^m$, and it is marked by the main configuration $\nu(1f_{7/2}) \times \pi(1d_{5/2})^{(m-2)}(1d_{3/2})^2$ in ^{139}Ba and ^{141}Ce. In ^{139}Xe and ^{141}Ba where stronger octupole correlations seem to be present [25], the $3/2^-$ ground state is distinguished by $\nu(1f_{7/2})^3 \times \pi(0g_{7/2})^{(m-2)}(1d_{5/2})^2$.

4. Magnetic dipole moment

The Electromagnetic properties constitute a severe test of our shell-model wave functions. thereby we have assembled in figure 10 the ground states magnetic dipole moments μ of some odd nuclei, where they are calculated using two sets of spin and orbital g factors, and compared to the scarce available data [26]. Making use of the *same* effective $g-$ factors used for even-even nuclei in our earlier works [14, 15] $(g_\pi^s, g_\pi^l) = (3.250, 1.069)$ and $(g_\nu^s, g_\nu^l) = (-1.506, 0.019)$ for protons and neutrons respectively, an agreement with the experimental data [26] is far more satisfactory than using the free $g-$ factors $(g_\pi^s, g_\pi^l) = (5.5857, 1.0)$ for protons and $(g_\nu^s, g_\nu^l) = (-3.8263, 0.0)$ for neutrons. As it was argued in [14, 15], the justification of this discrepancy is the missing of some spin-orbit partner in our model space $(r4h - r5i)$. However, the disagreement between the calculated value of $\mu(7/2^-)$ in ^{141}Ce and the experimental one remains unexplained.

5. Conclusions

Within the shell-model framework, we have deepened our survey of the ^{132}Sn mass region to odd-mass nuclei. The low-lying state energies of odd-even and even-odd nuclei are established, employing the *same* effective N3LOP interaction used before in the investigation of even-even nuclei [13, 14, 15]. The experimental spin with negative and positive parity states are well reproduced, and predictions are given in some cases. Also, a good description of the magnetic moments is obtained using effective $g-$factors. In addition, we have probed as well the position of $5/2^+$ state in ^{135}Sb, which is influenced by the splitting between $0g_{7/2} - 1d_{5/2}$ proton orbits.

Finally, at this step of calculations, we conclude that the inspection of the spectroscopic

IOP Conf. Series: Journal of Physics: Conf. Series **966** (2018) 012061 doi:10.1088/1742-6596/966/1/012061

properties of even-even, odd-even and even-odd nuclei is established with success employing the *same* N3LOP effective interaction. Moreover, the results are a good support for experimentalists by affirming their new measurements, and a starting point for theorists to develop other interactions. Therefore, these results and interaction provide a good starting point to extend our calculations to higher mass nuclei, where stronger deformations (quadrupole an octupole) are already observed experimentally in rare-earth nuclei.

References

[1] Bhattacharyya P Zhang C T Fornal B Daly P J Grabowski Z W 1997 *Phys. Rev* C **56** R2363
[2] Urban W Korgul A T. Rząca-Urban et al. 2000 *Phys. Rev* C **61** 041301(R)
[3] Fornal B Broda R Daly P J et al. 2001 *Phys. Rev* C **63** 024322
[4] Lui S H Hamilton J H Ramayya A V et al. 2010 *Phys. Rev* C **81** 014316
[5] Shergur J Brown A Fedoseyev V et al. 2002 *Phys. Rev* C **65** 034313
[6] Lozeva R Naïdja H Nowacki F et al. 2016 *Phys. Rev* C **93** 014316
[7] Sarkar S Sarkar M Saha 2004 *Eur. Phys. J* A **21**, 61
[8] Coraggio L Covello A Gargano A Itaco N (2005) *Phys. Rev* C **71** 057302
[9] Coraggio L Covello A Gargano A Itaco N (2009) *Phys. Rev* C **80** 021305(R)
[10] Simpson G S et al. 2014 *Phys. Rev. Lett* **113** 132502
[11] Naïdja H Nowacki F Sieja K 2015 *Journal. Phys. Conf. Series* **580** 012030
[12] Naïdja H Nowacki F Sieja K 2015 *Acta. Phys. Polonica* B **46** 669
[13] Naïdja H Nowacki F 2017 *Acta. Phys. Polonica* B **48** 587
[14] Naïdja H Nowacki F Bounthong B M. Czerwiński T. Rząca-Urban et al. 2017 *Phys. Rev* C **95** 064303
[15] Naïdja H Nowacki F Bounthong B 2017 *Phys. Rev* C **96** 034312
[16] Naïdja H Nowacki F accpeted in *Eur. Phys. J* web of conferences
[17] NNDC National Nuclear Data Center, Brookhaven National Laboratory, http://www.nndc.bnl.gov
[18] Urban W Kurcewicz W Nowak A Rzaca-Urban T et al. 1999 *Eur. Phys. J* **5** 239
[19] Andreozzi F Coraggio L Covello A Gargano A Kuo T. T. S Porrino A 1997 *Phys. Rev.* C **56** R16
[20] Caurier E, Martinez-Pinedo G, Nowacki F, Poves A and Zuker A P 2005 *Rev. Mod. Phys* **77** 427
[21] Caurier E and Nowacki F 1999 *Acta. Phys. Polonica* B **30** 705
[22] Entem D Machleidt R 2003 *Phys. Rev* C **68** 0411001
[23] Hjorth-Jensen M, Kuo T T S, Osnes E 1995 *Phys. Rep* **261** 125
[24] Shergur J Woöhr A Walters B et al. 2005 *Phys. Rev* C **72**, 024305
[25] Zhu S J et al. 1997 *J. Phys* G **23**, L77
[26] Stone N J 2005 *Atomic Data and Nuclear Data Tables* **90** 75

IOP Conf. Series: Journal of Physics: Conf. Series **966** (2018) 012062 doi:10.1088/1742-6596/966/1/012062

A STUDY ON 4 REACTIONS FORMING ^{46}Ti*

M. Cicerchia[1,2], T. Marchi[3,1], F. Gramegna[1], M. Cinausero[1], D. Fabris[4], G. Mantovani[1,2], M. Degerlier[5], L. Morelli[6], M. Bruno[6], M. DAgostino[6], C. Frosin[6], S. Barlini[7], S. Piantelli[7], S. Valdrè[7,8], M. Bini[7], G. Pasquali[7], G. Casini[7], G. Pastore[7], D. Gruyer[7], P. Ottanelli[7], A. Camaiani[7], N. Gelli[7], A. Olmi[7], G. Poggi[7], I. Lombardo[9], D. Dell'Aquila[9], N. Cieplicka-Orynczak[10]

[1] INFN Legnaro National Laboratories, Legnaro (PD), Italy
[2] Department of Physics and Astronomy, University of Padua, Padua, Italy
[3] KU Leuven, Department of Physics and Astronomy Institute voor Kern-en Stralingsfysica, 3001 Leuven, Belgium
[4] INFN of Padua, Padua, Italy
[5] Science and Art Faculty, Physics Department, Nevsehir Haci Bektas Veli University, Nevsehir, Turkey
[6] Department of Physics and Astronomy, University and INFN of Bologna, Bologna, Italy
[7] Department of Physics and Astronomy, University and INFN of Florence, Florence, Italy
[8] Grand Accelerateur National d'Ions Lourds, 14076 Caen, France
[9] Department of Physics, University and INFN of Naples, Naples, Italy
[10] Department of Physics and Astronomy, University and INFN of Milan, Milan, Italy and Institute of Nuclear Physics, Polish Academy of Sciences Krakow, Poland
Contact e-mail: cicerchia@lnl.infn.it

Abstract. The NUCL-EX collaboration is carrying out an extensive research program on pre-equilibrium emission of light charged particles from hot nuclei. The ultimate goal is to study how cluster structures affect nuclear reactions [1,2,3,4]. Indeed, a strong correlation between nuclear structure and reaction dynamics emerges when some nucleons or clusters of nucleons are emitted or captured [5]. At this purpose, the four reactions ^{16}O+^{30}Si, ^{16}O+^{30}Si, ^{18}O+^{28}Si and ^{19}F +^{27}Al have been measured at about 120 MeV projectile energy. Experimental data were collected at Legnaro National Laboratories, using the GARFIELD+RCo array, fully equipped with digital electronics [6]. Following an initial identification of particles and the energy calibration procedures, the complete analysis is being performed on an event-by-event basis. Experimental data are then compared to the theoretical predictions where events are generated by numerical codes based on pre-equilibrium and statistical models and then filtered through a software replica of the setup. Differences between the experimental data and the predicted data put into evidence effects related to the entrance channel and to the cluster nature of the colliding ions. After a general introduction on the experimental campaign, this contribution will focus on the preliminary results obtained so far.

1. Introduction

The idea of α-cluster structure in nuclei has been known since a long time [7]. The first α-cluster structure model of nucleus was proposed in the sixties by Ikeda, to describe the structure of the αconjugated nuclei (2Z=2N); the α-cluster structure appear at excitation energies close to the α-decay threshold. Nowadays, the nuclear clustering theories have gained new momentum due to the study of neutron-rich and exotic weakly-bound nuclei [8]. The clusters in nuclear systems could be pre-formed inside them or related to their dynamical formation. While for light nuclei the interplay between clusters emission with nuclear structure and reaction dynamics has been pointed out [7,5], in medium-mass nuclei the presence of clusters is much more complicated. Here, the interplay between preformation and coalescence plays a significant role. It is not evident whether the emission or capture of α-cluster implies that such an α-cluster is permanently present in the nucleus: it could be formed or dissolved at the instant of emission or capture.

Therefore, an interesting way to investigate the structural properties of medium mass systems is to search for the α-clustering effects in non-traditional observables, like those deriving from studies of pre-equilibrium process. In central collisions, the competition between evaporation and pre-equilibrium particles emission as a function of entrance channel parameters could provide information about the α-particles pre-formation probabilities in compound nucleus [5].

In this framework, since some years, the NUCL-EX collaboration has been engaged in an experimental campaign to study clustering structure effects in nuclei looking at the pre-equilibrium emission of light charged particles. The analysis of ^{16}O+^{116}Sn at 8, 12, 16 AMeV [3,4] brought to the observation of an over-production of α-particles emitted during the non-equilibrium stage of fusion nuclear reactions. The main hypothesis was to ascribe the α-particle excess to the effects induced by the cluster structure in the ^{16}O projectile nucleus. Following these results, a new experiment was proposed to obtain an evaluation of clustering effects in a model independent way: the comparative study of light charge particles emission from two different fusion reactions with different isospin ratio, ^{16}O+^{65}Cu and ^{19}F+^{62}Cu at 16 AMeV [2]. Although the analysis is still in progress, evidences of an α-particle fast production excess have been observed for both systems. A slight overproduction is moreover observed in the ^{19}F induced reaction with respect to the ^{16}O one. This might be ascribed to the lower energy of the predicted cluster state of the ^{19}F.

2. The Experiment

In order to continue the extensive experimental campaign, a comparative study of four nuclear reactions: ^{18}O+^{28}Si, ^{16}O+^{30}Si and ^{19}F+^{27}Al at 7 AMeV and ^{16}O+^{30}Si at 8 AMeV has been performed. The reactions main characteristics are reported in Table 1.

Table 1. The reactions main characteristics.

Entrance Channel	$E_{beam,lab}$ (AMeV)	$\vartheta_{grazing}$ (deg)	E^*_{CN} (MeV)	σ_{fus} (mb)	$v_{lab,res}$ (cm/ns)	E.R. Distrib. ϑ_{lab} (deg)
^{16}O+^{30}Si	7	10,1	88,0	1081	1,28	0-30
^{16}O+^{30}Si	8	8,8	98,4	1070	1,37	0-30
^{18}O+^{28}Si	7	9,0	98,5	1110	1,44	0-28
^{19}F+^{27}Al	7	8,9	103,5	1100	1,52	0-28

For central impact parameters and when the complete fusion occurs, all the studied reactions lead to the formation of the same compound nucleus, the ^{46}Ti*, even if with slightly different excitation energies (E^*_{CN}); therefore, a small difference in their de-excitation chain is expected. Otherwise, the choose of the same beam energy (7 AMeV) implies that the non-equilibrium processes should be almost the same [5]. The ^{16}O+^{30}Si reaction at 8 AMeV has been used to produce the compound nucleus at the same excitation energy of the ^{18}O+^{28}Si at 7 AMeV, hence the same statistical component.

The main goal of the experiment is to study the competition between fast emission and compound nucleus thermal evaporation and its possible relation to the structure of the colliding partners. For this purpose, the pre-equilibrium component of the particle emission spectra for these systems, especially the alpha-particle emission, has been measured and compared since it may be related either to the pre-formation of clusters in nuclei or to their formation during the collision dynamics. Both types of particles emission are produced in central events, which could be selected imposing the coincidence between an evaporation residue and one or more light charged particles.

In these medium-mass systems, the influence of possible structure effects, especially those related to the alpha-clustering, can be derived from the comparison of the different entrance channels, where the neutron and proton numbers of the colliding partners have been varied. Some information on the fast emission mechanism and its competition with statistical evaporation can be extracted from the comparison between statistical and pre-equilibrium light charged particle multiplicities and from the ratios of yields in specific exit channels.

2.1. The Experimental Details

The experiment has been performed at the Legnaro National Laboratories, with the $^{16,18}O$ and ^{19}F beams provided by the TANDEM-ALPI acceleration system. The GARFIELD plus RCo experimental setup, fully equipped with digital electronics [6], was used to detect light charged particles, in the whole apparatus, and heavier particles (like evaporation residues) in RCo. The performances of the apparatus permit the full event reconstruction and the study of many-body correlations. The apparatus has 488 detecting cells with a geometrical coverage of the order of 80% of 4π.

The GARFIELD array covers the polar angles $\vartheta = 30°-150°$. It consists of two drift chambers, filled with CF_4 gas at a pressure of 50 mbar, with gas microstrip read-out for the energy loss, ΔE, signals; they are divided in subsectors, each of which covering ~7° in azimuthal angle. In each sector four CsI(Tl) scintillators define four polar angle regions of about 14° and allow to detect the residual energy. The particles identification in charge and energy has been made via ΔE-E (μstrips-CsI(Tl)) technique and Fast vs. Slow pulse shape analysis of the CsI(Tl) signals technique; this last technique allows to obtain also light charged particle masses. The accuracy of energies is of the order of few percent with a 0.8-1 AMeV threshold.

The Ring Counter (RCo), covering $\vartheta = 7°-17°$, is a three-stage telescope consisting of an ionization chamber (IC), filled with CF_4 gas at a pressure of 25 mbar and divided in eight azimuthal sectors; behind each sector, an eight strip 300 μm thick silicon pad (Si) forms the second stage, followed by the third stage composed of 6 CsI(Tl) scintillators. Using the gas section, particles and fragments have been identified in charge with energy thresholds of 0.8-1 A MeV via ΔE-E (IC-Si) technique. Moreover, the mass determination for light fragments stopped in the silicon is obtained, using the pulse shape analysis (PSA) technique in Si, with a quite lower threshold (~2-3 AMeV). The lighter charge particles have been identified via ΔE-E (Si-CsI(Tl)) technique and/or through the Fast vs. Slow pulse shape analysis of the CsI(Tl) signals technique. A percent accuracy in the determination of the particle and fragment energy is possible in the RCo detector. Due to the high counting rate of elastic scattering, the smallest angles were shielded through an appropriate collimation system. The Evaporation Residues (ER) have been collected in an angular range $\vartheta = 8.6°-17°$, just beyond the grazing angles.

2.2. The Events Selection

To carry out a dedicated analysis, separate groups of events have been selected. The first one, which will be discussed in this paper, is related to central events; it was fulfilled asking for the detection of the evaporation residue in the forward direction (RCo) in coincidence with at least one light charged particle in the whole apparatus (GARFIELD+RCo). The coincidences with evaporation residue have been selected setting proper gates in the IC vs. Si ΔE-E energy spectra of the RCo and asking for any coincident particle in the entire apparatus. The evaporation residues are characterized by a velocity close to the center of mass velocity of the reaction (see Table 1). Since a high oxygen contamination of silicon target was observed, a very strict selection on experimental data was performed for the sub-sequent analysis. The only considered events are those for which the total charge has been at least the 70% of the expected total charge ($Z_{projectile}+Z_{target}=22$): therefore, $Z_{tot}>16$, which is the maximum possible charge related to the contaminated events. A correlation between the detected charge vs. the energy distribution

IOP Conf. Series: Journal of Physics: Conf. Series **966** (2018) 012062 doi:10.1088/1742-6596/966/1/012062

in the laboratory system for the reaction $^{16}O+^{30}Si$ at 8 AMeV is shown in Figure 1 for the total events (left panel) and for the selected central events (right panel) for the experimental data set.

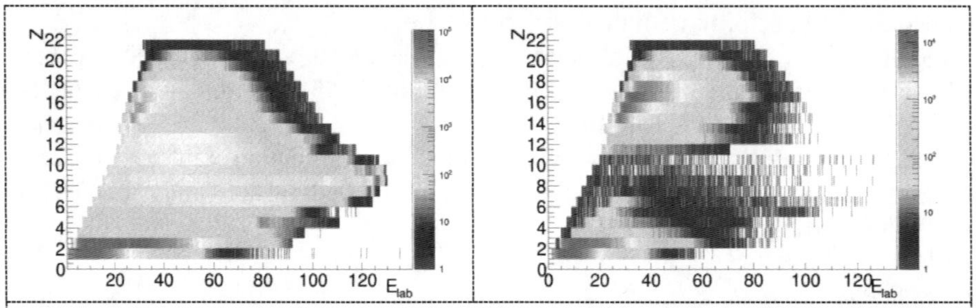

Figure 1. Correlation between the detected particles charge Z and their laboratory energy E_{lab} in the case of total events (left panel) and in the case of only central events (right panel) for the $^{16}O+^{30}Si$ at 8 AMeV.

3. Experiment vs. Simulation: Preliminary Results

A preliminary analysis has been performed on an event-by-event basis; the experimental observables have been studied and compared with those obtained by model predictions performed either with the statistical model GEMINI++ [9], based on a complete fusion hypothesis, and with some dynamical codes, like SMF [10] or AMD [11] codes, followed by GEMINI++ as afterburner. The effects of the experimental apparatus on the theoretical events have been taken in account through a software replica of the experimental array.

The GEMINI++ code [9] describes the thermalized emission from a compound nucleus; it is based on Hauser-Feshbach model, in which the Weisskopf description of evaporation mechanism is coupled with a quantum description of angular momentum and it gives information on the differences between the four reactions connected to the major part of the cross section, which is related to the de-excitation of a thermalized source.

The TWINGO code [10], based on the Stochastic Mean Field approximation, simulates the dynamical part of the reactions. Each nucleon is described by a set of test particles, arranged in a phase-space lattice and subject to a mean field. The time evolution of the one-body distribution function f(~r; ~p; t) is described using the Boltzmann-Nordheim-Vlasov (BNV) transport equation, that is substantially the Vlasov equation including the collision term. Related to initial impact parameter, this code produces the primary fragment distributions as a function of the impact parameter; these primary fragments are the hot ions produced in the earlier stage of the reaction and they can decay by emitting particles, gamma-rays and/or fissioning. To simulate the secondary particle emission, the GEMINI++ code has been used as afterburner.

The AMD (Antisymmetrized Molecular Dynamics) [11] is based on a Stochastic equation of motion for the Gaussian wave-packets representing the colliding nucleons and takes into account the particle-particle correlations with the possibility of describing and varying the degree of clusterization of the interacting partners.

3.1. Preliminary Results

The experimental energy distributions of the light charged particles in coincidences with residues have been compared with the predictions of the theoretical models. From the preliminary results, one can observe a slight difference in the four systems when compared to the complete fusion scenario.

Focusing on predictions of the statistical model simulated with GEMINI++, the first comparison between the single particle emission spectra and the simulated statistical component of the reactions, shows very similar experimental and simulated proton spectra on the whole angular range, and also α-particle spectra at backward angles; therefore, the complete fusion seems to account for almost all the central collision available cross section. On the contrary, a slight difference is observed in the α-particle spectra at the most forward angles, as signature of a small component of α-particle pre-equilibrium emission.

As shown in the Figure 2, for the angular region $8.6° \leq \vartheta \leq 13.2°$ (RCo), the results from the GEMINI++ complete fusion predictions (red lines in Figure 2) are in good agreement with the data for the proton, while a slight difference for α-particles is observed. From the comparison of the different systems with their own statistical predictions, one can observe a slight difference in fast emission for the different entrance channels. For example, looking at the three systems at 7 AMeV, a similar over-production compared to the statistical spectra are present for the ^{18}O and ^{19}F cases while almost no effect is observed for ^{16}O, even if they have the same projectile velocity. In ^{16}O+^{30}Si at 8 AMeV case, the higher the beam incoming energy, the higher the α-overproduction.

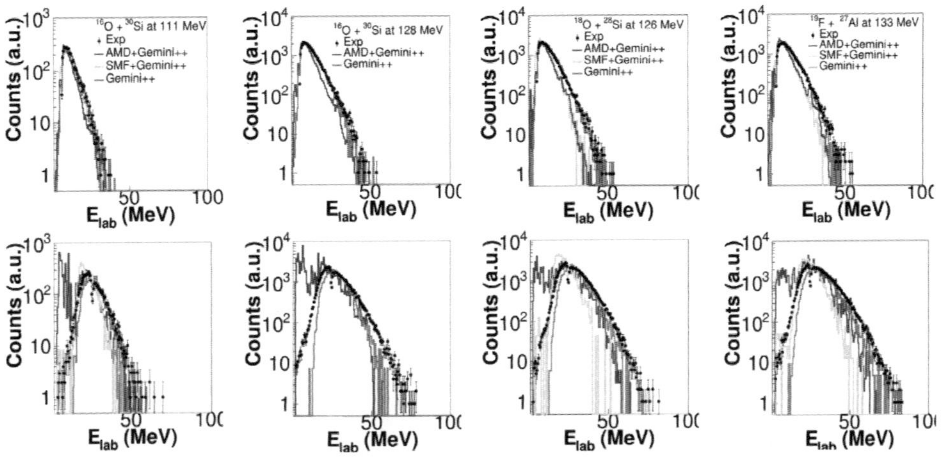

Figure 2. Experimental energy distributions (black full dots) collected at $8.6° \leq \vartheta \leq 13.2°$ for protons (top panels) and α-particles (bottom panels), for the four studied reactions, respectively from left to right: ^{16}O+^{30}Si at 7 AMeV, ^{16}O+^{30}Si at 8 AMeV, ^{18}O+^{28}Si at 7 AMeV and ^{19}F+^{27}Al at 7 AMeV. In all panels comparison to model predictions are provided: GEMINI statistical model (red line) and SMF+GEMINI++ (green lines) and AMD+GEMINI++ (blue lines). Model prediction are filtered through a software replica of the apparatus. The distributions have been normalized to the experimental maximum.

When we look at dynamical model simulations, Figure 2 shows some irregular trends. The SMF code (green lines in Figure 2), shows up some limitations in the description of the pre-equilibrium phenomenon: many nucleons emitted before the equilibrium cannot be properly reconstructed in the final event due to the impossibility to reconstruct them from scattered test particles; moreover, TWINGO, being based on Stochastic Mean Field approximation, is not even able to build up clusters like α-particles and also these fast particles are missing in the final spectrum; therefore, only statistical decay of the primary fragments can be taken in account. However, from this first comparison we observe a system which has anyway lost some nucleons and clusters in the early stage of the reaction and therefore results in colder emission spectra with respect to GEMINI++ alone. In this case, hence, the predictions on the fast emission can then be obtained only as a difference to the experimental spectra.

Finally, the AMD+GEMINI++ simulation (blue lines in Figure 2), being able to describe nucleon-nucleon correlations, can give several information. The simulated spectra reproduce quite well the experimental ones; apart from a secondary peak, lower in energy than the barrier which appear in the α-particles AMD-spectra and it is not present in the data. The physical meaning of this peak is under study.

Several calculations are planned to change and optimize the input parameters, especially related to the clusterization factor; unfortunately, the AMD code is very time expensive, therefore, the time needed to run the code for a sufficient number of events to compare to exclusive data is very long.

4. Conclusions and Perspectives

In order to probe pre-equilibrium and possible α-clustering effects in nuclei, the preliminary results for the $^{46}Ti^*$ compound system, formed through four different entrance channels: $^{16}O+^{30}Si$ at 8 A MeV and $^{16}O+^{30}Si$, $^{18}O+^{28}Si$ and $^{19}F+^{27}Al$ at 7 A MeV have been studied. Even if the analysis is still preliminary, from the proton and α-particle spectra of all the reaction we can infer that, as expected at this bombarding energies, most of the cross section is well described by fusion-evaporation. However, some small differences can be observed at forward angles in the α-particle spectra which can be ascribed to pre-equilibrium emission.

A deeper analysis is needed to understand the pre-equilibrium process; to better disentangle the pre-equilibrium emission, it is planned to use the Moscow Pre-equilibrium Model [4], based on the Hybrid model, which takes into account a pre-equilibrium stage before a second thermalized emission stage and allows to take into account possible clusters pre-formation in the colliding partners, especially the projectile.

A further analysis will be performed to get a better insight on the reaction interplay, studying the particle-particle correlations with selection of specific decay channels (1α, 2α, 3α ...).

References

[1] Morelli L *et al.* 2014 *Journ. of Phys. G* **41** 075107; Morelli L *et al.* 2014 *Journ. of Phys. G* **41** 075108.
[2] Fabris D *et al.* 2013 *PoS (X LASNPA) D* 061
[3] Kravchuk V L *et al.* 2010 *EPJ WoCs* **2** 10006
[4] Fotina O V *et al.* 2010 *Int. Journ. Mod. Phys. E* **19** 1134
[5] Hodgson P E, Běták E 2003 *Phys. Rep.* **374** 1-89
[6] Gramegna F *et al.* 2004 *Proceedings of IEEE Nucl. Symposium* 0-7803-8701-5/04/; Bruno M *et al.* 2013 *Eur. Phys. Jour. A* **49** 128
[7] Von Oertzen W *et al.* 2006 *Phys. Rep.* **432** 43
[8] Freer M *et al.* 2007 *Rep. Progr. Phys.* **70** 2149
[9] Charity R J 2010 *Phys. Rev. C* **82** 014610
[10] Colonna M *et al.* 1998 *Nucl. Phys. A* **642** 449
[11] Ono A 1999 *Phys. Rev. C* **59** 853

IOP Conf. Series: Journal of Physics: Conf. Series **966** (2018) 012063 doi:10.1088/1742-6596/966/1/012063

Nonnegative Tensor Factorization Approach Applied to Fission Chamber's Output Signals Blind Source Separation

M. Laassiri[1], **E-M. Hamzaoui**[2] **and R. Cherkaoui El Moursli**[1]

[1]LPNR, Faculty of Sciences, Mohammed V University, Rabat, Morocco

[2]National Centre for Nuclear Energy, Sciences and Techniques (CNESTEN), Rabat, Morocco

E-mail: mounia.laassiri@gmail.com

Abstract. Inside nuclear reactors, gamma-rays emitted from nuclei together with the neutrons introduce unwanted backgrounds in neutron spectra. For this reason, powerful extraction methods are needed to extract useful neutron signal from recorded mixture and thus to obtain clearer neutron flux spectrum. Actually, several techniques have been developed to discriminate between neutrons and gamma-rays in a mixed radiation field. Most of these techniques, tackle using analogue discrimination methods. Others propose to use some organic scintillators to achieve the discrimination task. Recently, systems based on digital signal processors are commercially available to replace the analog systems. As alternative to these systems, we aim in this work to verify the feasibility of using a Nonnegative Tensor Factorization (NTF) to blind extract neutron component from mixture signals recorded at the output of fission chamber (WL-7657). This last have been simulated through the Geant4 linked to Garfield++ using a ^{252}Cf neutron source. To achieve our objective of obtaining the best possible neutron-gamma discrimination, we have applied the two different NTF algorithms, which have been found to be the best methods that allow us to analyse this kind of nuclear data.

1. Introduction

Fission chambers (FC) are the most appropriate neutron detectors that are dedicated to deliver online neutron flux measurements for experimental fission reactors [1, 2, 3]. Those measurements have a wide range of applications, including characterization of experimental conditions, reactor monitoring and safety concerns [4, 5]. Depending on the application, such detector introduces a wide range of constraints, of several magnitudes with regard to neutron flux, gamma-ray flux and temperature. Hence, designing a specific fission chamber and measuring chain for a given application is a challenging task and can be achieved by a combination of experimental feedback and simulating tools. As thoroughly described in references [6, 7, 8], fission chambers can operate in three different modes: Pulse Mode, Campbelling Mode (also known as "fluctuation mode" or "mean square voltage mode") or Current Mode. The Current mode does not allow for pulse height discrimination between neutron and gamma radiations as do pulse and Campbelling Modes.

In an earlier work based on Geant4 linked to Garfield++ simulations [9], discrimination of gamma-rays from neutrons in WL-7657 FC detector was investigated by using the Nonnegative Tensor Factorization (NTF) algorithms. The objective was to extract independent components

(IC) from signals recorded at the output of WL-7657 FC, and to characterize obtained ICs in order to reach the discrimination goal [9].

Recently the used code was updated to be capable of simulating the output signals with ^{252}Cf neutron source [10]. The aim of the present work with a ^{252}Cf source was to check these methods and at the same time to investigate the differences in pulse shapes of neutron and gamma-ray radiations. As a continuation of that work, the present paper investigates the application of two NTF algorithms to achieve the discrimination task. The figure 1 below summarize the steps which will be followed in this project.

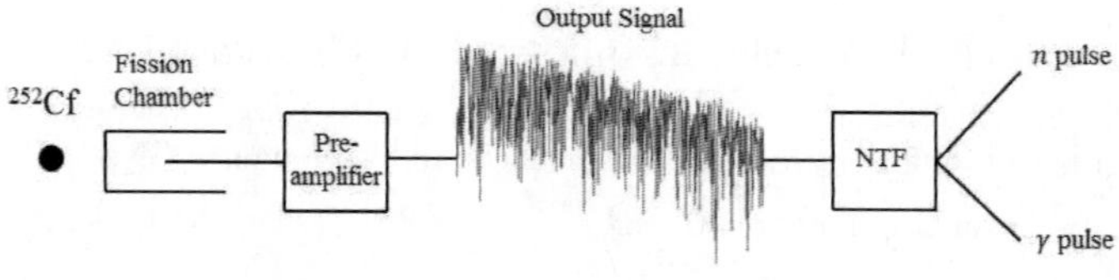

Figure 1. Block diagram of n-γ discrimination based on NTF approach.

2. Mechanical and Physical features in WL-7657 Fission Chamber

The WL-7657 fission chamber has been proposed for use as a neutron diagnostic system to measure the characteristics of neutrons from the TRIGA Mark II, research reactor of the Nuclear Studies Centre of Maâmora (CNESTEN-Morocco). The WL-7657 FC is a $\approx$ 1576.86 cm^3 volume and has two coaxial electrodes, the outer electrode (namely the cathode) have a diameter of about 76.2 mm, the inter-electrode space is filled with a 4% Nitrogen and 96% Argon at 1 atm pressure, with 90% enriched ^{235}U in U_3O_8 (1.68 g) deposited on the inner electrode (namely the anode) [11]. The mass of the fissile deposit coating is a key parameter in this simulation, as it impacts directly the amounts of signal within a given neutron flux. When a neutron crosses the fissile deposit, it is likely to induce a fission reaction that generates two charged particles (Fission Fragment) emitted in two nearly opposite directions. The fission fragment emitted out of the deposit ionizes the filling gas on its trajectory and consequently generates a high number of electron-ion pairs (Ar^+, e^-). Given that, a DC voltage of a few hundred volts is applied between the electrodes, the electrons and positive ions are separated and drift across the gas, generating current signal that can be amplified and processed. The figure 2 gives an outline of the neutron and gamma-ray interactions occurring in a fission chamber.

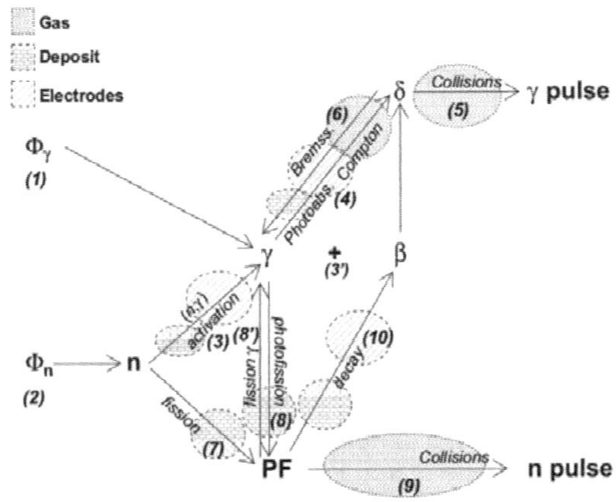

Figure 2. Outline of the neutron, gamma-ray interactions occurring in a fission chamber [12].

3. Nonnegative Tensor Factorization model

The basic 3D Nonnegative Tensor Factorization (NTF) model considered in this paper is illustrated in figure 3 [13]. The NTF is a generalization of Nonnegative Matrix Factorization (NMF) aimed at retrieving underlying components from high dimensional data. The basic formulation NTF is Parallel Factor Analysis Model (PARAFAC) with nonnegativity constraints and other possible natural constraints such as sparseness and/or smoothness on the basis of only the 3D tensor $\underline{Y}$. this permits to extract two common factors: a basis mixing matrix A and unknown components factors represented by a matrix X, which can be represented in slice factorization form:

$$Y_k = AD_kX + E_k, \quad (\text{k} = 1,2, \dots , \text{K}) \tag{1}$$

Where $Y_k = \underline{Y}_{:,:,k} \in \mathbf{R}^{I\times T}$, are frontal slices of a 3D tensor $\underline{Y} \in \mathbf{R}^{I\times T\times K}$, K is a number of frontal slices, in our application, Y is formed by recorded FC preamplifier's output signals, also referred to as "observations" or "data", $A = [a_{\text{ir}}] \in \mathbf{R}_{+}^{I\times R}$ is the basis (mixing matrix) representing common factors, $D_k \in \mathbf{R}_{+}^{R\times R}$ is a diagonal matrix that holds the $k-th$ row of the matrix $D \in \mathbf{R}_{+}^{K\times R}$, in its main diagonal, $X \in \mathbf{R}_{+}^{R\times T}$ is unknown matrix of independent components to be estimated, and $E_k = \underline{E}_{:,:,k} \in \mathbf{R}^{I\times T}$ is an additive noise for a $k-th$ frontal slice of the tensor $\underline{E} \in \mathbf{R}^{I\times T\times K}$ [13].

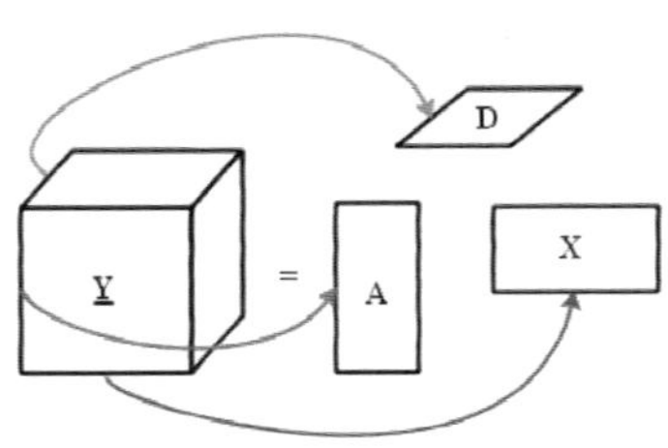

Figure 3. Decomposition of 3D tensor into three nonnegative matrices using the standard NTF model [14].

3.1. NTF1 model

In 3D NTF1 model, which is an extension of the NTF model (figure 4), a given data tensor $\underline{Y} \in \mathbf{R}^{I\times T\times K}$ (Observed data) is decomposed to a set of matrices A, D and $\{X_1, X_2, ..., X_K\}$ with nonnegative entries.

$$Y_k = AD_kX_k + E_k, \quad (\text{k} = 1,2, \dots , \text{K}) \tag{2}$$

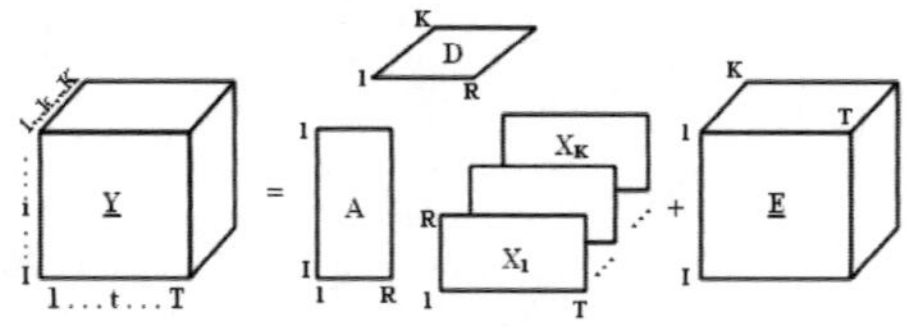

Figure 4. Illustration of NTF1 model.

Since the nonnegative diagonal matrices $D_k \in \mathbf{R}_+^{R\times R}$ are scalling matrices, they can usually be absorbed by the matrices $X_k \in \mathbf{R}_+^{R\times T}$ by introducing row-normalized matrices $X_k := D_kX_k \in \mathbf{R}_+^{R\times T}$, hence $Y_k = AX_k + E_k$ [13].

3.2. NTF2 model

In 3D NTF2 model, which can be illustrated as in figure 5, the observed data $\underline{Y} \in \mathbf{R}^{I\times T\times K}$ is decomposed to a set of matrices $\{A_1, A_2, ..., A_K\}$, D and X with nonnegative entries.

$$Y_k = A_kD_kX + E_k, \quad (\text{k} = 1,2, \dots , \text{K}) \tag{3}$$

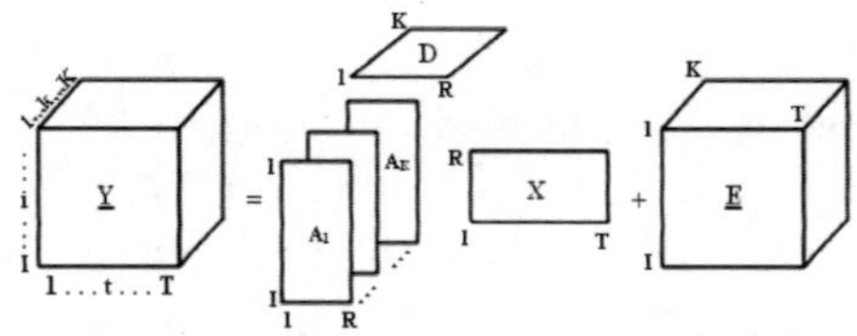

Figure 5. Illustration of NTF2 model.

The NTF2 model is similar to the NTF1 but in this case the nonnegative diagonal matrices $D_k \in \mathbf{R}_+^{R\times R}$, they can usually be absorbed by the matrices $A_k \in \mathbf{R}_+^{I\times R}$ by introducing column-normalized matrices $A_k := A_kD_k \in \mathbf{R}_+^{I\times R}$, hence $Y_k = A_kX + E_k$ [13].

4. Simulation results

In this paper the WL-7657 FC was simulated using Geant4 toolkit (version 4.10.02.p02) and Garfield++ (version 2015.1). A ^{252}Cf neutron source having an intensity of 2×10^3 n/s, was used to test the new neutron-gamma discrimination approach. In the first stage, Geant4 is used to simulate the FC geometry and the reaction models. We have included all necessary libraries in order to take all physical aspects of FC into account. In the second stage, we calculated via the Garfield++ software, the current signal generated by the drifting electrons and ions.

To test both NTF1 and NTF2 algorithms, we used the NTFLAB toolbox developed by Cichocki A. et al., which is implemented under MATLAB® environment [15]. To demonstrate the effectiveness of NTF algorithms, tensors of size $5 \times 600 \times 5$ were created from the simulated FC preamplifier's output signals (Observations), as shown in the figure 6. The algorithms were evaluated according to the values of their performance index of separability (PI) [16, 17]. Figure 6 shows the selected slices of mixed signals.

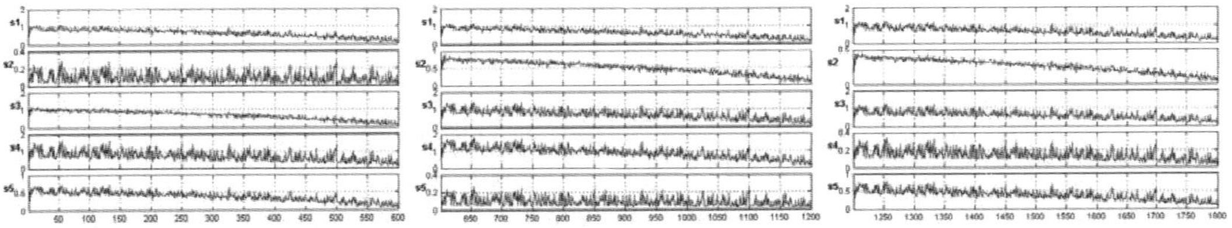

Figure 6. Selected slices of mixed signals.

The figure 7 and 8 show the sources recovered by the application of NTF1 and NTF2 algorithms respectively. The simulations results have been performed for neutron and gamma signals in which the nonnegative depend 10 hidden components or sources are collected in 5 slices $X_k \in \mathbf{R}_+^{2\times600}$ in case of NTF1 application and 2 hidden components or sources are collected in 1 slice $X \in \mathbf{R}_+^{2\times600}$ in case of NTF2 application, each representing 2 different kind of signals.

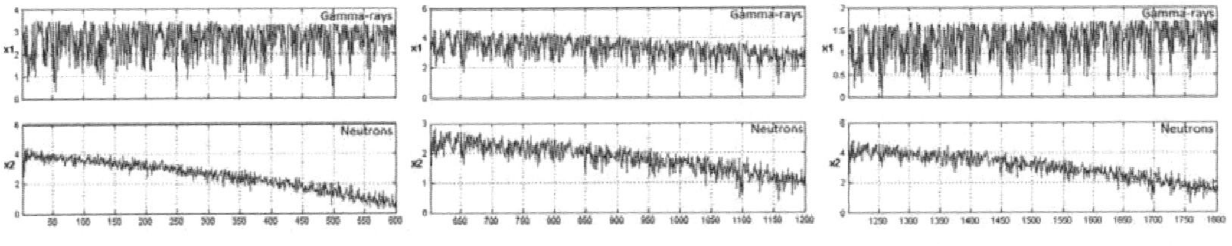

Figure 7. Spectra signals estimated with the NTF1.

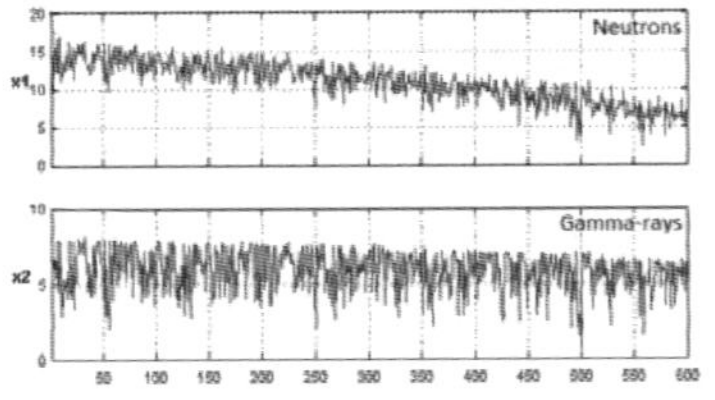

Figure 8. Spectra signals estimated with the NTF2.

It can be seen that the sources have been reasonably well separated. The computation of the Signal-to-interference ratio (SIR) of individual columns of the mixing matrix A, permits us to confirm that the WL-7657 FC output mixture signals are formed by two main independent components which may be corrupted by noise (figure 9). This leads to say that the neutron-gamma discrimination task has been well achieved.

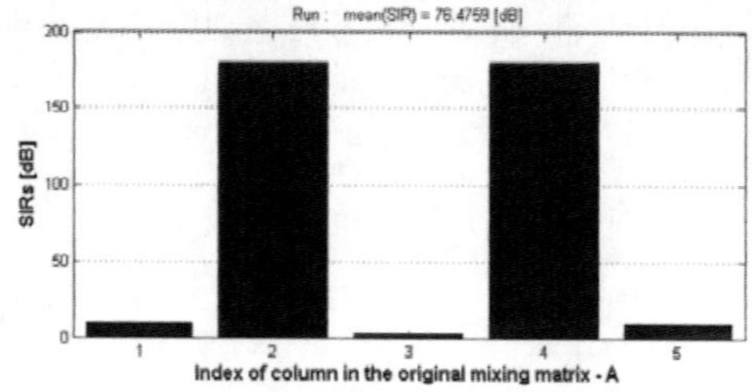

Figure 9. SIR evaluation according to the mixing matrix A.

5. Conclusion

The main objective and motivations of this study were to apply the Nonnegative Tensor Factorization algorithms to extract independent components that from the preamplifier's outputs of a simulated fission chamber. This last have been simulated through the Geant4 linked to Garfield++ using a ^{252}Cf source, which emits both neutrons and gamma-rays. The performance index values permit us to conclude that the NTF1 and NTF2 algorithms are the best blind source separation methods which can be applied to analyse such nuclear data. The plots of obtained independent components permit to define a qualitative criterion for the discrimination between neutron and gamma signals.

References

[1] Andriamonje S *et al.* 2006 *Nucl. Instrum. Methods Phys. Res. Sect. Accel. Spectrometers Detect. Assoc. Equip.* **562** 755-9
[2] Jammes C, Geslot B and Imel G 2006 *Nucl. Instrum. Methods Phys. Res. Sect. Accel. Spectrometers Detect. Assoc. Equip.* **562** 778-84
[3] Jammes C, Geslot B, Rosa R, Imel G and Fougeras P 2005 *Ann. Nucl. Energy* **32** 1131-45
[4] Duchene J 1962 *IRE Trans. Nucl. Sci.* **9** 123-127
[5] Filliatre P, Jammes C, Geslot B and Buiron L 2010 *Ann. Nucl. Energy* **37** 1435-42
[6] Boek H and Villa M 2004 *TRIGA reactor main systems* (AIAU - 24311)
[7] Geslot B, Berhouet F, Oriol L, Braud S, Jammes C, Filliatre P and Villard J F 2009 *Advancements in Nuclear Instrumentation Measurement Methods and their Applications (ANIMMA), First International Conference on (IEEE)* 1-4
[8] Elter Z, Jammes C, Pázsit I, Pál L and Filliatre P 2015 *Nucl. Instrum. Methods Phys. Res. Sect. Accel. Spectrometers Detect. Assoc. Equip.* **774** 60-7
[9] Laassiri M, Hamzaoui E-M and Cherkaoui El Moursli R 2017 *Results Phys.* **7** 1422-6
[10] http://wwwndc.jaea.go.jp/cgi-bin/FPYfig?iso=sCf252
[11] Mirion Technologies (ex: Imaging and Sensing Technology, Corp) WL-7657 Fission Counter and Control Datasheet
[12] Filliatre P, Vermeeren L, Jammes C, Geslot B and Fourmentel D 2011 *Nucl. Instrum. Methods Phys. Res. Sect. Accel. Spectrometers Detect. Assoc. Equip.* **648** 228-37
[13] Cichocki A 2009 *Nonnegative matrix and tensor factorizations: applications to exploratory multi-way data analysis and blind source separation* (Chichester, U.K: John Wiley)
[14] Cichocki A and Zdunek R 2006 *NTFLAB for signal processing* (Technical report, Laboratory for Advanced Brain Signal Processing, BSI, RIKEN, Saitama, Japan)
[15] Cichocki A and Zdunek R *NMFLAB for signal and image processing - NTFLAB for signal processing* available at: http://www.bsp.brain.riken.jp/ICALAB/nmflab. html (Visited on July 2017)
[16] Cichocki A and Amari S-I 2005 *Adaptive blind signal and image processing: learning algorithms and applications* (John Wiley & Sons)
[17] Hamzaoui E-M, Regragui F and Himmi M M 2011 *Contemp. Eng. Sci.* **4** 25-35

IOP Conf. Series: Journal of Physics: Conf. Series **966** (2018) 012064 doi:10.1088/1742-6596/966/1/012064

Pulse Shape Discrimination with EJ299 scintillators

A Muoio[1,2], F Cappuzzello[1,2], D Carbone[1], M Cavallaro[1], G Lanzalone[1,3], A Pappalardo[4], G Santagati[1], A Trifirò [5,6], S Tudisco[1]

[1]INFN - LNS, Via S. Sofia 62, 95123 Catania, Italy
[2]Dip. di Fisica e Astronomia, Univ. di Catania, Via S. Sofia 64, 95123 Catania Italy
[3]Univ. Degli Studi di Enna "Kore", Via delle Olimpiadi, 94100 Enna, Italy
[4]ELI-NP (Extreme Light Infrastructure-Nuclear Physics, Magurele, Romanian
[5]Dip. di Scienze MIFT, Univ. degli Studi di Messina, Viale F.S. D'Alcontres 31, 98166 Messina, Italy
[6]INFN - Sez. di Catania, Via S. Sofia 64, 95123 Catania, Italy

Abstract. Recently a new generation plastic scintillator PPO have been developed. They promise excellent performances in terms of neutron/gamma discrimination. In this work we will present the activity made at INFN-LNS on the plastic scintillator EJ299 in comparison with the most traditional liquid scintillator EJ301 used in several nuclear physics experiments.

In the past scintillation detectors have been used in several nuclear physics studies [1-3] and applications [4,5]. Recently, the possibility of manufacturing plastic scintillators with efficient neutron/gamma discrimination has been developed with a new polyvinyltoluene (PVT) polymer matrix loaded with a scintillating dye, 2,5-diphenyloxazole (PPO) [6].
Recently at INFN-LNS two research topics are fostering R&D on these scintillators, the study of Double Charge Exchange (DCE) in peripheral nuclear reactions[7] and Nuclear Reactions and astrophysics process in Plasmas (NRP) [8].
The exploration of detailed nuclear structure through DCE has historically got great benefits by the use of magnetic spectrometers. Recently large acceptance and high resolution spectrometers have been installed in leading European laboratories for fundamental nuclear physics with a major impact on the progress of knowledge in several branches of nuclear structure and reaction mechanisms.
Important upgrading of the existing facilities can be foreseen, especially in the view of emerging detection technologies, in order to extend the application of magnetic spectrometry to the more and more challenging experiments required by the present and future nuclear science. Among these, the MAGNEX spectrometer [9] of LNS has been coupled to organic scintillators neutron detector array, EDEN [10,11], to perform a better selection of reaction mechanisms. The improvement of PPO technology could be very useful to these topics.
NRP is a new field of studies related to recent laser technology advances, which in the near future can give the opportunity to investigate nuclear reactions and fundamental process in plasma under extreme conditions. Also for this contest, the "ideal" neutron detection module must have: i) high efficiency; ii) good discrimination of gammas from neutrons; iii) good timing performance for ToF neutron energies reconstruction.

IOP Conf. Series: Journal of Physics: Conf. Series **966** (2018) 012064 doi:10.1088/1742-6596/966/1/012064

In addition, laser-plasma experiments require detector able to work in hard environmental conditions. These requirements may be met by configuration based on plastic scintillators coupled with new solid state devices, Silicon Photomultiplier, SiPM [12][13][14], and a totally digital acquisition for the multi-hit signals expected from reactions events in plasma. In this work we found that the response of liquid scintillator is characterized by rice times of the order of 3-4 ns. One advantage of plastic scintillators include fairly high light output, with a rise time of 2-4 nanoseconds, but one of the major advantages is their flexibility. They are easily machined by normal means and shaped to desired forms. A special focus on the pulse shape discrimination capability of this material, which is characterized by low toxicity and low volatility, has been addressed. A possibility of manufacturing plastic scintillators with efficient neutron/gamma pulse shape discrimination is demonstrated using a system of a PVT polymer matrix loaded with a PPO. According to a commonly accepted mechanism [15] [16], both gamma and neutron induced pulses contain a fast decay (prompt) component and a slow decay (delayed) fluorescence one.
In order to characterize Pulse Shape Discrimination (PSD) for the two detectors the following experimental set-up was done.

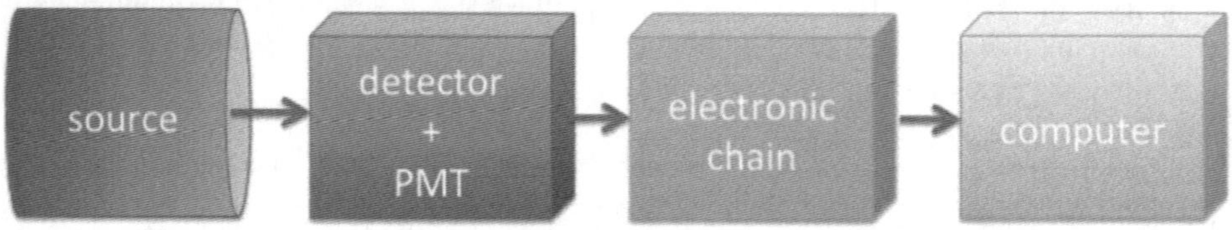

Figure 1 A sketch of experimental setup.

The first test on detectors was performed on liquid scintillator EJ301 + PMT (Photo Multiplier Tube) Hamamatsu R 1250. The liquid scintillator has been assembled and oriented towards the Am-Be source collimated through an aluminium plate 2 cm thick with a hole of 5 mm diameter, in order to keep less than 5% the difference in the path length of particles impinging in the detectors.

Figure 2. Scintillator EJ301 coupled with phototube.

IOP Conf. Series: Journal of Physics: Conf. Series **966** (2018) 012064 doi:10.1088/1742-6596/966/1/012064

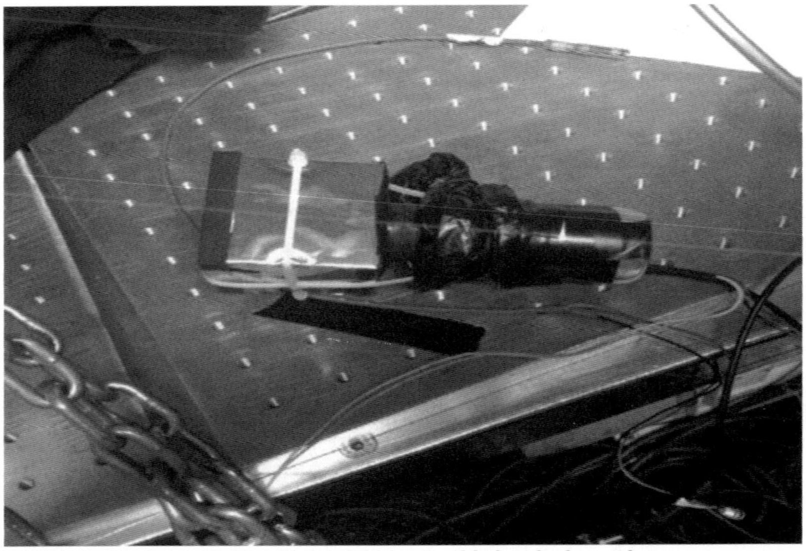

Figure 3. Scintillator EJ299 assembled with phototube.

The circuit, called BaF Pro [17] [18], is based on an analogic fast stretcher generating highly performing signals for PSD and timing purposes. A beam test has demonstrated a low detection threshold, a good pulse-shape discrimination even at low energies and a wide dynamic range for measurement of the neutrons energy. The voltage generated by the photon interaction of PMT, are send to BaF Pro.

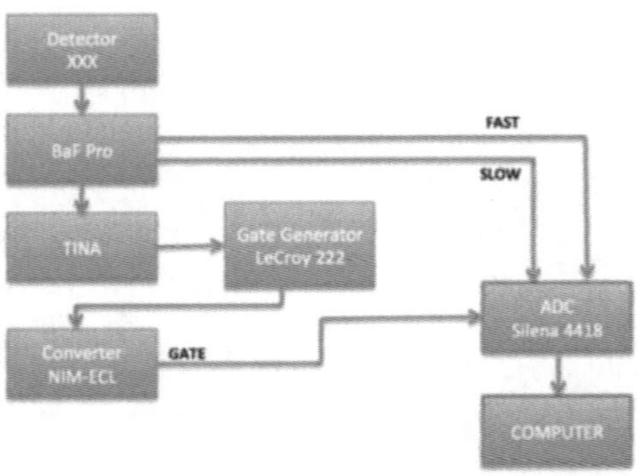

Figure 4. Block diagram of the electronic for the PSD.

The 16-channel BaF-pro module makes use of a sophisticated analog fast stretcher circuit to determine the fast component of the light output of the scintillator (hereafter indicated as ''fast'') and an integration section that provides the total energy of the signal (hereafter indicated as ''slow'').

Figure 5. BaF-Pro electronics.

Finally, a computer recorded the Fast and Slow components produced by the light collection system. Techniques for accurate discrimination between neutrons and photons interacting with scintillators have been established for a long time. The PSD techniques used to distinguish between the pulses from neutrons and the pulses from gamma rays rely on the differences in the pulse shapes produced. The goal of this research effort was to test the ability of a polyvinyltoluene research sample to produce recordable, distinguishable signals in response to gamma rays and neutrons. Pulse shape discrimination was performed to identify if the signal was generated by a gamma ray or a neutron.
The results of such separation are shown in fig 6 and 7 for EJ301 and EJ299 respectively (neutrons are bounded in a red polygon).

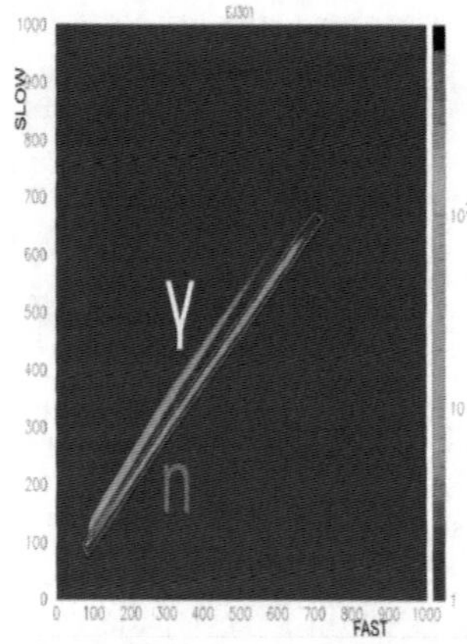

Figure 6. Pulse shape discrimination(PSD) performance for EJ301.

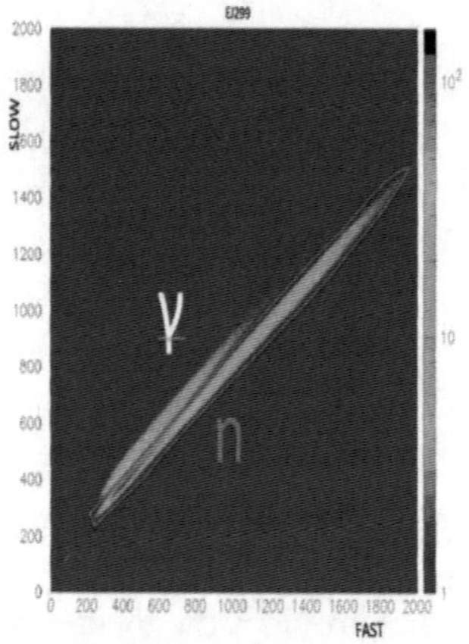

Figure 7. Pulse-shape discrimination(PSD) performance for EJ299.

Pulse shape analysis allowed the definition of a figure of merit as an indicative parameter for the neutron/gamma discrimination. Any detector with a Factor Of Merit (FOM) above 0.58, that is $FOM = \frac{S}{\delta_{neutrons}+\delta_{gammas}}$ where S is the separation between gamma and neutron peaks, and $\delta_{neutrons}$, δ_{gammas} are full width at half maximum (FWHM) of the corresponding peaks, can be considered to have adequate PSD capability for fast neutron detection in the presence of gamma rays; both detectors exceed this merit factor.

Acknowledgment
This project has received funding from the European Union's Horizon 2020 research and innovation programme under grant agreement No 654002.

References
[1] Cherubini S et al., 2004 *Eur Phys. J.A Vol. 20 355-358*
[2] Musumarra A et al., 1996 *NIM A 370 558-562*
[3] Amorini F et al., 1998 *Physical Review C 58 987-995*
[4] Tudisco S et al., 2003 Review of Scientific Instruments 74 4485-4490
[5] Lanzanò L et al., 2007 European Biophysics Journal with biophysics letters 336 823-829
[6] Zaitseva N et al., 2012 *NIM.* A 668 88-93
[7] Cappuzzello F et al., 2015 *Journal of .Phys Conference Series* 630 012018
[8] Negoita F et al., 2016 *Romanian Reports in Physics, Vol. 68, Supplement, P. S37–S144*
[9] Cappuzzello F et al., 2016 *Eur.Phys.J.A* 52: 167
[10] Cavallaro M et al., 2013 *NIM* A 700 65
[11] Cavallaro M et al., 2016 *Phys. Rev. C 93, 064323*
[12] Sciacca et al., 2006 *IEEE* 18 1633-1635
[13] Mazzillo et al., 2007 *Sensors and actuators A 138 306-312*
[14] Privitera S et al., 2008 *Sensors and actuators A* 8 4636-4655
[15] Birks J B, 1964 *The Theory and Practice of Scintillation Counting, McMillan Co., New York*
[16] Brooks F D, 1979 *NIM.* 162, 477
[17] Boiano C et al., 2006 *IEEE Transactions on Nuclear Science* NS53(2) 444
[18] Boiano C et al., 2008 *IEEE Nuclear Science Symposium Conference Record* N30-46

Institute of Physics
Dirac House, Temple Back
Bristol BS1 6BE UK

ISSN: 1742-6588
ISBN 978-1-5108-5896-1

9 781510 858961

CURRAN ASSOCIATES INC.
proceedings.com

2nd International Conference on Rheology and Modeling of Materials 2015 (IC-RMM2)

Journal of Physics: Conference Series Volume 790

Lillafured, Hungary
5 - 9 October 2015

IOP
Institute of Physics